T0310283

IN VIVO *GLUCOSE SENSING*

CHEMICAL ANALYSIS

A SERIES OF MONOGRAPHS ON ANALYTICAL CHEMISTRY
AND ITS APPLICATIONS

Series Editor
J. D. WINEFORDNER

Volume 174

A complete list of the titles in this series appears at the end of this volume.

IN VIVO *GLUCOSE SENSING*

Edited by

DAVID D. CUNNINGHAM

JULIE A. STENKEN

WILEY

A JOHN WILEY & SONS, INC., PUBLICATION

Published by John Wiley & Sons, Inc., Hoboken, New Jersey.
Published simultaneously in Canada.

For general information on our other products and services, or technical support, please contact our Customer Care Department within the United States at (800) 762-2974, outside the United States at (317) 572-3993 or fax (317) 572-4002.

Wiley also publishes its books in a variety of electronic formats. Some content that appears in print may not be available in electronic books. For more information about Wiley products, visit our web site at www.wiley.com

Library of Congress Cataloging-in-Publication Data

In vivo glucose sensing / edited by David D. Cunningham, Julie A. Stenken.
p. ; cm.
Includes bibliographical references and index.
ISBN 978-0-470-11296-0 (cloth)
1. Diabetes. 2. Blood sugar monitoring. 3. Biosensors. I. Cunningham, David D. II. Stenken, Julie A.
[DNLM: 1. Blood Glucose–analysis. 2. Blood Glucose Self-Monitoring. 3. Biosensing Techniques. WK 850 I35 2009]
RC660.A2I5 2009
616.4'62—dc22

2009025917

Printed in the United States of America
10 9 8 7 6 5 4 3 2 1

CONTENTS

PREFACE

New, reliable, user-friendly *in vivo* glucose sensing systems should have a significant impact on health care in the near future. The major motivation for sensing glucose *in vivo* is to help people with diabetes better manage their disease on a daily basis. The current widespread use of single-use glucose test strips has resulted in better day-to-day blood glucose control by many people with a significant delay in the onset of complications associated with diabetes.[1] To achieve additional improvements in the quality of life for people with diabetes, continuous sensing systems are needed to allow automated administration of insulin and other medications to even more tightly control the blood glucose. In addition, control of blood glucose levels in critically ill patients, even if they have not previously had diabetes, has recently been identified as decreasing the incidence of infections with an improvement in overall outcome.[2] Finally, reliable less-invasive technique sensing systems could find application in population screening tests to identify the large number of people with type 2 diabetes who remain undiagnosed for an extended period of time, and thus do not receive appropriate medical therapy. The reader should be aware of a 1997 edited volume by David Fraser that has complementary coverage to some chapters in our volume.[3] The aim of the present book is to inform the reader about the recently commercialized *in vivo* sensing systems and the major efforts underway to develop new as well as improved systems.

After a comprehensive introductory chapter, the book contains chapters describing the biological response of the body to sensing materials, commercially available systems, and systems under development. Two chapters then cover *in vivo* optical approaches, followed by chapters on noninvasive optical approaches, and finally a more recent effort using SERS. In general, the chapters describe both the analytical approach taken and the *in vivo* results to highlight the need for a deep understanding of the sensing technology, as well as the body's anatomy and physiology that frequently complicate the chemical analysis procedures.

In general, the systems described sense the glucose content of interstitial fluid that bathes the cells of the body. Systems that measure sweat,[4] saliva,[5,6] gingival crevicular,[7] or tear fluid[8] have generally not been included since the level of glucose is lower than that in interstitial fluid (and blood) and the glucose content in these fluids varies widely due to uncontrollable processes. NMR is widely used to measure glucose in laboratory research studies and the interested reader is referred to other sources.[9,10] Ideally, detailed information on commercial development of fully implanted systems would be included, however, most of the important technical details surrounding these systems remain confidential, at present. Besides waiting for publications in the literature, progress at small companies active in the area may be monitored by checking their Web sites. Examples given in patents also inform about technical progress, especially since U.S. patent applications are now published 18 months after the submission.

We hope you will find this book of use in your research in understanding the broad area of *in vivo* glucose sensing and in placing new developments into a sound technical perspective.

David D. Cunningham,
Abbott Park, IL, USA

Julie A. Stenken,
Fayetteville, AR, USA

REFERENCES

1. The Diabetes Control and Complications Trial Research Group. The effect of intensive treatment of diabetes on the development and progression of long-term complications in insulin-dependent diabetes mellitus. *New England Journal of Medicine* 1993, 329, 977–986.
2. van den Berghe G, Wouters P, Weekers F, Verwaest C, Bruyninckx F, Schetz M. Intensive insulin therapy in the critically ill patients. *New England Journal of Medicine* 2001, 345, 1359–1367.
3. Fraser DM. *Biosensors in the Body: Continuous In Vivo* Monitoring. John Wiley & Sons, New York, 1997.
4. Boysen TC, Yanagawa S, Sato F, Sato K. A modified anaerobic method of sweat collection. *Journal of Applied Physiology* 1984, 56, 1302–1307.
5. Reuterving CO, Reuterving G, Hagg E, Ericson T. Salivary flow rate and salivary glucose concentration in patients with diabetes mellitus influence of severity of diabetes. *Diabetes & Metabolism* 1987, 13, 457–462.
6. Prinz JF, deWijk RA. Sensory correlation of glucose levels in a starch-based semi-solid model system before and after alpha-amylase breakdown. *Archives of Oral Biology* 2007, 52, 168–172. Note that the amylase content of saliva from the mouth readily converts starch to glucose.
7. Yamaguchi M, Takada R, Kambe S, Hatakeyama T, Naitoh K, Yamazaki K, Kobayashi M. Evaluation of time-course changes of gingival crevicular fluid glucose levels in diabetics. *Biomedical Microdevices* 2005, 7, 53–58.
8. Baca JT, Feingold DN, Asher SA. Tear glucose analysis for the noninvasive detection and monitoring of diabetes mellitus. *Ocular Surface* 2007, 5, 280–293.
9. Seaquist ER, Tkac I, Damberg G, Thomas W, Gruetter R. Brain glucose concentrations in poorly controlled diabetes mellitus as measured by high-field magnetic resonance spectroscopy. *Metabolism* 2005, 54, 1008–1013.
10. deGraff RA. *In Vivo* NMR Spectroscopy: Principles and Techniques, 2nd edn. John Wiley & Sons, New York, 2008.

CONTRIBUTORS

Mark A. Arnold, Department of Chemistry and Optical Science and Technology Center, University of Iowa, Iowa City, IA, USA

Paul W. Barone, Department of Chemical Engineering, Massachusetts Institute of Technology, Cambridge, MA, USA

Kate L. Bechtel, George R. Harrison Spectroscopy Laboratory, Massachusetts Institute of Technology, Cambridge, MA, USA

Becky L. Clark, Department of Chemical Engineering, The Pennsylvania State University, State College, PA, USA

David D. Cunningham, Abbott Laboratories, Abbott Park, IL, USA

Heather M. Duman, Legacy Clinical Research and Technology Center, Portland, OR, USA

Michael S. Feld, George R. Harrison Spectroscopy Laboratory, Massachusetts Institute of Technology, Cambridge, MA, USA

Matthew R. Glucksberg, Biomedical Engineering Department, Northwestern University, Evanston, IL, USA

Marisha L. Godek, Department of Mechanical Engineering, Colorado State University, Fort Collins, CO, USA

Jared B. Goor, Department of Bioengineering, University of California, San Diego, CA, USA

David W. Grainger, Departments of Pharmaceutics and Pharmaceutical Chemistry and Bioengineering, University of Utah, Salt Lake City, UT, USA

Timothy Henning, Abbott Diabetes Care, Abbott Park, IL, USA

Olga Lyandres, Department of Biomedical Engineering, Northwestern University, Evanston, IL, USA

Milan T. Makale, Moores Cancer Center, University of California, San Diego, CA, USA

Mike McShane, Department of Biomedical Engineering, Texas A&M University, College Station, TX, USA

Jonathon T. Olesberg, Department of Chemistry and Optical Science and Technology Center, University of Iowa, Iowa City, IA, USA

Heather S. Paul, Department of Chemistry, University of North Carolina, Chapel Hill, NC, USA

Michael V. Pishko, Department of Chemical Engineering, Texas A&M University, College Station, TX, USA

Mark H. Schoenfisch, Department of Chemistry, University of North Carolina, Chapel Hill, NC, USA

Nilam C. Shah, Department of Chemistry, Northwestern University, Evanston, IL, USA

Wei-Chuan Shih, George R. Harrison Spectroscopy Laboratory, Massachusetts Institute of Technology, Cambridge, MA, USA

Gary W. Small, Department of Chemistry and Optical Science and Technology Center, University of Iowa, Iowa City, IA, USA

Erich Stein, Department of Biomedical Engineering, Texas A&M University, College Station, TX, USA

Julie A. Stenken, Department of Chemistry and Biochemistry, University of Arkansas, Fayetteville, AR, USA

Michael S. Strano, Department of Chemical Engineering, Massachusetts Institute of Technology, Cambridge, MA, USA

Richard P. Van Duyne, Department of Chemistry, Northwestern University, Evanston, IL, USA

Joseph T. Walsh, Department of Biomedical Engineering, Northwestern University, Evanston, IL, USA

W. Kenneth Ward, Divsion of Endocrinology, Diabetes and Clinical Nutrition, Oregon Health & Sciences University, Portland, OR, USA

George S. Wilson, Departments of Chemistry and Pharmaceutical Chemistry, University of Kansas, Lawrence, KS, USA

Jonathan M. Yuen, Department of Biomedical Engineering, Northwestern University, Evanston, IL, USA

Yanan Zhang, Departments of Chemistry and Pharmaceutical Chemistry, University of Kansas, Lawrence, KS, USA

CHAPTER 1

INTRODUCTION TO THE GLUCOSE SENSING PROBLEM

George S. Wilson and Yanan Zhang

In Vivo Glucose Sensing, Edited by David D. Cunningham and Julie A. Stenken

1.1 A SHORT HISTORY OF DIABETES AND GLUCOSE MEASUREMENT

Although the subject of some controversy, the Ebers papyrus (1550 BC) appears to be the earliest, largest, and most comprehensive reference to diabetes and describes one of the principal symptoms of the disease, excessive urination. Other scholars consider the discussion sufficiently vague that it may be regarded as a kidney disorder. In the second century AD, however, the condition was described in more detail by Areteus and the focus was on excessive urination, unquenchable thirst, and degradation of tissue. The name diabetes, taken from the Greek, siphon, was adopted, because fluid does not remain in the body.

There are various references to this condition also noting urine that is sweet, owing to the discharge of glucose when blood levels rise above a threshold for a particular patient. In China, the brilliant physician Zhang Zhongjing noted around AD 200 the "malady of thirst."[1]

In the second millennium AD, diabetes was diagnosed by "water tasters" who tasted the urine of patients to establish that it was sweet tasting. This resulted in a second general property, sweetness, and hence the term *mellitus*, coming from the Latin word for honey. Mathew Dobson, a British physician and chemist, suggested in 1766 that the sweetness in both urine and serum was due to sugar.[1] Throughout the seventeenth and eighteenth centuries, the sweetness of urine was used as an indicator of diabetes. It was observed that diabetes was fatal in less than 5 weeks in some (type 1) and a chronic condition (type 2) for others. It was also observed that urine glucose was reduced as the result of a high-protein, high-fat diet, whereas starchy food produced high sugar levels in the urine and blood. In the early to mid-nineteenth century, attention turned to consideration of diabetes as a metabolic disease and to reach such conclusions it was necessary to develop analytical tools that would enable glucose levels to be reliably monitored.

By the turn of the twentieth century, it was realized that diabetes was associated with the pancreas. In 1921, Banting and Best successfully isolated insulin from dog pancreas, and this was followed quickly by tests in humans. Eli Lilly began production of insulin in 1923.

1.1.1 Chemical Methods for Glucose Measurement

The advent of analytical methodology for glucose measurement began to show its influence through the work of Bernard, Bouchardat, and chemists Priestley, Lavoisier, Chevreul, and Wöhler.[1] They focused on chemical transformations linked to metabolism. As Claude Bernard, the eminent French physiologist, noted in his lectures at the College de France in the 1870s,[2] there were three methods commonly used at that time to detect glucose: polarimetry (rotation of polarized light), reduction of Cu(II) to Cu(I) by reducing sugars (Barreswill/Fehling), and the evolution of CO_2 resulting from the fermentation of a glucose-containing solution. These methods were first applied to the determination of glucose in urine, especially the work of Bouchardat, who is recognized as the first clinician to suggest regular monitoring of glucosuria (urine glucose) and also to specifically suggest that this should be the patient's responsibility.[3] Using the reducing sugar method (Fehling's solution) to measure glucose in blood (25 g of blood), Bernard was able to establish that glucose could be generated endogenously from glycogen and detected through its appearance in the blood that was not attributable to carbohydrate ingestion (gluconeogenesis). To make this discovery, it was necessary to develop an analytical method for the analysis of blood samples. He clearly realized and pointed out to his students[2] that the reduction of Cu(II) is not specific for glucose. In spite of the limitation of this method, it persisted for almost 100 years as a dominant glucose analysis approach along with the closely related Benedict's solution. In 1941, Miles Laboratories (now Bayer) developed the Clinitest®, essentially the Benedict's reagent in tablet form, which when added to the sample gave rise to an exothermic reaction necessary to facilitate the Cu(II) reduction. In 1956, a dip-and-read test for glucose in urine was developed by Bayer (Clinistix®). This represented a significant departure from previous technology since it employed glucose oxidase and peroxidase so that the peroxide formed from the reduction of oxygen could react with *o*-toluidine to give rise to a color development. The color was then compared with a color scale designated negative, light, medium, or dark. The test was useful for diabetic patients in determining whether glucose levels were above the renal threshold, but when the glucose levels were normal, little glucose was found in urine. In 1964, a test strip, based on the same technology (Dextrostix®), was developed for the measurement of glucose in blood by Anton (Tom) Clemens. Another related and significant development (1979) was the availability of a lancing system that simplified the blood sampling process, the Ames Autolet®, a fingerprick device.

1.1.2 Instrumental Readout for Glucose Strip Measurements

Rather than relying on comparison with a color scale, a reflectance meter was developed to read the Dextrostix strip called the Ames Reflectance Meter, primarily meant for use in doctors' offices. It was rather expensive ($495) but was used by a few patients. A version for patients that also provided a memory for results was introduced as the Glucometer® in 1985. A number of companies continued to produce reflectance meters, but in 1985 an electrochemically based test strip was described by Cass and coworkers.[4] In 1987, the Medisense ExacTech® sensor was launched. This device incorporated the use of an exogenous mediator as the electron acceptor, coupling the oxidation of reduced enzyme to the electrode. Since that time, the number of companies producing test strips has proliferated to more than 30. Both glucose oxidase and glucose dehydrogenase are employed. Electrochemical detection is the dominant technology for strips and improvements have come in the form of less blood (now around 0.2 μL) and less painful sampling, faster measurements, and data systems to help with diabetes management.

1.1.3 Glucose Biosensors for Clinical Applications

In the mid-1950s, Leland Clark developed an electrochemical method for oxygen measurements in biological fluids and made the discovery that if the Pt electrode used for detection could be separated from the biological medium by a gas permeable membrane, reliable measurements were possible. This approach was extended by immobilizing glucose oxidase on a membrane and measuring the reduction in oxygen levels as a result of the enzymatic reaction.[5] A miniaturized form of this sensor was proposed and referred to as the enzyme electrode.[6] In 1974, the Model 23 Yellow Springs Instruments glucose analyzer appeared in the market and is still used for clinical glucose measurements. The concept of an electrochemical biosensor has evolved into many different devices capable of continuous glucose measurements as well as a number of other analytes. Clemens at Miles was also responsible for developing a closed-loop system called the Biostator® that sensed blood glucose levels through an extracorporeal shunt and delivered insulin and/or glucose according to a control algorithm. The system was too massive and insufficiently reliable to serve ambulatory patients, but it did focus attention on glycemic control algorithms and on the importance of normoglycemia.[7–9]

1.1.4 Continuous Glucose Monitoring Systems (Electrochemical)

A report by Shichiri in 1982 involving the use of a subcutaneously implanted needle-type sensor in a pancreatectomized dog launched the quest to develop wearable systems for continuous glucose monitoring.[10] At the Central Diabetes Institute, Karlsburg, GDR, Fischer and coworkers examined subcutaneous monitoring in significant detail and demonstrated automated feedback control of subcutaneous glucose.[11–13] The possibility of sensor indwelling in the vascular bed was demonstrated in dogs by Gough and coworkers for sensors that survived for several months. Sensor failure was generally due to a failure of the electronics and not the sensor itself.[14] This technology was licensed to Medical Research Group, Inc. that later merged with Medtronic Minimed. A prototype intravenous catheter-type glucose sensor was developed. Clinical trials in type 1 diabetic patients were initiated in 2000 in France and the United States. Some sensors remained functional for up to a year based on a special nonlinear calibration algorithm. Beyond that, the longevity of the sensor was mainly compromised by a gradual loss of enzyme activity.

In 1993, the report of a 10-year study by the Diabetes Control and Complications Trial (DCCT) Study Group concluded that intensive insulin therapy (multiple injections of insulin daily and control of glycemia closer to the normal level) resulted in a 30–70% reduction in the complications of type 1 diabetes.[15] Coupled to this, however, was an increase of 300% in the incidence of hypoglycemia. Since this is a major concern for patients and their physicians, it was evident that if tighter control could be linked with hypoglycemia avoidance, the significant benefits could be realized. More recently, it has been concluded that the history of HbA1c measurements in patients does not adequately explain the risk for development of chronic complications. Glycemic excursions or variability may be as important as chronic hyperglycemia in the development of chronic complications.[16] This lends further support for the need of continuous monitoring systems.

In 1993, we reported on a study of nine normal subjects using a wearable continuous monitoring system.[17] This was the first application of error grid analysis

(EGA)[18] to demonstrate continuous monitoring sensor performance (see below). More than 99% of all values were demonstrated to fall in the A and B zones. By the end of the 1990s and the beginning of the twenty-first century, four systems have been approved by the FDA: CGMS Gold®/Guardian RT® (Minimed/Medtronic),[19,20] GlucoWatch Biographer (Cygnus/Animas),[21–23] DexCom STS (DexCom),[24] and FreeStyle Navigator® (TheraSense/Abbott).[25,26] All of these systems still exhibit instability over the approved 3–7-day period of implantation. Patients are accordingly advised to make as many as four fingerstick measurements per day. Each calibration is usually assumed to be valid for no more than 12 h. The time-dependent results are now generally available to the patient, but they are advised to use the continuous monitoring systems to detect "trends," while using the more reliable "fingerstick" systems to confirm results. The latest versions of four such monitoring systems are summarized in Table 1.1. Two systems have been developed in Europe that employ microdialysis sampling followed by detection with a conventional enzyme-based glucose sensor GlucoDay (Menarini Diagnostics)[20,27] and the Roche SCGM1.[28] These systems will be discussed in more detail elsewhere in this book. Despite the limitations of the continuous monitoring systems, valuable information has been obtained relating to insulin therapy. These include understanding the incidence of nocturnal hypoglycemia and arriving more quickly to improved metabolic control of patients.

1.1.5 Glucose Monitoring Systems (Nonelectrochemical)

All the above systems employ glucose oxidase and some electrochemical method for assessing the rate of the enzyme-catalyzed oxidation of glucose. There are two fundamental disadvantages of this approach: the sensing element must be in direct contact with biological fluid and the process of measurement requires the consumption of glucose. For these reasons, several alternative approaches have been developed that will be described in more detail elsewhere in this book. In general, however, they fall into two categories: first, spectroscopic measurements (optical rotation, near-infrared, Raman) for which a specific molecular signature has been identified, and second, measurements of glucose binding to specific agents where spectroscopic changes permit the binding process to be followed (fluorescent or other optical changes associated with glucose binding to boronic acid derivatives or binding to lectins such as concanavalin A).[29,30] Such methods have the advantage that they do not consume glucose in the course of making the measurement and this may prove to be an advantage in situations where the total amount of available glucose is limited. These methods are not automatically "noninvasive" as they may involve an implant that is externally interrogated or a device that enables interrogation of tear fluid via a contact lens.[31] Implanted devices can suffer from many of the same problems as the electrochemically based devices. Detection by near-infrared spectroscopy is probably the most advanced of the spectroscopic approaches and can, at least in principle, function as a truly noninvasive device.[32] Recently, a noninvasive system based on photoacoustic technology has been proposed (Aprise™, Glucon).[33]

TABLE 1.1 Continuous Glucose Sensors

Feature	Abbott FreeStyle Navigator®	MiniMed Paradigm® REAL-Time System	MiniMed Guardian® REAL-Time System	DexCom™ Seven™
Photos	Photo from Abbott	MiniMed Paradigm REAL-Time with new, smaller Mini-Link™ Transmitter; photo from Medtronic MiniMed	Guardian REAL-Time System with new, smaller MiniLink™ Transmitter; photo from Medtronic MiniMed	Photo from DexCom
FDA approval	March 13, 2008 for adults 18 +	Children 7–17 and adults 18 +	Children 7–17 and adults 18 +	March 2006 for adults 18 +
Accuracy	Error grid: 96.8–98.4% A + B	Consensus error grid: 98.9% A + B; MARD[a] (mean): 19.7%; MARD (median): 15.6%	Consensus error grid: 98.9% A + B; MARD (mean): 19.7%; MARD; (median): 15.6%	Clark error grid: 97% A B; MARD (mean): 15.7%; MARD (median): 11.4%
	GlucoWatch Biographer data for comparison: MARD: 17–21%; Clarke error grid A + B: 94%; Clarke error grid A: 60%			
Sensor life	Five-day wear indication	FDA approved for 72 h; users report longer wear times	FDA approved for 72 h; users report longer wear times	FDA approved for 7 days
Length of sensor probe	6 mm	0.5 in.	0.5 in.	13 mm

Gauge of sensor probe		23	23	26
Angle of sensor insertion	90°	45°	45°	45°
Insertion device available	Each sensor has a disposable inserter	Sen-serter®, manual insertion also possible	Sen-serter®, manual insertion also possible	DexCom SEVEN Applicator
Monitor size	3 in. × 2.5 in.	Displays on insulin pump, no separate monitor	3 in. × 2 in.	3 in. × 2.5 in.
Startup initialization time	10 h	2 h	2 h	2 h
Calibration	Calibrate at 10, 12, 24, and 72 h after insertion with no further calibration for the final 2 days of the 5-day wear	First calibration is 2 h after insertion. Second calibration within next 6 h after first, and then every 12 h. Will alarm if calibration value not entered	First calibration is 2 h after insertion. Second calibration within next 6 h after first, and then every 12 h. Will alarm if calibration value not entered	Must calibrate with One Touch Ultra—cannot be entered manually. Calibrate every 12 h, first calibration must have 2 done within 30 min of each other
Transmitter/sensor or body surface size	2 in. × 1 in. (combined)	Sensor the size of a nickel. Transmitter is 1.4 in. × 1.1 in. × 0.3 in. and attaches to the sensor	Sensor the size of a nickel. Transmitter is 1.4 in. × 1.1 in. × 0.3 in. and attaches to the sensor	2.5 in. (both combined)
Alarms on user—set low and high thresholds	Yes	Yes	Yes	Yes

Note: In clinical trials, some people never respond to alarms at night regardless of the volume. An alarming device (receiver or pump) that is under covers, a pillow, or underneath a body is almost impossible to hear. If you are considering a continuous glucose sensor, be sure to investigate the device's alarms to see if they will meet your needs

(continued)

TABLE 1.1 *(Continued)*

Feature	Abbott FreeStyle Navigator®	MiniMed Paradigm® REAL-Time System	MiniMed Guardian® REAL-Time System	DexCom™ Seven™
Displays glucose numbers	Every 1 min	Every 5 min	Every 5 min	Every 5 min
Displays directional trends	Yes, always has directional and rate of change arrow. Can view 2, 4, 6, 12, or 24 h glucose graph. Can go back 28 days	Yes, arrows that display how fast and in what direction, and 3 and 24 h graphs	Yes, arrows that display how fast and in what direction, and 3, 6, 12, and 24 h graphs	Yes, can display a 1, 3, or 9 h glucose graph
Displays rate of change	Yes, sideways arrow means dropping at less than 1 mg/dL/min. Up or down arrow means raising/dropping at over 2 mg/dL/min. 45° arrow means dropping/raising between 1 and 2 mg/dL/min	Yes, single and double arrows up or down communicate how fast glucose levels are falling or rising	Yes, single and double arrows up or down communicate how fast glucose levels are falling or rising	No
Predictive alarming	Yes, alarm on 10, 20, or 30 min before it thinks that you will hit that number based on the current trend. It estimates a future number by using algorithms and vector technology. (One parent has her child's alarm set at 20 min notice for highs and 30 min notice for lows)	No	Yes. Predictive alerts can be set to warn you 5, 10, 15, 20, 25, or 30 min before glucose limits have been reached. Rate of change alerts can be set to warn you when glucose levels are changing between 1.1 and 5 mg/dL/min, in 0.1 increments	No

Range of monitor to transmitter (factory stated)	10 ft (reports of significantly greater distance)	6 ft	6 ft	5 ft
Sensor storage: refrigerated or room temperature	Room temperature; 4 months life	Storage between 36 and 80°F without the need for refrigeration; 6 months life	Storage between 36 and 80°F without the need for refrigeration; 6 months life	Room temperature; 4 months life
Built-in BG monitor	Built-in Freestyle monitor	BD meter RF to pump or manually enter with other meters. With BD leaving market, future plans unknown	No, can use any meter and manually enter	Must calibrate with One Touch Ultra
Computer software	Freestyle CoPilot	Carelink™ Personal Software	Carelink™ Personal Software	DexCom DM Consumer Data Manager ($79)
Developing technology	Working on communications with a pump	Closed-loop sensor and pump in clinical trials (Yale, elsewhere)		Long-term (about 1 year/outpatient procedure) implantable sensor

From Children with Diabetes (www.childrenwithdiabetes.com). Used with permission.

[a] MARD is defined in equation (1.5).

1.2 SENSOR DESIGN

1.2.1 The Reactions

The discussion of sensors, especially those used for continuous monitoring, will be based on biosensors using glucose oxidase with electrochemical detection. The sensors generally adhere to the following sequence of reactions of enzyme, E, and mediator, M:

$$E_{ox} + \text{glucose} \rightarrow E_{red} + \text{gluconic acid} \tag{1.1}$$

$$E_{red} + O_2 \rightarrow E_{ox} + H_2O_2 \tag{1.2}$$

$$E_{red} + M_{ox} \rightarrow E_{ox} + M_{red} \tag{1.2'}$$

An exogenous mediator such as ferrocene or Os(III) may be employed to accept electrons from the reduced enzyme, but it must be realized that reaction (1.2) can still take place because oxygen will be present in any case. Oxygen can diffuse freely within the reaction layer but the mediator cannot because it will be anchored to prevent leaching out of the sensor. The enzymatic reaction obeys Michaelis–Menten kinetics according to what is called a "ping-pong" reaction, the sequence of reactions (1.1) and (1.2)/(1.2′). The objective in sensor design is to make reaction (1.1) the rate-determining step, meaning that its rate is proportional to the concentration of glucose. For this to be the case over a range of glucose concentrations between 2 and 20 mM, reaction (1.2)/(1.2′) must be very rapid with respect to reaction (1.1). Otherwise the sensor response becomes dependent on O_2 or mediator concentration and will, at higher concentrations, yield a response independent of glucose. Thus, if the conditions are properly arranged, the rate of reaction (1.1) can be determined by measuring the rate of disappearance of O_2, the formation of H_2O_2, or the formation of M_{red}. The concentrations of the electroactive products/reactants of reaction (1.2)/(1.2′) will be monitored as

$$O_2 + 2H^+ + 2e^- \rightarrow H_2O_2 \tag{1.3}$$

$$H_2O_2 \rightarrow 2H^+ + 2e^- + O_2 \tag{1.4}$$

$$M_{red} \rightarrow M_{ox} + 2e^- \tag{1.4'}$$

The mediator is normally chosen such that its oxidation (reaction (1.4′)) occurs at a potential lower than that for direct electrochemical oxidation of ascorbate and urate, two endogenous electroactive species. If reaction (1.4) is used as the basis for glucose monitoring, then some measure (usually a permselective membrane) must be taken to prevent endogenous interferences from reaching the electrode. Mediators can react directly with endogenous electroactive species. If reaction (1.4′) is used as a measure of the rate of the enzymatic reaction and oxygen is not excluded, then reaction (1.2) will have a parasitic effect on the response, leading to low results.[34] Heller and coworkers developed polymeric matrices in which the mediator could be immobilized and therefore would not diffuse out of the reaction layer. These are referred to as "wired" enzyme systems as they connect the redox chemistry of the enzyme to the electrode.[35,36] In some systems, the rate of the reaction is, in effect, measured by determining the charge necessary to oxidize the product of the two-reaction sequence (M_{red} or H_2O_2). This requires that the reaction is carried out in a fixed volume of sample for a defined time.

1.2.2 Control of Mass Transfer Using Membranes

The proper operation of a sensor based on the detection of peroxide (reaction (1.4)) can be controlled by the use of permselective membranes as shown in Figure 1.1. A membrane is located proximal to the electrode having the property of being permeable to peroxide, but not to endogenous electroactive species. Next is the enzyme layer, followed by an external membrane. In order to meet the condition noted above, this membrane is highly permeable to oxygen, but the permeability of glucose is significantly restricted. This property serves several useful purposes. First, as a consequence of reactions (1.1)/(1.2) above, glucose and oxygen should react in 1:1 stoichiometry, even though the tissue oxygen concentration is typically an order of magnitude lower than glucose. The membrane, in effect, creates a situation in the enzyme layer where oxygen is actually in excess. Second, the activity of the enzyme is made sufficiently high that glucose is immediately oxidized on its arrival in the enzyme layer. The rate of arrival (flux) and therefore the rate of the enzymatic reaction is defined by the concentration gradient between the outside and the inside of the outer membrane. Thus, as long as there is sufficient enzyme activity present, its exact activity does not matter and the sensor response will be limited by mass transfer and not by the rate of the enzymatic reaction. This is quite important because the activation energy for mass transfer is 3–4 kcal/mol as opposed to an enzyme-catalyzed reaction, which is ~11.8 kcal/mol at 300K.[37,38] This leads to temperature coefficients of 1.6–2.2%/°C and 7%/°C, respectively. The arrangement also accommodates small losses of enzyme activity without affecting the sensor sensitivity.

For blood glucose monitoring, it will generally be necessary to obtain linear sensor response in the range of 2–20 mM. In an air-saturated solution, the Michaelis constant (K_m) for glucose will be 5–6 mM. This corresponds to the substrate concentration yielding half the maximum reaction rate velocity. The maximum rate would occur at a concentration of about 15 mM, leading to mostly nonlinear

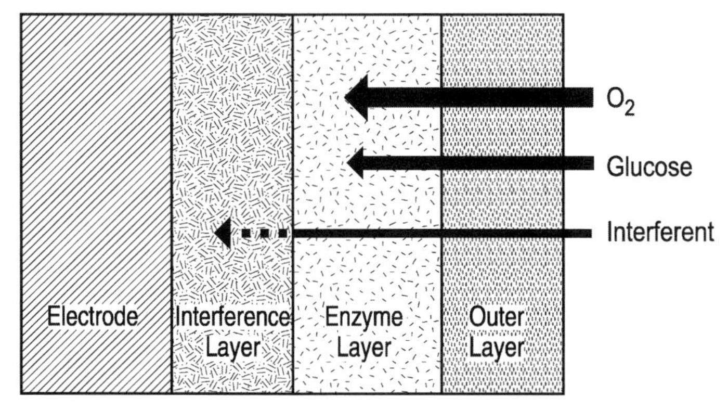

Figure 1.1 Multilayer sensor structure showing control of oxygen, glucose, and electroactive interferences.

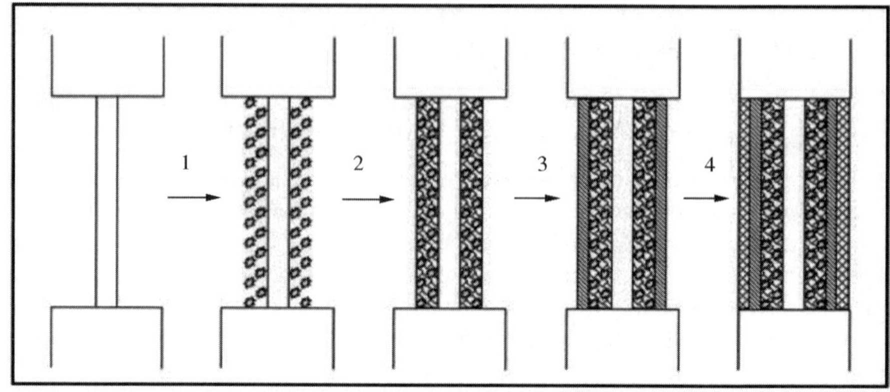

Step 1: Electrodeposition of GOx

Step 2: Electropolymerization of phenol

Step 3: Electrochemical cross-linking of (3-aminopropyl)trimethoxysilane

Step 4: Coating of polyurethane outer membrane

Figure 1.2 Electrochemical deposition of glucose oxidase (GOx) followed by electropolymerization of the polyphenol interference layer. Reprinted with permission from Ref. 40. Copyright 2002 American Chemical Society.

behavior starting at 10–11 mM. Thus, to obtain linear response over the required linear range, the K_m has to be much larger. There are two ways to do this. The first involves the lowering of the effective glucose concentration as noted above. The second approach involves increasing the rate of reaction (1.2)/(1.2′). Heller and coworkers have increased the effective K_m for the reaction to at least 40 mM using these two approaches.[39]

It is possible to combine the enzyme layer with the interference elimination layer through the electrochemical deposition of these layers in sequence. The enzyme layer is deposited first and the interference layer electropolymerized through it.[40,41] This is shown in Figure 1.2. The enzyme is first deposited on the electrode, and a key condition for the production of a smooth, compact enzyme layer is the presence of a detergent above its critical micelle concentration. Dynamic light scattering experiments show that the resulting micelle is approximately of the same diameter as glucose oxidase. The exact significance of this is unclear, but the detergent does prevent enzyme clumping during the deposition process. It is then possible to electropolymerize an interference reducing layer through the oxidation of phenol. The thickness of this layer can be conveniently controlled by controlling the electrodeposition conditions.

1.2.3 The Electrochemical Cell

The oxidation of hydrogen peroxide is actually quite complicated and depends upon the formation of a metal oxide on the electrode surface.[42–44] Consequently, there is no advantage to changing the applied potential, and a DC applied potential of 0.65 V versus a AgCl/Ag reference electrode will suffice. An often overlooked problem is the stability of the reference electrode. There are a number of species present in biological fluids that can dissolve the AgCl deposit off the electrode by forming soluble Ag(I) complexes. This could include endogenous amines including peptides and proteins. The surface must therefore be protected with a membrane that is Cl^- permeable. It is not necessary to use a conventional three-electrode system since the currents typically encountered are in the nanoampere range or less.

1.3 DATA ACQUISITION AND PROCESSING

1.3.1 Acquisition and Readout Device

In general, data acquisition involves digitization of analog signals by a microprocessor. The procedure of data acquisition itself is fairly straightforward. Data processing, however, is where most effort is required. Each device developer has spent substantial resources in dealing with abnormalities representing real-life challenges. This is the fundamental difference between data from a traditional lab instrument and a patient-wearable device. The wearable device may experience a strong electromagnetic field, a shower, moving from a warm to a cold environment, or impact on the monitoring unit or sensor that generates spurious signals. The device must be able to identify such results and handle them according to a predefined methodology. Therefore, error handling becomes an important task. Being able to identify such errors and correctly classify them is a continuing challenge to monitoring system developers. In some cases, sensor drift and calibration instability can be handled retrospectively, but this is not of much use for a real-time monitoring system.

Regardless of the detection mechanism, a typical data processing algorithm must include the following steps (with the sequence interchangeable):

- Error identification and handling
- Signal separation from background
- Filtering and noise reduction
- Signal drift adjustment (if any)
- Conversion of signal to the desired form and unit (glucose in mM or mg/dL)
- Further filtering and restrictions based on known physiological or medical facts
- Calculations of parameters of clinical significance: average, trend, rate of change, and so on
- Database management

The physical forms of readout and alarms depend on the status of the instrumentation. A handheld device with LCD display and input buttons/pads is adopted by most manufacturers. An audio alarm is also available. An on-screen menu-driven method is used for setting up user features such as display content and styles. A historical glucose graph is also available.

1.3.2 Management Software

The most desired outcome of a glucose sensor system is to prevent the occurrence of hyper- and hypoglycemia or, at least, reduce the severity of hypoglycemia. For a stand-alone monitor system, all it needs to do is to sound an alarm accurately and in a timely fashion. This, in concept, is an open-loop system. It requires the patient to decide how to manage the monitoring process. In reality, however, this seemingly simple task has been very difficult. "Sensitivity" and "specificity" are usually used to evaluate the effectiveness of the alarm methodology as well as the usefulness of the device.

$$\text{Sensitivity} = a/(a + c) = \text{true positive}/(\text{true positive} + \text{false negative})$$
$$\text{Specificity} = d/(b + d) = \text{true negative}/(\text{false positive} + \text{true negative})$$

	Patient with hypoglycemia	Patient without hypoglycemia
Alarm triggered	True positive	False positive
Alarm untriggered	False negative	True negative

Sensitivity tells how accurately the device sounds true alarms, while specificity defines how accurately the device avoids false alarms. The concept may appear vague, but it vividly describes real situations. A sensitivity of 80% means that there are two undetected hypoglycemia events every eight times the alarm sounds. A specificity of 80% indicates that in every 10 alarms there are 2 false alarms.

The rate of decrease in sensor glucose, in combination with a hypoglycemic threshold, can be used to predict an upcoming hypoglycemic event as we have demonstrated in rats.[45] The system is based on a finite-state machine, an algorithm that permits orderly passage between a safe state, a hypoglycemic state, and two intermediate states (Figure 1.3). Ideally, this is a good approach for forecasting potential hypoglycemia, thus permitting the patient to take corrective action before it is too late. The basis for the hypoglycemia alarm is to inform the patient that if glucose continues to decrease at the measured rate, a hypoglycemia threshold will be attained in 20 min. This affords sufficient time to permit corrective action. There are also indications in rat studies that early intervention, say at 150 mg/dL, requires less glucose infusion and avoids the rapid return of the glycemia to a hyperglycemic state. However, variation in patient physiology and status of disease (e.g., sensitivity to insulin) makes this task more complicated. A more reliable model must be developed to include numerous patient-specific parameters. When one tries to implement advanced triggers of alarms, or to be rightly termed "alerts," the risk of generating false alarms is significant. This is a condition that would not be tolerated by patients.

To date, there is no commercial glucose sensor system that provides complete diabetes management capability. Medtronic Minimed developed an "open-loop" system that utilized the data from CGMS Gold and the insulin pump. With historical insulin dosing parameters from individual patients, the new pump software can calculate a suggested insulin dosage. This suggested dosage is presented to the user as a reference.

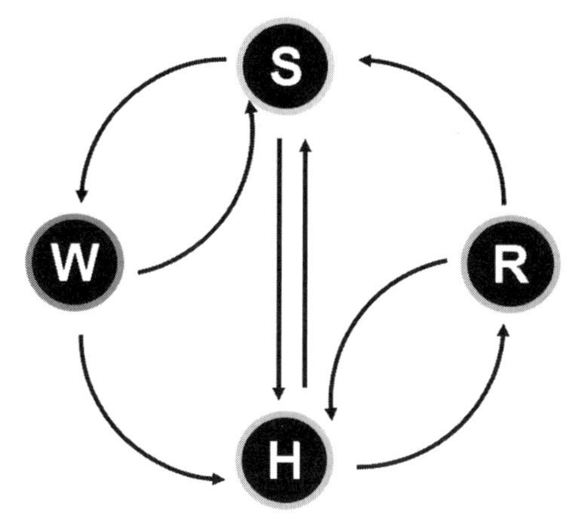

S: *Safety state*: normal blood glucose level, without risk for hypoglycemia

W: *Time-based warning state*: blood glucose level is decreasing and expected to reach a hypoglycemia threshold (HT) in less than m_h minutes

H: *Hypoglycemia state*: blood glucose level is lower than HT

R: *Recovery state*: blood glucose level is lower than HT but increasing and expected to be higher than HT in less than m_r minutes

Figure 1.3 Finite-state machine used to control monitor response in the course of avoiding hypoglycemic events.

1.4 GLUCOSE KINETICS

There is perhaps no subject more confusing in the development of glucose sensors than the relationship between the blood glucose, measured variously as capillary glucose, plasma glucose, or whole blood glucose. The calibration of continuous monitoring systems and the subsequent assessment of their performance depends on the assumption that the two are equal. There are three factors that can contribute to this discrepancy. First, the intrinsic response time of the sensor must be considered. If the response time exceeds about 5 min, then it can be a contributing factor. Second, signal processing techniques that remove noise from the signal can cause a delay. Finally, the physiological response must be taken into account. Bergman and coworkers have studied the interactions of glucose and insulin and have developed multicompartment models to explain glucose distribution and insulin resistance.[46] For the present discussion, we use a three-compartment model shown in Figure 1.4. The sensor is placed in the middle compartment and it is assumed to be measuring accurately the interstitial glucose concentration. There are two sources of glucose found in the blood: that derived by the conversion of glycogen into glucose (G_{end}) and glucose resulting from dietary intake (G_{ex}). When glucose reaches a relatively high threshold in the blood, it can be eliminated via the kidney (see above) (G_k). Insulin can also control the uptake of glucose by cells. Long-term osmotic equilibrium is assumed to be established between the capillaries and the interstitial fluid. Of interest, therefore, is the ratio of the blood glucose (BG) to the glucose of the interstitial fluid (IG). This was first addressed definitively by Fisher and coworkers using the so-called "wick" technique.[11] When BG concentrations are increasing rapidly due to the ingestion of carbohydrate, the BG/IG ratio is consistently greater than unity. On the other hand, decreases in blood glucose in diabetic subjects are typically triggered by the injection

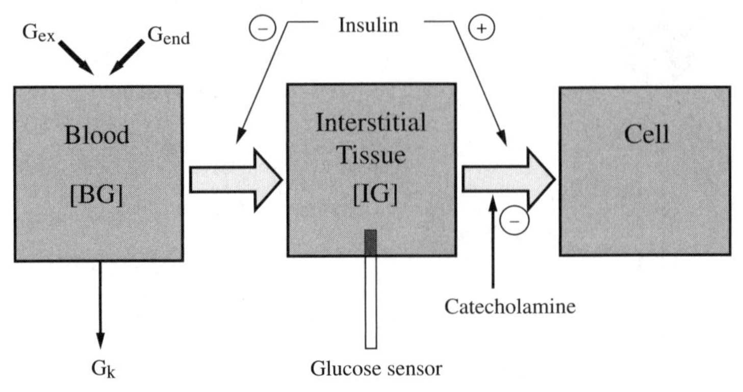

Figure 1.4 Glucose kinetics three-compartment model. G_{ex}, exogenous glucose; G_{end}, glucose produced by gluconeogenesis; G_k, glucose eliminated via the kidney; BG, blood glucose; IG, subcutaneous interstitial glucose. (See the color version of this figure in Color Plates section.)

of insulin, and the resulting response is not necessarily the second half of a "phase shift" or "time lag" as has frequently been suggested. During the decrease, the BG/IG ratio can be greater than, equal to, or less than unity. This question has been addressed in rats and in humans.[47,48] The effective ratio is determined by the placement of the sensor and by the extent of insulin resistance. The former condition (BG/IG > 1) is seen in young rats, especially when the sensor is implanted in adipose tissue, and the IG levels can remain low (in the hypoglycemic region) for extended periods of time even when the BG has recovered to normal levels. The latter (BG/IG < 1) condition is seen in old, obese, insulin-resistant rats. A ratio of BG/IG $= 1$ is achieved when the BG concentration is not changing and when insulin has not recently been administered. Thus, the ideal time for calibration is in the morning before breakfast.

The consequences of time-dependent nonunity ratios of BG/IG have also been manifested in the attempt to use fingerstick system sampling at alternate sites, that is, the arm or thigh rather than fingertips.[49] Once again, alternate sites show significant differences with respect to the capillary glucose value when glucose concentrations are changing rapidly.

1.5 EVALUATION OF SENSOR PERFORMANCE (*IN VITRO*)

Sensors should always be evaluated *in vitro*. While this is clearly not a substitute for *in vivo* testing, it is easier to diagnose fundamental problems without the complications that the biological milieu introduces. Furthermore, if they do not work reliably *in vitro*, they will not work *in vivo*. In addition to the linear dynamic range mentioned above, stability and reproducibility of characteristics in sensor production are very important. Linearity can be characterized by comparing the sensitivities (slope of the dose/response curve) at 5 and 15 mM glucose, assuming that they should not deviate by more than 10%. Stability can be measured in several different ways. Sensors can be stored dry and at room temperature between periodic sensitivity checks. This tends to yield more optimistic results than operating the sensor constantly in a buffer at 37°C. Storing in a refrigerator is not necessarily beneficial as the sensor may suffer thermal shock because the metal (electrode) and the associated polymer layers do not have the same thermal coefficient of expansion. Extended operation at body temperature can shorten the sensor lifetime.

If reaction (1.2) is used as the basis for the sensor response, it will be necessary to establish that the response does not depend strongly on oxygen concentration and also that the sensor does not respond to endogenous interferences such as ascorbate or urate. A useful way of making this determination is to measure the percent increase (decrease) in the signal corresponding to 5 mM glucose when the physiological concentration of the interference is added. Urate as an interferent to electrochemical sensors has never presented a serious challenge. It has a stable low concentration in the body (2–8 mg/dL, ~0.15–0.5 μM in serum). Most known sensor membranes are effective in blocking its diffusion. Ascorbate, with a base concentration of 0.4–1.0 mg/dL (0.02–0.06 mM) in serum, on the other hand, can vary over a wider range because it is present in most food substances. Ascorbate is also a popular food supplement for its role as an antioxidant. For all practical purposes, testing for ascorbate interference can be done with an addition of ascorbic acid to the test buffer to a level of 0.2 mM, which gives a maximum possible interfering level. The criterion for assessment can be that the overall signal caused by ascorbate be no more than 10% of the corresponding signal for 5 mM glucose. If a mediator is used instead of oxygen, it will still be necessary to verify that no chemical reaction occurs directly between the mediator and the endogenous reducing agents. If the mediator is oxidized at a potential lower than that of peroxide, a likely occurrence, then reaction (1.4) will compete with reaction (1.4′) with the result that the measured glucose concentration will be erroneously low. There have also been reports that oxidation of endogenous species such as ascorbate can foul the electrode.[50] There have been a number of reports of electropolymerized films serving to exclude electroactive interferences.[51,52] In our experience, many of these work well for short periods (1–2 days), but then the selectivity deteriorates rapidly. In addition to the electropolymerization of phenol,[53] we have had some success with sol–gels.[54]

1.6 EVALUATION OF SENSOR PERFORMANCE (*IN VIVO*)

1.6.1 *In Vivo* Sensitivity Loss

The literature is rather vague concerning the question of what happens to the sensitivity of a sensor when it is implanted in a biological fluid. In our hands, subcutaneous sensors immediately and rapidly lose sensitivity, a process that takes place in minutes. Despite losses in sensitivity of 10–30%, these sensors will function satisfactorily over periods in excess of 4 days. Indeed, the performance of the sensor frequently improves with time. The origin of this sensitivity loss has been studied in detail and some results are shown in Figure 1.5.[55] If the sensor is removed from the tissue and quickly calibrated in buffer solution (10 min), essentially the same *in vitro* sensitivity is obtained as for the *in vivo* value. Further incubation in buffer causes the sensitivity to rise until eventually the original *in vitro* value is obtained. The important

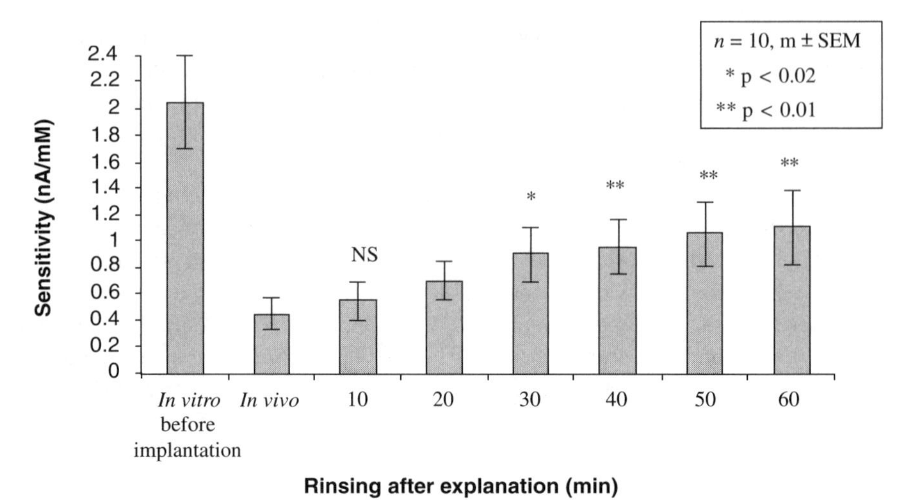

Figure 1.5 Loss of sensor sensitivity on implantation followed by regeneration on rising in glucose-containing buffer. Reprinted with permission from Ref. 55. Copyright 1996 Masson–Elsevier.

conclusion from these experiments is that the loss is specifically associated with the sensor itself and is retained even when the sensor is removed from the tissue. Since the process is overall reversible, this suggests that the cause of sensitivity loss is passive, meaning that the passage of glucose into the enzyme layer is being blocked. It is unlikely that biofilm formation or enzyme activity loss could be the cause of this very rapid sensitivity decrease and subsequent recovery on explantation.

Assuming that the regaining of the original *in vitro* sensitivity is caused by the leaching out of material from the interior of the sensor, we have examined the leachate using a proteomic approach. Rather than intact proteins, protein fragments dominate, suggesting that they are the result of proteolytic reactions at or near the sensor surface.[56] Studies in our laboratory have shown that incubating sensors with physiological concentrations of serum albumin, fibrinogen, and/or IgG, which are present in relatively high concentrations, produce very little change in the *in vitro* sensitivity.[57] On the other hand, incubation in serum can produce significant sensitivity losses.[58]

1.6.2 Calibration *In Vivo*

As noted above in the discussion of glucose kinetics, reliable calibration must eliminate the discrepancy between the tissue and blood glucose values. This suggests that calibration in the morning before breakfast will be the optimal solution. The performance of our glucose monitoring system in a diabetic patient is demonstrated in Figure 1.6,[59] where the performance of a sensor is evaluated over a period of 7 days. A key question is whether this calibration should be a one- or a two-point calibration. We have examined this question in considerable detail,[60,61] and this is also consistent with the finding of Heller and coworkers that a one-point calibration is to be preferred.[39]

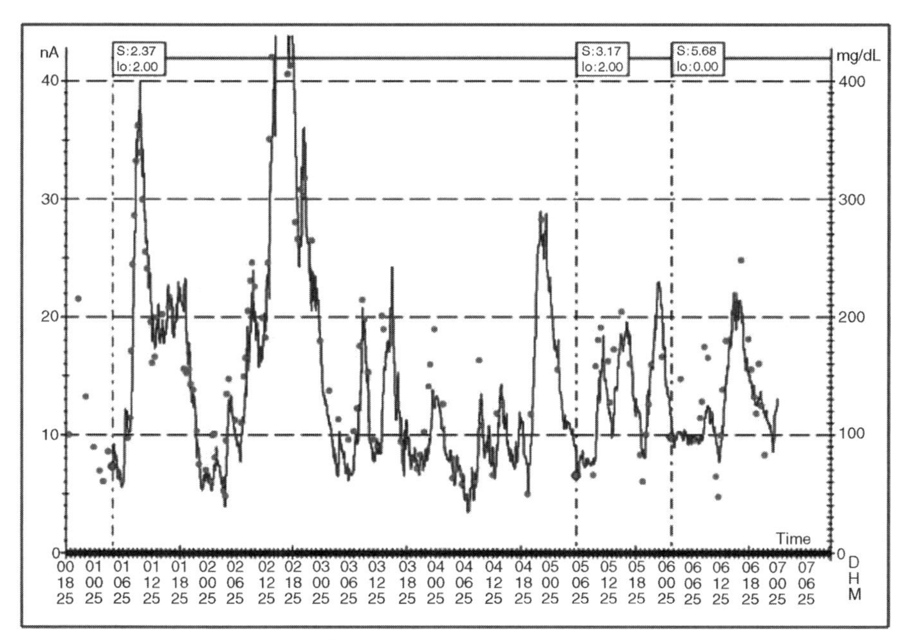

Figure 1.6 Seven-day monitoring of a diabetic patient. Points obtained from fingerstick method, solid trace — continuous sensor response. Three two-point calibrations are noted on days 1, 5, and 6.

At first glance, this conclusion seems counterintuitive. However, determining the slope of the calibration curve (the sensitivity) using two points means that the uncertainty of both points must be taken into account. In a one-point calibration, the I_0 value, that is, the current in the absence of glucose, is assumed to be a certain value with zero variance. Thus, the uncertainty of the slope of the calibration curve is determined only by one glucose measurement.

A nonstatistical method for *in vivo* sensor performance evaluation has come into use, the Clarke error grid analysis (EGA).[18] This approach was originally developed for the evaluation of fingerstick systems and is based on a conventional correlation plot of the performance of the test system with respect to a referee method (ideally a clinical analyzer). If the correlation were perfect, all points would fall on a 45° line. The area surrounding this line is divided into zones that predict the clinical consequences in terms of action taken by the patient, depending on where the measurements by the test system fall off the line. Zone A would yield a clinically accurate decision (take insulin, take glucose, or do nothing), zone B a clinically acceptable decision, and zone D a clinically erroneous decision. Figure 1.7 is an EGA of the data of Figure 1.6. It will be noted, for example, that a few points fall in zone D at a blood glucose concentration of less than about 5 mM. Such points are of concern because they represent a situation in which the patient believes that the BG value is in a safe region (greater than about 5 mM) when, in fact, it is in a hypoglycemic domain. This situation can arise if the blood glucose decreases more rapidly than the tissue glucose. Using this approach, sensors have been considered to perform adequately if the percentage of points falling in the A and B zones is at least 98–99%. Recently,

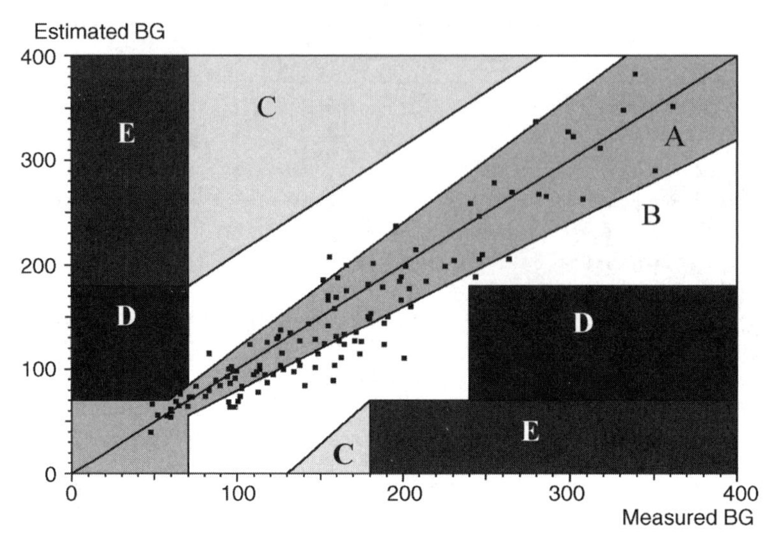

Figure 1.7 Clarke error grid analysis plot of the data of Figure 1.6.

Clarke and coworkers[25,62] have proposed an enhancement of the EGA called continuous glucose error grid analysis (CG-EGA). This approach seeks to account for the clinical consequences of the rate of change of blood glucose with respect to the rate derived from a continuous monitoring device. Two related analyses are performed: the conventional EGA, now called point error grid analysis (P-EGA) and rate error grid analysis (R-EGA). This approach was evaluated using clinical data provided by the TheraSense FreeStyle Navigator® continuous monitoring system.[25] Three regions were defined: hypoglycemia (BG $\leq$ 70 mg/dL), euglycemia (70 $<$ BG 180 mg/dL), and hyperglycemia (BG $>$ 180 mg/dL). P-EGA yields ~99% of values falling in the A and B zones. However, when the rates are analyzed as a function of the three regions, accuracy drops significantly in the hypoglycemic region. For this and related devices, rate accuracy is important, particularly if the continuous monitoring system is being used to avoid hypoglycemia. It is clear that making a series of independent measurements with a fingerstick system is not the same as time-correlated measurements generated by continuous monitoring systems.

In addition, all the paired points (n) including the concentrations measured from the glucose sensor, $[\text{glucose}]_{\text{sensor}}$, and the reference glucose measurement, $[\text{glucose}]_{\text{reference}}$, in the correlation plot are used to calculate the overall mean absolute relative difference (MARD). The median MARD is the median relative difference among all the measured values.

$$\text{MARD (mean)} = \left(\frac{\left(\sum\left(\left|[\text{glucose}]_{\text{sensor}} - [\text{glucose}]_{\text{reference}}\right|\right)/[\text{glucose}]_{\text{reference}}\right)}{n} \right) \times 100\%$$

$$\text{MARD (median)} = \text{median}\left(\frac{\left|[\text{glucose}]_{\text{sensor}} - [\text{glucose}]_{\text{reference}}\right|}{[\text{glucose}]_{\text{reference}}} \right) \times 100\% \qquad (1.5)$$

Most of the sensors currently marketed or under development have MARD values in the range of 10–20%. We feel strongly that as a real-time monitoring system a MARD of no greater than 15% should be adequate (see Table 1.1).

The correlation coefficient between the two methods is always reported. For clinical acceptance, a value of 0.85 or greater may be necessary. However, because correlation coefficient can be affected significantly by a single point at extreme values, or by a lack of dynamic range, one should use caution when looking at the numbers. It has further been suggested that there should be separate performance goals for glucose sensors in the various glycemia zones.[63] International Organization for Standardization (ISO) methods for defining accuracy have also been discussed.[64]

1.7 BIOCOMPATIBILITY

The word "biocompatibility" has been overworked to the point of exhaustion because there are few definitive criteria to define what it means. Several chapters in this book deal in some detail with the biology associated with implants. The present operational definition of biocompatibility will focus on whether interactions with the subcutaneous tissue materially affect the proper functioning of the sensor and whether the presence of the sensor influences adversely the biological environment. Studies of sensors indwelling in the vascular bed are less numerous. The intravascular sensor is less popular mainly because it requires surgery to implant and explant, and it cannot be percutaneous for a long term because of the risk of systemic infection. It is very troublesome if a sensor malfunctions and has to be taken out. Embolization of clots is generally not a risk because clots go to the lung where they are filtered and eventually dissolve. The implant is also placed in a vein rather than an artery, because it is less painful. When a clot adheres to the sensor, however, it becomes a major problem. The sensor is walled off from the blood and is no longer accessible to blood glucose. This is, in fact, a classic biocompatibility issue. This is the representative disadvantage for an intravascular sensor, and development of sensors capable of evolving NO may help to mitigate this problem (see below). The challenge of clot formation is ever present as long as the sensor is in the vascular bed. In contrast, a subcutaneous sensor can acquire a stable state once the acute interactions with the surrounding tissue subside.

Most of the sensors implanted in the subcutaneous tissue are 200–250 μm (33–31 ga) in diameter. They are usually implanted using a guide cannula whose outside diameter is 21–23 ga (813–635 μm). This insertion process will cause some tissue injury including breaking of capillaries. Experience in our laboratories has shown, however, that the damage is very slight and an edema around the implant, typically the size of a mosquito bite, disappears after about 24 h.

There are perhaps four sources of failure or apparent failure of subcutaneously implanted glucose sensors. The first is the passive loss of sensitivity resulting from uptake of species on or into the sensor.[55] It is quite likely that these species are of relatively low molecular weight (>15 kDa) and could include protein fragments, lipids, and a variety of small endogenous molecules. This process is relatively rapid and has generally been referred to as the "run-in" time, lasting 2–4 h. In addition, a second necessary phase of the initial run-in time is the rehydration of sensor

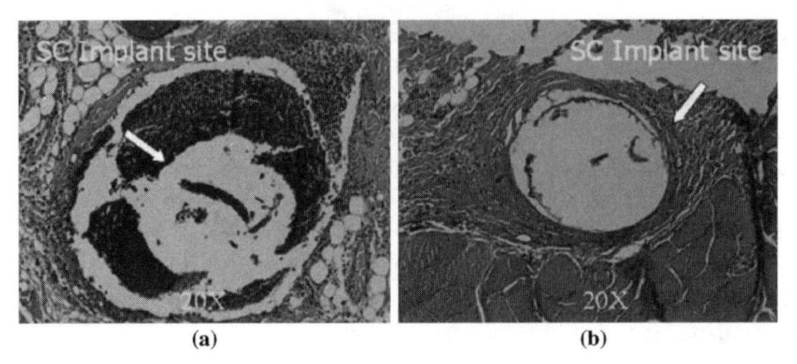

(a) (b)

Figure 1.8 Tissue section at implant site showing the effect of NO evolution on the acute inflammatory response. (a) Sensor with no NO evolution (control); (b) no evolution for about 18 h. Reprinted with permission from Ref. 65. Copyright 2005 John Wiley & Sons. (See the color version of this figure in Color Plates section.)

membranes that have been stored dry. It takes a finite amount of time for the membranes to hydrate and for the local microenvironment to establish the required mass transfer balance before stable operation can be achieved. The second is the perturbation of oxygen and glucose levels in the tissue surrounding the sensor resulting from tissue injury and the subsequent acute inflammatory response. This latter process, although complicated, has been studied in some detail and is described elsewhere in this book. It is typically manifested in sensor instability over the first 48 h but can result in cleanup of the small amount of capillary blood that may have accumulated. Tissue sections from a rat after 3 days show evidence of macrophage formation and angiogenesis (restoration of the damaged capillaries). We have demonstrated that sensors designed to evolve NO on implantation significantly reduce the acute inflammatory response as shown in Figure 1.8.[65] Experience in a number of laboratories has shown that subcutaneously implanted sensors can function reliably for 1–2 weeks if they survive the initial several days. For long-term implants, two other issues can become important. There will be a loss of enzyme activity that will eventually affect sensor response. Finally, the capsule formation caused by the foreign body reaction may limit the access of glucose and oxygen to the sensor. This latter issue has been addressed by encouraging angiogenesis as the sensor is implanted.[66] There is, in addition, the case already discussed and related to glucose kinetics, where the sensor correctly measures the glucose, but these values do not correspond to the blood value. This question can be resolved by calibrating the sensor under the appropriate conditions. Occasionally, sudden catastrophic failure of the sensor can occur. The reasons for this are not clear, but may be due to the fracture of a capillary owing to movement of the sensor in the tissue, thus releasing some blood.

1.8 FUTURE DIRECTIONS

The development of reliable, user-friendly glucose sensing systems remains a significant challenge. Applications can be diverse, including rapid screening of subjects for diabetes (say at a mall), self-monitoring of diabetes several times a day using test strips, continuous monitoring systems for bedside monitoring (especially appealing for monitoring children), wearable continuous monitoring systems, and monitoring of subjects (not necessarily diabetic patients) under trauma and critical care situations. The wide range of applications will demand a multiplicity of solutions very likely exploiting different technologies. It is to be hoped, for example, that truly noninvasive approaches, in which no component of the sensing system is in contact with a biological fluid, will become a reality.

ACKNOWLEDGMENTS

The authors gratefully acknowledge the support of the National Institutes of Health Grant No. NS047955. We thank Wayne V. Moore for helpful comments.

REFERENCES

1. Papaspyros NS. *The History of Diabetes Mellitus*, 2nd edn. Georg Thieme Verlag, Stuttgart, 1964.
2. Bernard C. *Leçons sur Le Diabète et La Glycogenèse Animale*. J.-B. Balbière et Fils, Paris, 1877, p. 95ff.
3. Joslin EP. Apollinaire Bourchardat. *Diabetes* 1952, 1, 490–491. See also: Bouchardat A. *De la glucosurie ou Diabète sucré son traitement hygiénique*. Libraire Germer Baillière, Paris, 1883.
4. Cass AEG, Davis G, Francis GD, Hill HAO, Aston WJ, Higgins IJ, Plotkin EV, Scott LD, Turner AP. Ferrocene-mediated enzyme electrode for amperometric determination of glucose. *Analytical Chemistry* 1985, 56, 667–671.
5. Clark LC, Lyons C. Electrode systems for continuous monitoring during cardiovascular surgery. *Annals of the New York Academy of Sciences* 1962, 102, 29–35.
6. Updike SJ, Hicks GP. The enzyme electrode. *Nature* 1967, 214, 986–988.
7. Clemens AH, Chang PH, Myers RW. The development of biostator, a glucose controlled insulin infusion system (GCIIS). *Hormone Metabolism Research Supplement Series (Blood Glucose Monitoring)* 1977, 7, 23–33.
8. Fogt EJ, Dodd LM, Jenning EM, Clemens AH. Development and evaluation of a glucose analyzer for a glucose-controlled insulin infusion system (Biostator). *Clinical Chemistry* 1978, 24, 1366–1372.
9. Heinemann L, Ampudia-Blasco FJ. Glucose clamps with the Biostator: a critical reappraisal. *Hormone Metabolism Research* 1994, 26, 579–583.
10. Shichiri M, Kawamori R, Yamasaki Y, Hakui N, Abe H. Wearable artificial endocrine pancreas with needle-type glucose sensor. *Lancet* 1982, 2, 1129–1131.
11. Fischer U, Ertle R, Abel P, Rebrin K, Brunstein E, Hahn von Dorsche H, Freyse EJ. Assessment of subcutaneous glucose concentration: validation of the wick technique as a reference for implanted electrochemical sensors in normal and diabetic dogs. *Diabetologia* 1987, 30, 940–945.
12. Rebrin K, Fischer U, von Woedtke T, Abel P, Brunstein E. Automated feedback control of subcutaneous glucose concentration in diabetic dogs. *Diabetologia* 1989, 32, 573–576.

13. Fischer U, Rebrin K, von Woedtke T, Abel P. Clinical usefulness of the glucose concentration in subcutaneous tissue: properties and pitfalls of electrochemical biosensors. *Hormone Metabolism Research* 1994, 26, 515–522.
14. Armour JC, Lucisano JY, McKean BD, Gough DA. Application of chronic intravascular blood glucose sensor in dogs. *Diabetes* 1990, 39, 1519–1526.
15. Diabetes Complications and Control Trial (DCCT) Group. The effect of intensive treatment of diabetes on the development and progression of long-term complications in insulin-dependent diabetes mellitus. *The New England Journal of Medicine* 329, 1993, 977–986.
16. Hirsch IB, Brownlee M. Should minimal glucose variability become the gold standard of glycemic control? *Journal of Diabetes and its Complications* 2005, 19, 178–181.
17. Poitout V, Moatti-Sirat D, Reach G, Zhang Y, Wilson GS, Lemonnier F, Klein JC. A glucose monitoring system for on line estimation in man of blood glucose concentration using a miniaturized glucose sensor implanted in the subcutaneous tissue, and a wearable control unit. *Diabetologia* 1993, 36, 658–663.
18. Clarke WL, Cox D, Gonder-Frederick LA, Carter W, Pohl SL. Evaluating clinical accuracy of systems for self-monitoring of blood glucose. *Diabetes Care* 1987, 10, 622–628.
19. Gross TM, Bode BW, Einhorn D, Kayne DM, Reed JH, White NH, Mastrototaro JJ. Performance evaluation of the MiniMed continuous glucose monitoring system during patient home use. *Diabetes Technology & Therapeutics* 2000, 2, 49–56.
20. Wentholt IM, Vollebrecht MA, Hart AA, Hoekstra JB, DeVries JH. Comparison of a needle-type and a microdialysis continuous glucose monitor in type I diabetic patients. *Diabetes Care* 2005, 28, 2871–2876.
21. Tamada JA, Garg S, Jovanovic L, Pitzer KR, Fermi S, Potts RO. Noninvasive glucose monitoring: comprehensive clinical results. Cygnus Research Team. *Journal of the American Medical Association* 1999, 282, 1839–1844.
22. The Diabetes Research in Children Network (DirecNet) Study Group. A randomized trial comparing the GlucoWatch Biographer™ with standard glucose monitoring in children with type 1 diabetes. *Diabetes Care* 2005, 28, 1101–1106.
23. Buckingham B, Block J, Burdick J, Kalajian A, Kollman C, Choy M, Wilson DM, Chase P. Response to nocturnal alarms using a real-time glucose sensor. *Diabetes Technology and Therapeutics* 2005, 7, 440–447.
24. Garg S, Zisser H, Schwartz S, Bailey T, Kaplan R, Ellis S, Jovanovic L. Improvement in glycemic excursions with a transcutaneous, real-time continuous glucose sensor. *Diabetes Care* 2006, 2, 44–50.
25. Kovatchev BP, Gonder-Frederick LA, Linda A, Cox DJ, Clarke WL. Evaluating the accuracy of continuous glucose-monitoring sensors: continuous glucose-error grid analysis illustrated by TheraSense Freestyle Navigator data. *Diabetes Care* 2004, 27, 1922–1928.
26. Wilson DM, Beck RW, Tamborlane WV, Dontchev MJ, Kollman C, Chase P, Fox LA, Ruedy KJ, Tsalikian E, Weinzheimer SA. The accuracy of the FreeStyle Navigator continuous glucose monitoring system in children with type 1 diabetes. *Diabetes Care* 2007, 30, 59–64.
27. Maran A, Crepaldi C, Tiengo A, Grassi G, Vitali E, Pagano G, Bistoni S, Calabrese G, Santeusanio F, Leonetti F, Ribaudo M, Di Mario U, Annuzzi G, Genovese S, Riccardi G, Previti M, Cucinotta D, Giorgino F, Bellomo A, Giorgino R, Poscia A, Varalli M.

Continuous subcutaneous glucose monitoring in diabetic patients: a multicenter analysis. *Diabetes Care* 2002, 25, 347–352.

28. Kapitza C, Lodwig V, Obermaier K, Wientjes K, Jan C, Hoogenberg K, Jungheim K, Heinemann L. Continuous glucose monitoring: reliable measurements for up to 4 days with the SCGM1 system. *Diabetes Technology and Therapeutics* 2003, 5, 609–614.
29. Yan J, Springsteen G, Deeter S, Wang B. The relationship between pK_a, pH and binding constants in the interaction between boronic acids and diols: it is not as simple as it seems. *Tetrahedron* 2004, 60, 11205–11209.
30. Aslan K, Lakowicz JR, Geddes CD. Tunable plasmonic glucose sensing based on the dissociation of ConA-aggregated dextran-coated gold colloids. *Analytica Chimica Acta* 2004, 517, 139–144.
31. Alexeev V, Das S, Finegold DN, Grabowski J, Somayajula K, Asher SA. Photonic crystal glucose sensing material for non-invasive monitoring of glucose in tear fluid. *Clinical Chemistry* 2004, 50, 2352–2366.
32. Arnold MA, Small GW. Noninvasive glucose sensing. *Analytical Chemistry* 2005, 77, 5429–5439.
33. Weiss R, Yegorchikov Y, Shusterman A, Raz I. Noninvasive continuous glucose monitoring using photoacoustic technology: results from the first 62 subjects. *Diabetes Technology and Therapeutics* 2007, 9, 68–74.
34. Mano N, Mao F, Heller A. On the parameters affecting the characteristics of the "wired" glucose anode. *Journal of Electroanalytical Chemistry* 2005, 574, 347–357.
35. Heller A. Electrical connection of enzyme redox centers to electrodes. *Journal of Physical Chemistry* 1992, 96, 3579–3587.
36. Heller A. Implanted electrochemical sensors for the management of diabetes. *Annual Review of Biomedical Engineering* 1999, 1, 153–175.
37. Krauss CJ, Spinks JWT. Temperature coefficients for self-diffusion in solution. *Canadian Journal of Chemistry* 1954, 32, 71–78.
38. Gutfreund H. *Kinetics for the Life Sciences: Receptors, Transmitters and Catalysts*, Cambridge University Press, New York, 1995, p. 234.
39. Csöregi E, Quinn CP, Schmidtke DW, Lindquist SE, Pishko MV, Ye L, Katakis I, Hubbell JA, Heller A. Design, characterization, and one-point *in vivo* calibration of a subcutaneously implanted glucose electrode. *Analytical Chemistry* 1994, 66, 3131–3138.
40. Chen X, Matsumoto N, Hu Y, Wilson GS. Electrochemically-mediated electrodeposition/electropolymerization to yield a glucose microbiosensor with improved characteristics. *Analytical Chemistry* 2002, 74, 368–372.
41. Matsumoto N, Chen X, Wilson GS. Fundamental studies of glucose oxidase deposition on a Pt electrode. *Analytical Chemistry* 2002, 74, 362–367.
42. Zhang Y, Wilson GS. Electrochemical oxidation of hydrogen peroxide on platinum and platinum/iridium electrodes in physiological buffer and its applicability to hydrogen peroxide-based biosensors. *Journal of Electroanalytical Chemistry* 1993, 345, 253–271.
43. Hall SB, Khudaish EA, Hart AL. Electrochemical oxidation of hydrogen peroxide at platinum electrodes. Part I. An adsorption-controlled mechanism. *Electrochimica Acta* 1998, 43, 579–588.
44. Hall SB, Khudaish EA, Hart AL. Electrochemical oxidation of hydrogen peroxide at platinum electrodes. Part V. Inhibition by chloride. *Electrochimica Acta* 45, 2000, 3573–3579.

45. Choleau C, Dokladal P, Klein JC, Ward WK, Wilson GS, Reach G. Prevention of hypoglycemia using risk assessment with a continuous glucose monitoring system. *Diabetes* 51, 2002, 3263–3273.

46. Bergman RN. Minimal model: perspective from 2005. *Hormone Research* 2006, 64 (Suppl. 3), 8–15.

47. Aussedat B, Dupire-Angel M, Gifford R, Klein JC, Wilson GS, Reach G. Interstitial glucose concentration and glycemia: implications for continuous subcutaneous glucose monitoring. *American Journal of Physiology. Endocrinology and Metabolism* 2000, 278, E716–E728.

48. Kulcu E, Tamada JA, Reach G, Potts RO, Lesho MJ. Physiological differences between interstitial glucose and blood glucose measured in human subjects. *Diabetes Care* 2003, 26, 2405–2409.

49. Ellison JM, Stegmann JM, Colner SL, Michael RH, Sharma MK, Ervin KR, Horwitz DL. Rapid changes in postprandial blood blouse produce concentration differences at finger, forearm, and thigh sampling sites. *Diabetes Care* 2002, 25, 961–964.

50. Lowry JP, O'Neill RD. Homogeneous mechanism of ascorbic acid interference in hydrogen peroxide detection at enzyme-modified electrodes. *Analytical Chemistry* 1992, 64, 453–456.

51. Geise RJ, Adams JM, Barone NJ, Yacynych AM. Electropolymerized films to prevent interferences and electrode fouling in biosensors. *Biosensors & Bioelectronics* 1991, 6, 151–160.

52. Sasso SV, Pierce RJ, Walla R, Yacynych AM. Electropolymerized 1,2-diaminobenzene as a means to prevent interferences and fouling and to stabilize immobilized enzyme in electrochemical biosensors. *Analytical Chemistry* 1990, 62, 1111–1117.

53. Craig JD, O'Neill RD. Electrosynthesis and permselective characterization of phenol-based polymers for biosensor applications. *Analytica Chimica Acta* 2003, 495, 33–43.

54. Chen X, Hu Y, Wilson GS. Glucose microbiosensor based on alumina sol–gel matrix/electropolymerized composite membrane. *Biosensors & Bioelectronics* 2002, 17, 1005–1013.

55. Thomé-Duret V, Gangnerau MN, Zhang Y, Wilson GS, Reach G. Modification of the sensitivity of glucose sensor implanted into subcutaneous tissue. *Diabète et Métabolisme* 1996, 22, 174–178.

56. Gifford R, Kehoe JJ, Barnes SL, Kornilayev BA, Alterman MA, Wilson GS. Protein interactions with subcutaneously-implanted biosensors. *Biomaterials* 2006, 27, 2587–2598.

57. Bindra DS. Development of potentially implantable glucose sensors. PhD dissertation, University of Kansas, 1990.

58. Gerritsen M, Jansen JA, Kros A, Vriezema DM, Sommerdijk NAJM, Nolte RJM, Lutterman JA, Van Hovell SWFM, Van der Gaag A. Influence of inflammatory cells and serum on the performance of implantable glucose sensors. *Journal of Biomedical Materials Research* 2000, 54, 69–75.

59. Reach G, Klein JC.Unpublished results, 1996.

60. Choleau C, Klein JC, Reach G, Aussedat B, Demaria-Pesce V, Wilson GS, Gifford R, Ward WK. Calibration of a subcutaneous amperometric glucose sensor. Part 1. Effect of measurement uncertainties on the determination of sensor sensitivity and background current. *Biosensors & Bioelectronics* 2002, 17, 641–646.

61. Choleau C, Klein JC, Reach G, Aussedat B, Demaria-Pesce V, Wilson GS, Gifford R, Ward WK. Calibration of a subcutaneous amperometric glucose sensor implanted for 7 days in diabetic patients. Part 2. Superiority of the one-point calibration method. *Biosensors & Bioelectronics* 2002, 17, 647–654.
62. Clarke WL, Anderson A, Farhy L, Breton M, Gonder-Frederick L, Cox D, Kovatchev B. Evaluating the clinical accuracy of two continuous glucose sensors using continuous glucose-error grid analysis. *Diabetes Care* 2005, 28, 2412–2417.
63. Klonoff DC. The need for separate performance goals for glucose sensors in the hypoglycemic, normoglycemic and hyperglycemic ranges. *Diabetes Care* 2004, 27, 834–836.
64. Klonoff DC. Continuous glucose monitoring: roadmap for 21st century diabetes therapy. *Diabetes Care* 2005, 28, 1231–1239.
65. Gifford R, Batchelor MM, Lee Y, Gokulrangan G, Meyerhoff ME, Wilson GS. Mediation of *in vivo* glucose sensor inflammatory response via nitric oxide release. *Journal of Biomedical Materials Research Part A* 2005, 75A, 755–766.
66. Updike SJ, Shults MC, Gilligan BJ, Rhodes RK. A subcutaneous glucose sensor with improved longevity, dynamic range, and stability of calibration. *Diabetes Care* 2000, 23, 208–214.

CHAPTER 2

THE MACROPHAGE IN WOUND HEALING SURROUNDING IMPLANTED DEVICES

Marisha L. Godek and David W. Grainger

2.1 INTRODUCTION

Biosensor–tissue compatibility is a significant obstacle in the development of viable, enduring implantable biosensors.[1,2] The clinical literature and product development files in industry are littered with biosensor failures in many different sites and

In Vivo Glucose Sensing, Edited by David D. Cunningham and Julie A. Stenken

situations. Problems with the continuous interfacing of the sensing device with host physiology arise from the complex immune response of the host to the chemical and physical properties of the implanted device. Phenomena surrounding this response arise from both acute and chronic inflammatory host reactivities, often characterized under the general term, "biocompatibility." Biocompatibility has been more formally defined as the ability of the implanted material to perform with an appropriate host response within a specific application or biomedical context and implies that the material must minimally perturb and be accommodated by the host at some acceptable level.[3,4] Implanted biosensors have an additional requirement: the host environment must not interfere with the sensor performance.[5]

Inflammation is a normal response to homeostatic disruption of multiple possible origins. Wounding produces inflammation as part of normal sequelae to healing. Since device implantation (or any invasive procedure) produces wounding, inflammation is a natural and known host response to a biomaterial. These events initiate immediately upon device contact with host physiology: with the adsorption of host proteins, surface activation of blood-derived elements including coagulation proteins, complement, and platelets, and progression to the eventual recruitment of inflammatory cells, unresolved wound healing and granulation tissue, and eventual fibrosis. The foreign body reaction (FBR)—the ubiquitous host inflammatory response to implanted biomaterials—is the consequence of these numerous physiological responses, including aberrant wound healing and inflammation reactions in the presence of an implanted device (e.g., sensors), grossly characterized by chronic inflammation, compromised healing at the implant site, and in many cases, biomedical device failure.[6,7] This response seems to be slightly dependent on device surface chemistry or topology, and more pronounced in soft tissues than in hard tissues, but nonetheless a recognized problem with substantial clinical implications and challenges in most device scenarios.[8] For implantable biosensors, FBR inflammatory events and ensuing fibrosis often translate to a loss of performance. Significantly, with multiple molecular, cellular, tissue-specific and intersecting, integrated, or synergistic kinetic timelines responsible for the exceedingly complex FBR, the host reaction is poorly understood.[9,10] Lack of understanding translates to lack of clinical control and inability to sustain reliable sensor reporting functions in many implant sites. Therefore, an understanding of the FBR at the cellular and molecular level is necessary for producing design parameters required for a new generation of devices that successfully suppress, avoid, or manipulate inflammation and associated fibrosis, and enhance desired tissue responses (e.g., angiogenesis and tissue integration). These design parameters must consider implant site-specific physiology, aspects of material-associated biocompatibility, and molecular and cellular components responding to the local trauma of sensor implantation.

2.2 THE IMPLANT SITE AS A WOUND SITE

Invasive or surgical placement of a biomedical device produces tissue trauma, resulting in the formation of a wound site. Normal wounds heal and remodel over time to restore tissue homeostasis in the absence of implanted materials. Interestingly, sites of long-term implants do not heal properly: the ensuing FBR prompted by the presence of an implant represents an aberrant form of wound healing.[7] Thus, despite surgical finesse, modern device design, sterility, and miniaturization, implant sites do not heal "normally," instead producing a state of chronic irritation and inflammation that resolves to an unusual state of tolerance in some cases and outright rejection in others. Examination of what is currently known about the progression of the essential "normal" wound healing response is important for understanding components that distinguish the FBR. Diverse research efforts currently focus on the FBR, crossing immunology, pathology, surgery, cellular and molecular biology, materials science, and biomedical engineering. One effort focuses on physical and chemical aspects of the unique biointerface between the implant material and the body, often distilled to studies of "cells on surfaces."[11–13] At the cellular level, this response can be examined in terms of key immune cells known to interact with materials and facilitate FBR development, in particular monocytes and elicited or activated macrophages primarily sourced from circulating monocytes.[7] These cells are implicated in acute and chronic inflammatory responses to foreign entities of various origins.[7,14,15] Importantly, these cells produce complex molecular signaling cascades using soluble proteins and peptides. Hence, these "cell on surfaces" studies rapidly evolve into molecular biochemistry studies in the detective work associated with implant FBR physiology. Furthermore, communication and other relationships between monocytes, macrophages, and other host cells (e.g., fibroblasts and endothelial cells) found in the implant site are significant to biocompatibility of implanted materials. Macrophage cells at implant site use soluble molecular signals to recruit and influence the behavior of other cell types at this site, contributing to the development of specific physiological outcomes (e.g., inflammation, thrombosis, fibrosis, angiogenesis, and wound stabilization).[7,10,15]

Given the molecular and cellular basis for wound healing and aberrant healing responses observed in implant sites, and the empirical correlation of macrophage activity in implant sites with FBR, this chapter focuses on the macrophage as a fundamental player in this context. Specifically, the chapter introduces basic monocyte and macrophage biology and currently accepted standards for classifying these cells within the mononuclear phagocyte system (MPS). This is important to understand their complex relationships with various cells present at a wound (implant) site. Further examination of aspects of monocyte and macrophage differentiation and macrophage activation responding to diverse wound-site stimuli, and the differences between classical, alternative, and innate activation are discussed. Elucidating the cellular basis for normal versus aberrant wound healing and similarities and differences between macrophages and dendritic cells (DCs) is critical, so roles of the macrophage in the FBR and origin of multinucleated cells—foreign body giant cells (FBGCs), osteoclasts, and Langhans cells—are also mentioned. The chapter closes with some review of the known effects of materials chemistry, topology, and porosity on cell response and the use of the cage implant system to examine *in vivo* components of the FBR.

2.2.1 The Mononuclear Phagocyte System as a Source of Wound-Site Monocytes and Macrophages

An appreciation for the complexity and diversity of cellular players believed to mediate the FBR at biomaterial implant sites is important to understanding the cellular and molecular basis of the FBR. Significant to this challenge is the concurrent evolution of new understanding of the MPS and its constituents: bone marrow progenitor cells and circulating monocytes and macrophages, cells intimately tied to the FBR.[16] Monocytes and macrophages are leukocytes that are part of a very adaptable, highly dynamic hematopoietically derived cellular response system, exhibiting vast heterogeneity with respect to maturity, morphology, and response to stimuli upon activation in peripheral tissue sites.[17] Characterizing a plastic and locally recruited population such as this is not trivial. Classification within the MPS involves cellular origin (from bone marrow progenitors), cellular morphology and ultrastructural features, protein expression profiling (e.g., nonspecific esterase and lysosomal hydrolase activities, and cytokine production), and cellular behaviors (e.g., phagocytosis).[18] The significance of each of these characteristics to local wound healing and also implant-related responses is even more complex.

More generally, MPS cells can be classified by location, morphology, and cell surface markers. Cell appearance and size both change with their progression to fully differentiated, mature states. Monoblasts and promonocyte (MPS progenitor cells in bone marrow) are smaller and less complex in both appearance and function than either more mature monocytes or macrophages.[18] Macrophage cells are derived from blood monocytes and bone marrow precursors, representing the chief differentiated MPS cell.[19] These cells are distributed throughout the body, exhibiting a broad range of structural and functional diversity despite originating from common progenitors. Monocytes and neutrophils, another important wound-site leukocyte, share a common myeloid marrow-based progenitor, the granulocyte–macrophage colony-forming unit (GM-CFU).[18] Monocytes likely remain in the bone marrow for less than 24 h before entering the peripheral blood, where they circulate for several days before entering tissues to replenish resident macrophage populations.[20,21] After transition, monocytes do not return to circulation, but reside in the organ, tissue, or implant site for several months as mature, terminally differentiated macrophages.[18] Previously, it was accepted that approximately 95% of tissue-resident macrophages were derived from monocytes, and the remaining 5% derived from local division of mononuclear phagocytes within the tissues that have yet to complete cell division.[22] However, recent studies suggest that significant local macrophage proliferation and renewal do occur, especially under steady-state conditions.[16–18,23] This circulating residency is species dependent: monocytes exhibit a half-life of 17.4 h in mice, up to 70 h in humans, and constitute between 5% and 10% of peripheral blood leukocytes in humans.[17,20,24] The circulating monocyte exhibits heterogeneity; monocyte morphology, size, nuclear morphology, and granularity vary. Interestingly, monocyte heterogeneity is conserved in humans and mice, and is believed to reflect monocyte developmental stages correlating to distinct physiological roles such as tissue homing or response to inflammatory signals.[17] This is important to many preclinical *in vivo* models of inflammation, wound healing, and modeling of the FBR, where equivalence of murine and human cell origins and phenotypes in tissue sites is important and where knockout murine models of wound healing are increasingly implemented experimentally. Circulating monocyte populations are frequently harvested as a source of progenitor inflammatory cells, and differentiated to macrophage *in vitro* for experimental exploitation, to serve as a "primary culture" in diverse experimental models.

Monocyte cells vary expression of their cell surface markers related to their physiological roles. "Classic" monocytes were described as $CD14^{hi}CD16^{-}$ cells, and differential expression of these cluster of differentiation (CD) markers and chemokine receptor (CCR) profiles allowed further classification of monocytes (for a detailed review of monocyte subsets and surface markers, see Ref. 17).[25] Generally, monocytes are $F4/80^{+}$ $CD11b^{+}$ cells that can also be classified based on the expression of specific surface receptors (i.e., CCR2, CD62L (L-selectin), and CX_3C chemokine receptor 1).[26] One subset, the "inflammatory monocytes," expresses CCR2, targeting migration toward CC-chemokine ligand 2 (i.e., $CCL2^{+}$ or $MCP\text{-}1^{+}$), important to attracting cells to wound sites.[27,28] Importantly, different subsets of monocytes are thought to differentiate to specific cell types: inflammatory monocytes are thought to differentiate to macrophages, aid in the clearance of pathogens and/or apoptotic cells, and facilitate inflammatory resolution, whereas *in vitro* experiments suggest that $CD14^{+}CD16^{+}$ monocytes are more likely to become dendritic cells.[14,17] Furthermore, monocyte subpopulations that share common traits with more mature cells have been described; $CD14^{+}CD16^{+}CD64^{+}CD86^{hi}HLA\text{-}DR^{hi}$ monocytes share characteristics with dendritic cells.[29]

Monocytes are one of the two cell types known to differentiate to dendritic cells (Figure 2.1), the most competent antigen-presenting cells known and efficient stimulators of B and T cells.[30,31] *In vitro*, this maturation process is facilitated by the addition of GM-CSF and interleukin-4 (IL-4), known to push $CD14^{+}$ monocytes toward a dendritic cell phenotype.[30,32,33] Like macrophages, dendritic cells exhibit phenotypic and functional heterogeneity (e.g., interstitial dendritic cells, lymphoid dendritic cells, and Langerhans dendritic cells); furthermore, dendritic cells and activated macrophages exhibit similar stellate morphologies.[34] Dendritic cells can be distinguished from macrophages by their ability to activate naïve T cells and by altered expression profiles of specific markers (e.g., the β_2 integrin CD11c).[16,35]

In vivo, dendritic cells are migratory cells found in the blood and lymph (veiled cells), in secondary lymphoid tissues (interdigitating dendritic cells), in organs (interstitial dendritic cells), and in the epidermis and mucous membranes (Langerhans cells). Collectively and constitutively, dendritic cells express more major histocompatibility complex (MHC) class II and costimulatory molecules required for antigen presentation to T cells than macrophages, and this is one of the main distinctions between these cell types. Similar to macrophages, dendritic cells are a plastic population modulated by cytokines present in the local environment. Dendritic cells play a role in wound healing by responding to tissue damage and directing effector cells to tissue damage sites. In addition, they are capable of secreting copious amounts of proinflammatory cytokines at their precursor stage.[34]

As monocytes emigrate from blood vessels to tissues, they develop (differentiate) into highly specialized cell types (e.g., dendritic cells and osteoclasts) and macrophages (Figure 2.1); examples include alveolar macrophages (lung), Kupffer cells (liver), Langerhans cells (epidermis), and brain microglia and osteoclasts (bone).[19] Although the development and differentiation pathways remain poorly understood *in vivo*, efforts to determine branch points within the developmental

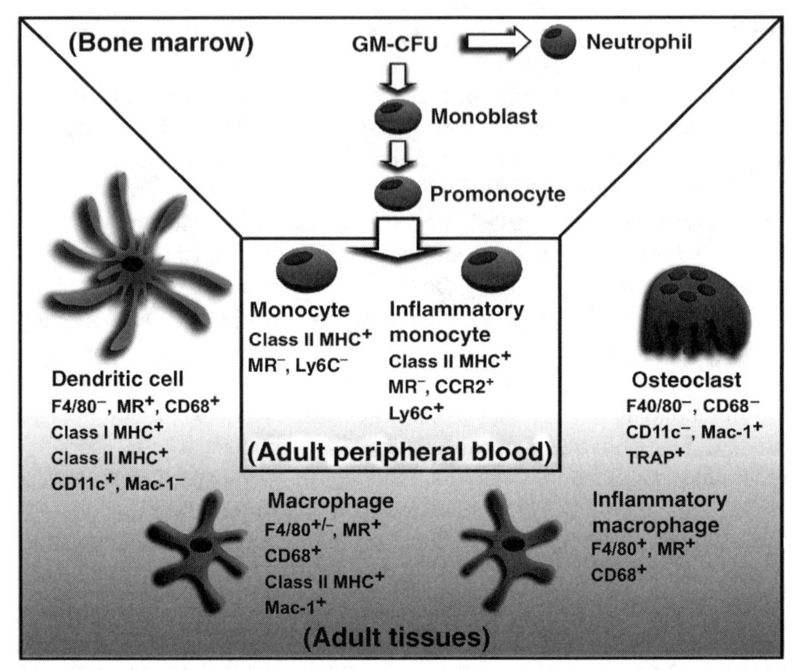

Figure 2.1 Monocyte (MC) derivation from precursors and maturation to highly specialized cell types: macrophages, dendritic cells, and osteoclasts. Neutrophils and monoblasts derive from a common precursor, the GM-CSF. Promonocytes precede MCs in maturity. MCs originate in the bone marrow and pass into the blood and tissues, where they mature to dendritic cells or macrophages (MΦs) in response to local environmental cues. Figure design by Dr. Marisha Godek; graphics by Dr. Gregory Harbers.

pathways have been undertaken.[16,17,19,35–37] Differentiated macrophage morphology is varied: cells typically possess a single, spherical concentric nucleus containing several nucleoli, large numbers of mitochondria and cytoplasmic vacuoles, a ruffled margin, and many fine cytoplasmic granules.[38] Macrophage heterogeneity is often associated with "immunocompetency," and specialization of function correlates to localization in different tissues.[18,19,39] Furthermore, small subpopulations of macrophages with distinct characteristics exist within specific tissues (e.g., spleen), suggesting a role for the microenvironment in phenotypic regulation.[17,19] Developing macrophages can be traced to early embryonic forms; these cells express macrophage markers including the β_2 integrin CD11b, mannose receptor (MR or MMR), and FMS (macrophage colony-stimulating factor 1 receptor precursor; CSF-1-R).[40–43] Importantly, it is still unclear whether tissue macrophages arise from lineage-committed precursors or randomly from the monocyte pool.[17]

Significantly, the implanted biomaterial provokes a spectrum of macrophage differentiation and activation profiles locally at the site, depending on this broad cell type heterogeneity, tissue site physiology, and device properties. It is also likely that local activation and acute wound healing at the implant site provide chemotactic cues that recruit other monocytes and immature MPS cells to the implant site.

2.3 MONOCYTE AND MACROPHAGE ACTIVATION AND INFLAMMATORY RESPONSE IN LOCAL TISSUE SITES

As a major immune system sentinel, macrophages play a critical role in host defense against a diverse group of environmental challenges.[44,45] Immune cell "activation" is frequently referred to, often described simply as an upregulation of a variety of signaling pathways resulting in altered gene expression and increased biochemical and functional activity, particularly receptor expression and macrophage phagocytic activity.[19] Much of this cell inflammatory activity is locally derived: implant sites provide considerable macrophage activation. Monocyte and macrophage heterogeneity converges in the host response to inflammation, where inflammatory monocyte-derived macrophages arise from a variety of stimuli.[17] Monocytes that enter tissues can mature into nonspecifically stimulated macrophages ("elicited" macrophages); alternatively, monocytes may be influenced by cytokines and mature into "activated" macrophages with enhanced functionality.[46] Generally, "activated" macrophages display more class II MHC, are more phagocytic, and release more cytokines (e.g., IL-1) than their unactivated counterparts.[47] Extreme activation leads to the overproduction of chemical mediators (e.g., cytokines, reactive oxygen intermediates (ROIs), and reactive nitrogen intermediates (RNIs)), which may in turn result in systemic imbalances often leading to tissue damage/necrosis, anaphylactic shock, and death. Once macrophage cells differentiate within a tissue, it is not known if they retain some degree of functional plasticity, or rather if they are "terminally differentiated."[17] However, tissue-resident macrophages are known to undergo "reactivation" in response to inflammatory stimuli, with enhanced expression of antigenic markers that are otherwise undetectable or present at very low levels.[46]

Inflammatory monocytes (defined as $CCR2^+Ly6C^+$) are recruited by cytokines and chemokines and undergo chemotaxis to inflammatory tissue sites where they differentiate to macrophages, likely under the local influence of further cytokine signals.[22] *In vitro* studies designed to simulate specific inflammatory stimuli have led to the delineation of different types of macrophage activation: classical, alternative, and innate. Furthermore, some stimuli (e.g., transforming growth factor, TGF-β, IL-4, IL-10) serve to "deactivate" macrophages (i.e., interfere with the process of macrophage activation or inhibit the cytotoxic functions of activated macrophages).[17,45] These processes are directly relevant to understanding roles of macrophage in the FBR: methods to reversibly activate or attenuate macrophage reactivity to implanted materials could prove fruitful in alleviating severity and duration of inflammation that leads to implant-based fibrosis complications.

2.3.1 Soluble Signaling Molecules Involved in Inflammatory Activation

A plethora of signaling molecules—acting in a tailored, concerted fashion—are involved in directing cellular activities and cell-to-cell communication, affecting a wide variety of cell behaviors including migration, adhesion, activation, proliferation, differentiation, and fusion—all important toward the establishment of a FBR *in vivo*. Chemokines are a superfamily of polypeptides and a subset of the larger class of cytokines that selectively control leukocyte activation, adhesion, and chemotaxis and typically have multiple effects on their target cells.[48,49] Cytokines are low molecular weight regulatory proteins or glycoproteins secreted by numerous cell types involved in cell-to-cell communication, but particularly potent macrophage activators capable of modulating macrophage function.[50] Infections, acute wounding, and tumors all elicit cytokine production as the basis for producing inflammation, but the types, kinetics, and cellular targets are important regulators of specific biological responses since most cytokines have short half-lives. Importantly, while cytokines direct macrophage behaviors, macrophages also produce many cytokines that act on various cell types including other macrophages. Inflammatory cytokine production may be the consequence of wounding, exposure to bacterial endotoxins (e.g., lipopolysaccharide (LPS)), microbes and viral particles, or priming factors (e.g., interferon-γ (IFN-γ).[51] A list of cytokines produced by macrophages and their biological actions can be found in Table 2.1. Importantly, cytokines are pleiotropic: their specific action depends upon the physiological context in which they are present and cells on which they might act. Significantly, macrophage cytokine secretory profiles are likely affected by cell maturity and local environmental cues. This is perhaps the most important challenge in controlling of the local tissue reaction to implanted biosensors: guidance of the macrophage cytokine expression and reactivity profiles in the presence of the implant.

In vitro, stimulants are routinely employed to push monocyte/macrophage cells to an "activated" state. These include LPS, a potent mitogen, and phorbol esters (e.g., phorbol 12-myristate 13-acetate (PMA)/12-*O*-tetradecanoylphorbol-13-acetate (TPA), and others) that act as differentiating agents.[52] LPS is an endotoxin present in Gram-negative bacterial cell walls, often associated with systemic bacterial infections, that leads to macrophage activation and subsequent production of inflammatory cytokines.[51,53] LPS induces cytokine gene expression via numerous transcription factors including members of the AP-1, C/EBP, Ets, and NF-κB/rel families.[54] Other bacterial components including peptidoglycans, trehalose diesters, lipoteichoic acid, and lipomannans are all recognized by and activating to macrophages.[54]

Phorbol esters are tumor promoters capable of binding to and activating protein kinase C (PKC), one critical component in T cell activation.[55,56] PMA is a structural analogue of diacylglycerol (DAG), an allosteric activator of PKC. PKC activation via phosphorylation leads to calcium release, resulting in a cascade of cellular responses including rapid proliferation.[56] Experimentally, phorbol esters have been used as chemoattractants and to differentiate nonadherent monocytes to adherent macrophage cultures.[55,57–60]

"Classical macrophage activation" can be achieved by stimulation with cellular cytokine IFN-γ and microbial activators (pyrogens, LPS), resulting in amplified microbicidal activity, production of reactive oxygen species (ROS) and proinflammatory cytokines, and increased antigen presentation. In contrast, "alternative activation" is the result of cell exposure to cytokines IL-4 and IL-13, associated with tissue repair and humoral immunity. Alternative activation enhances phagocyte endocytosis, heightened expression of MR, and increased cell growth and tissue repair. "Innate activation," typically associated with infections, involves stimuli (e.g., pathogen-derived LPS, peptidoglycans, and lipoteichoic acid) that act through

TABLE 2.1 Selected Macrophage Cytokine Products and Their Effects

Cytokine	Stimulus for production	Biological action	Induces production of	Cellular target(s)
G-CSF	IL-1, LPS	Granulocyte colony stimulation, terminal differentiation of myeloid cells, enhanced neutrophil function	—	Granulocytes, myeloid stem cells, neutrophils
GM-CSF	IL-1, TNF, LPS, retroviral infection	Granulocyte, eosinophil and macrophage colony stimulation, enhanced neutrophils/eosinophil function	IL-1, TNF, PGE_2, O_2	Macrophages, eosinophils, granulocytes
M-CSF	IL-1, LPS	Macrophage colony stimulation, antiviral	IL-1, IFN-γ, TNF-α, PGE_2, PA	Macrophages
TNF-α	IL-1, IL-2, GM-CSF, LPS	Tumor necrosis, endotoxic shock-like syndrome, cachexia, fever, acute-phase protein response	IL-1, IL-6, GM-CSF, MHC I, MHC II	Macrophages
IL-1	IL-2, TNF, GM-CSF, Ag presentation	Fever, acute-phase protein response, hypotension, increased ICAM-1 expression (endothelial cells)	IL-2, IL-4, IL-6, TNF, PGE_2, collagenase	Macrophages, endothelial cells
IL-6	IL-1, TNF, PDGF	Hemopoietic cell proliferation, fever, acute-phase protein response	IL-2R (T cells), IgG (B cells)	Macrophages, T cells, B cells
IL-10	T and B cell activation, LPS	Inhibitory to T cell proliferation and cytokine production	MHC II (B cells)	Macrophages, T cells, B cells
IL-12	T cell activation, LPS	Promotion of cell-mediated immunity, activated T and NK cell proliferation	IFN-γ	T cells
TGF-β	—	Fibrosis and wound healing *in vivo*, influences integrin expression and differentiation	—	Macrophages, T cells, B cells, epithelial cells
FGF	—	Angiogenesis *in vivo*, endothelial cell chemotaxis and growth	IFN-γ	Endothelial cells, myoblasts

(continued)

TABLE 2.1 *(Continued)*

Cytokine	Stimulus for production	Biological action	Induces production of	Cellular target(s)
PDGF	LPS, lectins, zymosan, thrombin coagulation	Neutrophil activation, collagen synthesis augmentation, mesenchymal cell proliferation and chemotaxis	IL-1, IL-1R, IFN-β, IFN-γ, PGE_2	Neutrophils, mesenchymal cells
EGF	—	Angiogenesis, wound healing, proliferation/ differentiation of basal epithelial cells	—	Epithelial cells
MIP-1α	—	Neutrophil recruitment, regulation of hemopoiesis	—	Neutrophils

Other macrophage cytokine products include IFN-α, IL-8, IL-15, IL-16, and IL-18. Abbreviations: colony-stimulating factor (CSF), interleukin (IL), transforming growth factor-β (TGF-β), fibroblast growth factor (FGF), platelet-derived growth factor (PDGF), epidermal growth factor (EGF), macrophage inflammatory protein-1α (MIP-1α), lipopolysaccharide (LPS), antigen (Ag), tumor necrosis factor (TNF), intracellular adhesion molecule-1 (ICAM-1), prostaglandin E_2 (PGE_2), major histocompatibility complex (MHC), and interferon (IFN). Sources for this information can be found in Refs 47,132–141.

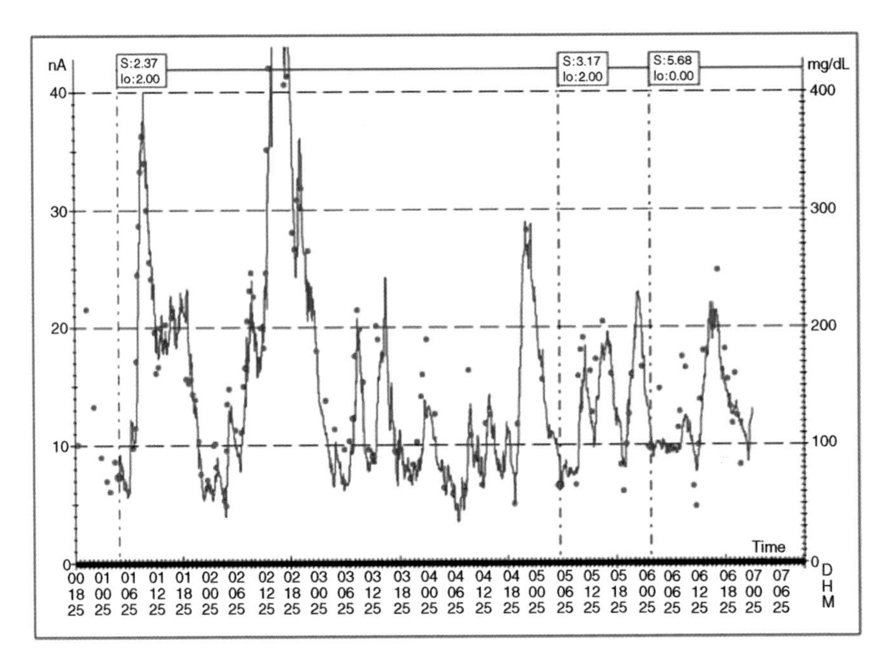

Figure 1.6 Seven-day monitoring of a diabetic patient. Points obtained from fingerstick method, solid trace. Three two-point calibrations are noted on days 1, 5, and 6.

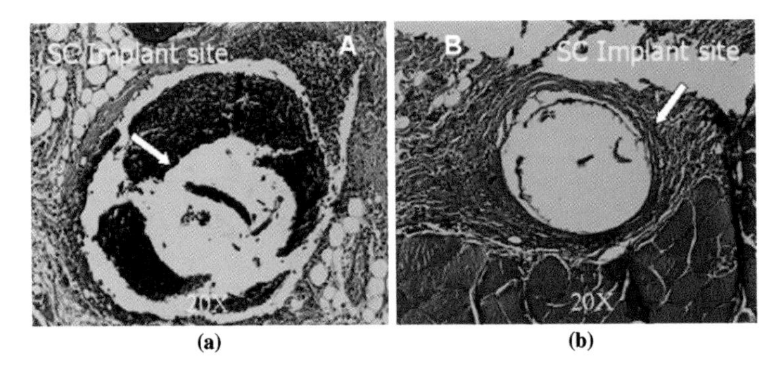

Figure 1.8 Tissue section at implant site showing the effect of NO evolution on the acute inflammatory response. (a) Sensor with no NO evolution (control); (b) no evolution for about 18 h. Reprinted with permission from Ref. 65. Copyright 2005 John Wiley & Sons.

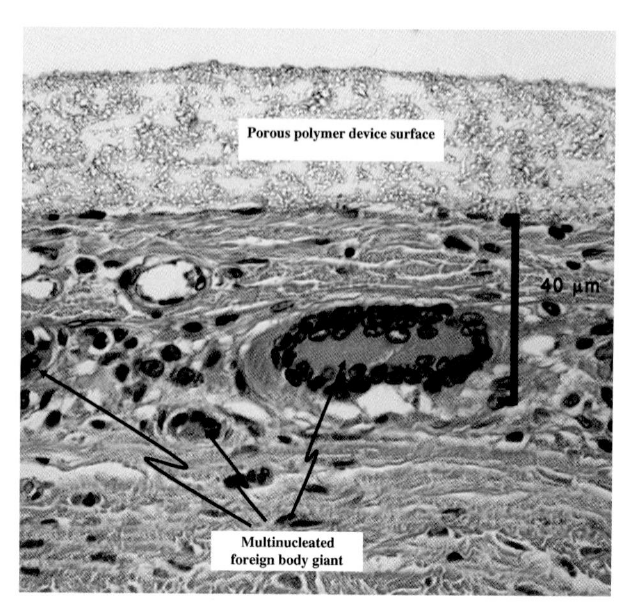

Figure 3.1 H&E section showing foreign body giant cells obtained 28 days after subcutaneous implantation of a porous polyvinyl alcohol implant in a rat. Reprinted with permission from Ref. 10. Copyright 2004 John Wiley & Sons.

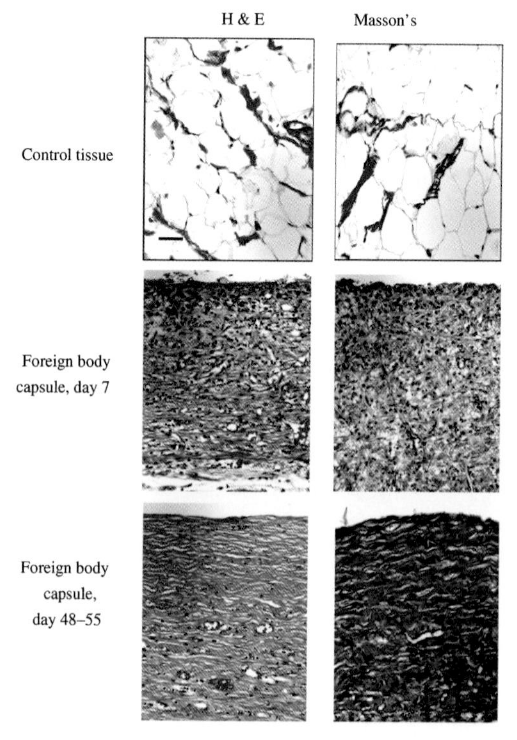

Figure 3.3 Histologic sections from foreign body capsules obtained at early and late time points after polyurethane implants in rats. The bar in the top left section represents 75 μm for all sections. Reprinted with permission from Ref. 16. Copyright 2007 John Wiley & Sons.

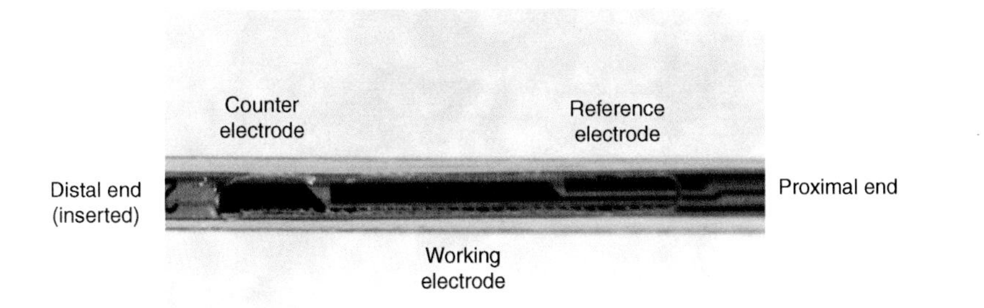

Figure 5.9 Close-up view of the Medtronic sensor showing the working, counter, and reference electrodes. Copyright 2008 Abbott. Used with permission.

Figure 6.8 Capsule differences between probes infused with or without MCP-1.

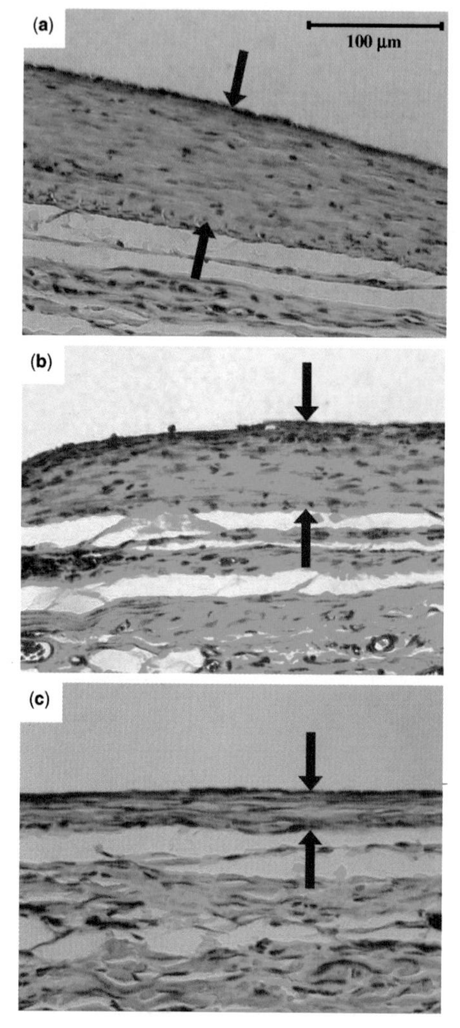

Figure 9.3 Optical micrographs of subcutaneous tissue showing the foreign body capsule (denoted by arrows) formed after 6 weeks at (a) bare silicone rubber elastomer, (b) xerogel-coated control, and (c) NO-releasing xerogel-coated implants. Reprinted with permission from Ref. 32. Copyright 2007 Elsevier.

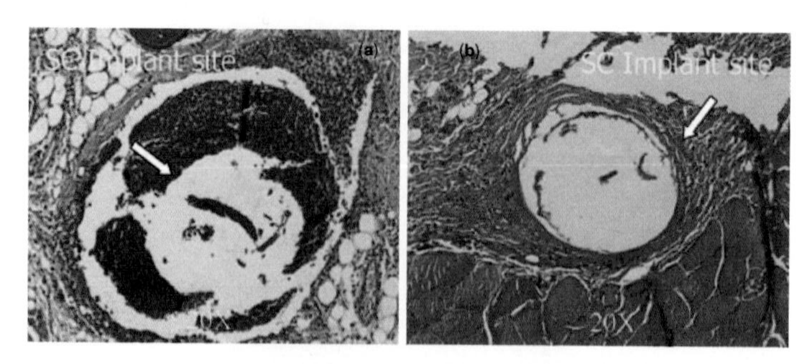

Figure 9.8 Tissue cross-section of (a) control and (b) NO-releasing sensor implant site. Reprinted from Ref. 48 with permission of John Wiley & Sons, Inc. Copyright 2005 John Wiley & Sons, Inc.

ligation of the toll-like receptors (TLRs), whereas cell deactivation can be achieved by IL-10, TGF-β, and steroids.[17]

In addition to cytokine and chemokine production, macrophages and other MPS cells are known to produce a wide range of substances (over 100) varying in size (32–440 kDa) and biological activity, affecting virtually every aspect of cell behavior from cell growth to cell death.[45] Enzymes such as lysozyme, proteases, and lipases, as well as enzyme inhibitors, complement components, adhesive proteins (e.g., fibronectin), ROIs and RNIs, and coagulation factors are all known secretory products of macrophages.[18,45] Furthermore, macrophages are known to modulate implant-associated angiogenesis and fibrogenesis by producing numerous factors (see below).

2.3.2 Macrophage Receptors as Phenotypic Markers of Activation States

Macrophage presence and reactivity in tissue sites are elicited though different combinations of effector molecules (e.g., cytokines, chemokines) acting on specific receptors; thus, the expression and presence of given membrane receptors on macrophage cell surfaces correlate to specific (acquired) functions and activation profiles.[18] Furthermore, macrophage maturity and/or differentiation state is/are often characterized based on the presence or absence of specific surface receptors (e.g., F4/80$^+$, Mac-1$^+$; Figure 2.1).[61] Receptor profiles vary between cells of different maturity, primary and secondary derivation, and between cell lines.[62]

Early membrane phenotypic markers identified on macrophages were receptors for the fragment constant (Fc) portion of IgG antibodies and for the complement protein C3.[63,64] Other important receptors include lipoprotein, fibronectin, laminin, fibrinogen, hormone, chemokine, lectin-like, adhesion and migration, and advanced glycosylation endproducts.[18,29] The macrophage mannose receptor (MMR) is a classic phagocytic receptor highly upregulated by IL-4 and implicated in IL-4-induced fusion to FBGC (see below).[65] Cytokine receptors include numerous interleukins (IL-1, IL-2, IL-3, IL-4, IL-6, IL-7, IL-10, IL-13, IL-16, IL-17), macrophage-colony stimulating factor (M-CSF/CSF-1) and GM-CSF, interferons (IFN-α, IFN-β, IFN-γ), and many others.[18] A diverse array of chemokine receptors have also been identified: CCR1, CCR2, CCR5, CCR8, CCR9, CXCR1, CXCR2, CXCR4, and CX3CR1.[18] Significantly, no single marker is known to reliably distinguish macrophages from dendritic cells.[16] Phenotypic characterization of macrophage in wound and implant sites has focused on determination of the qualitative (presence or absence) and semiquantitative expression levels (up- or downregulation) of these characteristic receptors using diverse methods including flow cytometry, immunoassays, and staining and genetic expression assays (e.g., polymerase chain reaction (PCR) and its variants, microarray assays).

2.4 ROLE OF THE MACROPHAGE IN NORMAL WOUND HEALING

The normal wound healing process is a concerted effort between multiple cell types including leukocytes (monocytes, macrophages, platelets, granulocytes, and lymphocytes), fibroblasts, endothelial cells, and keratinocytes.[14] Macrophages play an instrumental role in wound healing; the absence of wound-site macrophages has been correlated to impaired wound healing.[66] Under normal circumstances (i.e., the absence of a biomaterial) wound healing progresses through four stages: hemostasis, inflammation, proliferation, and resolution.[14] Hemostasis (vasoconstriction and clot formation) occurs initially, followed by the inflammatory stage, beginning within hours of injury and persisting for several days. The inflammatory stage is marked by the influx of leukocytes; tissue macrophages in particular play a critical role in wound healing.[67] "Proliferation" refers to the regenerative stage where protein synthesis, cellular differentiation or replication occurs, and the healing process begins. During the resolution stage, wound remodeling, accompanied by regression of newly formed capillaries, takes place.[14]

The sequence of events related to normal wound healing involves several essential cellular players. Neutrophils, comprising 60–70% of the circulating leukocyte population, arrive at the wound site first but are quickly replaced by monocytes.[18] Monocytes are attracted by a variety of chemokines, platelet-derived factors, and complement split products produced locally and transiently.[14] Monocytes are recruited throughout the inflammatory stage, and once at the injury site differentiate to macrophages, becoming the predominant cell type in the wound by day 5 and persisting for up to months.[7,18] Activated macrophages amplify the inflammatory response by producing chemoattractants for monocytes and other cell types, resulting in further cellular infiltration of the injury site. Macrophage-derived factors play a role in angiogenesis, extracellular matrix deposition, and fibroproliferation requisite for wound repair.[10,15,68] Furthermore, macrophage presence at the wound site is essential for removal of necrotic tissue and apoptotic cells.[67]

Primary macrophage functions in normal wound sites are (1) phagocytosis of apoptotic cells (e.g., neutrophils) and extracellular debris and the modulation of (2) angiogenesis and (3) fibrogenesis for wound-site homeostasis.[14,67] Phagocytosis of senescent neutrophils is believed to be mediated by specific subsets of lectin (monosaccharide specific), integrin (thrombospondin specific), and lipid (phosphatidylserine specific) receptors found on wound-site macrophages.[69] Numerous studies report wound-site macrophage stimulation of capillary growth (angiogenesis) required to supply nutrients for tissue regeneration.[14] *In situ* studies indicate that these macrophages produce various proangiogenic factors including transforming growth factors-α and -β (TGF-α, TGF-β), vascular endothelial growth factor (VEGF), platelet-derived growth factor (PDGF), IL-8, and basic fibroblast growth factor (bFGF), a potent endothelial cell proliferative agent that has been shown to be angiogenic *in vivo*.[14,70–75] In addition to producing proangiogenic factors, monocytes and macrophages are known to produce and secrete thrombospondins, as well as several other poorly defined antiangiogenic factors, suggesting an additional counter role in vascular regression and the resolution phase of wound healing.[76]

Similar to their modulation of angiogenesis, wound-site macrophages exhibit both pro- and antifibrotic functions required for acute and chronic wound stasis and remodeling.[77] Profibrotic functions include the ability of macrophages to produce extracellular matrix (ECM) components such as fibronectin, thrombospondins, and proteoglycans.[45] Furthermore, macrophage-derived growth factors such as TGF-β and PDGF directly stimulate collagen synthesis and fibroblast proliferation.[15] In contrast, macrophages produce numerous effectors of ECM degradation including metalloproteinases and proteinase activators.[78] These findings suggest a role for the wound macrophages in ECM remodeling. These same macrophage functions are also important in implant-associated fibrosis and encapsulation responses. Two chemokines that affect the participation of both monocyte and macrophage in wound healing—MCP-1 and macrophage inflammatory protein (MIP)-1α—are known to exhibit distinct patterned expression throughout the course of wound repair. MIP-1α levels peak concomitant with maximum macrophage levels within the wound site. Furthermore, MIP-1α antibody neutralization was shown to decrease macrophage numbers at the wound site, suggesting that MIP-1α is crucial for macrophage recruitment during wound repair.[79,80]

Limited studies have demonstrated that wound-site macrophages are phenotypically distinct from other macrophage populations, and these altered phenotypes have both temporal- and location-dependent components.[81,82] Based on the changing role of the wound macrophage in angiogenesis alone, a phenotypic shift from "reparative/proliferative" to "regression/antiproliferative" should occur.[14] However, a wound-specific phenotype is currently unknown and rigorous investigation toward this end has not been attempted; this type of study is complicated by the heterogeneous wound macrophage population, evolving phenotypes, the pleiotropic nature of wound macrophage function, and the overlapping phases of (and functional shifts assumed to accompany) wound healing.

2.5 THE FOREIGN BODY REACTION

The FBR is often described as an altered, abnormal, or aborted wound healing process. The most obvious distinction between normal and abnormal in this context is that wound healing in the presence of a foreign entity (e.g., a biomaterial or biomedical device) does not progress to or through the normal resolution stage (e.g., a scar). The FBR is often associated with unresolved local acute inflammatory events, chronic inflammation, continued and prolonged presence of monocytes, macrophages and lymphocytes, limited angiogenesis, and excessive fibrogenesis.[7] Chronic inflammation may also result from implant-associated infection, inherent chemical and physical properties, or motion of the implant.[7] Chronic inflammation is generally localized and usually short lived, but may persist for years in the presence of a foreign body.[7] Generally, prolonged, protracted inflammation, wound healing processes, and the ensuing FBR are all facets of the host response to tissue injury that occurs upon placement of a biomaterial or biomedical device.[7] In the context of implanted biomaterials, the FBR with the development of granulation tissue, chronic inflammation, and fibrosis is considered to be the normal wound healing response.[7]

The FBR to biomaterial implants physically comprises the implant, FBGC and surrounding granulation tissue comprising macrophages, fibroblasts, and capillaries, in varying amounts.[7] Surface chemical and physical properties of the implant (e.g., composition, surface roughness, general shape, and physiological placement) influence to some extent the intensity of the FBR reaction, its physiological structure, and thickness of the encapsulation tissue.[7] Notably, no materials or device designs reported to date completely attenuate the FBR, indicating a basal, generic level of host reactivity that may be impossible to eliminate. Most, but not all, implanted materials in soft tissue end up extensively fibrosed (abnormal formation of fibrous tissue) and physically isolated from the adjacent host tissue via an avascular fibrous capsular sheath (i.e., densely compacted collagen matrix).[83,84] Altered cellular phenotype(s) within the wound site, perhaps mediated by local inflammatory macrophage populations, may lead to overproduction of collagen and ECM components that contribute to fibrosis and fibrous encapsulation.[7] The FBR fibrous capsule provides an adherent, avascular, impermeable barrier often responsible for biomedical device failure or malfunction.[8,85] This constitutes one of the hallmarks of end-stage "healing" in response to a biomaterial, distinct from that of normal healing.[7] Relatively flat, uniform surfaces may have a FBR sheath containing only one to two macrophage-occupied strata, while rough or angular, edged surfaces exhibit a thicker cellular layer directly adjacent to the implant material, with varying degrees of granulation tissue subadjacent to the cellular layer.[7] End-stage healing (remodeling) at the implant site may involve either tissue regeneration (new tissue synthesis by parenchymal cells of the same type) or dysfunctional replacement (connective tissue, fibrous capsule formation).[86] The underlying tissue anatomical and physiological framework is critical for successful regeneration; without it, fibrosis occurs but normal function is not restored. The persistent inflammatory state surrounding implanted materials, elicited by several cellular players and correlated with macrophage presence, is likely to be responsible for the departure of wound healing mechanisms from normal prohealing and remodeling to excess matrix and fibrotic reactions observed around sensors. It is likely that macrophages within the biomaterial/implant wound site exhibit altered molecular repertoires compared to "normal" wound healing macrophage populations and that the molecular profile of biomaterials-associated macrophages lies somewhere between the "normal" wound healing macrophage populations and that of a giant cell. Unfortunately, none of these cell populations is well characterized at this time, and substantial effort must be undertaken to progress toward a thorough characterization of these cell types to understand the progression from normal to abnormal wound healing.

2.5.1 The Role of the Wound-Site Macrophages in the Foreign Body Reaction

In addition to terminal differentiation into macrophage phenotypes, MPS cells are known to exhibit an alterative terminal state in specific microenvironments within the host: the fusion product of multiple cells to produce a multinucleated giant cell (Figure 2.2).[18] The first reported observation of these cells dates back to 1868 by Langhans and later reports (1912, 1925, and 1927, respectively) by Lambert and Lewis.[10,87,88] Three types of giant cells are most commonly identified: osteoclast, Langhans cell, and FBGC. Each is distinct in location and appearance, but all share common traits: (1) derivation from MPS cells, (2) a role in tissue remodeling and/or immune defense, and (3) an association with disease, lesions, and tumors.[89]

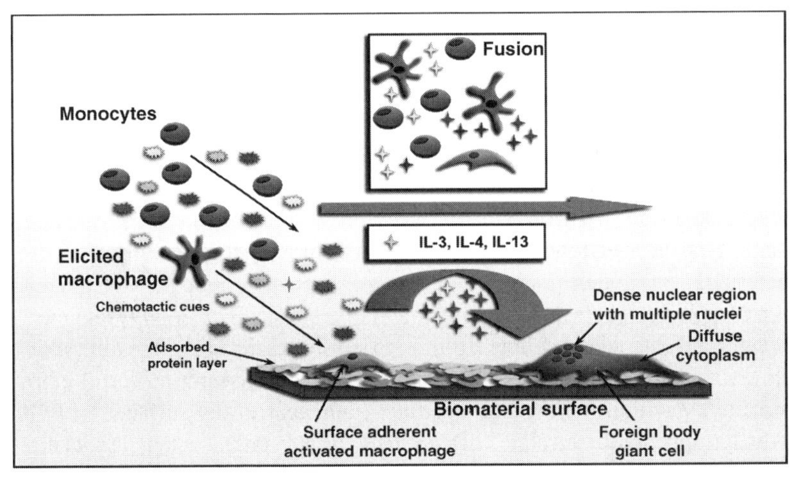

Figure 2.2 FBGC formation at the biomaterial surface. Circulating MCs and tissue-resident macrophages, MΦs, are elicited/attracted to the wound site (biomaterial) by chemotactic cues (e.g., MCP-1). MCs mature at the wound site to MΦs that may adhere to the protein adsorbed biomaterial. Adherent and/or activated MΦs produce soluble chemical signals (i.e., cytokines, chemokines) that mediate further influx of MCs, MΦs, and other cell types to the wound site. In the presence of specific cytokine signals (e.g., IL-3, IL-4, IL-13), MCs and MΦs may fuse into large multinucleated cells, FBGC, which may contain over 100 nuclei in their dense central nuclear region. *Note*: Cartoon is not to scale. Figure design by Dr. Marisha Godek; graphics by Dr. Gregory Harbers.

These cells are well known to be the products of fusion, not cell division in the absence of cytokinesis.[45,89–92]

Osteoclasts are multinucleated cells found attached to the bone that they actively resorb and critical for normal bone homeostasis—they play a key role in normal bone turnover and replacement. Normally, osteoclasts are recruited and stimulated by strain-induced signals coming from local regions of microdamaged, mechanically stressed bone. Osteoclastic activity resorbs small patches of bone as a normal prerequisite to bone neogenesis. However, they are also active in bone tumors, bone transplant rejection, and the pathogenesis of osteoporosis.[89] Langhans cells (not to be confused with the epithelial dendritic Langerhans cell) are examples of giant cells frequently encountered in diseased states involving granulomatous lesions such as tuberculosis, syphilis, sarcoidosis, and deep fungal infections.[87,93] These cells exhibit a nuclear arrangement that is horseshoe or circular in shape, and the nuclei are found at the periphery of the cells.

Foreign body giant cells, the large multinucleated cells characteristic of the FBR surrounding implant materials and devices, comprise many fused monocyte and macrophage cells in response to a diseased state and/or the presence of foreign entities such as implanted biomaterials.[7,10,87] Residing in direct proximity to the tissue–implant interface, FBGCs may persist for the lifetime of the implant.[7,10] The exact nature of their cellular state (metabolically active, phagocytic, quiescent, senescent) at extended times *in vivo* remains unknown.[7] Furthermore, the exact molecular signals that regulate FBGC formation and FBR development remain undefined.[10]

FBGCs may contain over a hundred nuclei, reflecting massive fusion of mononuclear MPS cells.[7,94] In contrast to the peripheral nuclear arrangement of the Langhans cell, the nuclei of the FBGC are found in a dense central location.[95] FBGCs are not highly phagocytic, possibly due to the relatively large size of the foreign entity that results in a FBR.[96] Despite the fact that FBGC do not phagocytize the implant directly, they do exhibit increased lysosomal and respiratory enzyme activity, and are reported to exhibit fewer cell surface and complement receptors, suggesting a change to a task-oriented phenotype directed toward dismantling and removing the insulting entity.[7,97] In fact, upon fusing to a giant cell phenotype, the macrophage is thought to redirect its functional role from endocytosis to establishment of a powerful external lysozome.[89] Both giant cells and osteoclasts adhere strongly to substrates through a "sealing zone" into which lysosomal enzymes and protons are secreted, allowing powerful extracellular bioactivity within a closed compartment adjacent to the surface.[98] FBGC presence always appears to correlate with an FBR. Importantly, such an association could be only correlative and not causative. That the FBGC contributes functionally to the FBR, further inflammatory mediation, or prolongation of abnormal fibrotic reactions remains to be proven.

Monocytes, the immature circulating progenitor form of the macrophage, differentiated macrophages, and activated macrophages all play important roles in the development of the FBR *in vivo*.[10,18]Figure 2.2 outlines key events associated with biomaterials placement in the body. The sequence begins with the rapid and immediate adsorption of wound-site host proteins from blood, lymph, or other extracellular fluids (complex biological milieu, thousands of possible proteins).[6] The biomaterials implant is likely to be completely covered with adsorbed proteins, many in nonnative conformations and also in dynamic states of exchange with other soluble species. Hence, the biomaterials surface chemistry is likely to be masked by a poorly controlled, poorly defined protein layer by the time host cells interrogate the implant site. Host neutrophils are the first cells to arrive at the site upon acute injury—their presence is very transient (neglected for simplicity in this diagram) as they are rapidly replaced by circulating monocytes that differentiate *in situ* in response to endogenous wound healing factors into mature, activated macrophages, often implant adherent. Subsequently, these macrophages provide soluble signals for further monocyte/macrophage infiltration, and, in response to cytokine cues, some macrophages may fuse to form FBGC. Whether only fully differentiated, mature, or inflammatory macrophage populations are exclusively involved in FBGC formation is not known. Furthermore, implant site-resident macrophages modulate angiogenesis and fibrogenesis as a wound healing response.[10,15]

Foreign body giant cell formation (i.e., macrophage fusion) is influenced and enhanced by specific cytokines including IL-3, IL-4, and IL-13.[95,99–101] IL-4 is known to upregulate expression of the MMR, also known as CD206, CCR2 (the receptor for monocyte chemoattractant protein-1, MCP-1), a classic phagocytic receptor implicated in fusion to FBGC.[65] MCP-1 has also been implicated in monocyte recruitment to biomaterials implanted in the peritoneal cavity and has been deemed requisite for macrophage fusion to FBGC.[10] Furthermore, MCP-1 has been detected on FBGC at implant sites 2–4 weeks after implantation, suggesting its role in recruitment of additional monocytes/macrophages and maintenance or progression of the FBR.[10] The fusion process is thought to be mediated by the interaction of transiently expressed cell surface proteins, although the detailed mechanism remains unclear.[89]

Several molecular markers that play a role in fusion have been identified: MMR, the macrophage fusion receptor (MFR), comprising CD44 and CD47, and the dendritic cell-specific transmembrane protein (DC-STAMP).[45,89,102,103] *In vivo* and *in vitro* macrophage studies performed by Vignery et al. have demonstrated that all cells tested, regardless of origin (species, organ/tissue), express the same functional markers, a characteristic of osteoclasts.[98,104–107] To date, no single molecule has been identified with restricted expression to either giant cells or osteoclasts, although it has been postulated that multinucleated macrophages acquire a tissue-specific molecular repertoire.[89]

In addition to altered or enhanced cell *surface* protein expression, distinct matricellular proteins (secreted macromolecules that interact with ECM proteins as modulators of cell–matrix interactions that serve no structural role) mediate elements of the FBR: inflammation, angiogenesis, and collagenous encapsulation. Members of the matricellular protein family (i.e., osteopontin, tenascins, thrombospondins, SPARC) are upregulated during inflammatory events and play a role in mediating cell–matrix interactions.[108–110] For example, Hevin and SPARC, two members of the SPARC family that are highly homologous, mediate different aspects of the FBR; Hevin plays a role in biomaterial-induced inflammation, whereas SPARC mediates collagen-based capsule formation.[110,111] Thrombospondin 2 (TSP2) has been shown to influence the development of the FBR capsule; experiments performed in TSP2-null mice demonstrated a highly vascularized thick capsule with abnormally shaped collagen fibers as compared to normal mice.[109]

2.5.2 Modulating the FBR *In Vivo*: Approaches to the "Pink Capsule"

The FBR appears to be a ubiquitous, graded response to all implanted materials at least in soft tissue, present to some extent regardless of device chemical composition or physiological placement. Frustration in addressing the clinical problems of the FBR with biomaterials chemistry alone has prompted alternative strategies. In this regard, both pharmacotherapies (i.e., drug delivery) and biomaterials interfacial physical modifications including topological and topographical cues, textures and surface-accessed porosity, and physical barrier strategies have been employed to evade or attenuate the FBR *in vivo*.[112] A primary goal has been to locally modulate the properties of resulting fibrosis and encapsulation *in vivo*—to achieve the "pink capsule"—a vascularized, perfused, and readily diffusive fibrous end point rather than the impenetrable barrier of the typical fibrous sheath encapsulating most devices. The "pink capsule" goal is a clinical resignation to the host tissue's persistent fibrotic response but with the compromise to produce angiogenesis sufficient to provide adequate blood flow for oxygen and analyte transport across the capsule to and from the device surface. This objective then necessarily has correlated production of the host inflammatory response with methods for prompting neovasculogenesis locally *in situ* adjacent to the implant. In fact, the correlated biology of wound healing and implant-associated inflammatory macrophages with follow-on angiogenesis can be exploited in this regard.[113]

Implant-associated coupling of local inflammation control and angiogenesis has been achieved with both local drug release and biomaterials physical cues, altering rates and histology of capsule formation, and associated angiogenesis. Dexamethasone, a well-known glucocorticoid anti-inflammatory drug active against macrophages, has been released for over a decade from commercial pacemaker leads in humans and more recently deployed in vascular stents to diminish inflammatory and fibrotic activities.[114–116] As dexamethasone might also attenuate angiogenesis, its local release is coupled with co-release of a vasculogenically active molecule (e.g., growth factors VEGF, PDGF, or bFGF) to induce new blood vessel production into the device coating or surface, resulting in a combination drug therapy at the implant site. Release of VEGF alone from implanted devices subdermally may not provide effective angiogenesis, as shown in the abnormal edematous vascular product from a subdermal VEGF releasing sensor coating in rats.[117,118] However, delivery of the VEGF transgene to an *ex ova* chorioallantoic membrane model provided new vascularity sufficient to improve acetaminophen biosensor performance for 8 days in this model.[119] Rats exposed to dual release of dexamethasone and VEGF from degradable microspheres exhibit attenuated inflammatory and fibrotic response and enhanced vessel growth.[120] However, another recent study has shown that combination local release of dexamethasone and VEGF from a subdermal sensor hydrogel coatings in rats alters several inflammatory responses but not in mutually additive, beneficial ways that might reliably produce the "pink capsule" or attenuated capsule performance objectives. Once drug release was exhausted, normal foreign body reactions recurred.[121]

Biomaterials surface texture, porosity, and microtopology have a long history of producing varied inflammatory, histological, and angiogenic responses *in vivo*. Intuitively, tissue inflammatory and angiogenic responses must be correlated effects from this influence but are largely empirically observed and without a clear and predictable mechanism or enduring FBR influence.[122–125] In one of the many examples, neovascularization at the interface of microporous polymer membranes and tissues was shown to occur in several implanted membranes with pore sizes sufficiently large to permit complete penetration by host cells (0.8–8 μm pore size). Larger pore membranes had 80–100-fold more vascular structures at the membrane–tissue interface for 5 μm pore size polytetrafluoroethylene (PTFE) membranes compared to 0.02 μm pore size PTFE membranes and maintained for 1 year subcutaneously in rats. Nonetheless, this strategy only transiently suppresses certain components of the FBR without complete, satisfactory resolution or clinical solution.[126] This suggests a cellular effort toward "biointegration" of these textured or porous membranes, advantageous to implant systems or devices where nutrient supply, oxygenation, and exchange of other soluble factors are critical to device function as well as tissue integration. How this might be promoted longer term to more permanently stabilize the implant site toward perfused, integrated tissue or a vascularized capsule remains to be achieved.

2.5.3 Foreign Body Giant Cell Formation *In Vitro*

To further characterize the role of the macrophage and significance of the FBGC in the FBR, much effort has been devoted to establishing reliable cell fusion protocols for *in vitro* studies. Most efforts focus on pooled human blood monocytes, matured in cultures using GM-CSF and subsequently stimulated with IL-4 and IL-13 to produce fused cells.[65,95,100,101] Numerous factors play into achieving successful *in vitro* fusion: cell source (primary versus secondary derived (immortalized) cells), cell maturity (monocyte versus macrophage lineages), serum source (autologous versus nonautologous, complement containing), cell age (passage number for secondary cells), and the cytokine cocktail employed to facilitate fusion. Human primary cells represent an expensive and short-lived cell source to form FBGCs *in vitro*; reliable use of secondary-derived immortalized monocyte–macrophage cell lines to produce relevant FBGC cultures would be greatly beneficial in terms of a readily available cost-effective cell source. To date, only limited fusion studies have been performed using cell lines, and the high fusion rates typical for primary-derived cells have not been achieved in cell lines.[65,127] This means that FBGC studies focus largely on *in vivo* implant models with both the benefits and the challenges of the full *in vivo* spectrum of cellular and molecular host response to dissect in order to tease out FBR mechanisms.

2.6 *IN VIVO* TECHNIQUES TO INTERROGATE CRITICAL COMPONENTS OF THE FBR

In vitro duplication of the FBR with much physiological fidelity is seemingly complicated by its relative simplicity compared to the *in vivo* implant wound healing scenario. Full cascades of cell–cell signaling responses, multiple cell types, and the phased kinetics of various cell-based inflammatory responses are completely missing from most models reported to date. Therefore, *in vivo* animal implant healing models have been developed in attempts to get more accurate information closer to the actual situation in human FBR.

2.6.1 The Cage Implant System

Historically, *in vivo* biocompatibility has been qualitatively assessed by histological evaluation of explanted devices.[6] The presence or absence of specific inflammatory and fibrotic cell types and the degree and duration of the incidence of these cells at the implant site can be correlated to a relative degree of success or failure in terms of tissue compatibility. However, this common "implant, sacrifice, explant, and evaluate" single-time-point-per-animal approach is quite limited in assessment of time-dependent phenomena (e.g., direct assay of molecular and cellular mediators) associated with the progression of the inflammatory response and the development of the FBR.

The "cage implant" model developed and described by Marchant et al.[128] was designed to examine synthetic materials and proximal physiological response(s) to them in real time *in vivo*. This system allows qualitative and quantitative assessments of the inflammatory response to materials of diverse composition *in vivo*.[128] Specifically, material changes and fibrous capsule formation/progression can be monitored, and associated inflammatory exudate can be collected and analyzed. Material changes can be further characterized by bulk physical analyses (molecular weight, stress–strain, FTIR) and by modern biomaterials relevant surface analytical techniques.[8] From inflammatory exudates, quantitative nonadherent leukocyte counts and phenotypic profiling can be obtained, as well as evaluation of extracellular exudates (e.g., alkaline and acid phosphatases, chemokines, cytokines, ECM production, etc).[128] The system is unique in its ability to internally control samples, provides real-time sampling and different types of inflammatory information from the implant site, and translation across different species and in different tissue sites.

2.7 FUTURE DIRECTIONS

Numerous previous studies of implantable biomaterials employ monocyte and macrophage cells to examine cytotoxicity, inflammatory response, and general "biotolerance." Primary deficiencies in these studies surround the lack of accuracy of the experiments compared to the problem, inability to collect the dimensionality of data necessary to accurately model MPS and inflammatory complexity, and lack of understanding of the cell mechanisms involved either *in vitro* or *in vivo*. The selection of the specific cells employed to this end is often limited by what is available, inexpensive, and practical, often excluding cells of primary derivation. Multiple immortalized model cell lines of monocyte and/or macrophage origin from human and murine sources are available, each with different morphologies, adhesion patterns, and growth requirements. No one has compared these cell sources using any basis for MPS fidelity. Thus, comparison of results for cell types arising from different species, of varied maturity and with different growth, subculture, and harvesting requirements is problematic. A thorough characterization of each cell type would support some comparison, but consensus is lacking in terms of what markers are appropriate for distinguishing maturity and differentiation, especially between closely related macrophages and dendritic cells. Characterization via surface marker detection is commonly employed, but often reliance is on relative levels of expression rather than presence or absence of markers to distinguish between cells of the MPS. In addition, most cellular and molecular phenotyping tools are currently focused on mouse markers, and cellular and *in vivo* inflammatory and healing model focus is therefore primarily murine derived. Attractive knock-in and knockout genetic models are also primarily murine. Many other animal molecular details are limiting, based on the narrow range of toolkits available. As murine physiology and healing mechanisms are likely distinct from larger animal models, the relevance of much of the data to human extrapolation might be suspected.

This begs the question—are these MPS cells truly different from each other, or rather, specialized adaptive cell states?[16] No single marker can be used to define if a cell is a macrophage. It has been suggested by Hume that, based on new insight provided by detailed characterizations of cells of the mononuclear system in recent years, a revised system for classifying immune cells of this lineage is overdue.[16]

Considering many different types of cell culture supports and their effects on the macrophage, it is evident that there is no unambiguous culture substrate to which other materials can be compared.[129] The activating nature of the surface may result from denatured or partially denatured proteins adsorbed to the surface that act as attractants and/or provide macrophage activating cues. Previous reports indicate that denatured non-ECM proteins serve as nonspecific attractants for granulocytes.[55] Recently, efforts to develop and employ new materials that mimic the three-dimensional environment more natural to cells have led to commercial synthetic support production (e.g., Ultra-Web® Synthetic ECM, SurModics/Donaldson®) for use in static cultures.[130] Culture methods have seen some advances, examples include roller bottle systems (designed to more closely represent cell growth in the presence of shear forces) and spinner flasks (designed to support suspended nonadherent cell growth). Other technique and technology advances seek to more accurately mimic *in vivo* conditions for *in vitro* studies.[131] Additional considerations for appropriate design of *in vitro* studies require the use of serum, given that serum proteins play determinant roles in cell interactions with surfaces *in vivo*. Macrophage studies can be also complicated by the presence of contaminating levels of endotoxins produced by bacteria. Therefore, all studies that employ macrophages should also describe measures taken to test for and reduce endotoxin contamination.

Advances in the quest for improved biocompatible materials rely on a thorough understanding of the FBR. This necessitates a detailed understanding of macrophage phenotype and function in this unique context. With the advent of new molecular characterization techniques and the likely discovery of novel macrophage markers, the relative relationships between cells of the MPS should be clarified. This should assist in elucidating relationships between these cells on biomaterials. In addition, understanding of the role of relatively new markers related to macrophage fusion (e.g., CD47 (MMR), MCP-1) and comparison to macrophage-specific (maturity) markers should provide a more detailed picture of the macrophage phenotype found at biomaterials-induced wound sites.[10,45,65]

Given that cell lines are certain to remain in use for *in vitro* purposes, improved phenotypic characterization should aid in the selection of specific cell lines that most faithfully represent primary cells within the experimental context. Specific validation guidelines for appropriate experimental design (serum inclusion, endotoxin testing, etc.) in the context of *in vitro* models for testing of biomaterials must be followed to collect meaningful data for comparisons and standardization. Relevant to *in vitro* work, characterization and comparison of immature and mature (aged) primary-derived cells to macrophage cell lines with respect to integrin expression on both standard tissue culture supports and model biomaterial surfaces would provide insight into the suitability of employing (aged, immortalized) adherent macrophage cultures that may be "prebiased" with respect to integrin expression based solely on their culture selection methods. *In vitro* studies must be carefully designed to be relevant to FBR progress *in vivo*; more relevant *in vitro* modeling of isolated aspects of the FBR requires an understanding of relationships between specific subsets of monocyte/macrophage and macrophage interaction with other cell types (e.g., fibroblasts, lymphocytes). Future work should address more thorough characterization of all macrophage phenotypes present at implant sites (adherent, activated, or fused macrophage) and duration of each phenotype using validated markers. This could be attempted *in situ* using specific markers directed toward surface antigens or soluble chemokine/cytokine products, or using techniques that allow cell harvest and sorting based on size and various surface markers. Prudent combinations of techniques with new methods for integration and interpretation of dynamic systems will likely yield the most convincing results in terms of characterizing FBR event sequelae. To conclude, better understanding of the molecular orchestration of events underlying the FBR should allow for design principles to provide improved device biocompatibility through modification and modulation of both materials and host inflammatory responses. New tools including mutant animal models, new molecular biomarkers, and new experimental techniques, including coculture and tissue cultures, should help provide more and better information. However, understanding the FBR will depend on new, coordinated improvements in molecular tools, wound healing characterization standards, cell phenotyping and standardized *in vitro* models, preclinical testing validation, new pharmacological control and modulating agents, and methods to integrate complex data sets from multiple cell types, signaling pathways, and feedback loops involved in inflammation.

ACKNOWLEDGMENTS

The authors gratefully acknowledge M. Howell, G. Callahan, M. Gonzalez-Juarerro (Colorado State University), and K. Ward (iSense) for guidance and input, and support of NIH grant EB 000894.

REFERENCES

1. Wisniewski N, Moussy F, Reichert WM. Characterization of implantable biosensor membrane biofouling. *Fresenius Journal of Analytical Chemistry* 2000, 366, 611–621.
2. Wilson GS, Gifford R. Biosensors for real-time *in vivo* measurement. *Biosensors & Bioelectronics* 2005, 20, 2388–2403.
3. Williams DF, Homsy CA. *Biocompatibility of Clinical Implant Materials*, Vol. 2. CRC Press, Boca Raton, FL, 1981, pp. 60–77.
4. Moussy F, Reichert WM. Biomaterials community examines biosensor biocompatibility. *Diabetes Technology & Therapeutics* 2000, 2, 473–477.
5. Sharkawy AA, Neuman MR, Reichert WM. Design considerations for biosensor based drug delivery systems. In: Park K (Ed.), *Controlled Drug Delivery: The Next Generation*. ACS Books, Washington, 1997, pp. 163–181.
6. Coleman DL, King RN, Andrade JD. The foreign body reaction: a chronic inflammatory response. *Journal of Biomedical Materials Research* 1974, 8, 199–211.
7. Anderson JM. Biological responses to material. *Annual Reviews of Materials Research* 2001, 31, 81–110.
8. Castner D, Ratner B. Biomedical surface science: foundations to frontier. *Surface Science* 2002, 500, 28–60.
9. Tang L, Eaton JW. Natural responses to unnatural materials: a molecular mechanism for foreign body reactions. *Molecular Medicine* 1999, 5, 351–358.

10. Kyriakides TR, Foster MJ, Keeney GE, Tsai A, Giachelli CM, Clark-Lewis I, Rollins BJ, Bornstein P. The CC chemokine ligand, CCL2/MCP1, participates in macrophage fusion and foreign body giant cell formation. *American Journal of Pathology* 2004, 165, 2157–2166.
11. Brodbeck WG, Nakayama Y, Matsuda T, Colton E, Ziats NP, Anderson JM. Biomaterial surface chemistry dictates adherent monocyte/macrophage cytokine expression *in vitro*. *Cytokine* 2002, 18, 311–319.
12. Collier TO, Thomas CH, Anderson JM, Healy KE. Surface chemistry control of monocyte and macrophage adhesion, morphology, and fusion. *Journal of Biomedical Materials Research* 2000, 49, 141–145.
13. Desai NP, Hubbell JA. Tissue response to intraperitoneal implants of polyethylene oxide-modified polyethylene terephthalate. *Biomaterials* 1992, 13, 505–510.
14. DiPietro LA, Strieter RM. Macrophages in wound healing. In: Burke B, Lewis CE (Eds.), *The Macrophage*. Oxford University Press, Oxford, 2002, pp. 434–456.
15. Hunt TK, Knighton DR, Thakral KK, Goodson WH, Andrews WA. Studies on inflammation and wound healing: angiogenesis and collagen synthesis stimulated *in vivo* by resident and activated wound macrophages. *Surgery* 1984, 96, 48–54.
16. Hume DA. The mononuclear phagocyte system. *Current Opinion in Immunology* 2006, 18, 49–53.
17. Gordon S, Taylor PR. Monocyte and macrophage heterogeneity. *Nature Reviews Immunology* 2005, 5, 953–964.
18. Hume DA, Ross IL, Himes SR, Sasmono RT, Wells CA, Ravasi T. The mononuclear phagocyte system revisited. *Journal of Leukocyte Biology* 2002, 72, 621–627.
19. Laskin DL, Weinberger B, Laskin JD. Functional heterogeneity in liver and lung macrophages. *Journal of Leukocyte Biology* 2001, 70, 163–170.
20. van Furth R, Cohn ZA. The origin and kinetics of mononuclear phagocytes. *The Journal of Experimental Medicine* 1968, 128, 415–435.
21. Volkman A, Gowans JL. The origin of macrophages from bone marrow in the rat. *British Journal of Experimental Pathology* 1965, 46, 62–70.
22. van Furth R, Cohn Z. Quantitative study on the production and kinetics of mononuclear phagocytes during an acute inflammatory reaction. *Journal of Experimental Medicine* 1973, 138, 1314–1330.
23. Pascual CJ, Sanberg PR, Chamizo W, Haraguchi S, Lerner D, Baldwin M, El-Badri NS. Ovarian monocyte progenitor cells: phenotypic and functional characterization. *Stem Cells and Development* 2005, 14, 173–180.
24. Whitelaw DM. The intravascular lifespan of monocytes. *Blood* 1966, 28, 455–464.
25. Passlick B, Flieger D, Ziegler-Heitbrock HW. Identification and characterization of a novel monocyte subpopulation in human peripheral blood. *Blood* 1989, 74, 2527–2534.
26. Palframan RT, Jung S, Cheng G, Weninger W, Luo Y, Dorf M, Littman DR, Rollins BJ, Zweerink H, Rot A, von Andrian UH. Inflammatory chemokine transport and presentation in HEV: a remote control mechanism for monocyte recruitment to lymph nodes in inflamed tissues. *Journal of Experimental Medicine* 2001, 194, 1361–1373.
27. Gu L, Tseng SC, Rollins BJ. Monocyte chemoattractant protein-1. *Chemical Immunology* 1999, 72, 7–29.
28. Newton RC, Vaddi K. Biological responses to C–C chemokines. *Methods in Enzymology* 1997, 287, 174–186.

29. Grage-Griebenow E, Zawatzky R, Kahlert H, Brade L, Flad H, Ernst M. Identification of a novel dendritic cell-like subset of CD64(+)/CD16(+) blood monocytes. *European Journal of Immunology* 2001, 31, 48–56.
30. Kiertscher SM, Roth MD. Human CD14 + leukocytes acquire the phenotype and function of antigen-presenting dendritic cells when cultured in GM-CSF and IL-4. *Journal of Leukocyte Biology* 1996, 59, 208–218.
31. Banchereau J, Steinman RM. Dendritic cells and the control of immunity. *Nature* 1998, 392, 245–252.
32. Romani N, Reider D, Heuer M, Ebner S, Kampgen E, Eibl B, Niederwieser D, Schuler G. Generation of mature dendritic cells from human blood. An improved method with special regard to clinical applicability. *Journal of Immunological Methods* 1996, 196, 137–151.
33. Zhou LJ, Tedder TF. CD14 + blood monocytes can differentiate into functionally mature CD83 + dendritic cells. *Proceedings of the National Academy of Sciences of the United States of America* 1996, 93, 2588–2592.
34. Banchereau J, Briere F, Caux C, Davoust J, Lebecque S, Liu Y-J, Pulendran B, Palucka K. Immunobiology of dendritic cells. *Annual Review of Immunology* 2000, 18, 767–811.
35. Ammon C, Meyer SP, Schwarzfischer L, Krause SW, Andreesen R, Kreutz M. Comparative analysis of integrin expression on monocyte-derived macrophages and monocyte-derived dendritic cells. *Immunology* 2000, 100, 364–369.
36. Miyamoto T, Ohneda O, Arai F, Iwamoto K, Okada S, Takagi K, Anderson DM, Suda T. Bifurcation of osteoclasts and dendritic cells from common progenitors. *Blood* 2001, 98, 2544–2554.
37. Witsell AL, Schook LB. Macrophage heterogeneity occurs through a developmental mechanism. *Proceedings of the National Academy of Sciences of the United States of America* 1991, 88, 1963–1967.
38. Douglas SD, Ho W-Z. Morphology of monocytes and macrophages. In: Lichtman MA, Beutler E, Kaushansky K, Kipps TJ, Seligsohn U, Prchal J (Eds), *Williams Hematology*. McGraw Hill, New York, 1990.
39. van Furth R. Current view on the mononuclear phagocyte system. *Immunobiology* 1982, 161, 178–185.
40. Hughes DA, Gordon S. Expression and function of the type 3 complement receptor in tissues of the developing mouse. *Journal of Immunology* 1998, 160, 4543–4552.
41. Hume DA, Monkley SJ, Wainwright BJ. Detection of c-fms protooncogene in early mouse embryos by whole mount *in situ* hybridization indicates roles for macrophages in tissue remodelling. *British Journal of Haematology* 1995, 90, 939–942.
42. Takahashi K, Donovan MJ, Rogers RA, Ezekowitz RA. Distribution of murine mannose receptor expression from early embryogenesis through to adulthood. *Cell and Tissue Research* 1998, 292, 311–323.
43. Shepard JL, Zon LI. Developmental derivation of embryonic and adult macrophages. *Current Opinion in Hematology* 2000, 7, 3–8.
44. Adams DO, Hamilton TA. The cell biology of macrophage activation. *Annual Review of Immunology* 1984, 2, 283–318.
45. Bogdan C, Nathan C. Modulation of macrophage function by transforming growth factor beta, interleukin-4, and interleukin-10. *Annals of the New York Academy of Sciences* 1993, 685, 713–739.

46. McKnight AJ, Gordon S. Membrane molecules as differentiation antigens of murine macrophages. *Advances in Immunology* 1998, 68, 271–314.
47. Goldsby RA, Kindt TJ, Osborne BA. Antigens. In: Kuby J (Ed.), *Immunology*. W.H. Freeman and Co., New York, 2000, pp. 61–81.
48. Goldsby RA, Kindt TJ, Osborne BA. Leukocyte migration and inflammation. In: Kuby J (Ed.), *Immunology*. W.H. Freeman and Co., New York, 2000, pp. 371–393.
49. Miller MD, Krangel MS. Biology and biochemistry of the chemokines: a family of chemotactic and inflammatory cytokines. *Critical Reviews in Immunology* 1992, 12, 17–46.
50. Goldsby RA, Osborne BA, Kindt TJ. Cytokines. Kuby J (Ed.), *Immunology*. W. H. Freeman and Co., New York, 2000, pp. 303–327.
51. Baer M, Dillner A, Schwartz RC, Sedon C, Nedospasov S, Johnson PF. Tumor necrosis factor alpha transcription in macrophages is attenuated by an autocrine factor that preferentially induces NF-kappaB p50. *Molecular and Cellular Biology* 1998, 18, 5678–5689.
52. Helinski EH, Bielat KL, Ovak GM, Pauly JL. Long-term cultivation of functional human macrophages in Teflon dishes with serum-free media. *Journal of Leukocyte Biology* 1988, 44, 111–121.
53. Rietschel ET, Brade H. Bacterial endotoxins. *Scientific American* 1992, 267(2), 54–61.
54. Sweet MJ, Hume DA. Endotoxin signal transduction in macrophages. *Journal of Leukocyte Biology* 1996, 60, 8–26.
55. Davis GE. The Mac-1 and p150,95 beta 2 integrins bind denatured proteins to mediate leukocyte cell-substrate adhesion. *Experimental Cell Research* 1992, 200, 242–252.
56. Teixeira C, Stang SL, Yong Z, Beswick NS, Stone JC. Integration of DAG signaling systems mediated by PKC-dependent phosphorylation of RasGRP3. *Blood* 2003, 102, 1414–1420.
57. Balsinde J, Balboa MA, Insel PA, Dennis EA. Differential regulation of phospholipase D and phospholipase A2 by protein kinase C in P388D1 macrophages. *Biochemical Journal* 1997, 321, 805–809.
58. Garcia JE, Lopez AM, de Cabo MR, Rodriguez FM, Losada JP, Sarmiento RG, Lopez AJ, Arellano JL. Cyclosporin A decreases human macrophage interleukin-6 synthesis at post-transcriptional level. *Mediators of Inflammation* 1999, 8, 253–259.
59. Gomez-Cambronero J, Chi-Kuang H, Yamazaki M, Wang E, Molski TFP, Becker EL, Sha'afi RI. Phorbol ester inhibits granulocyte-macrophage colony-stimulating factor binding and tyrosine phosphorylation. *American Journal of Physiology* 1992, 262, C276–C281.
60. Lai JM, Yu CY, Yang-Yen HF, Chang ZF. Lysophosphatidic acid promotes phorbol-ester-induced apoptosis in TF-1 cells by interfering with adhesion. *Biochemical Journal* 2001, 359, 227–233.
61. Leenen PJ, de Bruijn MF, Voerman JS, Campbell PA, van Ewijk W. Markers of mouse macrophage development detected by monoclonal antibodies. *Journal of Immunological Methods* 1994, 174, 5–19.
62. Yagnik DR, Hillyer P, Marshall D, Smythe CDW, Krausz T, Haskard DO, Landis RC. Noninflammatory phagocytosis of monosodium urate monohydrate crystals by mouse macrophages. Implications for the control of joint inflammation in gout. *Arthritis and Rheumatism* 2000, 43, 1779–1789.
63. Berken A, Benacerraf B. Properties of antibodies cytophilic for macrophages. *Journal of Experimental Medicine* 1966, 123, 119–144.

64. Lay WH, Nussenzweig V. Receptors for complement of leukocytes. *Journal of Experimental Medicine* 1968, 128, 991–1009.
65. McNally AK, DeFife KM, Anderson JM. Interleukin-4-induced macrophage fusion is prevented by inhibitors of mannose receptor activity. *American Journal of Pathology* 1996, 149, 975–985.
66. Subramaniam M, Saffaripour S, van de Water L, Frenette PS, Mayadas TN, Hynes RO, Wagner DD. Role of endothelial selectins in wound repair. *American Journal of Pathology* 1997, 150, 1701–1709.
67. Leibovich SJ, Ross R. The role of the macrophage in wound repair. A study with hydrocortisone and antimacrophage serum. *American Journal of Pathology* 1975, 78, 71–100.
68. Laustriat S, Geiss S, Becmeur F, Bientz J, Marcellin L, Sauvage P. Medical history of Teflon. *European Urology* 1990, 17, 301–303.
69. Savill J, Fadok V, Henson P, Haslett C. Phagocyte recognition of cells undergoing apoptosis. *Immunology Today* 1993, 14, 131–136.
70. Sunderkotter C, Steinbrink K, Goebeler M, Bhardwaj R, Sorg C. Macrophages and angiogenesis. *Journal of Leukocyte Biology* 1994, 55, 410–422.
71. Berse B, Brown LF, Van de Water L, Dvorak HF, Senger DR. Vascular permeability factor (vascular endothelial growth factor) gene is expressed differentially in normal tissues, macrophages, and tumors. *Molecular Biology of the Cell* 1992, 3, 211–220.
72. Nissen NN, Polverini PJ, Koch AE, Volin MV, Gamelli RL, DiPietro LA. Vascular endothelial growth factor mediates angiogenic activity during the proliferative phase of wound healing. *American Journal of Pathology* 1998, 152, 1445–1452.
73. Koch AE, Polverini PJ, Kunkel SL, Harlow LA, DiPietro LA, Elner VW, Elner SG, Strieter RM. Interleukin-8 as a macrophage-derived mediator of angiogenesis. *Science* 1992, 258, 1798–801.
74. Pugh-Humphreys RG. Macrophage-neoplastic cell interactions: implications for neoplastic cell growth. *FEMS Microbiology Immunology* 1992, 5(5–6), 289–308.
75. Klagsbrun M. Mediators of angiogenesis: the biological significance of basic fibroblast growth factor (bFGF)-heparin and heparan sulfate interactions. *Seminars in Cancer Biology* 1992, 3, 81–87.
76. Jaffe EA, Ruggiero JT, Falcone DJ. Monocytes and macrophages synthesize and secrete thrombospondin. *Blood* 1985, 65, 79–84.
77. Kovacs EJ, DiPietro LA. Fibrogenic cytokines and connective tissue production. *Journal of the Federation of American Societies for Experimental Biology* 1994, 8, 854–861.
78. Fukasawa M, Campeau JD, Yanagihara DL, Rodgers KE, Dizerega GS. Mitogenic and protein synthetic activity of tissue repair cells: control by the postsurgical macrophage. *Journal of Investigative Surgery* 1989, 2, 169–180.
79. DiPietro LA, Polverini PJ, Rahbe SM, Kovacs J. Modulation of JE/MCP-1 expression in dermal wound repair. *American Journal of Pathology* 1995, 146, 868–875.
80. DiPietro LA, Burdick M, Low QE, Kunkel SL, Strieter RM. MIP-1alpha as a critical macrophage chemoattractant in murine wound repair. *Journal of Clinical Investigation* 1998, 101, 1693–1698.
81. Nessel CC, Henry WL, Mastrofrancesco B, Reichner JS, Albina JE. Vestigial respiratory burst activity in wound macrophages. *American Journal of Physiology* 1999, 276, R1587–R1594.

82. Reichner JS, Meszaros AJ, Louis CA, Henry WL, Mastrofrancesco B, Martin BA, Albina JE. Molecular and metabolic evidence for the restricted expression of inducible nitric oxide synthase in healing wounds. *American Journal of Pathology* 1999, 154, 1097–1104.
83. Thomas CL, *Taber's Cyclopedic Medical Dictionary*, 16th edn. Davis Co., Philadelphia, PA, 1989.
84. Woodward SC. How fibroblasts and giant cells encapsulate implants: considerations in design of glucose sensors. *Diabetes Care* 1982, 5, 278–281.
85. Tang L, Eaton JW. Inflammatory responses to biomaterials. *American Society for Clinical Pathology* 1995, 103, 466–471.
86. Cotran RZ, Kumar V, Robbins SL. *Pathological Basis of Disease*, 6th edn. Saunders, Philadelphia, PA, 1999, pp. 50–112.
87. Langhans T. Uber Riesenzellen mit Wandstandigen Kernen in Tuberkeln und die fibrose Form des Tuberkels. *Virchows Arch Pathol Anat* 1868, 42, 382–404.
88. Lambert RA. The production of foreign body giant cells *in vitro*. *Journal of Experimental Medicine* 1912, 15, 510–515.
89. Vignery A. Osteoclasts and giant cells: macrophage–macrophage fusion mechanism. *International Journal of Experimental Pathology* 2000, 81, 291–304.
90. Murch AR, Grounds MD, Marshall CA, Papadimitriou JM. Direct evidence that inflammatory multinucleate giant cells form by fusion. *Journal of Pathology* 1982, 137, 177–180.
91. Saginario C, Qian HY, Vignery A. Identification of an inducible surface molecule specific to fusing macrophages. *Proceedings of the National Academy of Sciences of the United States of America* 1995, 92, 12210–12214.
92. Chambers TJ, Spector WG. Inflammatory giant cells. *Immunobiology* 1982, 161, 283–289.
93. Postlethwaite AE, Jackson BK, Beachey EH, Kang AG. Formation of multinucleated giant cells from human monocyte precursors. Mediation by a soluble protein from antigen- and mitogen-stimulated lymphocytes. *Journal of Experimental Medicine* 1982, 155, 168–178.
94. Mariano M, Spector WG. The formation and properties of macrophage polykaryons (inflammatory giant cells). *Journal of Pathology* 1974, 113, 1–19.
95. McNally AK, Anderson JM. Interleukin-4 induces foreign body giant cells from human monocytes/macrophages. Differential lymphokine regulation of macrophage fusion leads to morphological variants of multinucleated giant cells. *American Journal of Pathology* 1995, 147, 1487–1499.
96. Papadimitriou JM, Robertson TA, Walters MN. An analysis of the phagocytic potential of multinucleate foreign body giant cells. *American Journal of Pathology* 1975, 78, 343–358.
97. Papadimitriou JM, van Bruggen I. Evidence that multinucleate giant cells are examples of mononuclear phagocytic differentiation. *Journal of Pathology* 1986, 148(2), 149–157.
98. Baron R. Molecular mechanisms of bone resorption. An update. *Acta Orthopaedica Scandinavica. Supplementum* 1995, 266, 66–70.
99. McNally AK, Anderson JM. Foreign body-type multinucleated giant cell formation is potently induced by alpha-tocopherol and prevented by the diacylglycerol kinase inhibitor R59022. *American Journal of Pathology* 2003, 163, 1147–1156.
100. Kao WJ, McNally AK, Hiltner A, Anderson JM. Role for interleukin-4 in foreign-body giant cell formation on a poly(etherurethane urea) *in vivo*. *Journal of Biomedical Materials Research* 1995, 29, 1267–1275.

101. DeFife KM, Jenney CR, McNally AK, Colton E, Anderson JM. Interleukin-13 induces human monocyte/macrophage fusion and macrophage mannose receptor expression. *Journal of Immunology* 1997, 158, 3385–3390.

102. Cui W, Ke JZ, Zhang Q, Hua-Shu KE, Cecile C, Agnes V. The intracellular domain of CD44 promotes the fusion of macrophages. *Blood* 2006, 107, 796–805.

103. Yagi M, Takeshi M, Yumi S, Katsuya I, Naobumi H. DC-STAMP is essential for cell–cell fusion in osteoclasts and foreign body giant cells. *Journal of Experimental Medicine* 2005, 202, 345–351.

104. Vignery A, Niven-Fairchild T, Ingbar DH, Caplan M. Polarized distribution of Na+, K+-ATPase in giant cells elicited *in vivo* and *in vitro*. *Journal of Histochemistry and Cytochemistry* 1989, 37, 1265–1271.

105. Vignery A, Raymond MJ, Qian HY, Wang F, Rosenzweig SA. Multinucleated rat alveolar macrophages express functional receptors for calcitonin. *American Journal of Physiology* 1991, 261, F1026–1032.

106. Vignery A, Wang F, Qian HY, Benz EJ, Gilmore-Hebert M. Detection of the Na(+)-K(+)-ATPase alpha 3-isoform in multinucleated macrophages. *American Journal of Physiology* 1991, 260, F704–F709.

107. Vignery A, Wang F, Ganz MB. Macrophages express functional receptors for calcitonin-gene-related peptide. *Journal of Cellular Physiology* 1991, 149, 301–306.

108. Bradshaw AD, Reed MJ, Carbon JG, Pinney E, Brekken RA, Sage EH. Increased fibrovascular invasion of subcutaneous polyvinyl alcohol sponges in SPARC-null mice. *Wound Repair and Regeneration* 2001, 9, 522–530.

109. Kyriakides TR, Leach KJ, Hoffman AS, Ratner BD, Bornstein P. Mice that lack the angiogenesis inhibitor, thrombospondin 2, mount an altered foreign body reaction characterized by increased vascularity. *Proceedings of the National Academy of Sciences of the United States of America* 1999, 96, 4449–4454.

110. Puolakkainen P, Bradshaw AD, Kyriakides TR, Reed M, Brekken R, Wight T, Bornstein P, Ratner B, Sage EH. Compromised production of extracellular matrix in mice lacking secreted protein, acidic and rich in cysteine (SPARC) leads to a reduced foreign body reaction to implanted biomaterials. *American Journal of Pathology* 2003, 162, 627–635.

111. Barker TH, Framson P, Puolakkainen PA, Reed M, Funk SE, Sage EH. Matricellular homologs in the foreign body response: hevin suppresses inflammation, but hevin and SPARC together diminish angiogenesis. *American Journal of Pathology* 2005, 166, 923–933.

112. Patel ZS, Mikos AG. Angiogenesis with biomaterial-based drug- and cell-delivery systems. *Journal of Biomaterials Science. Polymer Edition* 2004, 15, 701–726.

113. Dagtekin G, Schiffer R, Klein B, Jahnen-Dechent W, Zwadlo-Klarwasser G. Modulation of angiogenic functions in human macrophages by biomaterials. *Biomaterials* 2003, 24, 3395–3401.

114. Bhattacharyya S, Brown DE, Brewer JA, Vogt SK, Muglia LJ. Macrophage glucocorticoid receptors regulate Toll-like receptor-4-mediated inflammatory responses by selective inhibition of p38 MAP kinase. *Blood* 2007, 109, 4313–4319.

115. Hua W, Mond HG, Strathmore N. Chronic steroid-eluting lead performance: a comparison of atrial and ventricular pacing. *Pacing and Clinical Electrophysiology* 1997, 20, 17–24.

116. Park SH, Lincoff AM. Anti-inflammatory stent coatings: dexamethasone and related compounds. *Seminars in Interventional Cardiology: SIIC* 1998, 3, 191–195.

117. Norton LW, Tegnell E, Toporek SS, Reichert WM. *In vitro* characterization of vascular endothelial growth factor and dexamethasone releasing hydrogels for implantable probe coatings. *Biomaterials* 2005, 26, 3285–3297.
118. Ward WK, Wood MD, Casey HM, Quinn MJ, Federiuk IF. The effect of local subcutaneous delivery of vascular endothelial growth factor on the function of a chronically implanted amperometric glucose sensor. *Diabetes Technology and Therapeutics* 2004, 6, 137–145.
119. Klueh U, Dorsky D, Kreutzer D. Use of vascular endothelial cell growth factor gene transfer to enhance implantable sensor function *in vivo*. *Journal of Biomedical Materials Research* 2003, 67A, 1076–1086.
120. Patil SD, Papadmitrakopoulos F, Burgess DJ. Concurrent delivery of dexamethasone and VEGF for localized inflammation control and angiogenesis. *Journal of Controlled Release* 2007, 117, 68–79.
121. Norton LW, Koschwanez HE, Wisniewski NA, Klitzman B, Reichert WM. Vascular endothelial growth factor and dexamethasone release from nonfouling sensor coatings affect the foreign body response. *Journal of Biomedical Materials Research. Part A* 2007, 81, 858–869.
122. Kellar RS, Kleinert LB, Williams SK. Characterization of angiogenesis and inflammation surrounding ePTFE implanted on the epicardium. *Journal of Biomedical Materials Research* 2002, 61, 226–233.
123. Menger MD, Walter P, Hammersen F, Messmer K. Quantitative analysis of neovascularization of different PTFE-implants. *European Journal of Cardio-Thoracic Surgery* 1990, 4, 191–196.
124. Sharkawy AA, Klitzman B, Truskey GA, Reichert WM. Engineering the tissue which encapsulates subcutaneous implants. III. Effective tissue response times. *Journal of Biomedical Materials Research* 1998, 40, 598–605.
125. Ward WK, Slobodzian EP, Tiekotter KL, Wood MD. The effect of microgeometry, implant thickness and polyurethane chemistry on the foreign body response to subcutaneous implants. *Biomaterials* 2002, 23, 4185–4192.
126. Brauker JH, Carr-Brendel VE, Martinson LA, Crudele J, Johnston WD, Johnson RC. Neovascularization of synthetic membranes directed by membrane microarchitecture. *Journal of Biomedical Materials Research* 1995, 29, 1517–1524.
127. Godek ML, Duchsherer NL, McElwee Q, Grainger DW. Morphology and growth of murine cell lines on model biomaterials. *Biomedical Sciences Instrumentation* 2004, 40, 7–12.
128. Marchant RE, Johnson SD, Schneider BH, Agger MP, Anderson JM. A hydrophilic plasma polymerized film composite with potential application as an interface for biomaterials. *Journal of Biomedical Materials Research* 1990, 24, 1521–1537.
129. van Kooten TG. Growth of cells on polymer surfaces. In: *Encyclopedia of Surface and Colloid Science*. Marcel Dekker, Inc., New York, 2004, pp. 1–19.
130. Schindler M, Ahmed I, Kamal J, Nur-E-Kamal A, Grafe TH, Chung HY, Meiners S. A synthetic nanofibrillar matrix promotes *in vivo*-like organization and morphogenesis for cells in culture. *Biomaterials* 2005, 26, 5624–5631.
131. Folch A, Toner M. Microengineering of cellular interactions. *Annual Review of Biomedical Engineering* 2000, 2, 227–256.
132. Bagby GC, Dinarello CA, Wallace P, Wagner C, Hefeneider S, McCall E. Interleukin 1 stimulates granulocyte macrophage colony-stimulating activity release by vascular endothelial cells. *Journal of Clinical Investigation* 1986, 78, 1316–1323.

133. Broudy VC, Kaushansky K, Segal GM, Harlan JM, Adamson JW. Tumor necrosis factor type alpha stimulates human endothelial cells to produce granulocyte/macrophage colony-stimulating factor. *Proceedings of the National Academy of Sciences of the United States of America* 1986, 83, 7467–7471.

134. Cannistra SA, Rambaldi A, Spriggs DR, Herrmann F, Kufe D, Griffin JD. Human granulocyte–macrophage colony-stimulating factor induces expression of the tumor necrosis factor gene by the U937 cell line and by normal human monocytes. *Journal of Clinical Investigation* 1987, 79, 1720–1728.

135. Dustin ML, Rothlein R, Bhan AK, Dinarello CA, Springer TA. Induction by IL 1 and interferon-gamma: tissue distribution, biochemistry, and function of a natural adherence molecule (ICAM-1). *Journal of Immunology* 1986, 137, 245–254.

136. Essner R, Rhoades K, McBride WH, Morton DL, Economou JS. IL-4 down-regulates IL-1 and TNF gene expression in human monocytes. *Journal of Immunology* 1989, 142, 3857–3861.

137. Jones ALaM, JL, *Clinical Hematology: Aplastic Anaemia*. Balliere Tindall, London, 1989.

138. Lee JD, Swisher SG, Minehart EH, McBride WH, Economou JS. Interleukin-4 downregulates interleukin-6 production in human peripheral blood mononuclear cells. *Journal of Leukocyte Biology* 1990, 47, 475–479.

139. Metcalf D. The molecular control of cell division, differentiation commitment and maturation in haemopoietic cells. *Nature* 1989, 339, 27–30.

140. Munker R, Gasson J, Ogawa M, Koeffler HP. Recombinant human TNF induces production of granulocyte-monocyte colony-stimulating factor. *Nature* 1986, 323, 79–82.

141. Wu CY, Demeure C, Kiniwa M, Gately M, Delespesse G. IL-12 induces the production of IFN-gamma by neonatal human CD4 T cells. *Journal of Immunology* 1993, 151, 1938–1949.

CHAPTER 3

STRATEGIES TO OVERCOME BIOLOGICAL BARRIERS TO BIOSENSING

W. Kenneth Ward and Heather M. Duman

In Vivo Glucose Sensing, Edited by David D. Cunningham and Julie A. Stenken

3.1 INTRODUCTION

An understanding of the foreign body reaction is becoming increasingly important as more biosensing implants are developed. Many of these devices undergo a reduction in accuracy, and some ultimately fail, because of the biofouling and the collagen deposition as part of the foreign body reaction that leads to device encapsulation. Biosensors are particularly susceptible to the effects of the foreign body reaction since even a very thin deposit of collagen or other protein can limit the influx of analytes into the device. In this chapter, we will review the foreign body reaction, especially with regard to subcutaneously implanted biosensors. In addition, we will review the known and potential targets for reducing the intensity of this response.

Before any discussion of potential targets or coping strategies to improve sensor performance, the foreign body reaction as understood by the scientific community must be reviewed. In this chapter, the foreign body reaction will be categorized into two broad segments: minutes to days and days to months.

3.2 THE EARLY FOREIGN BODY REACTION (MINUTES TO DAYS)

Much of the work on early events associated with the foreign body reaction comes from the work of Liping Tang and associates. This group found that soon after the implantation, fibrinogen becomes adhered to the entire surface of the implant. Upon binding, fibrinogen changes its conformation to resemble the structure of fibrin. This conformational change exposes epitopes P1 and P2, which are sites to which macrophages bind.[1–4] Studies from this group suggest that in humans and in rodents, histamine and the cells from which histamine is secreted (mast cells) play an important role in phagocyte recruitment.[5,6]

Macrophages are crucially important in organizing the subsequent events in the foreign body reaction. These cells are derived from circulating monocytes and enter the subcutaneous space through the endothelium of blood vessels. Macrophages are capable of phagocytizing foreign particles and necrotic tissue debris. They can cause chemical destruction of implanted materials or microbiologic organisms by releasing lysosomal enzymes and other compounds such as hydrogen peroxide, sodium hypochlorite, and superoxides.[7] To the extent that bleeding occurs during subcutaneous implantation, it is likely that platelets and blood-derived factors also contribute to the foreign body reaction. Although practical knowledge suggests that the presence of bleeding during implantation surgery has adverse effects, this issue has not been well studied.

Macrophages are guided to the site of the foreign body by chemoattractant cytokines (chemokines) (i.e., monocyte chemoattractant proteins (MCPs)). Once bound to fibrinogen, the macrophages activate, flatten, and begin to release cytokines and growth factors. These factors include platelet-derived growth factor (PDGF), basic fibroblast growth factor (bFGF), transforming growth factor-α (TGF-α), and tumor necrosis factor-α (TNF-α).

In some cases, macrophages fuse into foreign body giant cells (FBGCs). Such cells are composed of the nuclei of many macrophages, are found in abundance during intense foreign body reactions, and are thought to mediate the destruction of foreign materials. Compounds such as interleukin-4 (IL-4) are known to induce macrophage fusion *in vitro* in response to biomaterials.[8] Recently, osteopontin was found to inhibit the joining together of macrophages to form FBGCs.[9] Osteopontin is a glycoprotein found in bone and many other tissues and is thought to play an important role in the inflammatory response in many tissues. In Figure 3.1, a series of large FBGCs are noted in a tissue section taken 28 days after device implant. In support of its key role in the foreign body reaction, macrophages are attracted early to a foreign implant and persist in the area for as long as the device remains implanted.[10]

Another important aspect of the macrophage is its ability to secrete cytokines that control later aspects of the foreign body reaction (see Section 3.3.3). Since macrophages appear to persist for long periods of time in the tissue surrounding an implant, it is very likely that these cells are important acutely and chronically. As information continues to be discovered, the role of the macrophage is increasing in importance; this cell appears to orchestrate the response to foreign bodies. It responds to chemical signals and expressed other chemical signaling compounds in a complex, regulated system designed to encapsulate and wall off the implant.

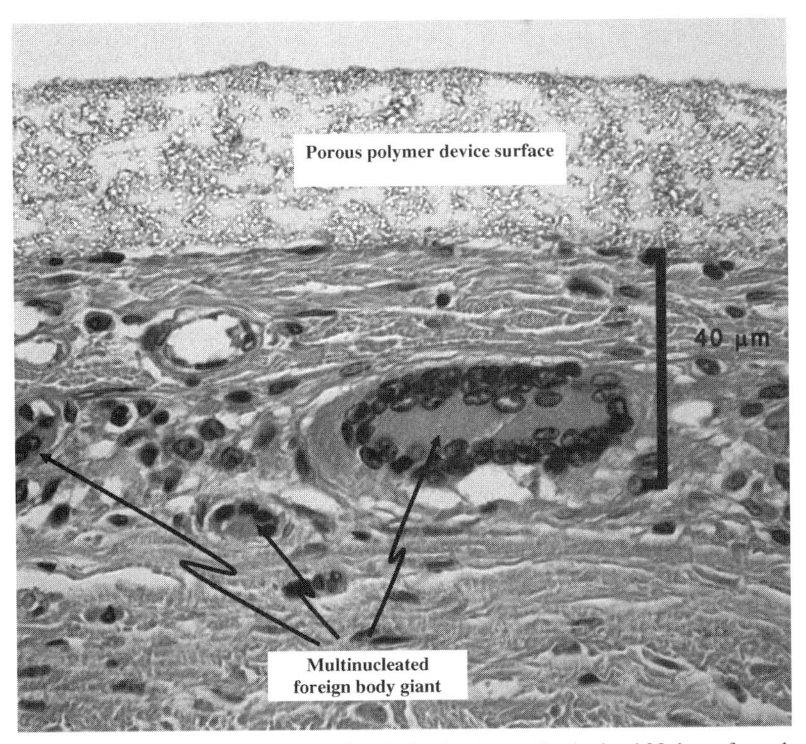

Figure 3.1 H&E section showing foreign body giant cells obtained 28 days after subcutaneous implantation of a porous polyvinyl alcohol implant in a rat. Reprinted with permission from Ref. 10. Copyright 2004 John Wiley & Sons. (See the color version of this figure in Color Plates section.)

Neutrophils are also rapidly attracted to the site of a foreign body reaction, but likely persist in large numbers for only a few days. They generally reside within blood vessels, but in response to signals such as histamine, which increases vascular permeability, neutrophils enter the extravascular space through a process known as diapedesis. A recent study by Wozniak found that nitric oxide produced by neutrophils plays an important role in loosening of total joint implants.[11]

3.3 SUBACUTE AND CHRONIC EFFECTS (DAYS TO MONTHS)

3.3.1 The Effect of the Foreign Body Reaction on Chronic Biosensor Function

The collagen encapsulation that occurs during the foreign body reaction represents a chronic fibrotic response and has been a major obstacle that reduces the useful life of implanted biomedical devices. A good example of the effect of encapsulation on a biosensor was found in our work with chronic amperometric glucose sensors implanted subcutaneously in rats with type 1 diabetes, as shown in Figure 3.2. The sensors were used to control insulin delivery in a closed-loop fashion every 2 weeks for several months. We found that within the first 2 weeks, there was very little lag between blood glucose values and sensed glucose values, nearly always less than 10 min. In addition, the sensitivity of the sensor to glucose was great and sensor accuracy was high. During this period of time, the ability of the artificial pancreas algorithm to control glucose levels by infusion of intravenous insulin was excellent. During closed-loop control, glucose was usually kept between 75 and 150 mg/dL.

These results should be contrasted with the situation with sensors that were implanted over 30 days. The lag in these devices rose to 15–30 min. There was also a decline in sensitivity of the sensors to glucose, sensor accuracy, and the degree of closed-loop glycemic control.[4] Based on the known effects and time course of collagen deposition that occurs as part of the foreign body reaction,[12,13] it is very likely that this decline in sensor function was a direct result of this process. The effect of the foreign body reaction on the function of biosensors has been comprehensively reviewed by Wisniewski and colleagues.[14,15]

3.3.2 Collagen

We found that at day 7 after implant of polyurethane devices in rats, collagen was relatively sparse, but by day 21, it was substantially more abundant. By days 48–55, the collagen was even more abundant, and, at this time, the collagen bundles were dense and oriented in a parallel fashion. Studies with immunofluorescence using antibodies against collagen confirmed these histologic results and showed that the fluorescence was greatest at days 48–55.[16] Figure 3.3 shows histology of rat tissue at the interface with the polyurethane foreign body, using H&E staining (left) and Masson's trichrome staining (right) at several time points after insertion. It can be seen that normal rat subcutaneous tissue has some collagen, but it is loosely woven and sparse. We have also investigated extracellular matrix proteins other than

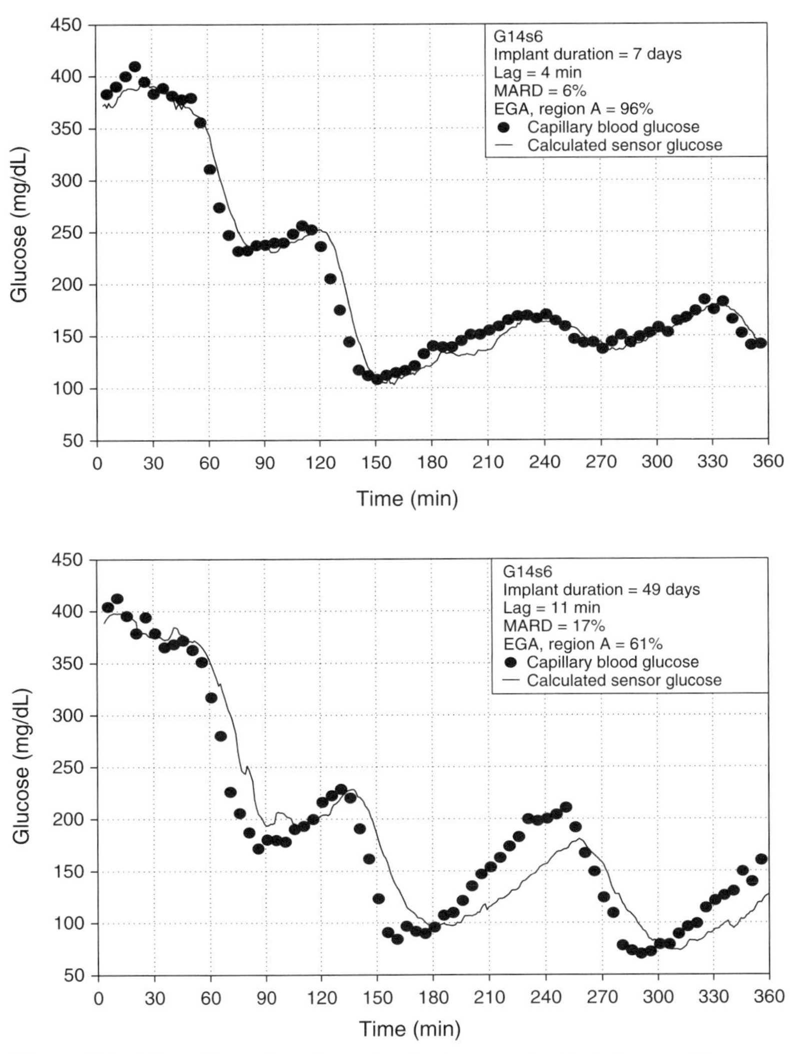

Figure 3.2 The effect of prolonged subcutaneous implantation on biosensor function. Blood glucose values shown in solid circles and glucose sensor values in the continuous lines. The early study (top panel), but not the late study (bottom), shows excellent sensor accuracy and minimal lag between blood glucose and sensed glucose values. MARD (mean absolute relative difference) refers to a sensor accuracy metric. EGA refers to the Clarke error grid analysis accuracy metric.

collagen. Using immunostaining, we found that the abundance of two other compounds, decorin and fibronectin, parallels that of collagen and, like collagen, increases over a period of several weeks.[17]

We also investigated the prevalence of the gene transcript for collagen 1 as measured by qRT-PCR on tissue harvested on days 7, 21, and 48–55 after implantation. These data revealed a marked increase in type 1 collagen mRNA, specifically an approximately 10-, 30-, and 5-fold increase at these time points. These results suggest

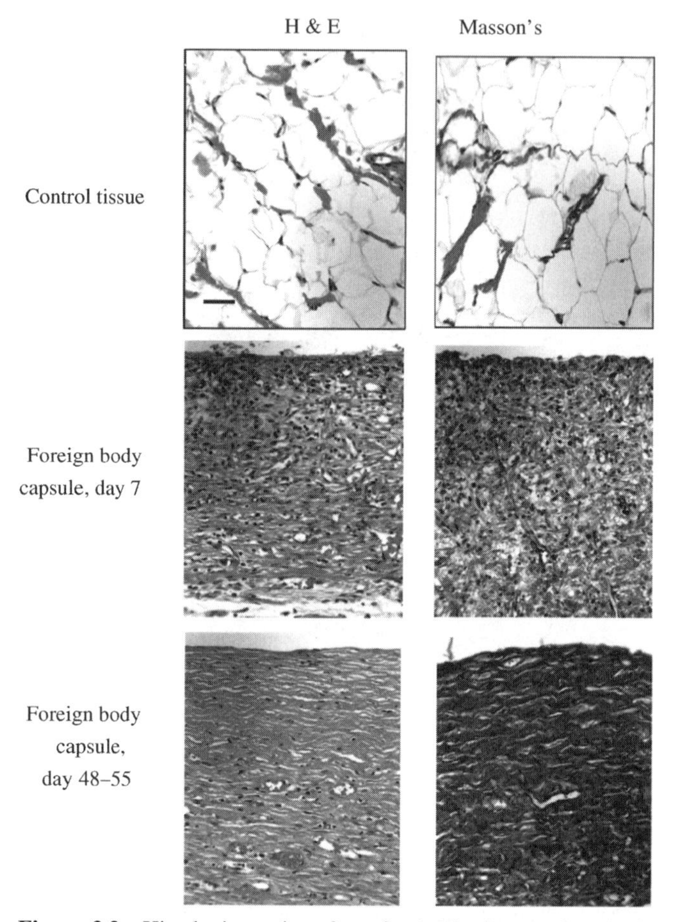

Figure 3.3 Histologic sections from foreign body capsules obtained at early and late time points after polyurethane implants in rats. The bar in the top left section represents 75 μm for all sections. Reprinted with permission from Ref. 16. Copyright 2007 John Wiley & Sons. (See the color version of this figure in Color Plates section.)

that, despite the collagen content being greatest at the final time point, the highest rate of collagen synthesis in the foreign body capsule (FBC) occurs during the intermediate stages of FBC formation. Collagen is secreted by activated fibroblasts; these actin-expressing cells were found in abundance within the encapsulated implant at many time points, especially during early and intermediate time points.[16] In summary, as verified by mRNA analysis and histologic staining, collagen has been identified as the major component of the FBC that encapsulates implants over weeks to months.

3.3.3 Transforming Growth Factor-β

Transforming growth factor-β (TGF-β) is a group of several isoforms that have several roles in wound healing and other aspects of cell growth. It is now believed that there are five members of the TGF-β family. These compounds, of which TGF-β_1 is the best studied, influence cellular proliferation, cellular differentiation, secretion of extracellular matrix proteins, apoptosis, and many other functions in many cell types. TGF-β_1 promotes collagen deposition in many tissues and has been associated with pathologic fibrosis of many organs.[18–21] There is substantial sequence homology among the five TGF-β isoforms and some of their roles overlap. For example, isoforms 1–3 all stimulate neutrophil chemotaxis.[22]

Considering its central role in many other fibrotic conditions, our group hypothesized that TGF-β (especially isoform 1) may play an important role in the formation of the collagen capsule during the foreign body reaction. For this reason, we implanted mock sensors in rats subcutaneously and excised foreign capsule samples at days 7, 21, and 48–55 after implantation. We found that the most abundant TGF-β isoform in all tissues was, in fact, TGF-β_1, which was expressed minimally in control tissue. We found that the expression of TGF-β_1 RNA was significantly increased in capsular tissues at all time points, with the highest level at day 7. Specifically, we observed an approximate 10-, 5–6-, and 3–4-fold increase of TGF-β_1 transcripts in day 7, 21, and 48–55 tissues, respectively, as compared to control tissue.

We also carried out ELISA assays to determine the TGF-β_1 protein levels in FBC tissues. By measuring acid-stable and acid-labile fractions, we were able to measure the latent TGF-β_1 protein and the active TGF-β_1 protein. The total TGF-β_1 protein in day 7, 21, and 48–55 FBC tissues was approximately 20, 13, and 10 times as much as in control tissues, respectively. These increases were reflected approximately equally in both the active and the latent TGF-β_1 protein. These data indicate that TGF-β_1 expression is elevated throughout FBC development and parallels closely the time course of the RNA transcripts. Specifically, we found that on day 7 both TGF-β_1 mRNA and protein levels were found to be significantly higher than those at the other two time points ($p < 0.05$).[16]

Our data found that TGF-β_1 is very highly expressed in the capsule that surrounds the implant due to the foreign body reaction. Like TGF-β expression, we also found that expression of TGF-β's primary signaling mediator, pSmad2, was greatest at day 7.[16] Brodbeck et al. from the Case Western Reserve group examined the expression of cytokine transcripts that developed in cellular material adherent to subcutaneously implanted cages that had been coated with biomaterials. Their study did not directly compare TGF-β expression at the implant site to normal subcutaneous tissue. Therefore, even though they found TGF-β expression in coated cages to be roughly similar to the control state (uncoated steel cages), their results cannot be directly compared to ours because we measured TGF-β expression directly from foreign capsule tissue and compared it to undisturbed subcutaneous tissue. Brodbeck et al. also found reduced inflammatory cytokine expression from hydrophilic surfaces.[23] Another group also investigated the role of TGF-β during the foreign body reaction. An increased expression of TGF-β was found in peritoneal washings in animals that were implanted with intraperitoneal alginate capsules but not in animals that received a saline control.[24] Although the overexpression was found in washings rather than directly from fibrotic tissue, their finding of increased TGF-β expression in response to a peritoneal foreign body is comparable to our finding in subcutaneous tissue.

TGF-β_1 has been thought to be an important cytokine in mediating fibrotic responses during normal wound healing in many tissues. Upon injury, TGF-β_1 is rapidly released from platelets to initiate the healing process and subsequently produced by tissue cells and inflammatory cells, including macrophages.[25] In response to increased TGF-β_1, fibroblasts rapidly become activated and differentiate to myofibroblasts, which are characterized by a high proliferative capacity and high capacity for collagen synthesis. Overexpression of TGF-β has been implicated in fibrotic disease in several tissues and organs.[18,20,26,27]

By using antibodies to alpha smooth muscle actin, we were able to identify activated fibroblasts (myofibroblasts). These contractile fibroblasts actively make collagen in contrast to quiescent fibroblasts. The increase in number of myofibroblasts after implantation was very similar to the increase in response to TGF-β,[16] and this finding is consistent with the concept that one of the major roles of TGF-β is to activate fibroblasts in part to stimulate the formation of collagen.

In summary, TGF-β_1 is a major promoter of fibrosis in many organs. Its intracellular mediators have been found to be very highly expressed in the tissue that surrounds the foreign body in the days to weeks after a subcutaneous implant.

3.3.4 Interleukin-13

Interleukin-13 is a product of the T helper-2 (Th2) cells, which affects many cells such as macrophages, cytotoxic T cells, and B-lymphocytes. IL-13 has been found to contribute to fibrosis in several organs and tissues.[28–30] It has also been reported that some of the profibrotic effects of IL-13 may be mediated by TGF-β.[31,32] There is also evidence that IL-13 can promote fibrosis independent of TGF-β.[33]

In our study of subcutaneous polyurethane implants into rats, we found IL-13 to be highly expressed at all time points, with greatest expression at day 21. The IL-13 expression was paralleled by increased numbers of T cells at all time points.[17] Our data thus suggest overexpression of IL-13 in foreign body reactions. However, the effect of IL-13 may not be entirely distinct from that of TGF-β, since IL-13 can lead to increased TGF-β expression.[31] Although we did not directly isolate Th2 cells in this study (we used the less specific CD3 T-cell stain), it is highly likely that the cells we observed in the capsular tissue were in fact Th2 cells. In summary, interleukin-13 is highly expressed in encapsulation tissue that surrounds a foreign body that develops around subcutaneous implants several weeks after their insertion.

3.3.5 Connective Tissue Growth Factor

In our study of the foreign body reaction to subcutaneously implanted devices in rats, the transcript for connective tissue growth factor (CTGF) was found to be elevated at all time points as compared to control tissue. However, it was difficult to quantify the exact increase versus control tissues because there were little to no CTGF transcripts in the control tissue.[17] Evidence from prior studies demonstrated that TGF-β is capable of increasing CTGF expression. In cardiac cells,[34] kidney tissue,[35] dermal scars,[36] and mouse embryonic fibroblasts,[37] TGF-β led to increased expression of CTGF. TGF-β is involved in a very large number of biochemical processes, including many effects on the immune system.[38] In contrast, CTGF has a much narrower range of effects; this specificity may be a mechanism by which some or all of the profibrotic effects of TGF-β are mediated. It will be interesting in further studies to determine whether IL-13 (like TGF-β) can induce CTGF expression, which could contribute to the fibrogenic effects of IL-13. In summary, CTGF is highly expressed in the subcutaneous tissue that surrounds a foreign body and thus may be a specific mediator of the TGF-β effect.

3.3.6 Extracellular Matrix Compounds

Decorin is a small, leucine-rich proteoglycan and one of the many extracellular matrix (ECM) compounds. It interacts closely with the D and E bands of type 1 collagen and increases collagen's tensile strength.[39,40] In rats with subcutaneously implanted devices, we found that the decorin was highly expressed at the middle and later stages of the foreign body reaction.[17]

Fibronectin is a cell-surface compound with domains that interact with many compounds, including integrins and extracellular matrix proteins such as collagen. Fibronectin expression can be stimulated by TGF-β and by CTGF,[41] although increases in fibronectin expression can also be mediated by TGF-β-independent means.[42] We found some evidence of increased expression of fibronectin in FBC tissue, but not to the degree in which TGF-β, CTGF, and decorin were expressed.[16,17]

3.3.7 The Issue of Causality

Although we have identified several molecules that appear to be involved in the subcutaneous foreign body reaction that produces fibrosis in mammals, it is quite possible that many other cytokines and messengers are also involved. Numerous growth factors, such as fibroblast growth factor and insulin-like growth factors, may play a part in foreign body reaction that leads to a fibrotic response. Participation of many compounds in this process would not be unusual given that fact that many biochemical processes are controlled by redundant pathways. In addition, it is always important to consider an "innocent bystander" effect when searching for causes of disease states. In other words, the direct cause of the collagen deposition that results from the foreign body reaction could be a molecule (such as thrombospondin-1 (TSP-1)) and the increased TGF-β expression that we observed could conceivably be simply an epiphenomenon. However, two lines of evidence suggest otherwise. First, it has been shown that in rodents, blockade of TGF-β with a neutralizing antibody successfully reduced skin fibrosis.[43,44] Second, it was recently found that an inhibitor of the ALK5 TGF-β receptor reduced the profibrotic signaling effects of TGF-β in dermal fibroblasts.[45] Thus, our data and those of others, taken together, suggest a primary role of TGF-β in the generation of subcutaneous encapsulation process that occurs due to the foreign body reaction.

3.4 SURFACE MICROGEOMETRY

Several decades ago, two studies showed that, as compared to dense vascular grafts, porous grafts enhanced the formation of a stable neointima as well as the growth of microvessels into the wall of the graft.[46,47] Later, porous polymers were examined during implantation into the subcutaneous space in laboratory rodents. Andrade and colleagues found that a polyester sponge led to the growth of capillaries into the sponge.[48] Soon thereafter, Brauker et al. published a widely cited study in which a great number of polymer materials were implanted subcutaneously in rats and the growth of microvessels into the implant interface was measured.[49] They found that of all the materials examined, a double layer of expanded polytetrafluoroethylene (ePTFE) had the greatest effect to promote angiogenesis. When this finding was extended to a model of implanted functioning glucose sensors, Updike and colleagues found that this material extended the life of subcutaneously implanted glucose sensors.[50]

The biomaterials research group from Duke University has examined the effect of the collagen capsule that results from the foreign body reaction on diffusion of glucose-like analytes into biosensors. They found reduced diffusion of such analytes caused by the hypocellular collagenous capsule.[12,13] Our research laboratory has extended the findings of the Brauker group and the Duke University group with regard to porous membranes and we have addressed the effect of different chemical compositions of polyurethanes. We examined the effect of dense and porous polymers on capillary formation and on the characteristics of the encapsulation tissue as a result of the foreign body reaction. We found that after being implanted for a period of 7 weeks, a dense, nonporous hydrophobic polyurethane membrane led to a relatively thick capsule, 170–230 μm in thickness. On the other hand, when a nonporous amphiphobic polyurethane with polyethylene oxide moieties was implanted, the resulting capsule was thinner, typically only 70–140 μm.[51] Our group also addressed the use of porous membranes implanted for 7 weeks. We found that a single layer of ePTFE led to a significant increase in the number of microvessels in the encapsulation tissue as compared to dense membrane implants. In addition, as compared to dense implants, a polyvinyl alcohol (PVA) porous sponge membrane led to similar increase in neovascularization.[52,53]

In terms of biomaterials in general, and porous materials in particular, it is important to understand that assessing the degree of biocompatibility can be difficult because the notion of biocompatibility is interpreted very differently by different workers. This issue, as it relates to the foreign body reaction to porous materials, was addressed elegantly by researchers from the Case Western Reserve University in a recent study. In this rat study, Voskerician et al. compared porous materials that could be used for abdominal wall repair. They found that if one used fibrous capsule thickness as the sole criterion of biocompatibility, then the materials were all similar. If one used tissue integration as the major criterion, ePTFE was the best, and if minimal inflammation was considered the main criterion, then condensed PTFE was the best. They concluded that their condensed PTFE, which had large mesh openings and low material thickness, would be expected to be associated with the most rapid wound healing.[54]

To summarize the concept of porous or geometrically patterned materials, it is generally accepted that such materials typically lead to better neovascularization and better tissue integration than solid materials. However, our laboratory experience with porous membranes suggests that such improvements are not sufficient to allow indefinite function of amperometric subcutaneous glucose sensors. Eventually (typically, after 1–2 months), such devices will lose sensitivity due to a collagen-rich capsule that reduces permeation of glucose and oxygen. It also should be mentioned that implants whose surfaces are textured or patterned may be more difficult to remove after implantation for several months, as compared to a smooth implant. Although removal has not been carefully studied, the ingrowth of vessels and collagenous tissue might make removal difficult.

Another important issue is that of implant size or thickness. In our opinion, this issue has not been studied extensively. Nonetheless, there is a general feeling that small or thin devices generate less of a foreign body reaction than larger or thicker devices. One study that dealt with this issue was reported by Joan Sanders and colleagues from the University of Washington. They conducted an *in vivo* study in rodents to assess the effect on fibrous capsule thickness and macrophage density from the diameter of polymer fibers. Single polypropylene fibers of diameters ranging from 2.1 to 26.7 μm were implanted subcutaneously. Results at week 5 demonstrated reduced fibrous capsule thickness for the smaller fibers. The reduced fibrous capsule thickness and macrophage density for very small fibers ($<6\,\mu m$) was thought to be due to reduced cell–material contact surface area or due to a curvature threshold effect that triggers cell signaling.[55] The Voskerician study referred to above also addressed this issue and their conclusions are consistent with those of the Sanders study.[54] In our laboratory, we addressed this issue by implanting differing thicknesses of polyurethane sheets into rats. In addition to the content of PEO, we found that the thickness of the polyurethane implant influenced the thickness of the encapsulation tissue, with lowest capsule thicknesses observed in the thinnest implants.[51]

3.5 SURFACE CHEMISTRY

Despite investigations of thousands of implanted biomaterials, no material that escapes host recognition has been found. There is no material that avoids the host foreign body reaction. From time to time, there has been a great deal of enthusiasm for "stealth" materials such as tetraglyme and diamond-like carbon, but further study reveals that these compounds cannot escape the foreign body reaction. Thus, in this regard, there is a commonality among all biomaterials studied to date. Even such "low-technology" substances as wood (that has excellent structural qualities) have been found recently to integrate well with bone, thus demonstrating the potential for use as a biological implant.[56,57]

Though there is some commonality among materials, not all materials lead to the same foreign body reaction. The chemical composition of an implant does have many effects on the tissue–implant interface, including the degree to which cells and proteins attach to the device, the degree to which material fragmentation occurs, and the degree to which new microvessels are formed in response to the implant. As mentioned in other areas of this chapter, polyethylene oxide and related hydrophilic moieties reduce the extent of protein biofouling. The presence of pores and interconnected channels increases microvessel growth.

The most common compounds used for implants are silicones, fluoropolymers, and polyurethanes. Although at one time silicone implants were thought to lead to autoimmune and rheumatic disorders, these studies were likely flawed and such compounds are now once again used for surgical implants.[58]

PTFE implants (especially expanded PTFE) have also been found to be clinically safe.[59,60] However, recently, there has been concern about potentially toxic effects of a processing component used during the manufacture of fluoropolymer materials, perfluorooctanoic acid (PFOA), a chemical that does not occur naturally in the environment. PFOA in low concentrations is present in most humans and even in cord blood of newborn infants.[61] Fluoropolymers are widely used in cookware and other materials to which the public is exposed. In rat pups, exposure to PFOA has been found to cause low birth weight, liver toxicity, and developmental delay,[62] leading to major concerns about the ubiquitous exposure of humans to this compound. The U.S. Environmental Protection Agency regards PFOA as a potentially toxic compound and hosts a discussion board regarding this compound on its Web site (http://www.EPA.gov).[63] The U.S. Food and Drug Administration also has concerns about PTFE implants with regard to this compound and device companies must demonstrate that devices containing PTFE are not toxic to test animals and do not leach out PFOA.

For most uses of medical devices, the degradation of a compound is an undesirable effect (a notable exception is biodegradable drug delivery systems such as polymers of lactic acid and/or glycolic acid). The fragmentation of implants is thought to be mediated in part by physical trauma (especially in the situation of orthopedic implants) and can be exacerbated by bacterial infection and exposure to chemicals released by host lysosomes. Since polyurethanes are so widely used as biomaterials, mechanisms of degradation have been addressed in some detail for this broad class of polymers. In terms of polyurethane implants, oxidation, rather than hydrolysis, is thought to be the main mechanism of polymer breakdown. This concept is supported by the recent finding that addition of an antioxidant inhibited the *in vivo* degradation of both polycarbonate urethanes and polyether urethanes.[64] In general, polycarbonate urethanes were found to be more resistant to the effects of chronic oxidation than polyether urethanes.[65] One method that appears to enhance the stability of polyetherurethanes is the substitution of polyether soft segment with polydimethylsiloxane (PDMS).[66] Another method is the inclusion of antioxidants into the polyurethane materials that make up the implant. These additives include Santowhite[64] and fluoroalkyl compounds derivatized with vitamin E.[67] Fluorinated surface-modifying compounds have also been found to reduce the binding of adhesive proteins such as fibrinogen and fibronectin to polyurethanes.[68] Recently, it has also been reported that the charge of compounds may affect the foreign body reaction to the materials and that, specifically, negatively charged surfaces may facilitate vessel ingrowth into fibro-porous mesh biomaterials.[69]

In terms of metals, avoidance of wear debris and corrosion resistance is important to achieve biocompatibility and to minimize cytotoxicity. For example, Koike et al. found that corrosion-resistant titanium alloys are biocompatible and were similar to the excellent results obtained from PTFE in terms of cytotoxicity.[70] Titanium–gold alloys are also corrosion resistant and relatively noncytotoxic.[71]

It has also been reported that superferritic grades of stainless steel (those with high chromium content) are corrosion resistant and might well serve as good biomaterials.[72]

A promising recent approach for understanding the biologic effects of biomaterials is the use of modern proteomics approaches. Such techniques have been used for the purpose of measuring protein expression profiles of macrophages (cultured on different biomaterials) to learn more about biocompatibility. Proteomics techniques were able to demonstrate that different types of polyurethane materials led to different intracellular and structural protein expression profiles.[73]

3.6 DEVICE MOTION

Another important (but little studied) issue is that of device motion. Although it is clear that micromotion has a major effect on skeletal implants (leading to increased bone formation),[74] the effect of motion on subcutaneous or intraperitoneal implants has not been systematically investigated. Informal impressions from our laboratory and others suggest that motion of an implant (from a device that is not well secured to surrounding tissue) leads to increased inflammation and, in some cases, formation of fluid masses around the implant. Motion was likely a contributing factor to fluid masses that developed around subcutaneous implants in dogs.[75] To avoid motion in devices that are fully implanted into subcutaneous tissue, it is often appropriate to secure the device to fascia or muscle using sutures or other means.

3.7 METHODS TO REDUCE THE FOREIGN BODY REACTION

There are a number of targets for the prevention or limitation of the foreign body reaction, as outlined in Figure 3.4. Unfortunately, the number of redundancies that occur during the complex immune response associated with the foreign body reaction makes the attempts to reduce the response very complex. This section covers some of the potential targets. In this section, we focus on pathways involving TGF-β and its associated upstream and downstream compounds.

3.7.1 Surface Modification to Reduce Fibrinogen and Macrophage Binding

The University of Washington Engineered Biomaterials group has studied the effect of tetraethylene glycol dimethyl ether (tetraglyme) applied by glow discharge plasma to hydrophobic tubing. They found that the tetraglyme greatly reduced fibrinogen adsorption and monocyte adhesion.[76] The same group also found that tetraglyme also inhibits platelet adhesion.[77] Since fibrinogen binding appears to be an important part of the foreign body reaction in subcutaneous tissue as well as in blood, this approach may hold promise to minimize protein fouling for subcutaneous implants. It has also been observed that certain polymer surfaces (hydrophilic and anionic surfaces) are able to inhibit monocyte adhesion.[78]

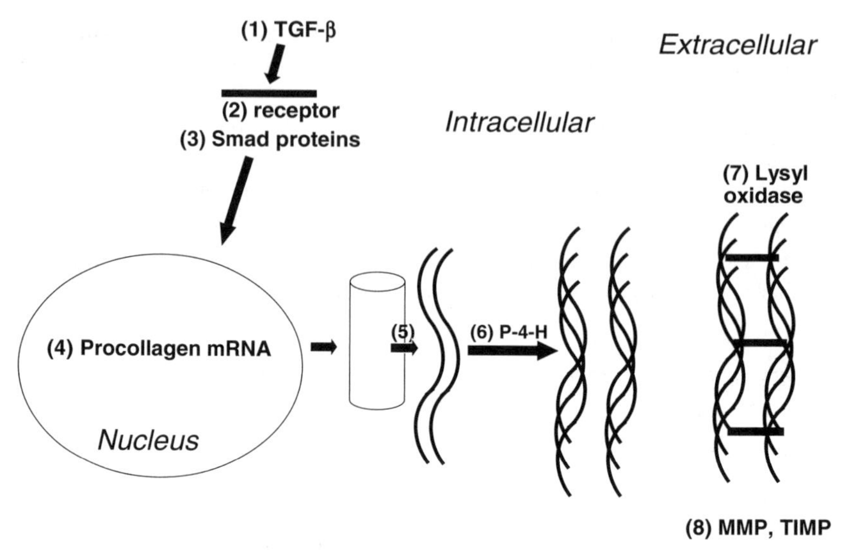

Figure 3.4 Factors affecting foreign body reaction and potential points of intervention at the level of the myofibroblast: (1) inhibit synthesis or release of TGF-β_1; (2) block stimulation by TGF-β of its membrane receptors on the activated fibroblast; (3) inhibit the Smad proteins, which transfer the TGF-β effect to the nucleus; (4) inhibit transcription of procollagen mRNA; (5) inhibit translation of the message to form procollagen; (6) inhibit prolyl-4-hydroxylase, which creates hydroxyproline and facilitates helix formation; (7) inhibit lysyl oxidase, which cross-links the collagen; (8) enhance the function of MMPs, which degrade collagen, or inhibit TIMPs, which degrade MMPs.

3.7.2 Phospholipids

Another method employed to minimize fouling is the use of phospholipid coating designed to mimic the normal mammalian cell membrane. The best studied of these compounds has been phosphorylcholine (PC). For example, Goreish et al. applied PC to polyethylene terephthalate substrate and found that PC decreased the number of inflammatory cells that bound to the substrate.[79] More important, they also implanted PC-coated substrates intramuscularly in rabbits and found that at week 13, the encapsulation tissue as a result of the foreign body reaction was much thinner in the PC-treated tissue samples.

Using the cage implant system, Kim et al. successfully polymerized a phospholipid on to a solid substrate and found reduced adsorption of proteins such as albumin, fibrinogen, and IgG and also reduced macrophage adhesion.[80] There has also been widespread interest in the use of PC coats for orthopedic joint implants as well as other biomedical applications.[81–83]

3.7.3 Passivating Proteins at the Implant Surface

It has been hypothesized that binding of certain proteins to the surface of an implant may inhibit key steps in the foreign body reaction such as the binding to fibrinogen. For example, Tang et al. found that albumin coating reduced the subsequent binding of macrophages.[1] Geelhood and Horbett, in collaboration with our group, found that hemoglobin was even better than albumin in terms of blocking fibrinogen adsorption to polystyrene.[84] However, when studied *in vivo*, passivating proteins are largely displaced by other serum proteins and it is thus very difficult to keep the passivating protein adhered to the device surface in animals. We also found that when the device is incubated *in vitro* with the passivating protein in the presence of human monocytes, the proteins are displaced much more quickly than when incubated in the presence of standard buffers. This finding suggests that monocytes release proteases that actively displace passivating proteins such as hemoglobin.[84] One potential avenue of exploration is to covalently bind passivating proteins (perhaps with cross-linking) such as hemoglobin to device surfaces in an attempt to block displacement or active removal.

3.7.4 Blockade of Collagen/Procollagen Production

Collagen is the most abundant protein in the human body and its synthesis will be briefly reviewed here. It is largely insoluble. There are 19 types of collagen, the most common of which is type 1 collagen. Its synthesis is stimulated by TGF-β. After TGF-β activates intracellular Smad proteins, the phosphorylated Smad moves to the nucleus and induces transcription of the procollagen gene.[85] The procollagen message moves to the ribosome, where procollagen is synthesized and subsequently modified by the addition of several carbohydrate moieties. Under the influence of prolyl 4-hydroxylase, proline residues in the procollagen molecule undergo transformation to hydroxyproline and are then assembled into a helical form. After the helical procollagen is secreted into the extracellular space, the molecule is further modified and then cross-linked by lysyl hydroxylase. The resulting protein matrix, now properly termed collagen, has great tensile strength. Collagen in the extracellular matrix is degraded by matrix metalloproteinases (MMPs). Heat shock protein-47 (HSP47),[86] α-ketoglutarate, and ascorbic acid are needed for normal collagen synthesis.

Chen and colleagues found that blockade of HSP47 with short interfering RNA chains is capable of reducing collagen content in keloid fibroblasts and suggested that such a treatment might be effective in fibroproliferative disorders.[87] Other mechanisms that have been suggested for reducing collagen formation include blockade of the stimulatory effect of TGF-β on procollagen synthesis,[88] blockade of prolyl hydroxylase,[89] blockade of lysyl hydroxylase,[90] or the use of topoisomerase 1 inhibitors, which has been shown to be effective in dermal cells from persons with systemic sclerosis.[91]

3.7.5 Matrix Metalloproteinases, Tissue Inhibitors of Metalloproteinases

Another possible strategy that, to our knowledge, has not been studied is to increase the concentration or effect of matrix metalloproteinases, compounds that are responsible for controlled, ongoing degradation of collagen. One such method would be to block tissue inhibitors of metalloproteinases (TIMPs). Such a method would have an indirect effect to decrease the amount of collagen in the extracellular matrix by inhibiting the natural pathway responsible for the degradation of MMP compounds.[92] It appears that one important physiologic method for regulation of endogenous MMPs is plasmin, which activates MMP-1 and thus leads to the degradation of collagen.[93] TIMP compounds also appear to be important in human fibrosis, as recently underscored by a very interesting study in patients with Dupuytren's disease, a fibroproliferative disorder. These authors found that individuals with this disorder had elevated serum concentrations of TIMP-1, accompanied by a reduced MMP-to-TIMP ratio,[94] suggesting an etiologic role for the TIMP compounds in this disorder.

We believe that further studies aimed at MMPs and TIMPs might hold tremendous promise for reducing the amount of dense collagen within the capsule surrounding the implant as a result of the foreign body reaction. Why do we say this? The relevant concept here is the notion of the final common pathway. One excellent example of a very complex pathway with involvement of multiple hormones and cytokines is secretion of acid by gastric parietal cells. Over the years, there have been multiple pharmacologic methods for inhibiting gastric acid secretion in attempts to treat gastrointestinal ulcers. In 1981, a landmark publication reported that substituted benzimidazones were capable of blocking the final step in parietal cell acid secretion, the proton pump.[95] This finding led to the development of proton pump inhibitors, which have made life better for millions of ulcer patients, and serves as an example of highly successful, focused pharmacologic research.[96] Prior to the development of proton pump inhibitors, there were many surgical and pharmacologic methods attempted for ulcer healing, none of which were as successful as these drugs. To our knowledge, MMP and TIMP compounds are to matrix collagen content as the proton pump is to secretion of hydrochloric acid, both examples serving as a final common pathway. In the setting of the foreign body reaction, it is possible that blockade of an early step such as fibrinogen adsorption might be overcome by redundant mechanisms, but that blockade of a later step (e.g., blockade of TIMPs) might not be able to be countered by other pathways.

3.7.6 Corticosteroids

Corticosteroids such as cortisol, dexamethasone, and prednisone have been used for decades as powerful anti-inflammatory agents that are active in virtually all organ systems. In addition to their immunosuppressive effects, these drugs have strong antifibrotic effects that include inhibition of normal wound healing. For example, in persons on large doses of corticosteroids, healing of surgical wounds is decreased, thus increasing the risk for serious postoperative complications such as wound dehiscence. Several workers have explored the effect of corticosteroids to decrease the foreign body reaction and the thickness and chemistry of the formed encapsulation tissue. Our group studied a drug release system in which dexamethasone was slowly released from the surface of a fully implantable glucose sensor in dogs. The animals underwent repeated glucose challenge studies over many months and those in which drug was given showed a trend toward prolonged sensor function.[75]

Moussy and colleagues have also been involved in work with corticosteroids in animals with implanted devices. This group found that dexamethasone released from

PLGA microspheres successfully reduced the acute and chronic inflammatory reaction. This group also found that a period of PLGA predegradation reduced the delay between implantation and steroid drug release from the PLGA matrix.[97,98] The major concern about the use of corticosteroids to block collagen deposition caused by the foreign body reaction is its systemic side effects, which include osteoporosis, glaucoma, diabetes, hypertension, central obesity, and psychological effects. To the degree that the steroid effect could be kept localized, these compounds may well hold promise to reduce the thickness of the resulting capsule surrounding the implant.

3.7.7 Potential Cytokine Targets: TGF-β and Other Pathways

Cytokines represent attractive targets to inhibit during the foreign body reaction. Evidence implicating TGF-β is presented in Section 3.3.3. Transforming growth factor-β was discovered in the early 1980s. It was found to increase neoplastic transformation of virus-transformed fibroblasts.[99] It was later found to have a more universal role in cell growth, not only in neoplastic tissue but also in many tissues during wound healing. In the late 1990s, it was found that TGF-β and its receptors (along with platelet-derived growth factor and its receptors) were present in the fibrogenic layer that surrounded facial titanium implants.[100] The suggestion that TGF-β was involved in the foreign body reaction was made by Hernandez-Pando et al. in 2000.[101] These workers found that in the chronic phase of the host response to injected nitrocellulose particles, multinucleated giant cells (MGCs) expressed TGF-β in tissue that was undergoing fibrosis. As discussed in Section 3.3.3, our research group found that TGF-β and its signaling factors were highly expressed at multiple time points after subcutaneous device implantation. TGF-β has also been found to be highly expressed in several seemingly unrelated fibrotic diseases, including disorders of the skin, kidney, and lung.

Taken together, these findings suggest that TGF-β (especially the β_1 isoform) might be an attractive target for inhibition of the foreign body reaction. One possible mechanism is the use of neutralizing antibodies, which Shah et al. found to be effective in blocking cutaneous fibrosis in rodents.[43,44,102] It may also be possible to block the effect of TGF-β by using inhibitors of TGF-β receptors.[45,103] Other potential blocking methods include the use of the natural TGF-β inhibitor, latency-associated peptide (LAP),[104] or blockade of compounds such as thrombospondin-1 that are known to activate TGF-β by interacting with LAP.[105] A possible indirect method of blocking the TGF-β effect to promote fibrosis would be inhibition of monocyte chemoattractant protein-1, which has been found to participate in the effect of TGF-β to stimulate fibroblast collagen production.[106]

Connective tissue growth factor appears to mediate some of the effects of TGF-β to promote collagen deposition. In a very interesting study in mammalian kidney fibroblasts, Duncan et al. were able to block the effect of TGF-β to promote collagen deposition by use of either an antisense CTGF gene or antibodies to CTGF.[107] This study is good evidence that CTGF is a major effector of TGF-β. Another study has verified the important role of TGF-β-mediated CTGF expression in kidney mesangial cells.[108] Although studies in cutaneous[36] and subcutaneous[17] tissues have verified the presence of TGF-β-mediated CTGF expression, to our knowledge, blocking studies in such tissues have not been published. For this reason, the importance of the putative role of CTGF in dermal and subcutaneous fibrosis cannot be verified at this point in time.

3.7.8 Other Cytokines and Hormones

In addition to TGF-β and its downstream effectors, there appear to be other cytokines whose inhibition may also lead to a decrease in the intensity of the foreign body reaction. We found that IL-13 and the T cells that synthesize and release IL-13 were overexpressed during subcutaneous implants in rats.[17.] Both Gretzer et al. and Brodbeck et al. found that IL-10 may have a role in the foreign body reaction.[78,109] One group found that the IL-13 effect could be blocked by gene silencing or blockade of the IL-13 α_2 receptor, formerly thought to be only a decoy receptor.[32] These authors conclude that blocking this receptor may be a logical means of inhibiting TGF-β-mediated fibrosis.

Another compound of interest is the hormone relaxin, which is secreted in large amounts during pregnancy in mammals and is known to create laxity in pelvic ligaments during late pregnancy and parturition. Relaxin is now known to have many more actions than this effect and is probably important as an antifibrosis compound in men and women. Workers in the field have found that relaxin can block fibrosis in skin and liver.[110,111] There has also been much interest in interferon-γ (IFN-γ). This compound has been found, by regulation of procollagen transcription, to decrease fibrosis[112] and to induce the formation and ingress of multinucleated giant cells.[113,114] It may have multiple effects to decrease aspects of wound healing.[115] It may also decrease fibronectin formation and smooth muscle actin activity, both of which play roles in the foreign body reaction.[116] Higashi and coworkers have some evidence that some of the role of IFN-γ to inhibit wound healing may be secondary to its effects to inhibit the expression of TGF-β signaling.[117]

In kidney tissue, there have been numerous studies examining the role of the renin–angiotensin system in fibrosis. In humans, inhibition of angiotensin II (Ang II) production at the converting enzyme step[118] or at the Ang II receptor[119] is well known to reduce the progression of kidney failure and albuminuria. Some studies suggest that one of the mechanisms by which inhibitors of Ang II production reduce kidney disease/fibrosis is by reduction of the concentrations and action of TGF-β.[120] From a pathological standpoint, kidney failure is characterized by kidney fibrosis, and there is good evidence that the renin–angiotensin system participates in chronic kidney fibrosis. Though studies are much fewer in number, there is some evidence that this system may also be important in cutaneous and subcutaneous tissues.[121,122]

3.7.9 Vascularizing Compounds

There has been a great deal of interest in creating neovessels within the capsule surrounding a foreign implant. As Updike and colleagues found, the encapsulation tissue surrounding an implant, despite being formed primarily of collagen, is not entirely devoid of capillaries.[123] Since capillaries are the source of glucose, oxygen, and other analytes that are measured by biosensors, some groups have administered blood vessel growth factors in an effort to increase the density of microvessels. For example, our group administered vascular endothelial growth factor-165 (VEGF-165) locally at the site of subcutaneous model sensor implants by the use of a slow mini-osmotic pump. We found a spatial effect of the VEGF; capsular tissue within 15 mm of the infusion port was seen to increase in capillary number, but tissue over 22 mm was not found to undergo such an increase.[52] In a second study in which functioning sensors were implanted, we found that VEGF delivery caused an increase in the ability of the sensor to respond to a glucose infusion in animals in which the VEGF had been slowly delivered to the local site of the sensor.[53] Other workers have also examined the effect of administration of VEGF or the VEGF gene. Klueh et al. found that vector transfer of the VEGF gene enhanced sensor function, using the chick chorioallantoic model.[124,125]

It should be mentioned that there would be risk associated with administration of VEGF to persons with diabetes. VEGF is known to have a role in generating the fragile retinal vessels that can undergo hemorrhage in diabetic retinopathy, and VEGF blockade may be helpful in this condition.[126] In addition, there are many well-studied associations of VEGF with neoplastic disease and VEGF antagonists are undergoing many trials in the field of cancer treatment.

Our belief is that vascularizing compounds can increase vessel growth within the encapsulation tissue surrounding a foreign body and probably improve sensor function modestly. Nonetheless, they have not been found to overcome the relentless process of collagen deposition and capsule formation caused by the foreign body reaction, which eventually blocks sensor function. In addition, they could have several major side effects.

3.8 SUMMARY

Within 1 week after subcutaneous implantation, a foreign body elicits a cascade of biochemical responses that ultimately encapsulate the implant in a matrix consisting of dense, parallel-oriented collagen accompanied by other proteins such as decorin and fibronectin. It appears that minimizing the size of a foreign implant reduces the intensity of the foreign body reaction, but there is a paucity of studies that address this specific issue. There is also a large body of evidence that TGF-β (primarily the β_1 isoform) promotes fibrosis in many cells and tissue, including the subcutaneous encapsulation with collagen deposition of a foreign body. IL-13 is probably a promoter of TGF-β expression and CTGF appears to be an important mediator of the TGF-β effect (perhaps one of the many mediators). The very first effect of TGF-β on the fibroblast is stimulation of Smad activity (especially, Smads 2, 3, and 4), which directly causes nuclear procollagen transcription. Procollagen is glycated after translation, modified to form a helix, and then secreted into the extracellular matrix along with other proteins that are also influenced by TGF-β. It is then cross-linked, which increases its tensile strength, and eventually enzymatically degraded by MMP compounds, which themselves are degraded by TIMP compounds. Because of the prominent role of TGF-β in this process, it is an attractive target. However, there might well be redundant pathways that lead to the resulting encapsulated object as a result of the foreign body reaction, and for this reason, stimulation of MMPs or blockade of TIMPs may be more successful since those compounds likely represent the final common pathway of collagen regulation.

ACKNOWLEDGMENTS

We thank the NIH (NIBIB and NIDDK) for grant support over the years and the Juvenile Diabetes Research Foundation for Grant 22-2006-1121. We thank Dr. Guanqun Allen Li for scientific collaboration regarding cytokine biochemistry and histology.

REFERENCES

1. Tang L, Eaton JW. Fibrin(ogen) mediates acute inflammatory responses to biomaterials. *Journal of Experimental Medicine* 1993, 178, 2147–2156.
2. Tang L, Ugarova TP, Plow EF, Eaton JW. Molecular determinants of acute inflammatory responses to biomaterials. *Journal of Clinical Investigation* 1996, 97, 1329–1334.
3. Tang L, Wu Y, Timmons RB. Fibrinogen adsorption and host tissue responses to plasma functionalized surfaces. *Journal of Biomedical Materials Research* 1998, 42, 156–163.
4. Ward WK, Wood MD, Casey HM, Quinn MJ, Federiuk IF. An implantable subcutaneous glucose sensor array in ketosis-prone rats: closed loop glycemic control. *Artificial Organs* 2005, 29, 131–143.
5. Zdolsek J, Eaton JW, Tang L. Histamine release and fibrinogen adsorption mediate acute inflammatory responses to biomaterial implants in humans. *Journal of Translational Medicine* 2007, 5, 31.
6. Tang L, Jennings TA, Eaton JW. Mast cells mediate acute inflammatory responses to implanted biomaterials. *Proceedings of the National Academy of Sciences of the United States of America* 1998, 95, 8841–8846.
7. Greenhalgh D. The role of monocytes/macrophages in wound healing. In: Robinson JP, Babcock GF (Eds), *Phagocyte Function: A Guide for Research and Clinical Evaluation.* Wiley-Liss Inc., 1998, pp. 349–357.
8. Kao WJ, McNally AK, Hiltner A, Anderson JM. Role for interleukin-4 in foreign-body giant cell formation on a poly(etherurethane urea) *in vivo*. *Journal of Biomedical Materials Research* 1995, 29, 1267–1275.
9. Tsai AT, Rice J, Scatena M, Liaw L, Ratner BD, Giachelli CM. The role of osteopontin in foreign body giant cell formation. *Biomaterials* 2005, 26, 5835–5843.
10. El-Warrak AO, Olmstead M, Apelt D, Deiss F, Noetzli H, Zlinsky K, Hilbe M, Bertschar-Wolfsberger R, Johnson AL, Auer J, von Rechenberg B. An animal model for interface tissue formation in cemented hip replacements. *Veterinary Surgery* 2004, 33, 495–504.
11. Wozniak W, Markuszewski J, Wierusz-Kozlowska M, Wysocki H. Neutrophils are active in total joint implant loosening. *Acta Orthopaedica Scandinavica* 2004, 75, 549–553.
12. Sharkawy AA, Klitzman B, Truskey GA, Reichert WM. Engineering the tissue which encapsulates subcutaneous implants. I. Diffusion properties. *Journal of Biomedical Materials Research* 1997, 37, 401–412.

13. Sharkawy AA, Klitzman B, Truskey GA, Reichert WM. Engineering the tissue which encapsulates subcutaneous implants. II. Plasma–tissue exchange properties. *Journal of Biomedical Materials Research* 1998, 40, 586–597.
14. Wisniewski N, Moussy F, Reichert WM. Characterization of implantable biosensor membrane biofouling. *Fresenius Journal of Analytical Chemistry* 2000, 366, 611–621.
15. Wisniewski N, Reichert M. Methods for reducing biosensor membrane biofouling. *Colloids and Surfaces B. Biointerfaces* 2000, 18, 197–219.
16. Li AG, Quinn MJ, Siddiqui Y, Wood MD, Federiuk IF, Duman HM, Ward WK. Elevation of transforming growth factor beta (TGFbeta) and its downstream mediators in subcutaneous foreign body capsule tissue. *Journal of Biomedical Materials Research A* 82, 2007, 498–508.
17. Ward WK, Li AG, Siddiqui Y, Federiuk IF, Wang X-J. Increased expression of interleukin-13 and connective tissue growth factor, and their potential roles during foreign body encapsulation of subcutaneous implants. *Journal of Biomaterials Science, Polymer Edition* 2008, 19, 1065–1072.
18. Border WA. Transforming growth factor-β and the pathogenesis of glomerular diseases. *Current Opinion in Nephrology and Hypertension* 1994, 3, 54–58.
19. Chen K, Wei Y, Sharp GC, Braley-Mullen H. Inhibition of TGFβ1 by anti-TGFβ1 antibody or lisinopril reduces thyroid fibrosis in granulomatous experimental autoimmune thyroiditis. *Journal of Immunology* 2002, 169, 6530–6538.
20. Atamas SP, White B. The role of chemokines in the pathogenesis of scleroderma. *Current Opinion in Rheumatology* 2003, 15, 772–777.
21. Leask A, Abraham DJ. TGF-β signaling and the fibrotic response. *Journal of the Federation of American Societies for Experimental Biology* 2004, 18, 816–827.
22. Parekh T, Saxena B, Reibman J, Cronstein BN, Gold LI. Neutrophil chemotaxis in response to TGF-β isoforms (TGF-β1, TGF-β2, TGF-β3) is mediated by fibronectin. *Journal of Immunology* 1994, 152, 2456–2466.
23. Brodbeck WG, Voskerician G, Ziats NP, Nakayama Y, Matsuda T, Anderson JM. *In vivo* leukocyte cytokine mRNA responses to biomaterials are dependent on surface chemistry. *Journal of Biomedical Materials Research A* 2003, 64, 320–329.
24. Robitaille R, Dusseault J, Henley N, Desbiens K, Labrecque N, Halle J-P. Inflammatory response to peritoneal implantation of alginate-poly-L-lysine microcapsules. *Biomaterials* 2005, 26, 4119–4127.
25. O'Kane S, Ferguson MW. Transforming growth factor-βs and wound healing. *International Journal of Biochemistry and Cell Biology* 1997, 29, 63–78.
26. McCormick LL, Zhang Y, Tootell E, Gilliam AC. Anti-TGF-β treatment prevents skin and lung fibrosis in murine sclerodermatous graft-versus-host disease: a model for human scleroderma. *Journal of Immunology* 1999, 163, 5693–5699.
27. Hetzel GR, Hermsen D, Hohlfeld T, Rettich A, Ozcan F, Fussholler A, Grabensee B, Plum J. Effects of candesartan and perindopril on renal function, TGF-β1 plasma levels and excretion of prostaglandins in stable renal allograft recipients. *Clinical Nephrology* 2002, 57, 296–302.
28. Oriente A, Fedarko NS, Pacocha SE, Huang S-K, Lichtenstein LM, Essayan DM. Interleukin-13 modulates collagen homeostasis in human skin and keloid fibroblasts. *Journal of Pharmacology and Experimental Therapeutics* 2000, 292, 988–994.
29. Jinnin M, Ihn H, Yamane K, Tamaki K. Interleukin-13 stimulates the transcription of the human α2(I) collagen gene in human dermal fibroblasts. *Journal of Biological Chemistry* 2004, 279, 41783–41791.

30. Joshi BH, Hogaboam C, Dover P, Husain SR, Puri RK. Role of interleukin-13 in cancer, pulmonary fibrosis, and other T(H)2-type diseases. *Vitamins and Hormones* 2006, 74, 479–504.

31. Lee CG, Homer RJ, Zhu Z, Lanone S, Wang X, Koteliansky V, Shipley JM, Gotwals P, Noble P, Chen Q, Senior RM, Elias JA. Interleukin-13 induces tissue fibrosis by selectively stimulating and activating transforming growth factor-β1. *Journal of Experimental Medicine* 2001, 194, 809–821.

32. Fichtner-Feigl S, Strober W, Kawakami K, Puri RK, Kitani A. IL-13 signaling through the IL-13α2 receptor is involved in induction of TGF-β1 production and fibrosis. *Nature Medicine* 2006, 12, 99–106.

33. Kaviratne M, Hesse M, Leusink M, Cheever AW, Davies SJ, McKerrow JH, Wakefield LM, Letterio JJ, Wynn TA. IL-13 activates a mechanism of tissue fibrosis that is completely TGF-β independent. *Journal of Immunology* 2004, 173, 4020–4029.

34. Chen MM, Lam A, Abraham JA, Schreiner GF, Joly AH. CTGF expression is induced by TGF-β in cardiac fibroblasts and cardiac myocytes: a potential role in heart fibrosis. *Journal of Molecular and Cellular Cardiology* 2000, 32, 1805–1819.

35. Yokoi H, Sugawara A, Mukoyama M, Mori K, Makino H, Suganami T, Nagae T, Yahata K, Fujinaga Y, Tanaka I, Nakao K. Role of connective tissue growth factor in profibrotic action of transforming growth factor-beta: a potential target for preventing renal fibrosis. *American Journal of Kidney Diseases* 2001, 38, S134–S138.

36. Colwell AS, Phan TT, Kong W, Longaker MT, Lorenz PH. Hypertrophic scar fibroblasts have increased connective tissue growth factor expression after transforming growth factor-β stimulation. *Plastic and Reconstructive Surgery* 2005, 116, 1387–1390; discussion 1391–1392.

37. Zhao Q, Chen N, Wang W-M, Lu J, Dai B-B. Effect of transforming growth factor-beta on activity of connective tissue growth factor gene promoter in mouse NIH/3T3 fibroblasts. *Acta Pharmacologica Sinica* 2004, 25, 485–489.

38. Li MO, Wan YY, Sanjabi S, Robertson A-K, Flavell L, RA Transforming growth factor-β regulation of immune responses. *Annual Reviews of Immunology* 2006, 24, 99–146.

39. Pringle GA, Dodd CM. Immunoelectron microscopic localization of the core protein of decorin near the d and e bands of tendon collagen fibrils by use of monoclonal antibodies. *Journal of Histochemistry and Cytochemistry* 1990, 38, 1405–1411.

40. Danielson KG, Baribault H, Holmes DF, Graham H, Kadler KE, Iozzo RV. Targeted disruption of decorin leads to abnormal collagen fibril morphology and skin fragility. *Journal of Cell Biology* 1997, 136, 729–743.

41. Weston BS, Wahab NA, Mason RM. CTGF mediates TGF-β-induced fibronectin matrix deposition by upregulating active α5β1 integrin in human mesangial cells. *Journal of the American Society of Nephrology* 2003, 14, 601–610.

42. Yung S, Lee CY, Zhang Q, Lau SK, Tsang RC, Chan TM. Elevated glucose induction of thrombospondin-1 upregulates fibronectin synthesis in proximal renal tubular epithelial cells through TGF-beta1 dependent and TGF-beta1 independent pathways. *Nephrology Dialysis Transplantation* 2006, 21, 1504–1513.

43. Shah M, Foreman DM, Ferguson MW. Control of scarring in adult wounds by neutralising antibody to transforming growth factor beta. *Lancet* 1992, 339, 213–214.

44. Shah M, Foreman DM, Ferguson MW. Neutralising antibody to TGF-β1,2 reduces cutaneous scarring in adult rodents. *Journal of Cell Science* 1994, 107, 1137–1157.

45. Mori Y, Ishida W, Bhattacharyya S, Li Y, Platanias LC, Varga J. Selective inhibition of activin receptor-like kinase 5 signaling blocks profibrotic transforming growth factor beta responses in skin fibroblasts. *Arthritis and Rheumatism* 2004, 50, 4008–4021.
46. Kusaba A, Fischer 3rd CR, Matulewski TJ, Matsumoto T. Experimental study of the influence of porosity on development of neointima in Gore-Tex grafts: a method to increase long-term patency rate. *American Surgeon* 1981, 47, 347–354.
47. Watanabe T. Experimental study on the influence of porosity on the development of the neointima in E-PTFE grafts. *Nippon Geka Gakkai Zasshi* 1984, 85, 580–591.
48. Andrade SP, Fan TP, Lewis GP. Quantitative *in-vivo* studies on angiogenesis in a rat sponge model. *British Journal of Experimental Pathology* 1987, 68, 755–766.
49. Brauker JH, Carr-Brendel VE, Martinson LA, Crudele J, Johnston WD, Johnson RC. Neovascularization of synthetic membranes directed by membrane microarchitecture. *Journal of Biomedical Materials Research* 1995, 29, 1517–1524.
50. Updike SJ, Shults MC, Gilligan BJ, Rhodes RK. A subcutaneous glucose sensor with improved longevity, dynamic range, and stability of calibration. *Diabetes Care* 2000, 23, 208–214.
51. Ward WK, Slobodzian EP, Tiekotter KL, Wood MD. The effect of microgeometry, implant thickness and polyurethane chemistry on the foreign body response to subcutaneous implants. *Biomaterials* 2002, 23, 4185–4192.
52. Ward WK, Quinn MJ, Wood MD, Tiekotter KL, Pidikiti S, Gallagher JA. Vascularizing the tissue surrounding a model biosensor: how localized is the effect of a subcutaneous infusion of vascular endothelial growth factor (VEGF)? *Biosensors & Bioelectronics* 2003, 19, 155–163.
53. Ward WK, Wood MD, Casey HM, Quinn MJ, Federiuk IF. The effect of local subcutaneous delivery of vascular endothelial growth factor on the function of a chronically-implanted amperometric glucose sensor. *Diabetes Technology & Therapeutics* 2004, 6, 137–145.
54. Voskerician G, Gingras PH, Anderson JM. Macroporous condensed poly (tetrafluoroethylene). I. *In vivo* inflammatory response and healing characteristics. *Journal of Biomedical Materials Research A* 2006, 76, 234–242.
55. Sanders JE, Stiles CE, Hayes CL. Tissue response to single-polymer fibers of varying diameters: evaluation of fibrous encapsulation and macrophage density. *Journal of Biomedical Materials Research* 2000, 52, 231–237.
56. Gross KA, Ezerietis E. Juniper wood as a possible implant material. *Journal of Biomedical Materials Research A* 2003, 64, 672–683.
57. Aho AJ, Rekola J, Matinlinna J, Gunn J, Tirri T, Viitaniemi P, Vallittu P. Natural composite of wood as replacement material for ostechondral bone defects. *Journal of Biomedical Materials Research B* 2007, 83, 64–71.
58. Smith HR. Do silicone breast implants cause autoimmune rheumatic diseases? *Journal of Biomaterials Science, Polymer Edition* 1995, 7, 115–121.
59. Godin MS, Waldman SR, Johnson Jr CM. The use of expanded polytetrafluoroethylene (Gore-Tex) in rhinoplasty. A 6-year experience. *Archives of Otolaryngology Head and Neck Surgery* 1995, 121, 1131–1136.
60. Hurst BS. Permanent implantation of expanded polytetrafluoroethylene is safe for pelvic surgery. United States Expanded Polytetrafluoroethylene Reproductive Surgery Study Group. *Human Reproduction* 1999, 14, 925–927.
61. Midasch O, Drexler H, Hart N, Beckmann MW, Angerer J. Transplacental exposure of neonates to perfluorooctanesulfonate and perfluorooctanoate: a pilot study. *International Archives of Occupational and Environmental Health* 2007, 80, 643–648.

62. Wolf CJ, Fenton SE, Schmid JE, Calafat AM, Kuklenyik Z, Bryant XA, Thibodeaux J, Das KP, White SS, Lau CS, Abbott BD. Developmental toxicity of perfluorooctanoic acid in the CD-1 mouse after cross-foster and restricted gestational exposures. *Toxicological Sciences* 2007, 95, 462–473.

63. http://www.epa.gov/opptintr/pfoa/pubs/pfoainfo.htm#concerns (accessed on August 19, 2007).

64. Christenson EM, Anderson JM, Hiltner A. Antioxidant inhibition of poly(carbonate urethane) *in vivo* biodegradation. *Journal of Biomedical Materials Research A* 2006, 76, 480–490.

65. Mathur AB, Collier TO, Kao WJ, Wiggins M, Schubert MA, Hiltner A, Anderson JM. *In vivo* biocompatibility and biostability of modified polyurethanes. *Journal of Biomedical Materials Research* 1997, 36, 246–257.

66. Wiggins MJ, MacEwan M, Anderson JM, Hiltner A. Effect of soft-segment chemistry on polyurethane biostability during *in vitro* fatigue loading. *Journal of Biomedical Materials Research A* 2004, 68, 668–683.

67. Ernsting MJ, Labow RS, Santerre JP. Surface modification of a polycarbonate-urethane using a vitamin-E-derivatized fluoroalkyl surface modifier. *Journal of Biomaterials Science, Polymer Edition* 2003, 14, 1411–1426.

68. Jahangir AR, McClung WG, Cornelius RM, McCloskey CB, Brash JL, Santerre JP. Fluorinated surface-modifying macromolecules: modulating adhesive protein and platelet interactions on a polyether-urethane. *Journal of Biomedical Materials Research* 2002, 60, 135–147.

69. Sanders JE, Lamont SE, Karchin A, Golledge SL, Ratner BD. Fibro-porous meshes made from polyurethane micro-fibers: effects of surface charge on tissue response. *Biomaterials* 2005, 26, 813–818.

70. Koike M, Lockwood PE, Wataha JC, Okabe T. Initial cytotoxicity of novel titanium alloys. *Journal of Biomedical Materials Research B* 2007, 83, 327–331.

71. Oh K-T, Kang D-K, Choi G-S, Kim K-N. Cytocompatibility and electrochemical properties of Ti–Au alloys for biomedical applications. *Journal of Biomedical Materials Research B* 2007, 83, 320–326.

72. Assis SL, Rogero SO, Antunes RA, Padilha AF, Costa I. A comparative study of the *in vitro* corrosion behavior and cytotoxicity of a superferritic stainless steel, a Ti-13Nb-13Zr alloy, and an austenitic stainless steel in Hank's solution. *Journal of Biomedical Materials Research B* 2005, 73, 109–116.

73. Dinnes DL, Marcal H, Mahler SM, Santerre JP, Labow RS. Material surfaces affect the protein expression patterns of human macrophages: a proteomics approach. *Journal of Biomedical Materials Research A* 2007, 80, 895–908.

74. Jasty MC, Bragdon C, Burke D, O'Connor D, Lowenstein J, Harris WH. *In vivo* skeletal responses to porous-surfaced implants subjected to small induced motions. *Journal of Bone and Joint Surgery* 1997, 79, 707–714.

75. Ward WK, Troupe JE. Assessment of chronically implanted subcutaneous glucose sensors in dogs: the effect of surrounding fluid masses. *American Society for Artificial Internal Organs* 1999, 45, 555–561.

76. Shen M, Pan YV, Wagner MS, Hauch KD, Castner DG, Ratner BD, Horbett TA. Inhibition of monocyte adhesion and fibrinogen adsorption on glow discharge plasma deposited tetraethylene glycol dimethyl ether. *Journal of Biomaterials Science, Polymer Edition* 2001, 12, 961–978.

77. Cao L, Sukavaneshvar S, Ratner BD, Horbett TA. Glow discharge plasma treatment of polyethylene tubing with tetraglyme results in ultralow fibrinogen adsorption and greatly reduced platelet adhesion. *Journal of Biomedical Materials Research A* 2006, 79, 788–803.
78. Brodbeck WG, Nakayama Y, Matsuda T, Colton E, Ziats NP, Anderson JM. Biomaterial surface chemistry dictates adherent monocyte/macrophage cytokine expression *in vitro*. *Cytokine* 2002, 18, 311–319.
79. Goreish HH, Lewis AL, Rose S, Lloyd AW. The effect of phosphorylcholine-coated materials on the inflammatory response and fibrous capsule formation: *in vitro* and *in vivo* observations. *Journal of Biomedical Materials Research A* 2004, 68, 1–9.
80. Kim K, Kim C, Byun Y. Biostability and biocompatibility of a surface-grafted phospholipid monolayer on a solid substrate. *Biomaterials* 2004, 25, 33–41.
81. Moro T, Takatori Y, Ishihara K, Konno T, Takigawa Y, Matsushita T, Chung U-I, Nakamura K, Kawaguchi H. Surface grafting of artificial joints with a biocompatible polymer for preventing periprosthetic osteolysis. *Nature Materials* 2004, 3, 829–836.
82. Moro T, Takatori Y, Ishihara K, Nakamura K, Kawaguchi H. Frank Stinchfield Award: grafting of biocompatible polymer for longevity of artificial hip joints. *Clinical Orthopaedics and Related Research* 2006, 453, 58–63.
83. Iwasaki Y, Ishihara K. Phosphorylcholine-containing polymers for biomedical applications. *Analytical and Bioanalytical Chemistry* 2005, 381, 534–546.
84. Geelhood SJ, Horbett TA, Ward WK, Wood MD, Quinn MJ. Passivating protein coatings for implantable glucose sensors: evaluation of protein retention. *Journal of Biomedical Research Part B* 2006, 81, 251–260.
85. Ulloa L, Doody J, Massague J. Inhibition of transforming growth factor-beta/SMAD signalling by the interferon-gamma/STAT pathway. *Nature* 1999, 397, 710–713.
86. Nagata K. Expression and function of heat shock protein 47: a collagen-specific molecular chaperone in the endoplasmic reticulum. *Matrix Biology* 1998, 16, 379–386.
87. Chen JJ, Zhao S, Cen Y, Liu X-X, Yu R, Wu D-M. Effect of heat shock protein 47 on collagen accumulation in keloid fibroblast cells. *British Journal of Dermatology* 2007, 156, 1188–1195.
88. Yamada H, Tajima S, Nishikawa T, Murad S, Pinnell SR. Tranilast, a selective inhibitor of collagen synthesis in human skin fibroblasts. *Journal of Biochemistry* 1994, 116, 892–897.
89. Kim I, Mogford JE, Witschi C, Nafissi M, Mustoe TA. Inhibition of prolyl 4-hydroxylase reduces scar hypertrophy in a rabbit model of cutaneous scarring. *Wound Repair and Regeneration* 2003, 11, 368–372.
90. Saika S, Ooshima A, Hashizume N, Yamanaka O, Tanaka S, Okada Y, Kobata S. Effect of lysyl hydroxylase inhibitor, minoxidil, on ultrastructure and behavior of cultured rabbit subconjunctival fibroblasts. *Graefe's Archive for Clinical Experimental Ophthalmology* 1995, 233, 347–353.
91. Czuwara-Ladykowska J, Makiela B, Smith EA, Trojanowska M, Rudnicka L. The inhibitory effects of camptothecin, a topoisomerase I inhibitor, on collagen synthesis in fibroblasts from patients with systemic sclerosis. *Arthritis Research* 2001, 3, 311–318.
92. Kahari VM, Saarialho-Kere U. Matrix metalloproteinases in skin. *Experimental Dermatology* 1997, 6, 199–213.
93. Pins GD, Collins-Pavao ME, Van De Water L, Yarmush ML, Morgan JR. Plasmin triggers rapid contraction and degradation of fibroblast-populated collagen lattices. *Journal of Investigative Dermatology* 2000, 114, 647–653.

94. Ulrich D, Hrynyschyn K, Pallua N. Matrix metalloproteinases and tissue inhibitors of metalloproteinases in sera and tissue of patients with Dupuytren's disease. *Plastic and Reconstructive Surgery* 2003, 112, 1279–1286.

95. Fellenius E, Berglindh T, Sachs G, Olbe L, Elander B, Sjostrand SE, Wallmark B. Substituted benzimidazoles inhibit gastric acid secretion by blocking ($H^+ + K^+$) ATPase. *Nature* 1981, 290, 159–161.

96. Spector R, Vesell ES. The power of pharmacological sciences: the example of proton pump inhibitors. *Pharmacology* 2006, 76, 148–155; discussion 156.

97. Hickey T, Kreutzer D, Burgess DJ, Moussy F. Dexamethasone/PLGA microspheres for continuous delivery of an anti-inflammatory drug for implantable medical devices. *Biomaterials* 2002, 23, 1649–1656.

98. Hickey T, Kreutzer D, Burgess DJ, Moussy F. *In vivo* evaluation of a dexamethasone/PLGA microsphere system designed to suppress the inflammatory tissue response to implantable medical devices. *Journal of Biomedical Materials Research* 2002, 61, 180–187.

99. Anzano MA, Roberts AB, Meyers CA, Komoriya A, Lamb LC, Smith JM, Sporn MB. Synergistic interaction of two classes of transforming growth factors from murine sarcoma cells. *Cancer Research* 1982, 42, 4776–4778.

100. Katou F, Ohtani H, Nagura H, Motegi K. Procollagen-positive fibroblasts predominantly express fibrogenic growth factors and their receptors in human encapsulation process against foreign body. *Journal of Pathology* 1998, 186, 201–208.

101. Hernandez-Pando R, Bornstein QL, Leon DA, Orozco EH, Madrigal VK, Cordero EM. Inflammatory cytokine production by immunological and foreign body multinucleated giant cells. *Immunology* 2000, 100, 352–358.

102. Shah M, Foreman DM, Ferguson MW. Neutralisation of TGF-β1 and TGF-β2 or exogenous addition of TGF-β3 to cutaneous rat wounds reduces scarring. *Journal of Cell Science* 1995, 108, 985–1002.

103. Okadome T, Oeda E, Saitoh M, Ichijo H, Moses HL, Miyazono K, Kawabata M. Characterization of the interaction of FKBP12 with the transforming growth factor-β type I receptor *in vivo*. *Journal of Biological Chemistry* 1996, 271, 21687–21690.

104. Young GD, Murphy-Ullrich JE. Molecular interactions that confer latency to transforming growth factor-beta. *Journal of Biological Chemistry* 2004, 279, 38032–38039.

105. Ribeiro SM, Poczatek M, Schultz-Cherry S, Villain M, Murphy-Ullrich JE. The activation sequence of thrombospondin-1 interacts with the latency-associated peptide to regulate activation of latent transforming growth factor-β. *Journal of Biological Chemistry* 1999, 274, 13586–13593.

106. Gharaee-Kermani M, Denholm EM, Phan SH. Costimulation of fibroblast collagen and transforming growth factor beta1 gene expression by monocyte chemoattractant protein-1 via specific receptors. *Journal of Biological Chemistry* 1996, 271, 17779–17784.

107. Duncan MR, Frazier KS, Abramson S, Williams S, Klapper H, Huang X, Grotendorst GR. Connective tissue growth factor mediates transforming growth factor beta-induced collagen synthesis: down-regulation by cAMP. *Journal of the Federation of American Societies for Experimental Biology* 1999, 13, 1774–1486.

108. Chen Y, Blom IE, Sa S, Goldschmeding R, Abraham DJ, Leask A. CTGF expression in mesangial cells: involvement of SMADs, MAP kinase, and PKC. *Kidney International* 2002, 62, 1149–1159.

109. Gretzer C, Emanuelsson L, Liljensten E, Thomsen P. The inflammatory cell influx and cytokines changes during transition from acute inflammation to fibrous repair around implanted materials. *Journal of Biomaterials Science, Polymer Edition* 2006, 17, 669–687.
110. Unemori EN, Beck LS, Lee WP, Xu Y, Siegel M, Keller G, Liggitt HD, Bauer EA, Amento EP. Human relaxin decreases collagen accumulation *in vivo* in two rodent models of fibrosis. *Journal of Investigative Dermatology* 1993, 101, 280–285.
111. Williams EJ, Benyon RC, Trim N, Hadwin R, Grove BH, Arthur MJP, Unemori EN, Iredale JP. Relaxin inhibits effective collagen deposition by cultured hepatic stellate cells and decreases rat liver fibrosis *in vivo*. *Gut* 2001, 49, 577–583.
112. Granstein RD, Flotte TJ, Amento EP. Interferons and collagen production. *Journal of Investigative Dermatology* 1990, 95, 75S–80S.
113. Fais S, Burgio VL, Silvestri M, Capobianchi MR, Pacchiarotti A, Pallone F. Multinucleated giant cells generation induced by interferon-gamma. Changes in the expression and distribution of the intercellular adhesion molecule-1 during macrophages fusion and multinucleated giant cell formation. *Laboratory Investigation* 1994, 71, 737–744.
114. Khouw IM, van Wachem PB, De Leij LFMH, Van Luyn MJA. Inhibition of the tissue reaction to a biodegradable biomaterial by monoclonal antibodies to IFN-gamma. *Journal of Biomedical Materials Research* 1998, 41, 202–210.
115. Miles RH, Paxton TP, Zacheis D, Dries DJ, Gamelli RL. Systemic administration of interferon-gamma impairs wound healing. *Journal of Surgical Research* 1994, 56, 288–294.
116. Strutz F, Heeg M, Kochsiek T, Siemers G, Zeisberg M, Muller GA. Effects of pentoxifylline, pentifylline and gamma-interferon on proliferation, differentiation, and matrix synthesis of human renal fibroblasts. *Nephrology Dialysis Transplantation* 2000, 15, 1535–1546.
117. Higashi K, Inagaki Y, Fujimori K, Nakao A, Kaneko H, Nakatsuka I. Interferon-gamma interferes with transforming growth factor-beta signaling through direct interaction of YB-1 with Smad3. *Journal of Biological Chemistry* 2003, 278, 43470–43479.
118. Lewis EJ, Hunsicker LG, Bain RP, Rohde RD. The effect of angiotensin-converting-enzyme inhibition on diabetic nephropathy. The Collaborative Study Group. *New England Journal of Medicine* 1993, 329, 1456–1162.
119. Perico N, Ruggenenti P, Remuzzi G. Losartan in diabetic nephropathy. *Expert Review of Cardiovascular Therapy* 2004, 2, 473–483.
120. Sharma K, Eltayeb BO, McGowan TA, Dunn SR, Alzahabi B, Rohde R, Ziyadeh FN, Lewis EJ. Captopril-induced reduction of serum levels of transforming growth factor-beta1 correlates with long-term renoprotection in insulin-dependent diabetic patients. *American Journal of Kidney Diseases* 1999, 34, 818–823.
121. Abiko M, Rodgers KE, Campeau JD, Nakamura RM, Dizerega GS. Alterations of angiotensin II receptor levels in full-thickness excisional wounds in rat skin. *Wound Repair and Regeneration* 1996, 4, 363–367.
122. Sun Y, Weber KT. Angiotensin-converting enzyme and wound healing in diverse tissues of the rat. *Journal of Laboratory and Clinical Medicine* 1996, 127, 94–101.
123. Gilligan BJ, Shults MC, Rhodes RK, Updike SJ. Evaluation of a subcutaneous glucose sensor out to 3 months in a dog model. *Diabetes Care* 1994, 17, 882–887.
124. Klueh U, Dorsky DI, Kreutzer DL. Use of vascular endothelial cell growth factor gene transfer to enhance implantable sensor function *in vivo*. *Journal of Biomedical Materials Research A* 2003, 67, 1072–1086.

125. Klueh U, Dorsky DI, Kreutzer DL. Enhancement of implantable glucose sensor function *in vivo* using gene transfer-induced neovascularization. *Biomaterials* 2005, 26, 1155–1163.

126. Spaide RF, Fisher YL. Intravitreal bevacizumab (Avastin) treatment of proliferative diabetic retinopathy complicated by vitreous hemorrhage. *Retina* 2006, 26, 275–278.

CHAPTER 4

A WINDOW TO OBSERVE THE FOREIGN BODY REACTION TO GLUCOSE SENSORS

Milan T. Makale and Jared B. Goor

In Vivo Glucose Sensing, Edited by David D. Cunningham and Julie A. Stenken

4.1 INTRODUCTION AND BACKGROUND

Progress in the field of implantable glucose biosensors would be considerably accelerated by the development of a model system allowing real-time imaging of local tissue conditions and microvascular flow around an active sensor. Such a system would help to characterize the relationship between sensor performance and various sensor membrane surfaces in terms of local host tissue effects, for example, collagen encapsulation of the sensor and microvascular deterioration (Figure 4.1). Accordingly, we have adapted the rodent dorsal skinfold tissue window chamber so that local tissue and blood flow conditions can be imaged simultaneously with the acquisition of tissue glucose measurements using an implanted planar sensor array.

Clearly, the successful development of an implantable glucose sensor that can operate dependably for many months to years has been a broadly recognized and much sought after goal for almost four decades. The ability to tightly control blood glucose based on a continuous or semicontinuous glucose monitoring system would have the potential to revolutionize diabetes management and reduce the incidence of diabetic complications.[1] While considerable effort and expense has yielded a number

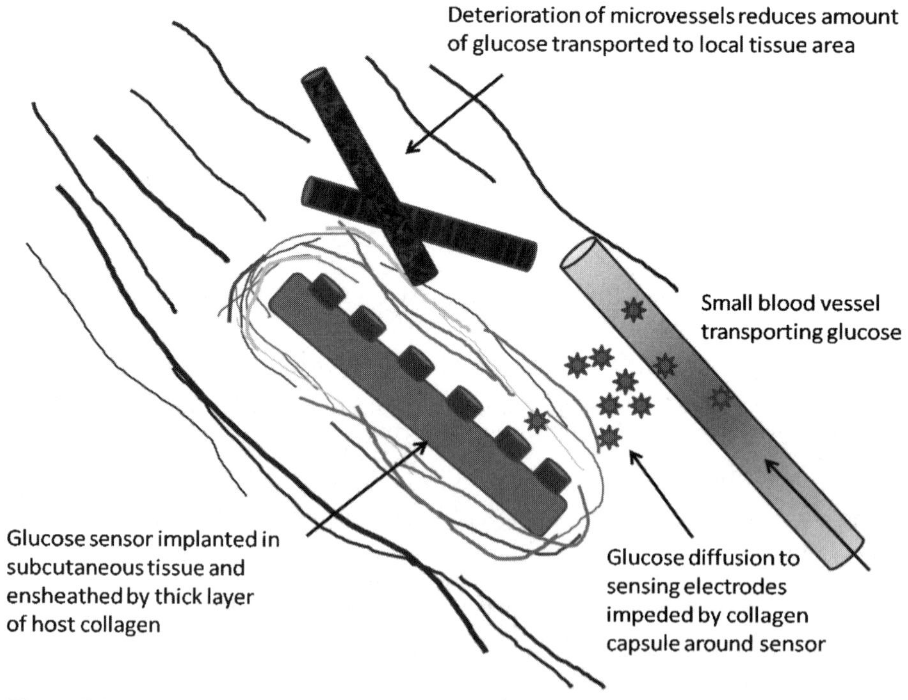

Figure 4.1 Schematic illustration showing encapsulation of a glucose sensor by host collagen and deterioration of the local microvessels adjacent to the implant. Fewer vessels deliver less glucose to the local tissue and the diffusion of glucose to the sensor is inhibited by the presence of the collagen capsule.

of potentially viable glucose sensing technologies, the accuracy and stability of the subcutaneously implanted glucose sensors produced thus far have been inadequate as clinical devices for diabetes management.[2] For example, a recent study using needle electrodes in the same subject found that the two sensors at times recorded opposite trends in glucose concentration.[3] Typically, *in vivo* sensors have failed after several hours or days.[4–7] Some sensors have lasted months, but have required frequent recalibration due to declining performance.[8]

The primary difficulty with implanted sensors is the host's foreign body reaction, which prevents the maintenance of a working interface between local microvasculature and the sensing surface, thereby impeding the transfer of sufficient quantities of analyte for reliable detection. This limitation has thus far proved to be a formidable barrier to any further progress in the development of implantable glucose sensors.[9] The essential components of the host reaction that diminish glucose mass transfer include the following: fouling of the implanted sensor with proteins and other biomolecules, the local accumulation of inflammatory fluid exudates, encapsulation with immune cells and collagen, and insufficient local blood supply resulting from rarefaction of blood vessels.[10,11]

Sensor performance may be even more problematic in diabetics than in normoglycemics as long-standing hyperglycemia is frequently associated with an aggravated and chronic host reaction. There are significant diabetes-induced changes in wound healing, inflammation, microvascular function, and microvascular permeability and density, and each of these alone could significantly impact biosensor signal generation.[12–14] Moreover, a possible complication in the interpretation of the effect of the host tissue reaction on implanted sensor performance is that the amount of analyte diffusing to the sensors may vary significantly according to normal local metabolic conditions. It is well known that the proportion of patent microvessels and the rate of blood flow in a given area oscillate in accordance with local tissue needs and the general physiologic state of the host subject.[12,15]

Several pioneering papers have described a histological approach to systematically characterize the foreign body reaction to an implant, and some have addressed the impact of implant size, material composition, surface features, and porosity.[16–21] A very few investigators have attempted the conceptually important experiment of implanting an operational sensor, then manipulating the host tissue reaction over an extended time period, and measuring the effects on the signal quality.[22] However, no studies have attempted to directly image and quantify the salient components of the foreign body reaction during the actual performance of a biosensor. Furthermore, experiments in which subcutaneously implanted sensors have been evaluated in diabetic subjects appear to be lacking in the open literature. This lack of data from both healthy individuals and diabetics is to a large extent a consequence of the difficulty in obtaining direct information about which tissue factors are influencing sensor performance at a particular time period.

Several noninvasive methods have been proposed to study the tissue conditions around implants, including high-resolution magnetic resonance imaging (MRI) approaches.[4,23,24] However, at present, MRI does not have adequate resolution for detailed imaging of the microvasculature, collagen, and protein deposits adjacent to a sensor in the living subject. In addition, the use of MRI is somewhat incompatible with the simultaneous operation of an electronic glucose sensing system containing metal parts, as the MRI magnetic fields are locally disrupted by metals, and the electric currents in the sensor and its connections would be affected by magnetic fields and RF imaging pulses.

In response to the perceived need for a more dynamic, yet technically feasible model system to directly evaluate the factors affecting an implanted sensor, we have adapted the rodent dorsal skinfold tissue window chamber to allow a clear tissue window and a planar sensor array to be installed together in the same subject.[25,26] This arrangement allows the tissues and vessels adjacent to biosensor electrodes (glucose and/or oxygen) to be imaged along with the simultaneous recording of the output of each sensor electrode in the unanesthetized subject. The imaging is nondestructive, can be performed periodically over several weeks, and can be done in normoglycemic and diabetic hamsters, rats, and other small animals. Moreover, the recent development of new blood and tissue labeling methods in living subjects, along with the introduction of the confocal and multiphoton laser scanning microscopes that can image deeply into tissues, make the window chamber–planar biosensor system a powerful new technology with which to address previously inaccessible questions related to *in vivo* sensor function.

This chapter provides a synopsis of the salient host tissue responses that relate to implanted sensor performance and describes in detail the development and current adaptation of the window chamber–biosensor method to address host tissue factors. The use of normoglycemic and diabetic animal models, specifically the Syrian hamster and Fat Sand Rat, to study the role of the microvasculature on sensor function is described. Limitations of the window chamber–biosensor method are also outlined. Finally, some important and useful features of standard brightfield, fluorescence, confocal, and multiphoton imaging as they apply to the window chamber are noted.

4.2 SALIENT FEATURES OF THE FOREIGN BODY REACTION TO AN IMPLANTED SENSOR

The transfer of glucose from the blood to an implanted biosensor is dictated by the proximity of the sensing surface to the host tissue, by the local tissue vascularity and perfusion, and by the response of the tissue to the implant surface.[11] All of these parameters change with the length of time after implantation. The primary components of the foreign body reaction to an implant along with the potential to observe these in a tissue window chamber preparation are listed in Table 4.1.

The major factors impacting sensor performance, whatever the physiological basis, are the degree of local vascularity and the loss of functional microvessels, together with the eventual presence and thickness of a fibrous capsule. Continued inflammation and collagen deposition eventually reach an equilibrium state, and the thickness of the fibrous capsule has been proposed as an index of biocompatibility.[32] The thickness and vascularity of the capsule depend on the size, surface texture, and porosity of the implant.[33–35]

TABLE 4.1 Foreign Body Reaction to an Implant in a Tissue

Tissue reaction	Potentially observable in window
Inflammation and edema	Yes (Ref. 27)
Fluid pocket formation	Possible—tissue geometry may not be amenable
Retraction of microvessels	Yes—window system very amenable for this (Ref. 28).
Chronic inflammation/edema	Yes
Accumulation of neutrophils and macrophages	Yes—fluorescent markers may be used (Ref. 29)
Establishment of a collagenous/ cellular capsule	Yes—invasion and coating with fibroblasts has been observed

Salient features of host reaction derived from Refs 10, 30, and 31.

4.3 PREVIOUS STUDIES OF THE FOREIGN BODY REACTION TO AN IMPLANTED SENSOR AND THE NECESSITY TO EVALUATE ACTIVE SENSORS *IN SITU*

Previous studies of the foreign body reaction have focused on histological methods and have not used active, working sensors in direct contact with the local tissue. In several innovative quasi functional studies, passive materials were variously implanted, then excised at various times, and the encapsulating tissue was evaluated in terms of its mass transfer properties.[18–21,36] Ward and his colleagues infused vascular endothelial growth factor (VEGF) near the surface of implanted sensors and were able to modulate the local foreign body reaction, most likely angiogenic in nature, to enhance sensor performance over several weeks.[22] These kinds of studies provide a great deal of useful information on the nature and progression of the foreign body reaction to an implant and on the potential to improve sensor function by manipulating this response. They do not, however, have the ability to compare sensor output with static and dynamic tissue conditions adjacent to the sensing surface, such as capsule thickness, microvascular density and perfusion, proximity of the sensing surface to the tissue, and the blood concentration of analyte. To directly assess the role of such variables, the site of sensor implantation needs to be imaged in real time together with the acquisition of signals from an active sensor.

In addition, the ability to optimize biosensor design is of central importance and initially depends on the determination of what aspects of the foreign body reaction and biosensor surface properties are critical to the success of the implanted biosensor. To accomplish this efficiently, it would be very beneficial if active sensors could be imaged *in situ*. Thus, sensor performance could be quantified relative to the manipulation of local tissue and microvascular conditions in response to various implant properties. Some important implant features include surface texture, porosity, and surface material composition. Surface texture of the implant has been observed to affect the extent of collagen formation. Smooth implant surfaces, which the local tissues cannot invade, elicit a greater fibrotic encapsulating response.[34] The secretion of cytokines, such as interleukins and tumor necrosis factor, and complement activation may also be related to the degree of roughness, surface irregularities, and surface area. Porosity in the range of tens of microns can affect the activation of microvasculature to initiate angiogenesis.[18,37,38] Macrophages cannot enter pores less than 50 μm, while larger pores allow them to enter and release angiogenic factors.[33] Surfaces with pores greater than 50 μm and variously pillared surfaces can encourage tissue ingrowth and fixation, thereby reducing movement of the implant and local irritation.[39]

4.4 TISSUE AND MICROVASCULAR DIFFERENCES BETWEEN NORMOGLYCEMIC AND DIABETIC INDIVIDUALS THAT MAY BE RELEVANT TO SENSOR RESPONSE AND THE NEED FOR A DYNAMIC IMAGING MODEL

The majority of *in vivo* glucose sensor evaluations have been done with normoglycemic animal or human subjects. The fact that in diabetics local tissue reactions are exacerbated and are more complex than those seen in normoglycemics has not been directly addressed in the context of signal generation of an implanted biosensor. In diabetic individuals, healing can be greatly protracted and the subcutaneous tissues exhibit an inflammatory response to injury that frequently does not resolve nor proceed to the next stage of wound resolution.[40,41] Therefore, events developing in response to sensor implantation, such as the accumulation of inflammatory exudates, microvessel function, and fibrous encapsulation, could be significantly different in diabetics compared with normoglycemics.

The inability of diabetics to heal competently is thought to a large extent to be due to microvascular defects.[40] Microvascular changes induced by diabetes could be expected to significantly influence the transport of glucose and oxygen to an implanted sensor. Human and animal studies have shown that there are clear differences in the vasculature between normoglycemics and individuals with types 1 and 2 diabetes.[42–44] Diabetic microvascular abnormalities have been seen in adult humans and in children.[45] Hyperglycemia of even a few weeks duration has been shown to compromise the structure and function of the microvasculature with the cellular endothelium as the primary target.[46–49] The damage to the microvasculature is variable but global and has been documented for heart, kidney, skin, intestinal smooth muscle, skeletal muscle, and the eye.[14,46,48–50] Vascular endothelial cell function is compromised in the presence of elevated blood glucose, possibly by disrupting cell-to-cell tight junctions and intracellular vacuolar transport.[51]

The above-mentioned effects of diabetes on the gross characteristics of the microvasculature, together with changes in the thickness, ultrastructure, and charge characteristics of the capillary basement membrane (CBM), result in significantly increased permeability to fluids and solutes, leading to local tissue edema and eventually architectural changes to the microvasculature.[47,52–54] The microvessels in hyperglycemia are often reduced in density within tissue beds, and the remaining vessels along with newly formed vessels exhibit unusual tortuosity and are fragile with multiple aneurysms.[55] Under normal conditions, the endothelial lining of the microvessels responds to local metabolic conditions by releasing nitric oxide (NO), a potent vasodilator that modulates the rhythmic constriction and dilatation of vessels within microvascular beds, a phenomenon known as vasomotion.[14,50] Vasomotion influences the rate and pattern of local microvascular flow. Endothelial cell damage results in a loss of vasomotion, and blood flow becomes inefficient and slow. Moreover, the peripheral neuropathy that often occurs in diabetes affects autonomic nerves that control blood flow, so that both microvascular and macrovascular functions are often abnormal in people with diabetes: vessels show an impaired response to heating, sodium nitroprusside, and acetylcholine.[12,14,56,57] Altered vessel morphology and patterns result, in part, from the changed blood flow and inefficient perfusion of local tissue. Moreover, because hyperglycemic individuals have diminished angiogenic capacity, the ability of their microvascular beds to support wound healing is compromised.[58]

All the aforementioned diabetic microvascular defects, including changes in the pattern, density, distribution, permeability, and reactivity of microvessels could greatly affect the effective delivery of glucose to an implanted sensor. However, as mentioned earlier, there has been no systematic comparative study of the operation of subcutaneously implanted glucose or oxygen sensors in diabetic individuals. The majority of previous work with subcutaneously implanted sensors has focused more on immediate applications rather than on mechanisms of signal production *in vivo*.[3–7,59] A critical question is whether diabetes and its attendant vascular deterioration and loss of functionality, together with the predilection for an aggravated inflammatory response, impact the viability and accuracy of an implanted sensor. This question certainly imposes the need to be able to image the tissue and microvasculature adjacent to an implanted, operational sensor on a real-time, continuous basis and to relate such observations to sensor signal quality.

4.5 SINGLE SENSORS VERSUS SENSOR ARRAYS AND THE IMPETUS FOR THE TISSUE WINDOW CHAMBER–BIOSENSOR SYSTEM

The frequent failure rate and the variability exhibited by sensors *in vivo* provides a rationale for testing multiple, closely spaced sensors in the same subject. A reasonable assumption is that multiple, closely spaced sensors within the same individual will be more likely to be exposed to the same general plasma concentrations of glucose and overall physiological conditions. Therefore, it will be easier to discern significant tissue effects that develop on a small spatial scale and that lead to variability between sensors. Figure 4.2, derived from an implanted multiple oxygen sensor array, illustrates how even the calibrated and adjusted tissue oxygen readings from several carefully manufactured sensors can vary. Each working electrode, that is, oxygen sensor, in this test was separated by only 1–2 mm. The use of multiple sensor arrays allows the comparison of sensor output from many different sensor electrodes, facilitating assessment of how tissue conditions on a small scale can affect sensor output. This concept, in addition to the needs described in the foregoing sections,

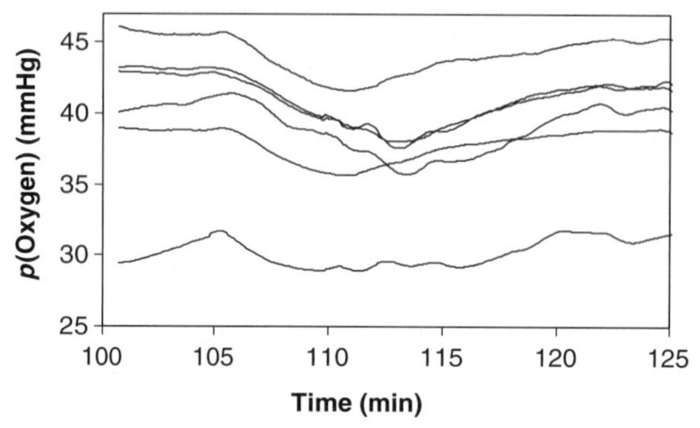

Figure 4.2 Graphical recording of oxygen levels measured from several different oxygen electrodes of an implanted planar sensor array. The oxygen levels inhaled by the subject were varied, and despite the fact that the sensors were fabricated to precise specifications and the output current calibration was adjusted, the oxygen levels measured across tissue distances of 1–2 mm varied considerably.

prompted us (the UCSD Biosensors Lab) to develop a planar sensor array that could be implanted together with a transparent tissue window as a modification of the rodent dorsal skinfold chamber. This unique preparation allows high-resolution imaging of the tissue adjacent to 18 working oxygen sensor electrodes along with the recording of the output of each electrode. Four or more of the oxygen electrodes may be adapted for glucose detection and measurement.

The dorsal skinfold window chamber from which the Biosensors Lab apparatus is derived was developed almost 60 years ago for tumor studies in mice.[60] Window chambers have been used extensively to chronicle the behavior of transplanted cancers and to analyze microvessels, healing wounds, and the image of the internal viscera.[61–65] Tissue chambers have been placed on the human arm, rabbit ear, and the mouse, rat, and hamster dorsal skinfold.[62,67–68] The current basic design of the rodent dorsal skinfold tissue window chamber was described in 1980 and comprises two parallel thin metal plates sandwiching a longitudinal fold of dorsal skin.[69] An approximately 1 cm diameter round area of skin is removed from one side of the skinfold and a round glass coverslip is secured over the site. For simultaneous installation of a sensor array, the contralateral skin of the fold also has a round area of skin removed. The UCSD Biosensors Lab has modified the contralateral metal plate so that a 1.2 cm diameter, round, planar sensor array can be held against the exposed subcutaneous tissue, and this thin tissue layer (approximately 200 μm thick) is sandwiched between the glass coverslip and the planar sensor (Figure 4.3). The vasculature of the subcutaneous layer can be clearly seen, as shown in Figure 4.4, and the individual sensor electrodes can be definitively imaged with conventional brightfield microscopy (Figure 4.4b) and fluorescence microscopy (Figure 4.4c).

The window chambers designed for the hamster, the diabetic desert Sand Rat (*Psammomys obesus*), and the mouse consist of two titanium alloy frames, each having a 12 mm diameter circular opening fitted with a ring for attachment of the window or sensor array (Figure 4.3). The standard chamber plates for the mouse have the same size opening but are approximately 50% smaller in surface area. However, as

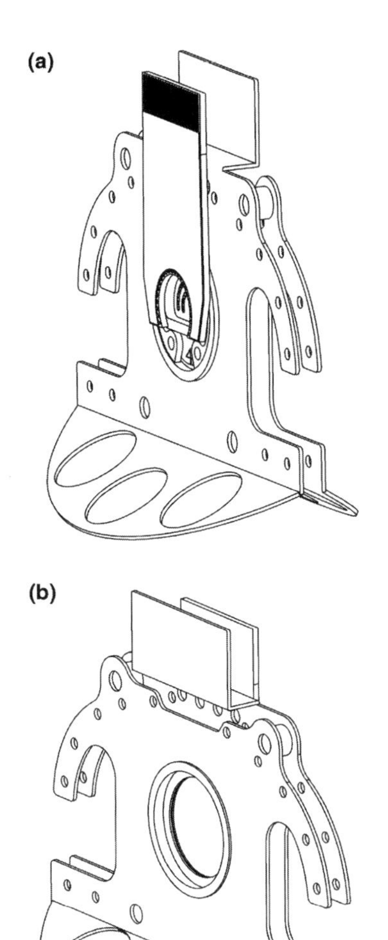

Figure 4.3 Schematic drawings illustrate two mated window chamber plates, one containing a planar sensor and its connector (a), while the companion plate has an open ring for the placement of a glass coverslip (b).

of this writing, sensors have not been implanted in mouse chambers, although this would likely be feasible if the sensor connector were to be made lighter and more flexible. Each attachment ring projects 0.95 mm toward the tissue and, in conjunction with appropriate spacers to separate the frames, creates sufficient separation to allow vigorous perfusion of the microvasculature in the thin sheet of tissue. The edges of the coverslip and sensor disk are coated with silicon adhesive before insertion into the ring to provide a critically important seal. In some experiments, an additional 100 μm thick plastic disk is glued, using clear silicone, to the internal surface of the window to minimize the thickness of the fluid layer between the tissue and sensor array surfaces. The frames are held together with four 10 mm screws and 4 mm hexagonal nuts. Prior

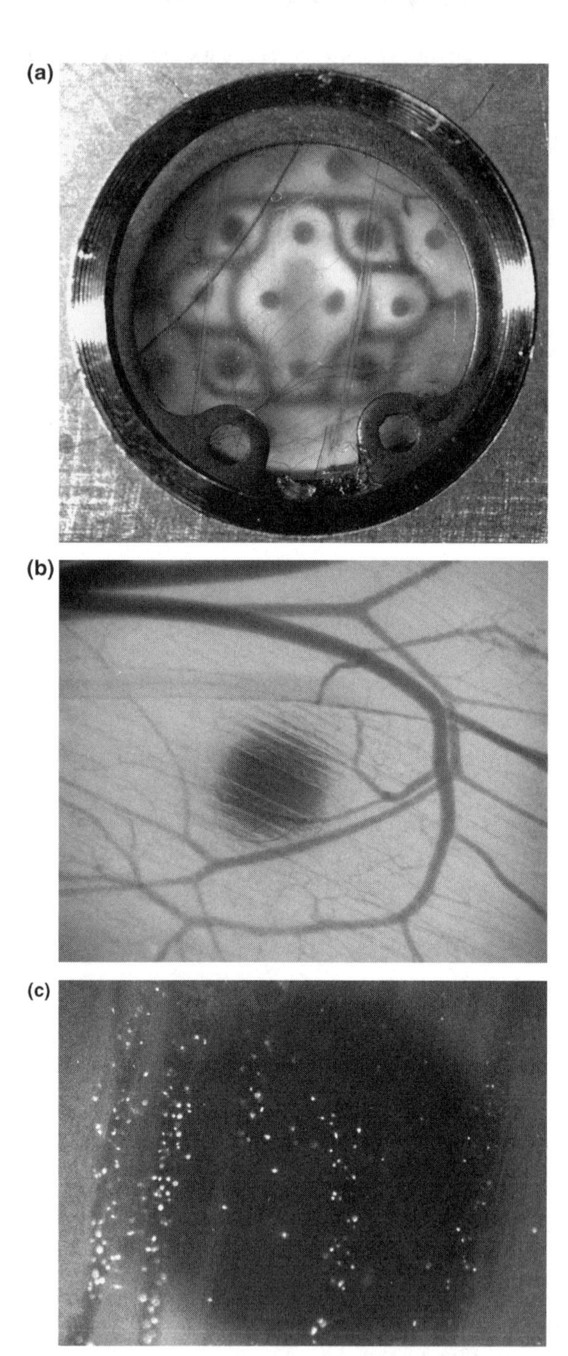

Figure 4.4 Panel (a) shows an implanted chamber with the skin retractor muscle and its vasculature clearly visible, and through the transparent retractor muscle, the planar sensor array can be seen. Panel (b) depicts the transparent hamster skin retractor muscle and its feeding arterioles and capillaries overlying an oxygen electrode (dark circle, 125 μm in diameter) belonging to a planar array. Note the clarity of the preparation. Panel (c) is an image of fluorescently labeled red blood cells coursing through the skin retractor muscle vessels over an adjacent sensor electrode. The blood flow across this sensor (125 μm in diameter) can be determined with accuracy.

to implantation, the frame surfaces, coverslip, and screws are autoclaved and rinsed extensively with sterile saline. The sensor array is sterilized by soaking in 6% glutaraldehyde solution, followed by rinsing in several changes of sterile saline solution over a 24 h period. A properly implanted skinfold tissue window chamber–planar sensor apparatus allows clear imaging for about 2 weeks.

4.6 USE OF OXYGEN SENSORS AS A SURROGATE GLUCOSE SENSOR FOR *IN VIVO* TESTING AND IMPORTANT ISSUES RELATED TO *IN VITRO* SENSOR CALIBRATION

The calibration of glucose sensors is relatively involved and the manufacture of glucose sensors is expensive. For testing purposes, it is simpler to measure oxygen than glucose. Oxygen has low intrinsic solubility and does not diffuse to long distances in aqueous solution. This makes oxygen a good candidate for testing tissue restrictions on mass transfer using an active biosensor.[25,70] As a practical approach, a planar oxygen sensor array in conjunction with a tissue window can be used to determine guiding principles. Then the initial findings can be validated using glucose sensor arrays.

Sensors implanted in experimental subjects need to be carefully calibrated prior to testing *in vivo*. Immediately following testing, calibration validation needs to be performed to demonstrate that the calibration was not lost during the testing procedure. It is important to maintain proper calibration to eliminate any variability produced by the sensors themselves. If sensors are not appropriately calibrated prior to experimentation, it will not be possible to get an accurate measure of the variability due to tissue effects.

To avoid the inconsistencies and ambiguity generated by liquid boundary layers, sensors should be calibrated in the gas phase whenever feasible. This will preclude the presence of a liquid boundary layer immediately adjacent to the sensor surface. This boundary layer is relatively static and does not effectively transfer changes in analyte concentration. Determination of sensitivity to analyte prior to sensor implantation and after sensor explantation allows the separation of tissue mass transfer effects from sensor variance and drift. The use of oxygen sensors in the same physical configuration and housing as the glucose sensor allows the gas-phase calibration and determination of boundary layer effects. Then the boundary layer in stirred glucose solutions is scaled by the previously determined oxygen boundary layer revealed by comparing oxygen sensors calibrated in a liquid with the same oxygen sensors calibrated in the gas phase where the oxygen boundary layer is absent. This approach provides a basis for estimation of glucose concentration gradients when the sensor is implanted in tissues.[25,70]

4.7 GENERAL DESIGN AND FABRICATION OF THE PLANAR SENSOR ARRAY

The configuration of the dorsal tissue window chamber plates used as described here allows the installation of a circular, planar sensor array 1.2 cm in diameter (Figure 4.5).

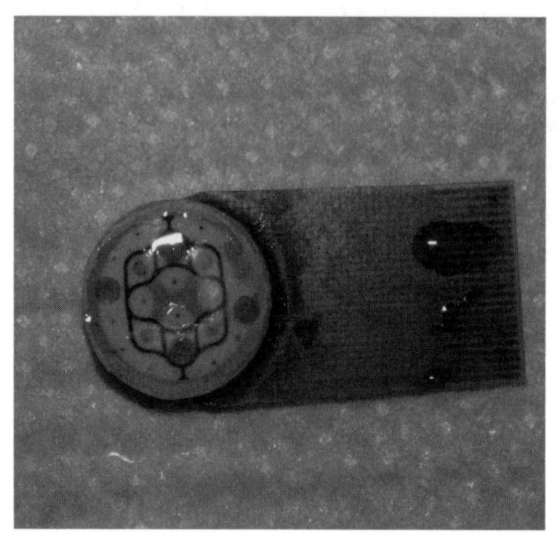

Figure 4.5 A planar 1.2 cm diameter oxygen–glucose sensor array. The small (125 μm) dots are the oxygen electrodes, the ribbon-like counter electrode is visible, and the Ag reference electrodes can be seen. Four of the oxygen electrodes have circular pockets of glucose oxidase integrated into the overlying membrane.

Any sensor type and surface membrane can be used. The sensor arrays are fabricated by patterned thick film deposition of platinum paste on an alumina disk that is then baked at 700°C.[25,71] Eighteen disk platinum working oxygen electrodes of 125 μm diameter are 1–2 mm apart, and there are six common platinum disk counter electrodes approximately 875 μm in diameter. A common potential silver reference electrode is electro-deposited on a platinum electrode base in the form of a ribbon, and then is chloridized to create the Ag/AgCl junction. A 25 μm layer of conductive electrolyte and a 25 μm layer of polydimethylsiloxane are deposited on the alumina disk. Immobilized glucose oxidase is deposited in a pocket of the sensor membrane directly over each four of the oxygen electrodes. The optimal shape of the enzyme pocket was predicted using computer models. The various sensor electrodes are physically linked to a pin connector that then communicates via a ribbon cable with a multichannel potentiostat. Data are displayed using a custom Labview program on a PC computer.

4.8 CHAMBER VARIANTS AND SIMULTANEOUS RECORDING OF SENSOR FUNCTION AND IMAGING

The surgical preparation of the dorsal skinfold window chamber can be varied to allow the sensor and window to be placed next to either subcutaneous fat, the skin retractor muscle, or subcutaneous fascia.[26] A plastic restraining device comfortably holds the conscious or lightly sedated animal subject in a position on the microscope stage that allows imaging of the window and simultaneous connection of the sensor to the

(a)

(b)

Figure 4.6 Panel (a) depicts the subject restrainer and the skinfold chamber that is bolted to a stainless steel stabilizing plate. Panel (b) shows the subject holder mounted onto the stage of a confocal microscope.

electronic recording apparatus (Figure 4.6). The test subject is usually acclimated to the plastic restrainer over several days prior to testing, so that during the actual experiments the animal is more likely to quietly rest or sleep. A small cone for delivery of air containing various physiological concentrations of oxygen can be attached to the nose end of the restraining cylinder and animals can be catheterized for the delivery of glucose and for blood sampling.

4.9 SENSOR BIOMATERIALS AND MEMBRANE TESTING

Our laboratory began to use the dorsal skinfold window chamber–biosensor system to evaluate sensor membrane materials and textures in terms of the host reaction and relate these to the function of an implanted sensor. Figure 4.7a is a brightfield side view of a silastic implant etched with 15 μm wide longitudinal grooves and Figure 4.7b shows a window chamber with a silastic implant sandwiched between the cover glass and the subcutaneous retractor muscle. The microvasculature can be imaged through the silastic, highlighting the fact that extremely thin, namely, tenths of microns, implants of many materials are transparent or semitransparent. The confocal and multiphoton laser scanning microscopes would in particular be expected to image effectively through scattering samples such as thin biomaterials implants.

(a)

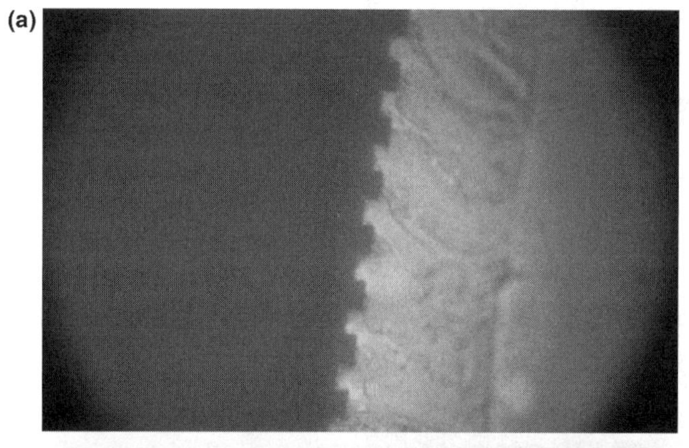

(b)

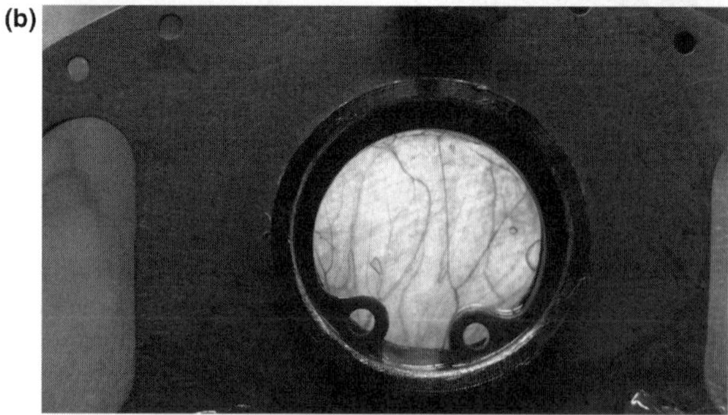

Figure 4.7 Panel (a) is a high-resolution brightfield side view of silastic etched with 15 μm wide longitudinal grooves. Panel (b) conveys the concept that very thin biomaterial implants, such as a microthin sheet of silastic etched with 15 μm wide grooves, implanted between the subcutaneous retractor muscle and glass in the window chamber shown here do not adversely affect the optical quality of the preparation.

A variety of material types and textures of sensor membrane surfaces have been evaluated using the window chamber–biosensor approach and these have been combined in various configurations, including the half and half arrangement in which 50% of a working sensor is coated with one type of material and 50% is coated with some other material. Figure 4.8a shows a planar oxygen sensor array that is half covered with 15 μm groove-etched silicone and the remaining half is coated with smooth silicone. Figure 4.8b shows that the smooth side (right to the viewer) is optically clear, while the etched side of the sensor is experiencing a significant foreign body reaction and appears considerably more opaque. The explanation for this relates to the development of markedly different host tissue and immune reactions in response to textured surfaces having different surface patterns and surface feature dimensions. The next step with sensors coated with variously textured surfaces would be to compare sensor electrode output in terms of amplitude, response time, and how long the signal persists over days and weeks.

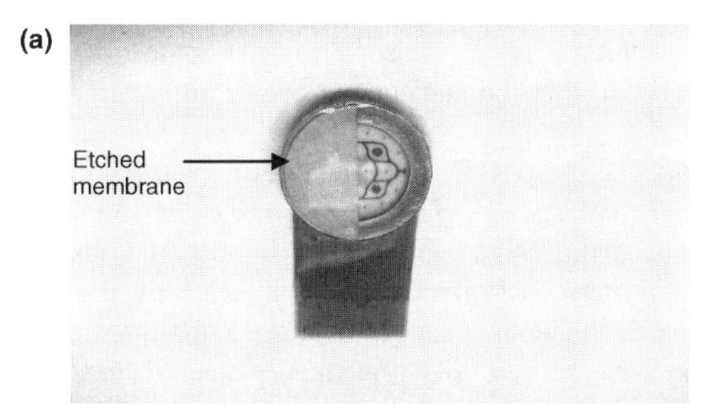

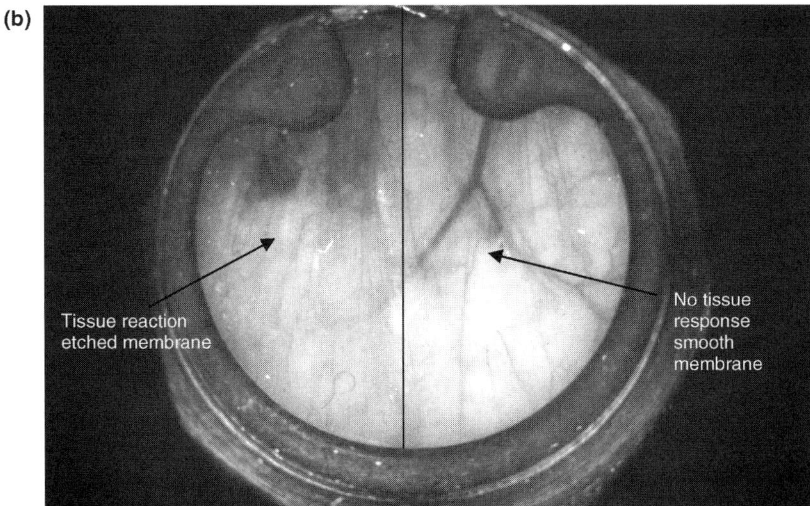

Figure 4.8 Panel (a) shows a planar oxygen sensor array half coated with etched silicone and half coated with smooth silicone. Panel (b) shows that the etched side (left to the viewer) is experiencing a host response and appears cloudy, while the smooth material on the right remains optically clear.

4.10 NORMAL AND DIABETIC ANIMAL MODELS COMPATIBLE WITH THE WINDOW CHAMBER–BIOSENSOR SYSTEM

The dorsal skinfold window chamber was initially designed to be used with rats, mice, and hamsters. Juvenile and adult Syrian hamsters tolerate the chamber–sensor apparatus very well and they can be easily catheterized in both the jugular and carotid vein for the systematic infusion of glucose solution as well as for blood sampling. The Syrian (APA) hamster will develop diabetes after a single dose of streptozotocin.[72] Even the laboratory rat can tolerate the dorsal skinfold window chamber, but the apparatus must be modified to accommodate the rather thick subcutaneous skin retractor muscle of this species. An advantage of the rat is that several genetic variants exist, such the BB rat, which develops the complete diabetic syndrome.[73] However, these diabetic animals require considerable attention and maintenance. At present, the mouse can tolerate a window chamber without an attached sensor, which is useful for visualization of the vascular and permeability studies, but to include a planar sensor array the connector stub must be reduced in size and weight so that the subject can comfortably carry the apparatus.

The Fat Sand Rat is a member of the Gerbil family and develops a syndrome quite comparable to human type 2 diabetes.[74–76] They are also relevant to type 1 (insulin dependent) diabetes as diabetic Sand Rats eventually become insulin dependent. This model is comparatively cost-effective as it is relatively easy to maintain under standard conditions, the diabetes is simple to induce naturally, and the incidence of diabetes is high. We have successfully implanted planar oxygen and glucose sensors using the dorsal skinfold chamber in both normoglycemic and diabetic Sand Rats. Glucose and oxygen readings have been obtained with these animals (Figure 4.9).

The diabetes that develops in Fat Sand Rats is polygenic in origin and a spectrum of diabetic phenotypes is observed, a characteristic that parallels diabetes in humans. This contrasts with many other rodent models, such as the BB Wistar rat and NOD mouse, in which the diabetes results from a single gene with a generally uniform phenotypic expression. The natural range of the Fat Sand Rat occupies the Middle East, and the diet of this species is rich in vegetable matter and high in complex carbohydrates. Conversion to a standard commercial rodent diet that is relatively high in simple carbohydrates causes a large proportion (90%; V.M. Chenault, U.S. FDA, personal communication) of the animals, over several weeks, to develop obesity and exhibit hyperinsulinemia, hyperglycemia, markedly decreased glucose tolerance, and insulin resistance. This diabetic syndrome is very similar to maturity–onset (type 2) diabetes in humans.[74–76] After several months on a rich diet, the animals will develop insulin-dependent diabetes and those animals not treated with insulin die in diabetic coma.[75]

In diabetes the disruption of the normal vascular architecture and loss of capillaries may affect the efficiency of glucose transport to a sensor surface. The Fat Sand Rat may be a good model for such effects, as this species was reported by

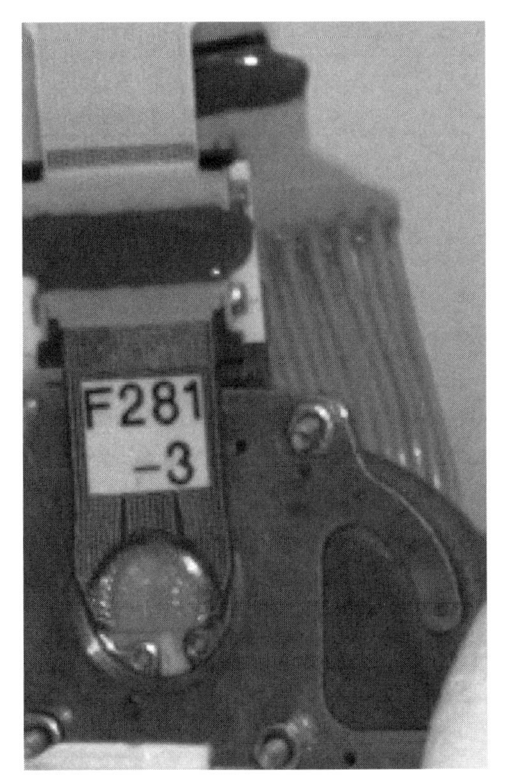

Figure 4.9 Photograph of the dorsal skinfold tissue chamber mounted on a Fat Sand Rat. The sensor array is connected to the data acquisition system via a ribbon cable.

Marquie and coworkers to develop capillary basement membrane thickening in the skin, microangiopathy in skeletal muscle, and myocardium.[77] This is comparable to types 1 and 2 human diabetics that exhibit microvascular deterioration that includes excessive capillary permeability, capillary basement membrane thickening, microaneurysms, and areas of poor perfusion.[53] In the kidneys of insulin-dependent Fat Sand Rats, Marquie and his colleagues found glomerulosclerosis, tubular changes, and marked basement membrane thickening in the Bowman's capsule and glomerular capillaries.[77]Table 4.2 summarizes microvascular pathology seen in the diabetic Fat Sand Rat, the diabetic BB rat, and the NOD mouse.[73]

Retinopathy, which is generally rare in rodents, does not appear in the hyperglycemic Jerusalem type A strain of Fat Sand Rat.[78] The French strain of Fat Sand Rat showed angiopathy and cataract formation after hyperglycemia of 4 months, but no retinopathy.[77] It has been suggested that retinopathy in the Sand Rat may be influenced by the duration of hyperglycemia, which may need to be at least 6 months (V.M. Chenault, U.S. FDA, personal communication). An advantage of using the Sand Rat is that it can be maintained in a diabetic state for up to 1 year, while maintenance is comparatively difficult in other rodent models of diabetes. Moreover, Sand Rats live about 3 years, so diabetes of 6 months and 1 year duration in the Sand Rat could be expected to be roughly equivalent to diabetes of about 12.5 and 25 years,

TABLE 4.2 Microvascular Pathology in the Diabetic Fat Sand Rat and Diabetic BB Rat

Pathology/species	Diabetic human	Diabetic Sand Rat	Diabetic BB rat	Diabetic NOD mouse
Vessel permeability increase	√	√**	√	ND
Basement membrane thickened	√	√	√	√
Endothelial damage	√	√	√	√
Altered vessel reactivity	√	√	√	√
Wound healing inhibited	√	√	√	√

√ means pathologic feature has been documented in literature. ND = not documented. √** = evidence for increased permeability from authors' unpublished work.

respectively, in the human. Therefore, given the high metabolic rate and pace of aging of this animal, it would be feasible to test for the microvascular effects of long-standing diabetes on sensor function after 6 months to 1 year of hyperglycemia. The window chamber–biosensor apparatus can be installed on Sand Rats of any age and on both normoglycemic and diabetic individuals.

4.11 LIMITATIONS OF THE WINDOW CHAMBER–BIOSENSOR SYSTEM

Although the dorsal skinfold window chamber–biosensor model system is versatile and allows considerable insight into tissue conditions associated with sensor function, as is the case with any methodology, it has certain limitations. For example, the window chamber does not recreate a three-dimensional tissue–sensor implant environment. The sensor is pressed against the subcutaneous tissue and the entire arrangement is planar and physically supported. This is in contrast to the subcutaneous environment that would apply to a clinically used sensor. Moreover, the window chamber stretches the skin of the dorsum and this exerts mechanical stress that would not typically exist in the subcutaneous environment. Another limitation of the window chamber is that eventually the cutaneous tissue begins to heal, so that by approximately 4 weeks after implantation, fibroblasts can be observed invading the chamber *en masse* and the subcutaneous tissues can no longer be imaged with clarity. The chamber is generally viable for at least 2 weeks, and a well-prepared chamber can last much longer, although this should not be routinely expected. Sensors developed for clinical use will be designed to remain in place for various lengths of time. Some will remain implanted for several days, "disposable sensors," while other types are being developed to reside in the patient for several months or even years. The window chamber can be used to learn much about early tissue and microcirculatory effects on sensor function, for example, those processes and local events appearing several days and weeks after sensor implantation, but imaging after several months will not be feasible with this methodology.

4.12 SOME NOTES ON THE IMAGING OF THE WINDOW CHAMBER–BIOSENSOR MODEL

4.12.1 Standard Brightfield and Fluorescence Microscopy

This approach can be used successfully with the dorsal skinfold window chamber to assess tissue condition (brightfield) and net tissue flow (fluorescence). The subject is placed in a specially designed holder that comfortably secures the animal on the microscope stage. The holder can be used for different microscopes (Figure 4.6). Blood flow can be imaged following intravenous injection with FITC-dextran (fluorescein isothiocyanate-dextran, 2×10^6 MW, Sigma). The FITC-dextran is made up in sterile water to a concentration of 20–50 mg/mL and 100–300 μL of this solution is infused, usually via a jugular catheter. The capillaries and blood flow can be easily visualized with a good quality CCD video camera, using a dry 20× objective. The blood vessels can also be highlighted with the injection of FITC-lectin or FITC-rhodamine, infused intravenously in 100–300 μL volumes of sterile normal saline.

4.12.2 Confocal and Multiphoton Laser Scanning Microscopy

Confocal microscopy greatly reduces the image blurring effect of scattered light from biological tissues by imposing a pinhole aperture at the conjugate focus of the objective. This pinhole only permits light from the focal plane to enter the detection system, thereby largely eliminating scattered light rays. Although the net signal amplitude reaching the detection system, usually comprised of photo multiplier tubes (PMTs), is considerably diminished, the resultant image is much clearer due to the absence of poorly focused scattered light. Such a system is easily adapted to accommodate an animal bearing a window chamber, and with the use of FITC-dextran and FITC-lectin, along with fluorescent antibody and reflected light imaging of collagen fibers, the microvasculature, microvascular flow, and collagenous deposition adjacent to each sensor electrode can be clearly imaged (Figure 4.10). Thin confocal slices or image planes can be acquired axially in the vicinity of each sensor, thereby allowing a three-dimensional rendering of the tissue to be produced later. Confocal microscopy offers the advantage of imaging several fluorophores simultaneously.

Multiphoton laser scanning microscopy is based on the principle that two temporally coincident photons, each with half the energy and wavelength of a single photon that would normally excite a given fluorophore, will together be sufficient to excite the fluorophore and cause light to be emitted.[79] The fluorescence generated by multiphoton events depends on the incident laser intensity and the illuminated area and falls off as a quadratic function of the distance from the focal plane.[80,81] Therefore, two photon excitation events occur only in a subfemtoliter volume within the focal plane and virtually no excitation develops outside this space. This means that very little energy is transferred to the tissue outside the multiphoton excitation volume, and tissue phototoxicity is very much reduced. In contrast to confocal microscopy, most of the light from the focal plane is acquired due to the absence of a

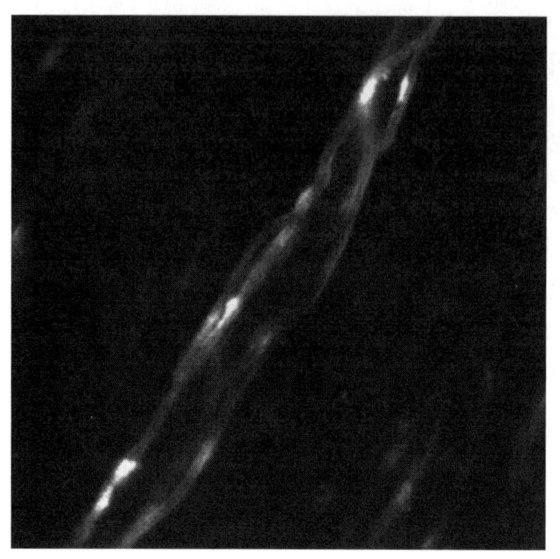

Figure 4.10 A confocal laser scanning microscope image taken through the dorsal skinfold window chamber on a tie-2 GFP (green fluorescent protein) mouse. Two capillaries in which the endothelium is expressing GFP are shown in the image.

pinhole, thereby improving the signal-to-noise ratio. Moreover, with multiphoton microscopy, tissue penetration is greatly increased as IR light suffers from considerably less scattering in tissue than do shorter wavelengths. There is virtually no interference from light generated above and below the focal plane as excitations do not occur in these regions. The depth penetration of confocal microscopy is approximately 50–100 μm, while multiphoton microscopy can image in scattering tissue to a depth of 500–700 μm and in some cases as far as 1000 μm.[80] The dorsal skinfold window chamber retractor muscle can be 200 μm thick and if both folded layers are left intact, the tissue thickness can be 400 μm, depending on the age and species of the animal subject. Multiphoton microscopy, together with intravenously injected fluorophores and reflectance imaging, should yield in a clear window chamber excellent three-dimensional images of the vasculature and tissue adjacent to the working electrodes of an implanted planar sensor array (Figure 4.11).

4.13 CONCLUSIONS AND FUTURE DIRECTIONS

The implantable biosensors field is facing a daunting obstacle in the form of the robust host foreign body reaction to an implant. Moreover, the local tissue reaction to implants is exacerbated in the diabetic subject. The development of a practical, reliable implantable glucose sensor will be very difficult unless the foreign body reaction can be manipulated and mitigated by implant properties. The design and testing of features, such as surface texture, implant shape, surface chemistry, and the inclusion of inhibitors and growth factors, needs to be evaluated in terms of direct

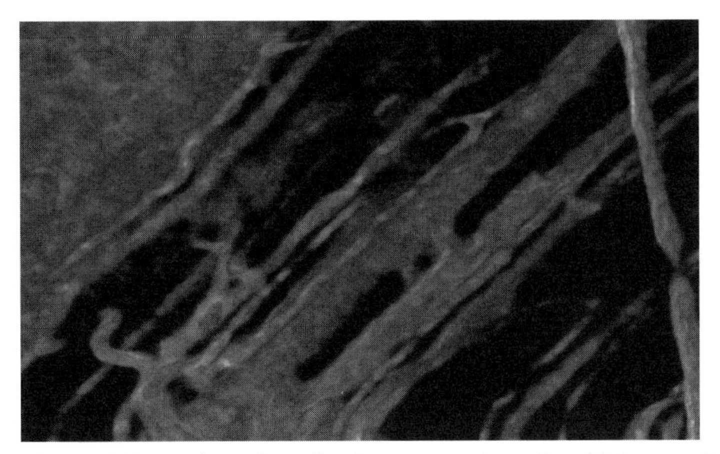

Figure 4.11 A three-dimensional reconstruction of multiphoton microscope acquired image slices of a series of small vessels in the dorsal skinfold window chamber. Taken with a 40× water immersion objective and digitally zoomed to approximately 60×.

imaging of the foreign body reaction and the simultaneous measurement of sensor performance *in vivo*. We have developed a tissue window system that allows real-time imaging of local tissue and vascular conditions together with the acquisition of tissue glucose measurements. This system has been tested with both normoglycemic and diabetic rodent models. Future studies with the window chamber ought to involve systematic modifications of sensor surface properties, and the corresponding sensor output should be recorded in normoglycemics and diabetic subjects, not over a few days but during several weeks. The findings of such experiments need to be complemented and extended by studies lasting months, in which the best candidate sensor variants are implanted subcutaneously as active glucose sensors, and evaluated.

So far decades of effort and large expenditures directed toward the development of a long-term implantable glucose sensor have failed to meet the expectations of medical researchers and the lay public. What should be apparent from this experience is that the complex foreign body reaction by the host dominates, and thus methodical, stepwise research using relevant and powerful *in vivo* model systems cannot be bypassed. This approach is virtually requisite if we are to ever have a clinically useful, subcutaneously implantable glucose sensor.

ACKNOWLEDGMENTS

The authors wish to thank Prof. David A. Gough of the UCSD Bioengineering Department Biosensors Laboratory for his teaching and for sharing his scientific insights with younger colleagues and also Prof. David Kleinfeld, Dr. Philbert Tsai, and Mr. Benjamin Migliori of the UCSD Physics Department for collaborating with the authors to obtain multiphoton images of the tissue window chamber vasculature. The authors also wish to thank Mr. Rick Calou for his drawings of the window chamber–biosensor assembly.

REFERENCES

1. Research Group DCCT. The effect of intensive treatment of diabetes on the development and progression of long-term complications in insulin-dependent diabetes mellitus. *New England Journal of Medicine* 1993, 329, 977–986.
2. Gerritsen M. Problems associated with subcutaneously implanted glucose sensors. *Diabetes Care* 2000, 23, 143–145.
3. Metzger M, Leibowitz G, Wainstein J, Glasser B, Raz I. Reproducibility of glucose measurements using the glucose sensor. *Diabetes Care* 2002, 25, 1185.
4. Johnson K, Mastrototaro J, Howey D, Brunelle R, Burden-Brady P, Bryan N, Andrew C, Rowe H, Allen D, Noffke B, McMahan WC, Morff RJ, Lipson D, Nevin RS. *In vivo* evaluation of an electroenzymatic glucose sensor implanted in subcutaneous tissue. *Biosensors & Bioelectronics* 1992, 7, 709–714.
5. Moatti-Sirat D, Capron F, Poitout V, Reach G, Bindra D, Zhang Y, Wilson G, Thevenot D. Towards continuous glucose monitoring: *in vivo* evaluation of a miniaturized glucose sensor implanted for several days in rat subcutaneous tissue. *Diabetologia* 1992, 35, 225.
6. Kerner W, Kiwit M, Linke B, Keck FS, Zier H, Pfeiffer EF. The function of hydrogen-peroxide-detecting electroenzymatic glucose electrode is markedly impaired in human subcutaneous tissue and plasma. *Biosensors & Bioelectronics* 1993, 8, 473.
7. Thome-Duret V, Gangerau M, Zhang Y, Wilson G, Reach G. Modification of the sensitivity of glucose sensor implanted into subcutaneous tissue. *Diabetes and Metabolism* 1996, 22, 174.
8. Updike S, Shults M, Gilligan B, Rhodes R. A subcutaneous glucose sensor with improved longevity, dynamic range, and stability of calibration. *Diabetes Care* 2000, 23, 208–214.
9. Friedl K. Corticosteroid modulation of tissue response to implanted sensors. *Diabetes Technology & Therapeutics* 2004, 6, 898–901.
10. Colton CK. Engineering challenges in cell encapsulation technology. *Trends Biotechnology* 1996, 14, J58–J62.
11. Stenken J, Reichert W, Klitzman B. Magnetic resonance imaging of a tissue implanted device biointerface using *in vivo* microdialysis sampling. *Analytical Chemistry* 2002, 74, 4849–4854.
12. Morris SJ, Shore AC, Tooke JE. Responses of the skin microcirculation to acetylcholine and sodium nitroprusside in patients with NIDDM. *Diabetologia* 1995, 38, 1337–1344.
13. Affonso FS, Cailleaux S, Pinto LFC, Gomes MB, Tibirica E. Effects of high glucose concentrations on the endothelial function of the renal microcirculation of rabbit. *Arquivos Brasileiros de Cardiologia* 2003, 18, 156–165.
14. Lefrandt JD, Bosma E, Oomen PHN, Hoeven JH, Roon AM, Smit AJ, Hoogenberg K. Sympathetic mediated vasomotion and skin capillary permeability in diabetic patients with peripheral neuropathy. *Diabetologia* 2003, 46, 40–47.
15. Riihimaa P. Markers of microvascular complications in adolescents with type 1 diabetes. PhD dissertation, University of Oulu, Finland, 2003.
16. Woodward SC. How fibroblasts and giant cells encapsulate implants: considerations in design of glucose sensors. *Diabetes Care* 1982, 5, 278–281.
17. Rebrin K, Fischer U, Hahn von Dorsche H, von Woetke T, Abel P, Brunstein E. Subcutaneous glucose monitoring by means of electrochemical sensors: fiction or reality? *Journal of Biomedical Engineering* 1992, 14, 33–40.

18. Salzmann DL, Kiemer LB, Berman SS, Williams SK. The effects of porosity on endothelialization of ePTFE implanted in subcutaneous and adipose tissue. *Journal of Biomedical Materials Research* 1997, 34, 463–476.
19. Sharkawy AA, Klitzman B, Truskey GA, Reichert WM. Engineering the tissue which encapsulates subcutaneous implants. I. Diffusion properties. *Journal of Biomedical Materials Research* 1997, 37, 401–412.
20. Sharkawy AA, Klitzman B, Truskey GA, Reichert WM. Engineering the tissue which encapsulates subcutaneous implants. II. Plasma-tissue exchange properties. *Journal of Biomedical Materials Research* 1998, 40, 586–597.
21. Sharkawy AA, Klitzman B, Truskey GA, Reichert WM. Engineering the tissue which encapsulates subcutaneous implants. III. Effective tissue response times. *Journal of Biomedical Materials Research* 1998, 40, 598–605.
22. Ward WK, Wood MD, Casey HM, Quinn MJ, Federiuk IF. The effect of local subcutaneous delivery of vascular endothelial growth factor on the function of a chronically implanted amperometric glucose sensor. *Diabetes Technology & Therapeutics* 2004, 6, 137–145.
23. Qui H, Hedlund L, Neuman M, Edwards C, Black R, Cofer G, Johnson GA. Measuring the progression of foreign body reaction to silicone implants using *in vivo* microscopy. *IEEE Transactions on Biomedical Engineering* 1998, 45, 921–926.
24. Ramponi S, Rebaudengo C, Cabella C, Grotti A, Vultaggio S, Aime S, Morisetti A, Lorusso V. Contrast-enhanced MRI of murine sponge model for progressive angiogenesis assessed with gadoteridol (ProHance) and gadocoletic acid trisodium salt (B22956/1). *Journal of Magnetic Resonance Imaging* 2008, 27, 872–880.
25. Makale MT, Lin JT, Calou RE, Tsai AG, Chen PC, Gough DA. Tissue window chamber system for validation of implanted oxygen sensors. *American Journal of Physiology: Heart and Circulatory Physiology* 2003, 284, H2288–H2294.
26. Makale MT, Chen PC, Gough DA. Variants of the tissue–sensor array window chamber. *American Journal of Physiology: Heart and Circulatory Physiology* 2005, 289, 57–65.
27. Devoisselle JM, Begu S, Tourne-Peteilh C, Desmettre T, Mordon S. *In vivo* behavior of long-circulating liposomes in blood vessels in hamster inflammation and septic shock models: use of intravital fluorescence microscopy. *Luminescence* 2001, 16, 73–78.
28. Stern SJ, Flock ST, Small S, Thomsen S, Jacques S. Photodynamic therapy with chloroaluminum sulfonated phthalocyanine in the rat window chamber. *American Journal of Surgery* 1990, 160, 360–364.
29. Unal C, Sen C, Iscen D, Dalcik H. *In vivo* observation of leukocyte–endothelium interaction in ischemia reperfusion injury with the dorsal window chamber and the effects of pentoxifylline on reperfusion injury. *Journal of Surgical Research* 2007, 138, 259–266.
30. McNally AK, DeFife KM, Anderson JM. Interleukin-4-induced macrophage fusion is prevented by inhibitors of mannose receptor activity. *American Journal of Pathology* 149, 1996, 975–985.
31. Burczak K, Gamian E, Kochman A. Long-term *in vivo* performance and biocompatibility of poly(vinyl alcohol) hydrogel macrocapsules for hybrid-type artificial pancreas. *Biomaterials* 1996, 17, 2351–2356.
32. Von Recum AF, Optiz H, Wu E. Collagen types I and III at the implant/tissue interface. *Journal of Biomedical Materials Research* 1993, 27, 757–761.
33. Costantino PD, Friedman CD, Lane AG. Synthetic biomaterials in facial plastic and reconstructive surgery. *Facial Plastic Surgery* 1993, 9, 1–15.

34. Bobyn J, Jacobs J, Tanzer M, Urban R, Aribini R, Sumner D, Turner T, Brooks C. The susceptibility of smooth implant surfaces to periimplant fibrosis and migration of polyethylene wear debris. *Clinical Orthopaedics and Related Research* 1995 311, 21–39.
35. Saltzman WM, Langer R. Transport rates of proteins in porous materials with microgeometry. *Biophysical Journal* 1989, 55, 163–171.
36. Pfeiffer EF. The glucose sensor: the missing link in diabetes therapy. *Hormone and Metabolic Research Supplement* 1990, 24, 154–164.
37. Brauker J, Carr-Brendel V, Martinson L, Crudele J, Johnston W, Johnson R. Neovascularization of synthetic membranes directed by membrane microarchitecture. *Journal of Biomedical Materials Research* 1995, 29, 1517–1524.
38. Padera R, Colton C. Time course of microtexture driven neovascularization. *Biomaterials* 1996, 17, 277–284.
39. Picha G, Drake R. Pillared-surface microstructure and soft-tissue implants: effect of implant site and fixation. *Journal of Biomedical Materials Research* 1995, 30(3), 305–312.
40. Nagato H, Umebayashi Y, Wako M, Tabata Y, Manabe M. Collagen-poly glycolic acid matrix with basic fibroblast growth factor accelerated angiogenesis and granulation tissue formation in diabetic mice. *Journal of Dermatology* 2006, 33, 670–675.
41. Arul V, Kartha R, Jayakumar R. A therapeutic approach for diabetic wound healing using biotinylated GHK incorporated collagen matrices. *Life Sciences* 2007, 80, 275–284.
42. Klein R. Hyperglycemia and microvascular and macrovascular disease in diabetes. *Diabetes Care* 1995, 18, 258–268.
43. Lawson SR, Gabra BH, Nantel F, Battistini B, Sirois P. Effects of a selective bradykinin B1 receptor antagonist on increased plasma extravasation in streptozotocin-induced diabetic rats: distinct vasculopathic profile of major key organs. *European Journal of Pharmacology* 2005, 514, 69–78.
44. Flyvbjerg A, Orskov H. Diabetic angiopathy: new experimental and clinical aspects. *Hormone and Metabolic Research* 2005, 37, 1–3.
45. Cheung AT, Price AR, Duong PL, Ramanujam S, Gut J, Larkin EC, Chen PC, Wilson DM. Microvascular abnormalities in pediatric diabetic patients. *Microvascular Research* 2002, 63, 252–258.
46. Algenstaedt P, Schaefer C, Biermann T, Hamann A, Scwarzloh B, Greten H, Ruther W, Hansen-Algenstaedt N. Microvascular alterations in diabetic mice correlate with level of hyperglycemia. *Diabetes* 2003, 52, 542–549.
47. Carlson EC, Audette JL, Veitenheimer NJ, Risan JA, Laturnus DI, Epstein PN. Ultrastructural morphometry of capillary basement membrane thickness in normal and transgenic diabetic mice. *Anatomical Record Part A: Discoveries in Molecular, Cellular, and Evolutionary Biology* 2003, 271, 332–341.
48. Feletou M, Boulanger M, Staczek J, Broux O, Duhalt J. Fructose diet and VEGF-induced plasma extravasation in hamster cheek pouch. *Acta Pharmacologica Sinica* 2003, 24, 207–211.
49. Reusch JEB. Diabetes, microvascular complications, and cardiovascular complications: what is it about glucose? *Journal of Clinical Investigation* 2003, 112, 986–988.
50. Dogra GK, Hermann S, Irish AB, Thomas MAB, Watts GF. Insulin resistance, dyslipidaemia, inflammation and endothelial function in nephritic syndrome. *Nephrology Dialysis Transplantation* 2002, 17, 2220–2225.

51. Harhaj NS, Antonetti DA. Regulation of tight junctions and loss of barrier function in pathophysiology. *International Journal of Biochemistry and Cell Biology* 2004, 36, 1206–1237.
52. Hainsworth DP, Katz ML, Sanders DA, Sanders DN, Wright EJ, Sturek M. Retinal capillary basement membrane thickening in a porcine model of diabetes mellitus. *Comparative Medicine* 2002, 52, 523–529.
53. Stitt AW. The role of advanced glycation in the pathogenesis of diabetic retinopathy. *Experimental and Molecular Pathology* 2003, 75, 95–108.
54. Stolar MW, Chilton RJ. Type 2 diabetes, cardiovascular risk, and the link to insulin resistance. *Clinical Therapeutics* 2003, 25, B4–B31.
55. Bates DO, Harper SJ. Regulation of vascular permeability by vascular endothelial growth factors. *Vascular Pharmacology* 2003, 39, 225–237.
56. Caballero AE, Arora S, Saouaf R, Lim SC, Smakowski P, Park JY, King GL, LoGerfo FW, Horton ES, Veves A. Microvascular and macrovascular reactivity is reduced in subjects at risk for type 2 diabetes. *Diabetes* 1999, 48, 1856–1862.
57. Correa RC, Alfieri A. Plasmatic nitric oxide, but not Willebrand factor, is an early marker of endothelial damage, in type 1 diabetes mellitus patients without microvascular complications. *Journal of Diabetes and Its Complications* 2003, 17, 264–268.
58. Rojas JD, Sennoune SR, Martinez GM, Bakunts K, Meininger CJ, Wu G, Wesson DE, Seftor EA, Hendrix MJC, Martinez-Zauilan R. Plasmalemmal vacuolar H^+ -ATPase is decreased in microvascular endothelial cells from a diabetic model. *Journal of Cellular Physiology* 2004, 201, 190–200.
59. Chiu NF, Wang JM, Yang LJ, Liao CW, Chen CH, Chen HC, Lu SS, Lin CW. An implantable multifunctional needle type biosensor with integrated RF capability. *Conference Proceedings of the IEEE Engineering in Medicine and Biology* 2005, 2, 1933–1936.
60. Algire GH. An adaptation of the transparent chamber technique to the mouse. *Journal of the National Cancer Institute* 1943, 4, 1–11.
61. Jain RK, Munn LL, Fukumara DD. Dissecting tumor pathophysiology using intravital microscopy. *Nature Reviews Cancer* 2002, 2, 266–276.
62. Menger MD, Laschke MW, Vollmer B. Viewing the microcirculation through the window: some twenty years experience with the hamster dorsal skinfold chamber. *European Surgical Research* 2002, 34, 83–91.
63. Vranckx JJ, Slama J, Preuss S, Perrez N, Svensjo TS, Breuing K, Bartlett R, Pribaz J, Weiss D, Eriksson E. Wet wound healing. *Plastic and Reconstructive Surgery* 2002, 110, 1680–1687.
64. Bertera S, Geng X, Tawdrous Z, Bottino R, Balamurugan AN, Rudert WA, Drain P, Watkins SC, Trucco M. Body window-enabled *in vivo* multicolor imaging of transplanted mouse islets expressing an insulin–dimer fusion protein. *Biotechniques* 2003, 35, 718–722.
65. Chen DC, Agopian VG, Avansino JR, Lee JK, Farley SM, Stelzner M. Optical tissue window: a novel method for optimizing engraftment of intestinal stem cell organoids. *Journal of Surgical Research* 2006, 134, 52–60.
66. Branemark P-I. *Intravascular Anatomy of Blood Cells in Man.* Karger, Basel, 1971, pp. 4–10.
67. Arfors KE, Jonsson JA, McKenzie FN. A titanium rabbit earchamber: assembly, insertion and results. *Microvascular Research* 1970, 2, 516–518.

68. Kerger H, Torres Filho IP, Rivas M, Winslow RM, Intaglietta M. Systemic and subcutaneous microvascular oxygen tension in conscious Syrian golden hamsters. *American Journal of Physiology: Heart and Circulatory Physiology* 1995, 268, H802–H810.
69. Endrich B, Asaishi K, Gotz A, Messmer K. Technical report: a new chamber technique for microvascular studies in unanesthetized hamsters. *Research in Experimental Medicine* 1980, 177, 125–134.
70. Makale MT, Jablecki MC, Gough DA. Mass transfer and gas-phase calibration of implanted oxygen sensors. *Analytical Chemistry* 2004, 76, 1773–1777.
71. Lin JT. Testing of planar thick film fabricated oxygen sensor with galvanostatic techniques. PhD thesis, University of California San Diego, La Jolla, CA, 2000.
72. Horiuchi K, Takatori A, Inenayo T, Ohta E, Ishii Y, Kyuwa S, Yoshikawa Y. Histopathological studies of aortic dissection in streptozotocin-induced diabetic APA hamsters. *Experimental Animals* 2005, 54, 363–367.
73. Sima AF, Shafrir E. *Animal Models of Diabetes: A Primer*. Harwood Academic Publishers, Amsterdam, 2001.
74. Kalderon B, Gutman A, Levy E, Shafrir E, Adler J. Characterization of stages of development of obesity–diabetes syndrome in Sand Rat (*Psammomys obesus*). *Diabetes* 1986, 35, 717–724.
75. Marquie G, Duhault J, Jacotot B. Diabetes mellitus in Sand Rats (*Psammomys obesus*). Metabolic pattern during development of the diabetic syndrome. *Diabetes* 1984, 33, 438–443.
76. Lewandowski PA, Cameron-Smith D, Jackson CJ, Kultys ER, Collier GR. The role of lipogenesis in the development of obesity and diabetes in Israeli Sand Rats (*Psammomys obesus*). *Journal of Nutrition* 1998, 128, 1984–1988.
77. Marquie G, Duhault J, Hadjiisky P, Petrov P, Bouissou H. Diabetes mellitus in Sand Rats (*Psammomys obesus*): microangiopathy during development of diabetic syndrome. *Cellular and Molecular Biology* 1991, 37, 651–667.
78. Shafrir E, Gutman A. *Psammomys obesus* of the Jerusalem colony: a model for nutritionally induced, non-insulin dependent diabetes. *Journal of Basic Clinical Physiology and Pathology* 1993, 4, 83–99.
79. Schenke-Layland K, Riemann I, Damour O, Stock UA, Konig K. Two-photon microscopes and *in vivo* multiphoton tomographs: powerful diagnostic tools for tissue engineering and drug delivery. *Advanced Drug Delivery Reviews* 2006, 58, 878–896.
80. Tsai PS, Nishimura N, Yoder E, Dolnick E, White GA, Kleinfeld D. Principles, design and construction of a two-photon laser-scanning microscope for *in vitro* and *in vivo* brain imaging. In: Frostig RD (Ed.), *In Vivo Optical Imaging of Brain Function*. DCRC, Boca Raton, FL, 2002.
81. Masters BR, So PTC. Multi-photon excitation microscopy and confocal microscopy imaging of *in vivo* human skin: a comparison. *Microscopy and Microanalysis* 1999, 5, 282–289.

CHAPTER 5

COMMERCIALLY AVAILABLE CONTINUOUS GLUCOSE MONITORING SYSTEMS

Timothy Henning

In Vivo Glucose Sensing, Edited by David D. Cunningham and Julie A. Stenken

5.1 INTRODUCTION

5.1.1 Why Do We Need Continuous Glucose Monitoring?

This chapter will explain the technology used in the three commercially available continuous glucose monitoring (CGM) systems that are approved by the FDA in 2008. The history of how the technology was developed, the technical hurdles that had to be overcome, and the performance of these technologies will be covered. The future development and uses of the technology will also be discussed. Although the technology will undoubtedly evolve in the coming years, the underlying principles of how the technology works and the challenging problems that were overcome in developing the technology will serve as a foundation for future technological developments of CGM.

People with types 1 and 2 diabetes face the daily challenge of maintaining a stable glucose level in their body. Home glucose testing has allowed people with diabetes to frequently test their blood and obtain an accurate reading of their blood glucose at that point in time. The person with diabetes can then act on the glucose reading by injecting insulin if their glucose reading is too high or by eating if their glucose level is too low. Although home glucose testing is very valuable, it can only provide the person with diabetes a glucose reading at that specific moment in time. Because of the inconvenience and cost of pricking your finger and making a glucose measurement, the person with diabetes often performs testing less than twice per day. The glucose levels of people with diabetes are however very labile and periodic testing cannot provide them enough information to properly control their glucose level. A clinical study was done where the diabetic subject performed on average nine blood glucose measurements per day.[1] Despite the great number of tests that were performed daily, the subjects still spent on average 4.8 h per day in hyperglycemia (blood glucose above 180 mg/dL) and 2.1 h in hypoglycemia (blood glucose below 70 mg/dL). Hyperglycemia is known to contribute to long-term vascular damage and hypoglycemia can lead to mental confusion and ultimately death. The time spent outside euglycemia by people with diabetes should not be blamed on the person because they cannot measure frequently enough with home glucose testing to maintain control. Maintaining euglycemia during sleep is beyond the control of the person with diabetes. The only solution to this problem is a device that can provide an almost continuous measurement of blood glucose, so that person with diabetes can obtain glucose values frequently enough to adjust their treatment accordingly to maintain euglycemia. CGM can also provide alarms to alert the person with diabetes while they are sleeping. The technology is being developed that will allow the CGM to work with an insulin pump and provide an automated system to maintain euglycemia.

Home glucose test strips were able to reach the market two decades before the first continuous glucose monitors. This points to the tremendous challenges faced in the development of CGM. Following the long and colorful development of the technology will make it easier to understand the present systems.

5.1.2 Why Electrochemical and Not Optical Systems?

Glucose is a monosaccharide that provides the fuel for all the cells in the body. Nature has evolved enzymes that react with glucose producing by-products that can be used to generate an analytical signal. The details of the enzyme reaction are covered in the subsequent section. The enzyme glucose oxidase is used in the current CGM systems and was used for the home glucose test strips. The home glucose test strips have now evolved and use the enzyme glucose dehydrogenase that does not require oxygen as a coreactant. YSI, formerly Yellow Springs Instruments, introduced a blood analyzer in 1972 that used an immobilized glucose oxidase membrane on top of an electrode to generate hydrogen peroxide. The same YSI glucose electrode is used to measure many blood samples, which differentiates the electrode from single-use glucose strips. Hydrogen peroxide is electrochemically active and can be oxidized at a precious metal electrode to produce a current that is proportional to glucose. This instrument is still in production today and serves as the reference method for most of the clinical trials involving CGM. The YSI instrument has many properties that make the technology ideal for use in CGM: glucose oxidase is inexpensive and relatively stable and is very selective for glucose, the enzyme is easily immobilized, the electrochemical detection of the hydrogen peroxide produced from the glucose oxidase reaction can be done with inexpensive electronics, and the electronics can be located far away from the electrodes requiring only small wires to transmit the current.

Optical systems have certainly been constructed that can measure the reaction of glucose with glucose oxidase. The term optical systems refers to systems that measure a change in an optical property of a reaction zone, for example, change in color, in response to a reaction of glucose with another chemical or biological agent not normally present in the body. The first home use glucose test strips were based on color changes that resulted from the glucose oxidase reaction (for a review of how home glucose test strips work, see Ref. 2). The color change requires a dye that is irreversibly changed in the reaction. A reservoir of fresh dye would be needed to run the reaction continuously as would be needed for a continuous glucose sensor. The optical means of measuring glucose must be located in proximity to the color formation. This would require that the optical reader be placed inside the body or possibly outside the skin just above a color-forming reaction that is located just underneath the skin. As with most optical systems, alignment of the instrument with the reaction center is critical. The optical system based on enzymes would, therefore, have the following disadvantages over electrochemical systems: requires a reservoir of dye, the optical reading system needs to be near the color change, optical readers need to be aligned with color change, and optical readers are typically more expensive than electrochemical circuits. To get around the problem of consuming reagent, these systems have found other means of reacting with glucose. Inorganic boranate compounds that can reversibly bind glucose and naturally occurring glucose binding proteins have been discovered.[3–8] These binding events between glucose and the binding molecule can be used to trigger an optical signal. The details of these reactions are not the subject of this chapter, so the reader should refer to the reference for further details. Because of the many technical challenges facing optical systems for continuous glucose monitoring, their progress has been slow but deliberate. Several companies have shown results in animals that look promising and the author expects that commercial continuous glucose monitors based on optical systems will be available in the next decade.[9]

Much research has been done on the direct detection of glucose in the body using electromagnetic radiation with no chemicals placed inside the body.[10,11] Light impinges on the body and the reflected or transmitted light is changed based on the interaction of the body with the light and that change possibly contains information about glucose levels in the body. This technique is called noninvasive glucose monitoring, but since electromagnetic radiation does enter the body, the measurement is not truly noninvasive. Several companies are actively working on continuously shining light into the body so that a continuous measurement of glucose can be made. None of these systems are commercially available, so they will not be covered in this chapter except for the preceding comments.

5.1.3 Operation of Electrochemical Systems

5.1.3.1 Enzyme Reactions at Electrodes The reaction of glucose oxidase with glucose is explained in many sources. An elementary understanding of this reaction, discussed in the following sections, is necessary to comprehend the various generations of biosensor technology. Since the current commercial CGMs encompass two generations of biosensor technology, it is necessary to provide a broader understanding of these key fundamental points. This and the following sections are particularly relevant because most of the differences between the commercial systems arise from the distinction in generations of biosensor to which they belong.

Glucose oxidase belongs to a class of enzymes that has an electroactive FAD site at which their specific reaction occurs. Glucose reacts with the FAD site in glucose oxidase and two electrons are transferred from glucose to the FAD site. Glucose is oxidized to gluconic acid. The reduced FAD site must be oxidized before the enzyme can oxidize another glucose molecule. In nature, the FAD site is oxidized by oxygen dissolved in the sample and the oxygen is reduced to hydrogen peroxide (a description of the glucose reaction can be found in Section 1.2.1). Hydrogen peroxide is electroactive and can be oxidized at an electrode. The current produced by the hydrogen peroxide will be related back to the amount of glucose in the sample that reacts with glucose oxidase. The hydrogen peroxide generated from the naturally occurring oxygen allows the reaction of the enzyme and substrate to be sensed at the electrode and can be characterized as a mediator between the enzyme reaction and the electrode. The discovery was made that electroactive molecules could be added to the reaction that would perform the same function as oxygen.[12] Unlike oxygen that is naturally present in most samples, these compounds would need to be added to a sample because they are not naturally occurring compounds. The electroactive molecules also function to mediate the reaction between the enzyme reaction and the electrode, hence they are called mediators. Replacing oxygen with a mediator has many advantages, which will be described in greater detail in the following sections.

5.1.3.2 Electrode Configurations The term "electrode" has different meanings depending on the context of the use of the term. Electrode could mean a single piece of metal at which one reaction occurs. Electrode also has the meaning of a device that senses a specific compound, such as a "combination pH electrode." In reality, a combination pH electrode is a pH sensor that is composed of a pH electrode and a reference electrode. In this chapter, the complete device for detecting glucose will be referred to as a glucose sensor. The glucose sensor is composed of at least two electrodes, which will be explained in this section.

Electrochemical detection of glucose requires that an enzyme that reacts with glucose produce a change in the oxidation state of a reactant: oxygen or a mediator. The change in oxidation state must then be sensed at a metal surface by an exchange of electrons between the reactant and the metal surface. The electrode where this occurs is called the working electrode. The working electrode then must contain at a minimum the enzyme needed for the reaction and a metal surface for the exchange of electrons. The major characteristic required for a good working electrode is that the reaction with the reactant should be fast and should remain so for the whole duration of the measurement. Metal surfaces have a tendency to adsorb other molecules that may physically block the exchange of electrons, which results in a "poisoning" or "fouling" of the working electrode, as described in Section 1.6.1. Many designs of working electrodes contain a membrane that prevents the poisoning of the metal surface.

At a minimum, the working electrode must be electrically coupled to a second electrode that "completes the circuit." The second electrode must allow a current to flow into or out of it that is directly opposite to the current flowing through the working electrode. A battery will not work if only one end of it is attached to the device that it is intended to power. Both ends of the battery are required to be attached so that a current can flow out from one end of the battery into the other end. The electrode that performs this function is called the counter electrode. The main requirement for the counter electrode is that it must always pass a current that is larger than the working electrode so that the current at the working electrode is not limited by the current output at the counter electrode.

A third electrode function, called the reference electrode, also needs to be performed for a reliable electrochemical measurement of glucose. The reference electrode serves the purpose of providing a known electrochemical potential for the sensor. A reference electrode has associated with it a chemical reaction that occurs at a known potential. At this point, the concept of potential will be explained in a manner as simple as possible to fully comprehend how the various electrodes operate. All CGM systems use a silver/silver chloride reference electrode. The reduction of silver chloride to silver occurs at a known potential. The potential is the energy of the electrons in the metal part of the electrode. The potential of the silver/silver chloride reaction is known and constant as long as the chloride level in the sample is known and constant. The chloride level in the body varies over a narrow range, 98–108 mM, and so it is relatively constant for each person. The importance of having a defined and stable reference electrode cannot be underestimated. The electronics that control the sensor apply a potential difference between the working electrode and the reference electrode. Since the potential of the reference electrode is fixed, the potential to which the working electrode is driven by the electronics is also known and fixed. When an oxidation occurs at an electrode, the electrochemical potential of the electrons in the reactant can find a lower electrochemical potential in the metal electrode, so they transfer to the electrode. The potential of the working electrode *must* be at such a level where the reactant will react when it arrives at the metal surface, so that a current is produced that is proportional to the amount of glucose in the sample (for a basic explanation of electrochemical reactions, see Chapter 1 of Ref. 13). In some commercial CGM systems, the function of the counter electrode and the reference electrode are combined into one electrode. Sensors that have only one working electrode and one reference/counter electrode are called "two-electrode" sensors, and sensors with separate working, counter, and reference electrodes are called "three-electrode" sensors. The advantages and disadvantages of each type will be discussed in Section 5.5.

5.1.3.3 First/Second/Third-Generation Sensors The technology of electrochemical sensors can be described in terms of first-, second-, and third-generation sensors.[14,15] The term generation does not mean that one generation was developed prior to the other, but rather it is strictly used to identify the chemical reaction used to make the glucose measurement. The commercially available CGM systems are either the first- or the second-generation sensors, and future systems are likely to be third-generation sensors. A first-generation glucose sensor uses oxygen to react with glucose oxidase and produce hydrogen peroxide (Figure 5.1). Hydrogen peroxide is then oxidized at the metal surface of the working electrode. The potential at which the metal surface must be maintained to enable the oxidation of hydrogen peroxide to occur is 600–700 mV when using a silver/silver chloride reference electrode. First-generation sensors must find a way to cope with the large concentration difference between oxygen and glucose. In people with diabetes, glucose is 2–30 mM and the oxygen concentration is typically 0.1–0.3 mM. The desire is to have the enzyme reaction be limited by the concentration of glucose and not by the concentration of oxygen, so all first-generation systems face this problem.

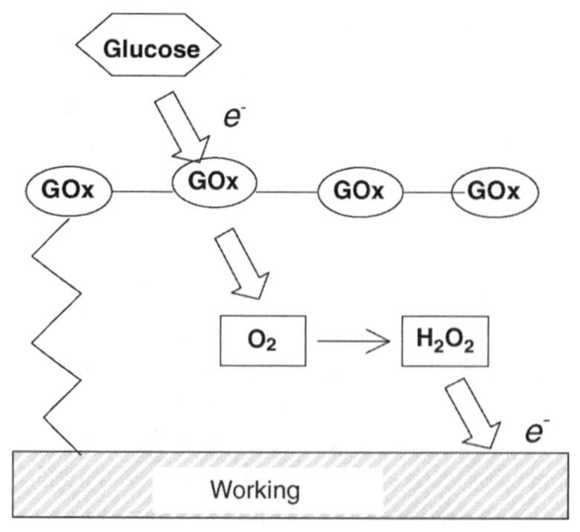

Figure 5.1 Electrochemical reaction of glucose with glucose oxidase immobilized on a working electrode. Copyright 2008 Abbott. Used with permission.

First-generation systems require that a diffusion barrier be used that slows down the diffusion of glucose more than that of oxygen so that the oxygen concentration is not rate limiting. Polymers with high oxygen solubility, such as polyurethanes and polysiloxanes, are good choices (for further discussion of membrane properties, see Section 1.2.2). Necessarily, when the glucose diffusion is restricted, the current generated is also reduced, which can make the miniaturization of such systems problematic. Another solution to the oxygen limitation is to dilute the sample containing glucose, which lowers the glucose concentration but not the oxygen level. The YSI glucose analyzer was the first electrochemical glucose analyzer to be sold and it relied on dilution of samples. Of course, sensors implanted in the body cannot rely on diluted samples.

A second-generation sensor uses an exogenous mediator (i.e., a mediator other than hydrogen peroxide) to react with glucose oxidase instead of oxygen (Figure 5.2). The mediator can be added at a higher concentration than oxygen to eliminate the problem of low oxygen concentrations. All single-use electrochemical test strips sold today are second-generation sensors. The potential at which the mediator can react at the metal surface of the working electrode can be manipulated based on the chemical structure of the mediator. The commercially available CGM system based on second-generation technology uses a potential of approximately 40 mV versus a Ag/AgCl reference electrode. The second-generation sensor does not require as restrictive a membrane as the first-generation sensor to slow down the diffusion of glucose, so the currents can be higher. The challenge of using a second-generation system inside the body is that the mediator must remain attached to the sensor. The mediator must also maintain its ability to react with the enzyme after it is coupled to the working electrode. Attempts were made to physically entrap the mediator over the electrode using a semipermeable membrane, but the mediator can leak out and the construction of this type of sensor is complex.[16–18] Immobilization of the mediator in a “redox polymer” has been more successful.[19] Development of a second-generation glucose sensor that can be used inside the body has been challenging and only recently it has been approved by the FDA and commercialized.

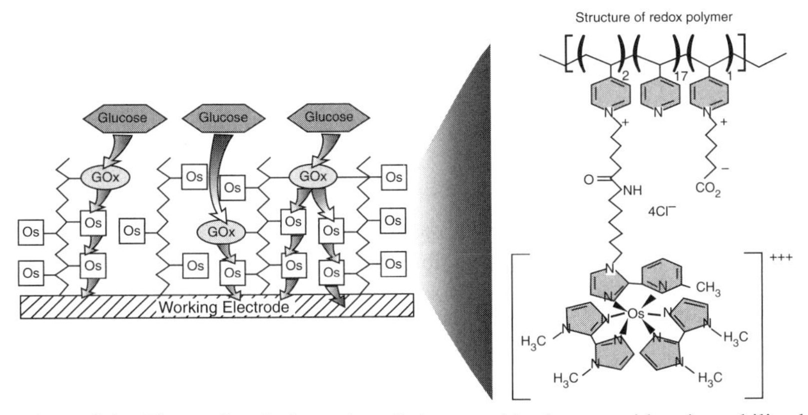

Figure 5.2 Electrochemical reaction of glucose with glucose oxidase immobilized within a redox polymer on a working electrode. Copyright 2008 Abbott. Used with permission.

Home use glucose test strips do not require that the mediator be confined to the working electrode. As a result, the single-use, second-generation home use glucose test strips have been commercially available for 20 years prior to the implanted second-generation glucose sensor being approved.

A third-generation sensor will not require oxygen or a mediator to react with the enzyme. In a third-generation sensor, the enzyme itself is in electrical communication with the metal surface of the electrode. Electrons can be transferred back and forth between the metal surface and the enzyme. Making such an electrical connection has proven difficult, but some recent success has been reported.[20–23] The active site of the enzyme is typically buried inside the enzyme so that exchange of electrons with the metal surface is difficult. If such a system is developed, it would have the advantages of second-generation system because it would not be oxygen dependent and would have the simplicity of a first-generation sensor because it would not require a mediator. The third-generation systems developed to date show a sluggish electron exchange with the enzyme.[24] The electron exchange can be increased by applying a larger potential to the metal surface, but applying a larger potential makes the electrode more susceptible to electrochemical interferences that can be found inside the body.

5.2 HISTORY OF CGM

5.2.1 Doctor Updike and Generation 1 (Dexcom)

One of the first researchers in the field of continuous glucose monitoring and enzyme electrodes is Dr. Stuart Updike of the University of Wisconsin. Dr. Updike published an article in 1967 on the electrochemical detection of glucose based on the depletion of oxygen from the reaction of glucose with immobilized glucose oxidase.[25] As a practicing nephrologist, Dr. Updike noted the utility of such a sensor to provide continuous glucose results for the care of patients with diabetes. In 1980, Dr. Updike was granted a patent for a system that would automatically extract blood out of a patient, mix the blood with a diluent, and then read the glucose concentration using a sensor.[26] The sensor had glucose oxidase immobilized on the surface that reacted with glucose and caused a local reduction in the oxygen concentration. The blood samples were oxygenated to a constant level. The patent proposed that samples could be extracted as often as every 2 min. Overall, the device is large and contains a pump, multiple valves, diluent solution, calibration solution, heparin solution, as well as the glucose sensor and electronics. The device could be used in a hospital setting but would not be practical for everyday use by a person with diabetes. In 1982, Dr. Updike and colleagues recognized that hydrophobic membranes would restrict glucose diffusion much more than oxygen diffusion.[27] If the membrane was effective, the reaction of glucose and glucose oxidase would not be limited by oxygen but by glucose. Using this special membrane, the sample did not need to be diluted, which raised the possibility of making a measurement directly in the body and the possibility of making an artificial pancreas was raised. The membrane was also found to stabilize glucose oxidase, which would be required for a glucose sensor to operate in the body.

The Updike group applied for a patent in 1985 for a glucose meter with a reusable glucose oxidase membrane that could measure glucose directly in blood.[28] The meter was eventually sold commercially by Eli Lilly under the name Direct 30/30.[29] The meter was different from all the other glucose meters designed for home use because the glucose sensor was reusable. It remains the only commercially available home use glucose monitoring system that was based on the detection of hydrogen peroxide, generation 1, rather than an exogenous mediator, generation 2. The glucose sensor could be used by a person with diabetes to run as many whole-blood fingerstick samples as they wished in a 30 day period. The membrane had to be kept wet between samples and calibrated once per day by the user using an aqueous calibrator. The analytical performance of the meter was good. The device was recalled because the user interface was not robust enough, but the Direct 30/30 demonstrated a significant improvement in glucose membrane technology. Previous membranes were often a series of very porous materials, for example, nylon netting. Updike's new membrane was a polyurethane membrane derivatized with anionic carboxylate groups and hydrophilic polyether segments.[28] This membrane had no defined pores and relied on the solubility of glucose and oxygen in the membrane to control the reactant flux to the enzyme. This new approach was able to slow down glucose diffusion much more than oxygen, so undiluted whole-blood samples could be analyzed. This membrane technology would also make it possible to operate a glucose sensor inside the body. At this same time, John Mastrototaro worked for Lilly on glucose sensors.[30] Mastrototaro moved to MiniMed and started their continuous glucose monitoring program.

Dr. Updike founded a company, Markwell Medical, which used the membrane technology to implant continuous glucose sensors in dogs.[31–33] The implants contained all the sensor electronics, telemetry, and batteries to use the devices for many months. Dr. Updike pioneered the use of different materials to promote the regrowth of blood capillaries around the implanted device. The body naturally tries to wall off any implanted device because the body sees it as a foreign body. Using Velcro-like materials, Updike's group claimed to be able to promote the ingrowth of blood capillaries. The fully implanted sensor was rather large, 7 cm long and 1.2 cm high, because of the state of electronics miniaturization at that time. The device for its size was not attractive for human implantation and the company received very little funding for many years. In 1999, Markwell Medical received substantial funding and it was renamed DexCom. The objective of DexCom was to further develop the totally implanted continuous glucose sensors that Markwell Medical was using. Eventually, they were able to run a human clinical trial with mixed success.[34]

5.2.2 Professor Heller and Generation 2 (Abbott)

The typical electrochemical glucose sensor has a problem with the low level of oxygen found in biological samples compared to the high level of glucose. The earliest electrochemical glucose analyzers diluted the sample to decrease the glucose concentration. The problem of low oxygen concentration was overcome in generation 2 sensors by the use of exogenous mediators that take the place of oxygen in the enzyme reaction. These mediators are soluble in blood, which makes them ideal for use in blood sampled from the body but would not be appropriate for use inside the body. In the early 1990s, Dr. Adam Heller at the University of Texas developed polymers that contained mediators and enzymes chemically attached to the polymers.[19,35] "Redox Polymers" could then react with glucose and did not require oxygen. Electrons are exchanged between the mediator on the polymer and the enzyme. The mediator is then able to exchange its electrons with the other mediators in the polymer until the electrons are eventually exchanged with the metal surface of the electrode. Because the process does not require oxygen, the need to decrease the glucose concentration is much less, so higher signals or smaller electrodes are possible. Exogenous mediators are also typically oxidized at lower potential than the endogenous mediator hydrogen peroxide, which is an advantage because fewer oxidizable interferences can be oxidized at low potentials. Although the theory is simple, the practical challenges of designing such a membrane are formidable. The mediator must be able to rapidly exchange electrons with the enzyme. Some home use glucose strips use mediators that do not exchange very rapidly with the enzyme, but they can overcome this limitation by using a large excess of mediator. There are limited attachment sites in a typical polymer, so large excesses of mediator cannot be created. If the mediator is too loosely attached to the polymer, the mediator will be lost and if the mediator is too tightly attached to the polymer, it may not be loose enough to react with the enzyme.

Dr. Heller and his son licensed the redox polymer technology from the University of Texas in 1991 and formed the company E. Heller and Company. The company became TheraSense in 1997 and was bought by Abbott Laboratories in 2004. Along the way of developing a subcutaneous implanted glucose sensor, TheraSense decided to also develop a glucose test strip for use outside the body. TheraSense's glucose test strip business was very successful and it continued the development of an implanted sensor. The long time required to commercialize a product is an indicator of the tremendous challenges that had to be overcome to commercialize a second-generation sensor.

5.3 PERFORMANCE REQUIREMENT OF CGM

5.3.1 Performance Measures

5.3.1.1 Clinical Use of CGM CGM systems are at present used by doctors and patients in one of the two ways that are described in this section and some of the future applications of the technology are covered in Section 5.7. The systems are approved to be worn for short periods, 3–7 days, The goal of these short wear sessions is to provide insight into the glucose patterns found within a person's day. These patterns can be interpreted by a doctor to change the treatment regime of a person with diabetes. The first Medtronic CGM system approved did not even provide a live glucose result to the user. The data had to be downloaded at the doctor's office and were viewed only retrospectively. The data, however, are still very valuable for the adjustment of insulin dosage and insulin type. One of the unique aspects of diabetes is that the person with diabetes has the control through the use of insulin, diet, and exercise to make a significant difference in their disease. Even with a live continuous display, the results may need to be interpreted by a trained professional to see the correct pattern. Custom software is also expected to improve this interpretation by at a minimum speeding up the analysis. As the number of people with diabetes increases, the workload on primary care physicians and endocrinologist will become worse. Technologies like CGM are needed to allow doctors to make correct treatment decisions and more effectively use their time with the patient. Although the ability to see trends and adjust insulin dosing is important, the next use of CGM is as the lifesaver. If a person with diabetes chooses to wear a CGM device essentially everyday, then the information can be used by the person with diabetes in their everyday treatment of their disease. Again, the person with diabetes, especially those using insulin, decides how to treat their disease several times a day. The more information they have that can be easily interpreted, the better they can treat themselves. Using conventional blood glucose meters, they know what their glucose level is at the time when they made a glucose measurement. However, they neither know what their glucose level was 10 min before they made the blood glucose measurement nor do they have any means of estimating what it will be 10 min after. CGM devices enable them to know their glucose level every minute of the day. The CGM device can be used to set alarm levels to warn the person with diabetes when they reach a dangerous glucose level. Two of the devices will also project ahead and warn a user that they are heading for a potentially dangerous glucose level. The alarm features, however, are only important if the CGM device is accurate. An inaccurate device will lead to many false alarms and the user will eventually turn off the alarms because they are annoying. One CGM device that was not commercially successful suffered from this problem.[36]

5.3.1.2 Accuracy Accuracy is the most important criteria used to judge the performance of an analytical measurement. A measurement of a body component, however, is judged differently than your typical laboratory analysis. Accuracy will not only be defined in terms that a scientists is normally accustomed to, but it will also be defined by the diagnostic requirements of the disease that the measurement is directed at. Accuracy can then become a curious mix of both analytical performance and medical necessity. The methods used to judge the accuracy of blood glucose meters are well established and have been applied to CGM systems. The accuracy methods used for glucose meters will be described along with the limitations of these methods when they are applied to CGM as well as the new methods that have been developed just for CGM.

The accuracy of home glucose meters can be measured like any other analytical technique by comparing the results of samples run on both a blood glucose meter and a reference instrument. The paired points are plotted with goals of a slope of one, intercept of zero, and minimal scatter about the unity line as measured by the correlation coefficient (r). The International Standards Organization (ISO) has chosen to define accuracy using a plot of paired points, but in a way that includes precision and takes into account the diagnostic significance of the points. Because glucose varies over such a wide range, the accuracy is defined differently for two different glucose ranges. Above 75 mg/dL glucose, 95% of the blood glucose measurements must be within +/−20% of the reference instrument. Below 75 mg/dL glucose, 95% of the blood glucose measurement must be within +/−15 mg/dL of the reference instrument. If the specification for above 75 mg/dL was applied to lower glucose levels, then the accuracy, absolute difference between blood glucose meter and reference instrument, of the measurement would have to improve so that at 20 mg/dL the allowed error would be +/−4 mg/dL. Such accuracy would be hard to obtain with the reference instrument and impossible with an inexpensive and disposable blood glucose strip. Using a range of +/−15 mg/dL is clinically acceptable at low glucose levels because the treatment is the same once a low level is reached, so the absolute accuracy is not needed. If the criteria of +/−15 mg/dL were applied to high glucose levels, the accuracy of the whole-blood glucose would have to be very good. This 15 mg/dL is only 5% of 300 mg/dL, which is significantly tighter than the 20% allowed. The treatment decision would be the same if the glucose value is within 20%, so this level of accuracy is acceptable at higher glucose levels. This method of defining accuracy inherently includes a measurement of precision. If a glucose meter was not precise, then 95% of the values would not fall within their ranges. The ISO method can be applied to the evaluation of continuous glucose meters by comparing the continuous blood glucose measurement to a blood sample taken at the same time and measured on a reference instrument. No ISO standards have been developed for CGM.

Dr. William Clarke wanted to know if his patients could accurately predict their glucose level based solely on how they felt.[37,38] He asked patients to predict their glucose level based on how they felt and then measured their blood glucose level. Their predictions were not very accurate, but Dr. Clarke looked for a method of determining accuracy from relatively poor data. He decided to look at the data from the standpoint of whether the glucose prediction would be so far off so as to cause the patient harm or was it close enough to not to harm the patient. He developed what is known today as the Clarke error grid. The grid has several zones with different results for the patient depending on which zone the data are located. The grid is used by plotting data pairs from the device being evaluated and a reference instrument (see Figure 1.7). The grid has many of the same constraints as the ISO standard discussed above. Above 75 mg/dL, the results are good if they fall within +/−20% of the reference value. Below 75 mg/dL, the grid is more tolerant than the ISO standard, which reflects the clinical reality that any low glucose value needs to be treated. The ISO standard requires 95% of the paired points to be within certain limits, but does not comment on the 5% of the points that can fall outside those limits. The grid seeks to classify these points in terms of the hazard to the patient, so it looks at the clinical significance of all the points. The grid became so popular that several variations have been debated to possibly improve the grid. The consensus error grid sought to fix some of the deficiencies of the Clarke error grid. Similar to the ISO standard, the grid measures both accuracy and precision.

The Clarke error grid and subsequent variations are better suited for the determination of the accuracy of new techniques like CGM than the ISO standard. The grid does not require the high degree of analytical accuracy that is necessary to pass the ISO standard. Many companies developing noninvasive glucose measurements have relied solely on the grid. The statistics from the grid can however be deceiving. A high percentage of data in the A and B zones of the grid does not mean the device being tested is accurate. Data in the B zone represent those data that are not analytically very accurate but would also not harm the patient. In the opinion of the author, use of the ISO standard, possibly with relaxed requirements, and the grid together are the best ways to judge new technology.

Another parameter commonly used to measure the accuracy of glucose measurements is to report the differences between the CGM device and the reference instrument. The mean or median difference can be reported for all values. The median difference tends to be lower than the mean if there are any large outlier values. The sign of the difference measurement will show if there is any bias to the measurement. However, positive and negative deviations will cancel each other out, so this is not the standard parameter used. Taking the absolute value of all the differences eliminates the problem of differences canceling each other out. The difference will have the units of mg/dL. Because of the wide physiological range of glucose, a 20 mg/dL difference at high glucose level would be of no concern, while the same difference at low glucose level would be a concern. Since there is such a wide range of glucose values, the differences can be normalized by dividing it by the reference value. The new statistic is dimensionless and is often expressed as a percentage. The median absolute relative difference (MARD-median) and mean absolute relative difference (MARD-mean) are probably the most reported number for accuracy (equation (5.1)). Because the number is based on the absolute value of the difference, the MARD does not tell you if any bias is present.

$$\begin{aligned}\text{MARD(mean)} &= \left(\left(\sum \frac{\left|[\text{Glucose}]_{\text{sensor}} - [\text{Glucose}]_{\text{reference}}\right|}{[\text{Glucose}]_{\text{reference}}}\right) \Big/ n\right) \times 100\% \\ \text{MARD(median)} &= \text{MEDIAN}\left(\frac{\left|[\text{Glucose}]_{\text{sensor}} - [\text{Glucose}]_{\text{reference}}\right|}{[\text{Glucose}]_{\text{reference}}}\right) \times 100\%\end{aligned} \tag{5.1}$$

The following provides a practical example of how the different statistical parameters used to judge the accuracy of glucose monitoring relate to each other. A person with type 2 diabetes using insulin to control the disease performed 176 glucose measurements using a blood glucose meter over a 23 day period. The average of all the glucose values was 205.6 mg/dL, which is a higher average than would be desirable, but this subject may be representative of a large number of the population with diabetes who are not in good control of their glucose. Each of the 176 glucose values was randomly assigned an offset that would result in a specified standard deviation for the whole data set. The various statistical parameters were then calculated for the encoded level of variation in the glucose values. If the process were repeated on the same data set, the results would have been slightly different because of the random assignment of the deviations. The results are summarized in Table 5.1 and a series of Clarke error grids are shown in Figure 5.3. The mean absolute difference (MAD) tends to change proportionally with increasing error while the Clarke error grid zones exhibit a more unusual behavior. The Clarke error grid zones change significantly when the errors cause certain boundaries in the grid to be crossed, as can be best seen in Figure 5.3. There is no deliberate offset of the slope or Y-intercept encoded into the results. If errors were introduced into the slope and intercept, some of

TABLE 5.1 Simulated Error in Glucose Measurements

(a)

Error at 1 SD mg/dL	Slope	Intercept mg/dL	r	MAD mg/dL	MARD	SE mg/dL	SRE
10	1.002	0.4	0.99	8.7	4.2%	10.7	5.2%
20	1.004	0.8	0.96	17.4	8.5%	21.4	10.4%
40	1.008	1.6	0.85	34.8	16.9%	42.8	20.8%
60	1.011	2.4	0.74	52.1	25.4%	64.2	31.2%
80	1.015	3.2	0.63	69.5	33.8%	85.6	41.7%

(b)

	Distribution of data points in error grid					
Error at 1 SD mg/dL	A	B	C	D	E	Outside
10	99.4%	0.1%	0.0%	0.6%	0.0%	0.0%
20	96.0%	3.4%	0.0%	0.6%	0.0%	0.1%
40	76.8%	21.8%	0.1%	1.0%	0.0%	0.3%
60	58.5%	36.0%	1.8%	2.2%	0.6%	1.0%
80	46.5%	40.0%	5.7%	3.5%	1.5%	2.6%

MAD, mean absolute difference; MARD, mean absolute relative difference; SD, standard deviation; SE, standard error; SRE standard relative error.

the parameters would be affected and others would not. The standard error is not normally reported for CGM, but is a popular metric for noninvasive glucose monitoring systems. The standard error and mean absolute difference are very similar for measurements without significant outliers. The standard error is significantly higher than the mean absolute difference when significant outliers are present in the data set.

The different methods of measuring accuracy were all developed for blood glucose meters that provide a reliable glucose measurement at one point in time. Since CGM provides a continuous stream of glucose results, are the same accuracy measures appropriate for CGM? Any single glucose result reported by a CGM device is preceded by many glucose measurements and followed by many more glucose measurements. Does the wealth of data provided by CGM devices mean that they do not need to meet the same accuracy requirements of glucose meters? This question is still being debated. In the near future, the common accuracy metrics used for blood glucose meters will be applied to CGM devices because no other methods have been established as industry standards. When are the CGM results good enough so that the person with diabetes does not need to use blood glucose fingerstick measurements and can rely solely on the CGM results? This is still an unanswered question because no CGM device has been approved that does not require calibration with blood glucose measurements.

CGM devices have the ability to alarm the patient if their blood glucose level is beyond the preset limits or even if the device projects that the patient's blood glucose level will be beyond the preset limits. The accuracy of the alarms can be measured based on the number of true alarms given, false alarms given, and alarms that should

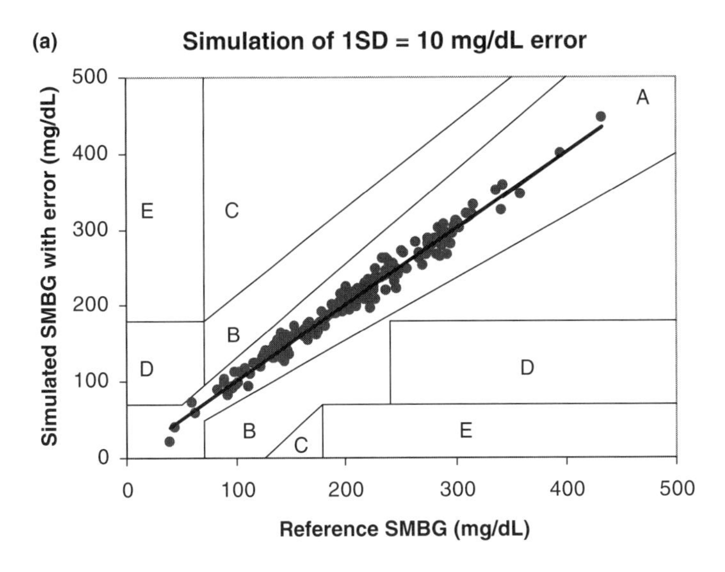

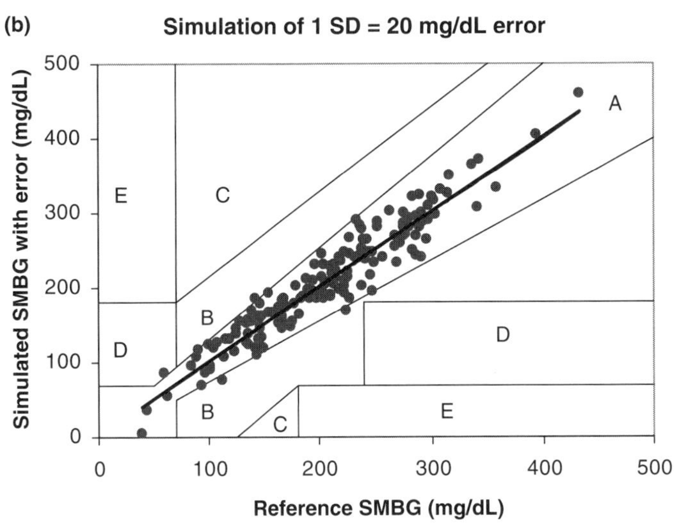

Figure 5.3 Simulation of the incorporation of different amounts of error into self-monitoring blood glucose (SMBG) readings: (a) −10 mg/dL, (b) −20 mg/dL, (c) −40 mg/dL, and (d) −80 mg/dL. Copyright 2008 Abbott. Used with permission.

have been given but were missed. The subject is so complicated that another chapter would be needed just to cover this subject (see additional information in Section 1.3.2.1). Since blood glucose meters do not alarm, they do not provide any history to establish acceptable criteria. Does a false alarm in the middle of the day matter less than a false alarm in the middle of thc night that wakes the person up? If the CGM reads 79 mg/dL and alarms and the blood glucose is 80 mg/dL, above the alarm threshold of 79 mg/dL, is that a false alarm? Everyone would agree that the CGM is

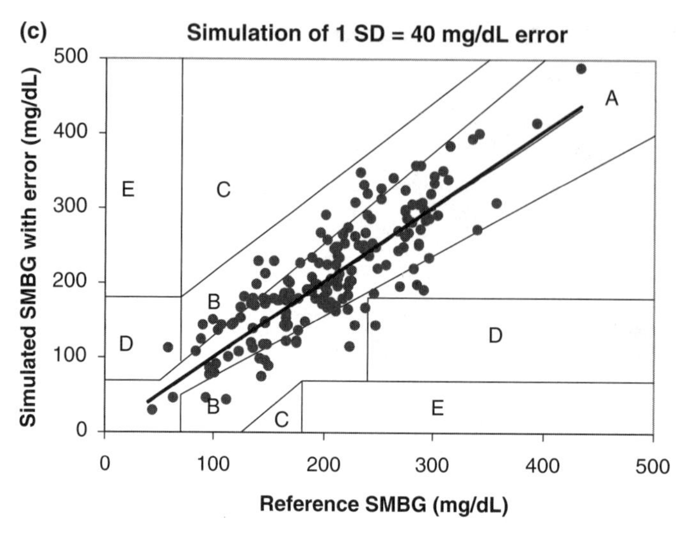

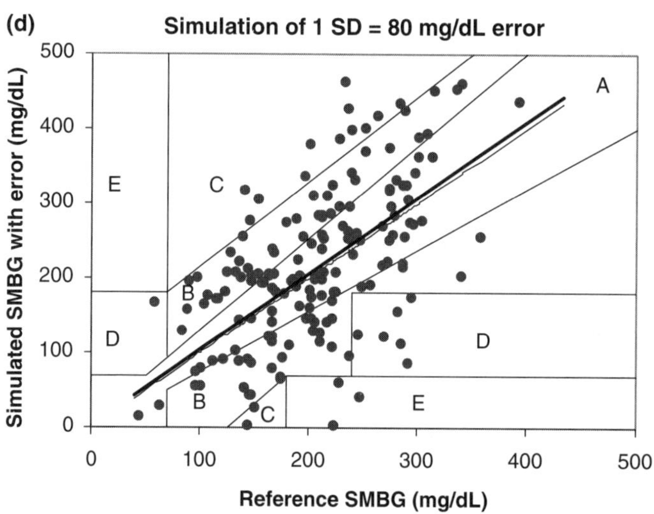

Figure 5.3 *(Continued)*

giving an accurate reading when it reads 79 mg/dL because it is so close to the blood reading of 80 mg/dL. Several methods of reporting alarm accuracy have been proposed, but no consensus exists today on which method is correct.

5.3.1.3 Precision Measuring the precision of blood glucose strips is very easy and involves testing a number of strips with the same sample with a fixed glucose level. The standard deviation and relative standard deviation (also called percent coefficient of variation) can then be easily calculated for the strips. There is no such easy precision measurement for CGM devices. The manufacturers have placed two CGM devices in a person and then compared the glucose readings made by each device at the same point in time. Because the device measures glucose as frequently as every minute and can be worn for several days, a comparison of two devices can yield thousands of paired data points. The differences between the paired points can be calculated and a standard deviation of the differences calculated to indicate the precision. However, the precision only compare two devices. It would not be practical to put 20–30 devices in one person to get a better indication of the device to device variability. Two devices could be worn by a number of people, which adds the person to person variability to the device to device variability. Precision is, therefore, a rarely quoted statistic for CGM devices. As mentioned in Section 5.3.1.2, precision is a component of most of the accuracy measurements, so it is not totally neglected.

5.3.2 User Requirements

5.3.2.1 Short-Term Versus Long-Term Systems All the CGM devices on the market are at present approved for either 3, 5, or 7 days of wear. These short-term systems fulfill the needs of the person with diabetes for the monitoring of glucose patterns, as described in Section 5.3.1.1. These systems can also be worn continuously by changing the systems every 3–7 days. A long-term system, greater than 6 months, would at a minimum have a device implanted in the body and a display outside the body. Such systems are being developed and will be discussed in Section 5.7. Both types of systems require some type of calibration using a fingerstick blood glucose measurement. The frequency of calibration of the short-term systems is approximately daily, while the frequency of calibration of the long-term systems has not been determined but is likely to be less frequent than short-term systems.

The short-term systems are the only systems on the market. Because the devices are inserted by the user at home, they allow a user to choose how often and at what times they want to wear the device. The person with diabetes may not want to wear the device at times for personal reasons. Short-term devices are flexible and the diabetic can ease into the use of the device. Long-term systems require a long-term commitment to the use of the device. Inserting the long-term device and retrieving it from the body will likely be done in a physician's office and some of the larger devices require surgical procedures. The long-term devices do eliminate the need to change the device every few days as is required for the short-term devices. The long-term devices will also be more discrete than short-term devices because a transmitter or wires will not have to be worn by the diabetic on the skin. Long-term systems, however, are years away from reaching the market.

On-market CGM devices are limited to 3–7 day. There are several technical reasons that will make extending the lifetime beyond 7 days very difficult. The most daunting task is to get the devices to stick to the skin for more than 7 days. The skin continually expels dead cells, so its surface is not stable. The lifetime can be extended by placing bandages over the device to help it adhere, but this secondary bandage will still have many of the same limitations of the primary adhesive used to stick the devices. The longer the device remains in the skin, the greater the chance of an infection developing around the site. Bacteria can grow along transcutaneous devices and form biofilms that can lead to infection. No serious infections have been reported with the current 3–7 day devices. The longer the device remains in the body, the greater the chance that it will evoke an immune system response that could alter the system performance. The recommendation is to replace insulin cannulas every 3 days, but this may be because of the irritating effect of the insulin on the local tissue provoking an inflammation response. The CGM sensors seem to evoke much less tissue irritation. A final technical challenge will be sensor lifetime for extended periods of time. The robustness of the sensors for extended insertion times has not been reported.

5.3.2.2 Types 1 and 2 Diabetes The term diabetes covers a wide range of conditions and each of these conditions has different needs for CGM. Type 1 diabetes is defined as a complete loss of insulin secretion that typically happens at a young age. People with type 1 diabetes depend completely on the injection of insulin to live. These people must pay attention to the glucose level throughout each day or risk serious short- and long-term health consequences. A major threat to the life of a person with type 1 diabetes is to pass out while operating machinery such as driving a car because of low glucose (hypoglycemia). The consequences of low glucose are threatening enough that the person with type 1 diabetes will spend large portions of the day with high glucose levels (hyperglycemia) to protect themselves from passing out. High glucose levels over a long period of time will damage the eyes, kidneys, and nerves as well as cause cardiovascular disease. The type 1 diabetic, therefore, will be a key user of CGM devices. The person with type 1 diabetes has already shown a desire to seek new technology to treat their disease by adopting insulin pumps. The CGM device needs to protect the diabetic by setting of alarms if their glucose is low or high. The alarm feature is especially important during sleep when the person with diabetes is not aware of the typical physiological clues that warn them of problems. The CGM device will also allow tighter glucose control by allowing the person with diabetes to safely maintain a glucose level that is not too high. A distinct subset of type 1 diabetics is younger children where the parents are primarily responsible for the patient's treatment. This small population will likely be one of the firsts to adopt CGM because the parents need information to manage their child's disease as the child may not be capable of communicating effectively.

Type 2 diabetes is a very complex disease where the cells of the body are not able to utilize the insulin produced by the body. The disease is usually associated with aging and obesity. Describing the disease as the body has too many cells (fat) for the amount of insulin produced is a gross simplification. Losing weight no doubt typically helps the type 2 diabetic regain glucose control but the disease is progressive. The disease is treated by weight loss, better nutrition, and drugs that help the body better utilize the insulin present. Patients at this stage of the disease do not suffer from the extreme glucose levels as seen in type 1 diabetes and would not need the advantages of CGM. The loss of glucose control becomes worse with time and eventually some people with type 2 diabetes start using insulin injections to help maintain glucose control. The person with type 2 diabetes at this stage has not completely lost their insulin production, but the loss is bad enough to require insulin injections. At this point in the disease, measuring glucose levels more frequently is required. The injection of too much insulin can produce hypoglycemia and hyperglycemia becomes more prevalent. The disease begins to resemble type 1 diabetes but the patients are typically older. Because of their age, these people with type 2 diabetes may not be as technologically savvy as the typical person with type 1 diabetes and therefore less able to handle CGM devices. However, the large aging population in the United States, known as "baby boomers," now facing type 2 diabetes is much better prepared to handle technology than previous generations. A paradigm shift is possible where aging baby boomers will adopt some of these newer technologies to achieve a better quality of life. The baby boomers are also more likely to able to afford the new technology than past generations.

5.3.2.3 Ease of Use The use of a CGM device involves a number of skills that may not be present in all people with diabetes. A patch is applied to the skin and a sensor inserted across the skin. The procedure is similar to inserting an insulin catheter for an insulin pump, so it should be simple enough for any insulin pump user to do. Applying the adhesive patch is similar to applying a Band-Aid and should be simple enough for most people with diabetes to accomplish. The sensor insertion needs to be simple because it will be done on the outer part of the forearm or on the stomach. Performing the insertion on the arm means that it will have to be done with one hand. The insertion should not involve any complicated twisting, turning, or pulling. Adults certainly have enough available skin space to wear the devices on a continuous basis, which would involve rotating to different sites as the previous wear site heals. Small children who wear the devices continuously may have to use alternate body sites such as the buttocks. One of the on-market devices is at present approved for use by children.

The CGM devices will be used by patients at home on a regular basis, so the process of using the device should be simple to learn and simple to do. Placement on the body and insertion of the device were discussed and another potentially complex operation is the calibration of the device. Once the devices are placed on the body, they all require a minimum of one blood glucose value for calibration. Ideally, the blood glucose value would be automatically read by the CGM device and used for calibration. Requiring a separate blood glucose meter for calibration requires that the data be transferred in some manner from the blood glucose meter to the CGM device. The transfer can involve cables and additional steps that make the process more complex. One manufacturer has placed a blood glucose meter inside their display device, so the transfer of information is done automatically for the user. In this case, the display device can also be used to perform blood glucose measurements to check the accuracy of the CGM device and a separate blood glucose meter does not have to be carried around by the user. The device that displays the glucose results should be easy to read in different lighting conditions and have buttons and menus that are easy to navigate. Because of the proliferation of handheld electronics such as cell phones, the design of user interfaces is a well-known skill and most devices are intuitive to operate.

5.4 HOW TO MAKE A MEASUREMENT

5.4.1 Components

All the on-market CGM devices have similar components and the details of each manufacturer's device will be explained in the subsequent section. An adhesive holds a patch to the skin. The sensor is inserted across the skin using a sharp device to make a cut into the skin and the sharp device is then retracted leaving the sensor in the skin. The sensor has a glucose-oxidase-based working electrode located under the skin and electrical contacts located above the skin. Electrical contact is made from the sensor to an electrical measuring apparatus, the potentiostat. The electrical measuring apparatus is powered by batteries. The electrical measuring apparatus is reusable and may be integrated into the sensor patch or separately attached to the skin and connected to the sensor patch using wires. The electrical measuring apparatus communicates with a device that displays the glucose results using telemetry. The display device is also used to input calibration information. Besides displaying the most current glucose value, the display device will also show plots of the glucose values over time, alarm settings, and various software menus. The display device will have a large memory typically capable of recording glucose results of months.

5.4.2 The Signal

The working electrode of the sensor contains the enzyme glucose oxidase. The enzyme reacts with glucose and the enzyme is reduced. The reduced enzyme then reacts with either oxygen, the first-generation sensor (Figure 5.1), or exogenous mediator, the second-generation sensor (Figure 5.2). The hydrogen peroxide produced by the reaction with oxygen or the reduced exogenous mediator is oxidized at the metal surface of the working electrode that is poised by the potentiostat at a potential sufficient to cause the oxidation of the hydrogen peroxide or the mediator. Electrons flow into the metal surface and an equal number of electrons flow out of the counter electrode (or combination counter and reference electrode if the system is two electrode). The technique of measuring the current while the potential is held constant is called chronoamperometry. More sophisticated potential profiles, such as sweeping the potential over a fixed range, are often used in electrochemical measurements. However, since the current measurement will be made continuously for several days, keeping a constant potential is an advantage because it allows a constant glucose diffusion profile to be established across the enzyme-containing membrane that covers the metal conductor. The flux of reactant to the metal surface then becomes a function of the flux of glucose across the membrane. The concentration of glucose at the metal surface is assumed to be zero because sufficient enzyme is present to oxidize all the glucose. The concentration of glucose at the membrane surface in contact with the body is assumed to be the concentration inside the body. The change in glucose concentration from the surface in contact with the body to the surface in contact with the metal conductor is the concentration gradient that drives glucose diffusion into the membrane. Over a defined period of time, the amount of glucose diffusing into the membrane is the flux of glucose. If the glucose concentration in the body increases, the flux of glucose entering the membrane will be higher and the amount of reactant reaching the metal surface will be higher leading to a higher current. The thicker the membrane, the longer it will take for the change in flux outside the membrane to affect the flux inside the membrane. As a generalization, thicker membranes will react more slowly than thinner membranes. The diffusion coefficient of the glucose in the membrane will also affect how quickly the current changes. The change in response is often called the "response time" of the electrode and can be reported as the time for the current to change 90% of the final change.

An important consideration in the design of the electrode is the thickness of the enzyme membrane. The membrane needs to contain an excess of glucose oxidase so that all the glucose arriving at the membrane will be oxidized within the membrane. The amount of enzyme in the membrane will be determined by both the concentration of the enzyme in the membrane and the thickness of the membrane. The membrane controls the flux of glucose and oxygen into the membrane. Because it plays a pivotal role in determining the current, any changes in the solubility or diffusion rate of the reactants will affect the current. Changing membrane properties can be partially compensated for by the calibration process that will be explained in the subsequent section. An ideal electrode would require very few calibrations. A sensor that does not require calibration has not been commercialized. The membrane will also play a key role in determining how sensitive the sensor is to motion within the body. An ideal sensor would consume so little glucose that there would not be any glucose gradient in the body outside the membrane surface. If such a gradient existed, then it could be disturbed by movement of the sensor or by movement of the fluid around the sensor. The generation 1 sensor produces hydrogen peroxide that is not naturally present in the body. The hydrogen peroxide will diffuse not only to the metal surface of the electrode but also into the body. The generation 2 sensor has immobilized mediator, so the reactant does not diffuse into the body.

The current flows continuously through the electronics but is typically digitized at a fixed frequency. A number of digitized current measurements can be averaged to provide a single current value or converted to a single glucose value that will be displayed to the person with diabetes using the device. The method of converting the current to a glucose value is called "calibration" and is explained in the following section.

5.4.3 Calibration

The current produced by the continuous glucose sensor must be converted into a glucose value in order for it to be useful to the user. A device that comes precalibrated, such as is done with glucose test strips, would be ideal because it would not need any calibration. This has challenges because each insertion of the sensor into the body produces a unique response. Calibration requires that the current produced at a point in time is matched to a blood glucose measurement made with a glucose strip. The current can be divided by the glucose level determined by the glucose strip to produce a glucose sensitivity factor. This simplified explanation assumes the current is zero when the glucose concentration is zero or is predictable. Future glucose levels can be calculated by dividing the current by the sensitivity factor. This method of calibration looks to be very straightforward and should not be a cause for concern. In reality, a number of factors make calibration a challenging process that requires a detailed discussion to understand all the possible implications. This section will briefly review some of the more interesting challenges; however, a whole chapter could be devoted just to the calibration of CGM devices.

The calibration is done with one or more blood glucose measurements using blood glucose strips. The blood measurement will then set the glucose sensitivity factor and affect the accuracy of all future sensor results until another calibration is accomplished. Blood glucose strips do not have the accuracy of the more sophisticated glucose analyzers used in central laboratories (see Section 5.3.1.2 for glucose strip accuracy). It would not be practical for people with diabetes to carry around with them large clinical analyzers, so glucose strips are the only possibility for calibration. The inaccuracies of the blood glucose strips will then affect the accuracy of the CGM results and in essence CGM measurements can be no more accurate than the technique used to calibrate them. Once the CGM sensor is calibrated, the question becomes when does it need to be recalibrated? The manufacturers of on-market CGM systems all employ different recalibration schemes, which points to the complexity of the question. An empirical approach involves calculating accuracy based on using different numbers of calibration points and at different frequencies. With the increase in the number of calibration points and the frequency of calibration, the accuracy will reach a level at which additional calibrations do not measurably improve the accuracy. The frequency of calibration must also be tolerable to the user otherwise the device would not be used, so a compromise is struck between increasing the number of calibration points and making a device that can be used. It is also a matter of fact that a user would not want to do calibrations in the middle of the night.

Some researchers have suggested that CGM sensors need to be calibrated by taking blood glucose measurements at two points in time when the blood glucose measurements are very different. By pairing the two blood glucose measurements to two current readings of the sensors, the two data pairs can be used to calculate a line that describes the response of the sensor. This approach is important if at zero glucose the current is not zero or is not predictable. If at zero glucose the current is zero or a known stable value, then this can be used as one point for defining the calibration line and a blood glucose measurement, or measurements taken substantially at the same time can be used as a second point to define the calibration line. Since all the on-market CGM systems do not require a pair of glucose values with differing values, they either have zero current at zero glucose or assume or predict what the current is at zero glucose. The calibration can also be limited by the blood glucose reading. The manufacturers limit the range of acceptable blood glucose values so that low glucose and high glucose values cannot be used for calibration. The manufacturers presumably do this because they have found that extreme blood glucose values are not as reliable for calibration as more moderate glucose values. Calibration when glucose is rapidly rising, such as immediately after eating, is not recommended. The limitations applied to calibration would seem to be related to the dynamics of equilibration between the blood and the interstitial space where the sensor is placed in. As mentioned previously, the subject of calibration is both complex and controversial and is one of the continuing challenges to improve the accuracy of CGM sensors.

5.5 COMMERCIALLY AVAILABLE CGM DEVICES

5.5.1 Medtronic

MiniMed obtained the first FDA approval for a CGM device in 1998 and was purchased by Medtronic in 2001. The first CGM devices from MiniMed did not

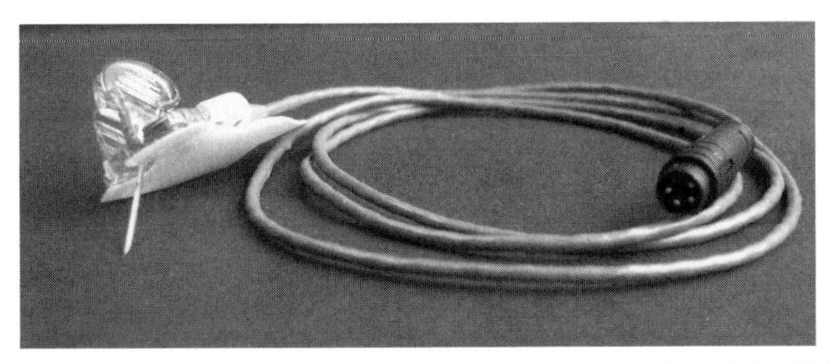

Figure 5.4 Medtronic glucose sensor with cable. Copyright 2008 Abbott. Used with permission.

display the glucose reading to the user. The sensor was inserted in a doctor's office and the patient carried a device that recorded the output of the sensor. When the patient returned to the doctor's office after 3 days, the data were downloaded into a computer along with a minimum of four blood glucose measurements per day that were used to calibrate the output of the sensor. The system made a glucose measurement every 10 s and every 5 min it stored the glucose data in its memory. This type of nondisplay device did not allow the user to manage their diabetes while wearing the device. The results were used by the doctor to observe the normal daily variations in glucose of the user and recommend insulin adjustments and possible lifestyle changes to the user. Improvements were made to the sensor performance and a discontinuity in the calibration algorithm that occurred at midnight was fixed. The device was sold under the name CGMS® System Gold.™ As shown in Figure 5.4, the sensor is attached to a cable and the cable would be attached to a monitor that could be worn on a belt. Since the glucose result is not displayed on the device, the monitor could also be hidden under clothing.

Medtronic has now developed a system that displays the glucose results to the user in real time. The system is called Guardian® REAL-Time. The sensor is again placed in the skin but now a MiniLink™ Real-Time transmitter is also attached to the sensor (Figure 5.5). The transmitter wirelessly sends the glucose information to a monitor that displays the glucose result. The system requires a blood glucose measurement every 12 h for calibration. The monitor can also display arrows showing the direction of change in glucose and the number of arrows displayed indicates how rapidly the glucose is changing. In addition, the user can program in warnings that will project future glucose readings and warn the user if their glucose is projected to be dangerously high or low. Alternatively, the monitor can be a Medtronic insulin pump that is equipped with the correct technology to receive and display the signal. This system is called the Paradigm® REAL-Time 522 or 722 pump.

The sensor that attaches to the skin is shown in Figure 5.6. A needle containing a slot is used to insert the sensor (Figure 5.7). The slot holds part of the sensor and allows the needle to be withdrawn but leaves the sensor in the body. The sensor is inserted at an approximately 45° angle and the length of the sensor in the body is 12 mm. The sensor consists of a plastic housing that contains within it a thin flexible substrate that contains working, counter, and reference electrodes (Figure 5.8). The sensor is a three-electrode sensor that does not require a large reference electrode because

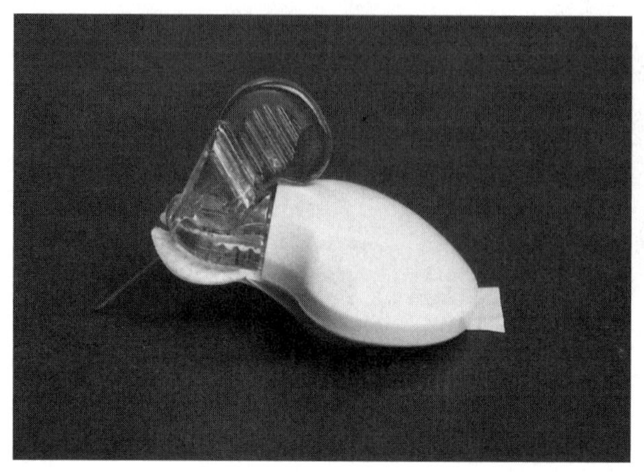

Figure 5.5 Medtronic transmitter and glucose sensor. Copyright 2008 Abbott. Used with permission.

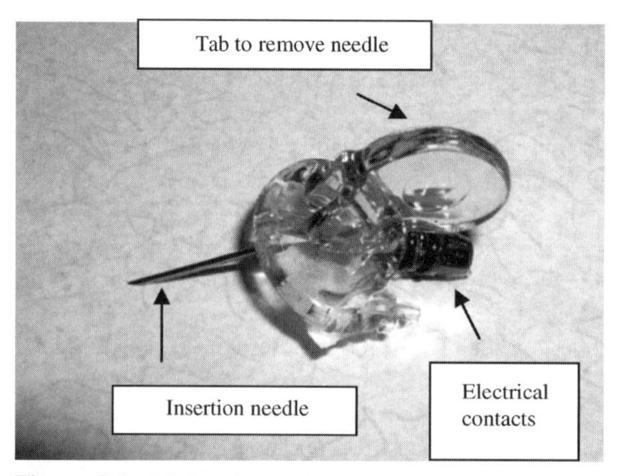

Figure 5.6 Medtronic continuous glucose sensor. Copyright 2008 Abbott. Used with permission.

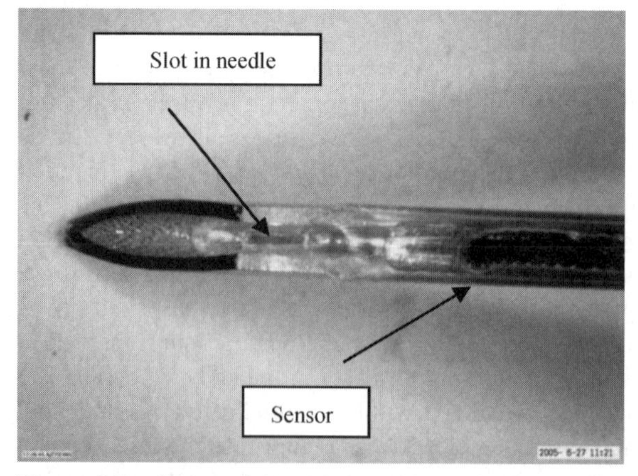

Figure 5.7 Close-up view of the metal needle with a slot where the Medtronic sensor sits within. Copyright 2008 Abbott. Used with permission.

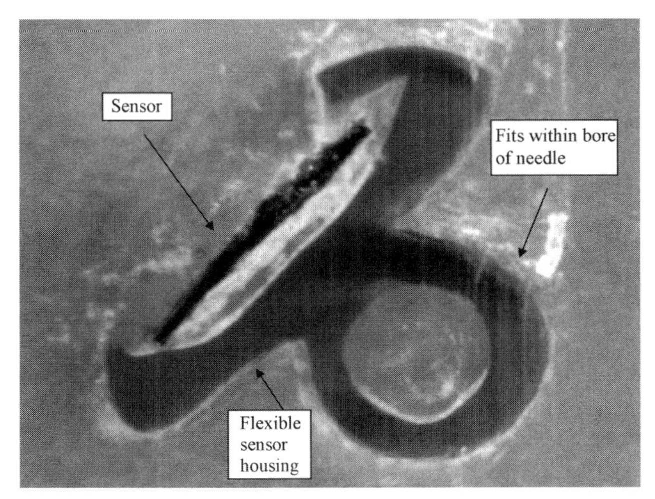

Figure 5.8 Cross section of the thin Medtronic sensor contained within a plastic housing and also showing the portion of the housing that fits within the bore of the needle used for the insertion. Copyright 2008 Abbott. Used with permission.

current does not flow through the reference electrode. A glucose oxidase membrane is applied to the working electrode. The membrane will react with glucose to produce hydrogen peroxide that is oxidized at the working electrode to produce a current. The sensor is, therefore, a first-generation sensor because it uses oxygen to react with the glucose oxidase. The construction of the sensor is proprietary, but some general observations can be made. The electrodes are formed on a thin plastic substrate using what appears to be a thin film process used in the semiconductor industry for patterning metals (Figure 5.9). Platinum is plated on the surface of the working electrode and gives the electrode a black appearance. The outer plastic housing appears to protect the thin flexible plastic piece that the electrodes are printed on during the insertion process into the body. The thin plastic piece with the printed electrodes has a wider portion that resides outside the body (Figure 5.10). The wider portion acts as points of electrical contact to a cable leading to the CGMS System Gold

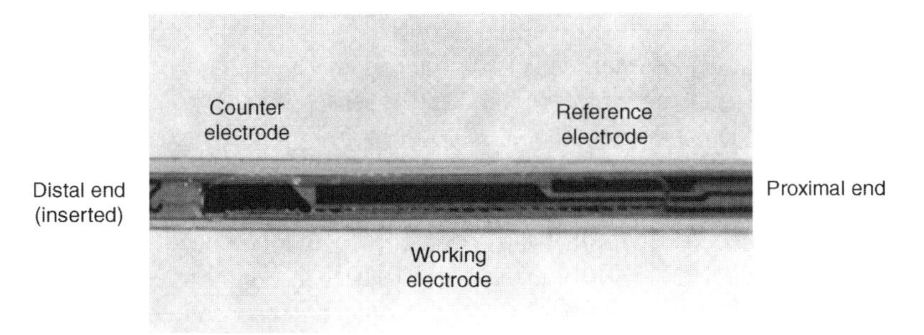

Figure 5.9 Close-up view of the Medtronic sensor showing the working, counter, and reference electrodes. Copyright 2008 Abbott. Used with permission. (See the color version of this figure in Color Plates section.)

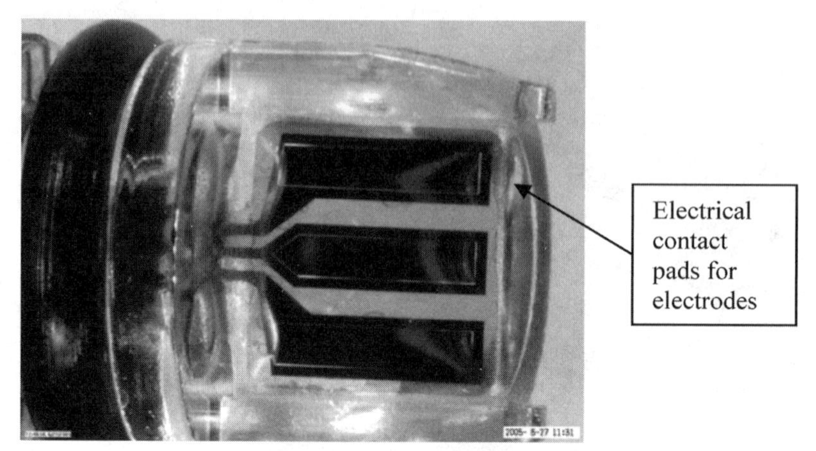

Figure 5.10 Close-up view of the Medtronic sensor's electrical contact pads for the working, counter, and reference electrodes. Copyright 2008 Abbott. Used with permission.

monitor or to the Guardian REAL-Time transmitter. Adhesive patches keep the sensor attached to the body for 3 days.

5.5.2 Dexcom

The company developed by Stuart Updike at the University of Wisconsin and under the name of Markwell Medical became DexCom in 1999. Although the company was originally started to commercialize a long-term implanted continuous glucose sensor, it later became focused on a subcutaneous sensor. Dexcom received FDA approval in 2006 to sell their 3-day continuous glucose monitoring system and in 2007 to sell their 7-day sensor. The 3-day device is named the DexCom short-term continuous glucose monitoring system (STS™) to differentiate it from the long-term system (LTS™) and the 7-day device is named the SEVEN™ STS®. The device was approved to be used with blood glucose strips, so the users do not solely rely on the STS glucose result for their treatment. This is referred to as an adjunctive use of the device. The device requires a prescription.

The general use of the system will be described and then a more detailed description of the sensor and how it is inserted into the skin will be provided. The sensor comes to the patient in a sterile package that contains a patch for adhering the STS sensor and STS transmitter to the skin and a sensor insertion device, STS applicator (Figure 5.11). The patch is applied to the skin by removing plastic tabs that cover the adhesive in a manner similar to an adhesive bandage. The user then pushes on the inserter and then pulls on the inserter, thus driving an approximately 24 gauge needle into the skin with the sensor contained within the needle and then removing the needle leaving the sensor in the skin. The used inserter is removed from the patch and discarded (Figure 5.12). The STS transmitter is inserted into the patch that causes it to make electrical contact with the sensor (Figure 5.13). The transmitter is powered by a nonreplaceable battery. The transmitter is in radio frequency

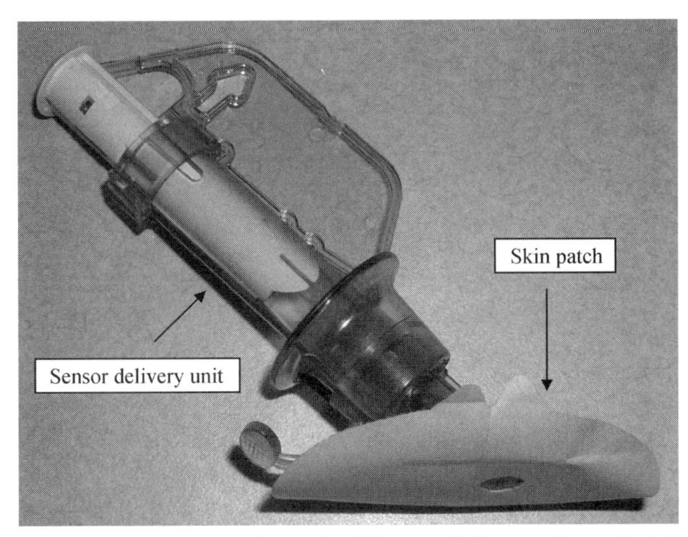

Figure 5.11 Dexcom skin patch with sensor delivery unit attached. Copyright 2008 Abbott. Used with permission.

communication with the STS receiver (Figure 5.14). The glucose results are displayed on the receiver.

Once the sensor is inserted, the user is instructed to wait 2 h before they are prompted to calibrate the sensor. The user must run duplicate blood glucose measurements using a specific brand of glucose meter and download the fingerstick data into the receiver. The calibrated sensor will transmit data from the transmitter to the receiver every 5 min. The data are stored inside the receiver and can be displayed

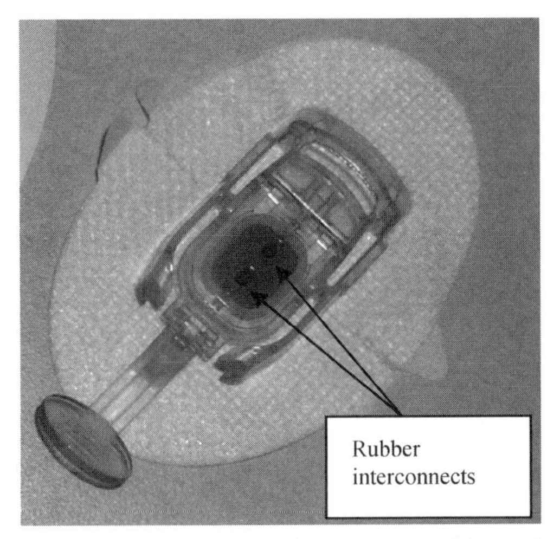

Figure 5.12 Top view of the Dexcom skin patch with the sensor delivery unit removed. Copyright 2008 Abbott. Used with permission.

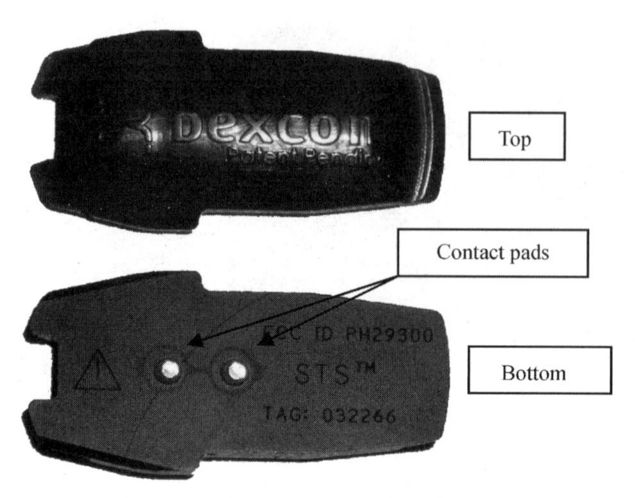

Figure 5.13 Top and bottom views of the Dexcom transmitter that is inserted into the skin patch. Copyright 2008 Abbott. Used with permission.

on the receiver's screen in a number of ways. The sensor must be recalibrated using a blood glucose measurement at least every 12 h. After 3 or 7 days depending on the version of the product, the sensor and patch are removed and discarded but the transmitter is saved and used for the next insertion.

The electrodes inserted under the skin consist of a thin platinum wire and a silver wire with a silver chloride coating on the outside (Figure 5.15). The end of the platinum wire that is inserted under the skin is coated with an insulating coating except at the tip. At the tip of the wire some of the insulating coating has been removed and a membrane is visible. The exposed membrane contains glucose oxidase and is the site where the reaction occurs with glucose. The exact composition of the membrane is proprietary. A small distance further up the insulated platinum wire, the silver wire is wrapped around the platinum wire. The ends opposite to those in the skin of the platinum and silver wires do not go into the skin. At this end, the platinum wire does

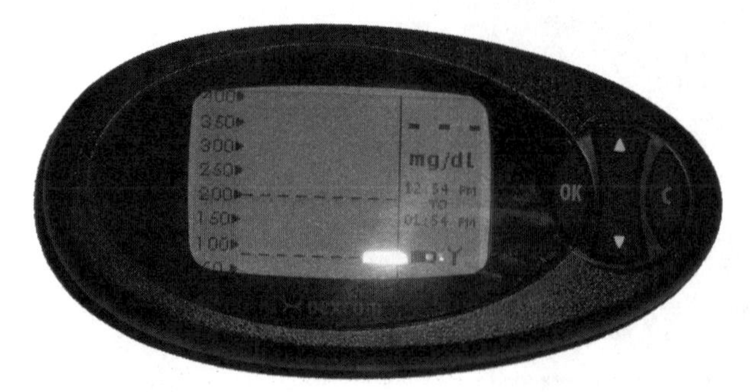

Figure 5.14 Top view of the Dexcom receiver unit. Copyright 2008 Abbott. Used with permission.

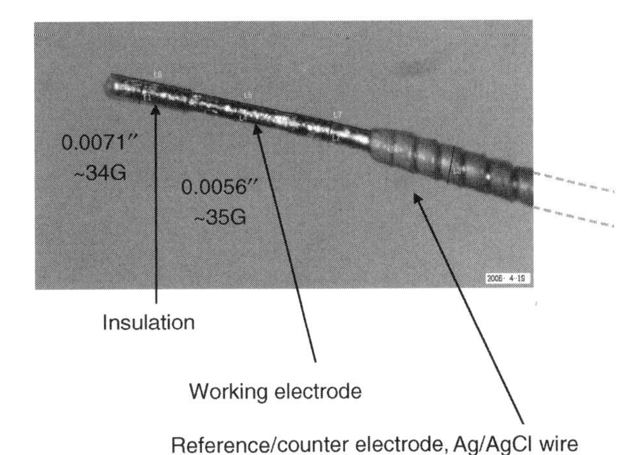

Figure 5.15 Close-up view of the tip of the Dexcom sensor that is inserted under the skin. Copyright 2008 Abbott. Used with permission.

not have an insulting coating and the silver wire does not have a silver chloride coating (Figure 5.16). The ends of these wires are used to make electrical contact with the STS transmitter.

The STS inserter places the platinum and silver electrodes into the skin and makes electrical contact between the electrodes and the adhesive patch. A detailed explanation is required to understand how the STS inserter works (Figure 5.17). In the unused inserter, the two electrodes are inside the barrel of the needle. When the user presses on the inserter, the needle, with electrode inside, first pierces through two conductive rubber pads housed in the patch. The insertion occurs at an approximately 45° angle to the skin and the conductive rubber pads are also angled at 45°. The needle continues to pierce through the skin to a depth of approximately 1.2 cm. The user then pulls back the inserter that withdraws the needle. The electrodes remain in the skin when the needle is withdrawn because a cylinder fills the bore of the needle upon retraction preventing the electrodes from retracting with the needle. Finally, the cylinder is also fully removed. When the needles and cylinder are removed, the conductive rubber pads that were pierced by the needle collapse around the ends of the electrodes. The ends of the electrodes have different lengths so that each end makes electrical contact with only one of the conductive

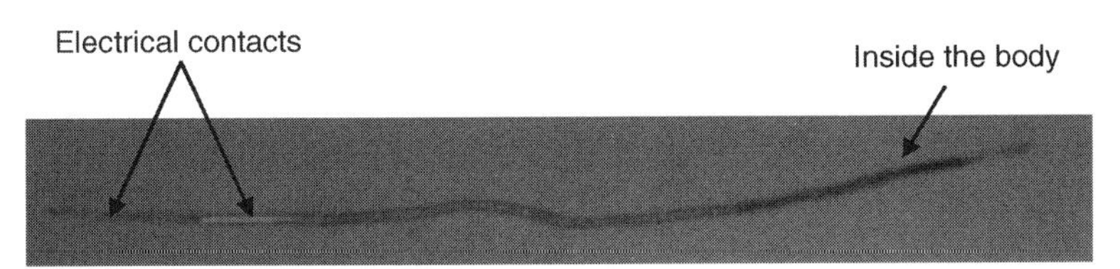

Figure 5.16 Complete view of the Dexcom sensor. Copyright 2008 Abbott. Used with permission.

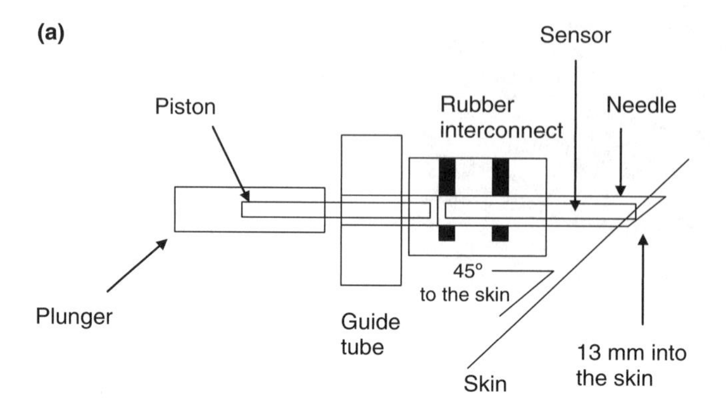

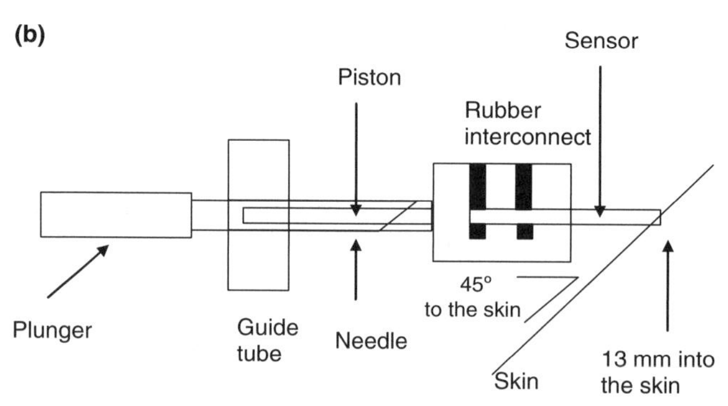

Figure 5.17 Schematic representation of the Dexcom inserter. (a) Inserting the sensor into the skin within the needle. (b) Retracting the needle leaving the sensor in the skin and in contact with rubber interconnects. Copyright 2008 Abbott. Used with permission.

rubber pads. The inserter is removed from the patch and the conductive rubber pads move to a position horizontal to the skin. Since the electrodes are thin wires, they easily bend to accommodate the movement of the pads and are not pulled out of the skin. When the transmitter is placed into the patch, it makes electrical contact with the rubber pads, which means the electrodes are now in electrical contact with the transmitter (Figure 5.18).

The DexCom membrane contains glucose oxidase that reacts with glucose in the presence of oxygen to produce hydrogen peroxide. The hydrogen peroxide is oxidized at the metal surface of the working electrode, which means it is a first-generation sensor. The transmitter applies a potential that is generally 0.5–0.7 V between the working electrode and the silver/silver chloride wire. An equal number of electrons flowing into the working electrode flow out of the silver/silver chloride wire and cause the reduction of silver chloride to silver. The electrode measurement is a two-electrode system. Because the silver chloride coating is consumed, there must be enough silver chloride to react with all the electrons being produced at the working electrode for the lifetime of the sensor. This necessitates the use of a large surface area

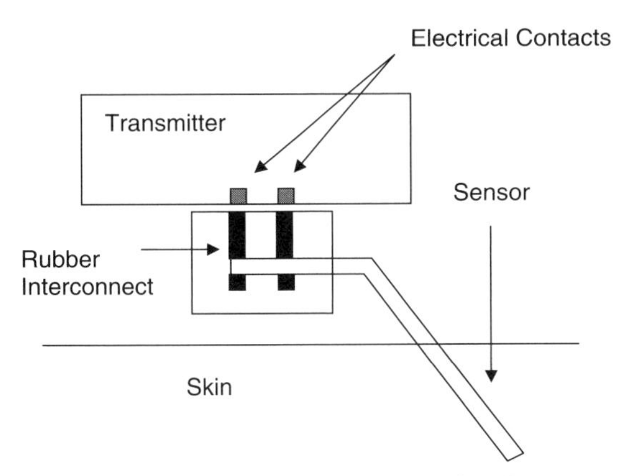

Figure 5.18 Schematic representation of the Dexcom sensor and transmitter mounted on the skin. Copyright 2008 Abbott. Used with permission.

silver chloride coating when compared to the surface area of the working electrode. Depletion of the silver chloride layer would mean that the reference counter electrode would not have a defined potential and would then not act as a proper reference electrode for the system. The Dexcom system carries a warning about the use of acetaminophen (active ingredient in Tylenol) when using the device. This seems to indicate that the system operates at a potential where acetaminophen can be oxidized at the working electrode and that the working electrode cannot prevent acetaminophen from reaching the metal surface of the working electrode. Preventing acetaminophen from reaching the metal surface is difficult because acetaminophen is uncharged and so it cannot be stopped by ionic membranes. Cation exchange membranes are often used in glucose electrodes because the fixed anionic sites of the membrane repel anionic interferents found in the body, primarily urate and ascorbate, but will not stop acetaminophen.

5.5.3 Abbott

TheraSense was founded in 1996 to develop the glucose monitoring technology invented by Dr. Adam Heller at the University of Texas. TheraSense was purchased by Abbott Laboratories in 2004 and combined with their MediSense business unit to form Abbott Diabetes Care Inc. The FreeStyle Navigator® continuous glucose monitoring system was approved by the FDA in 2008. The FreeStyle Navigator system is available by prescription to display glucose values for up to 5 days. The FreeStyle Navigator receiver continuously displays a glucose result, which with fingerstick blood glucose measurements can be used by people with diabetes to guide their treatment.

The sensor comes to the patient in a sterile package that contains a sensor support mount for adhering the sensor and transmitter to the skin and a sensor insertion device. The mount is applied to the skin in a similar manner to apply an

(a)

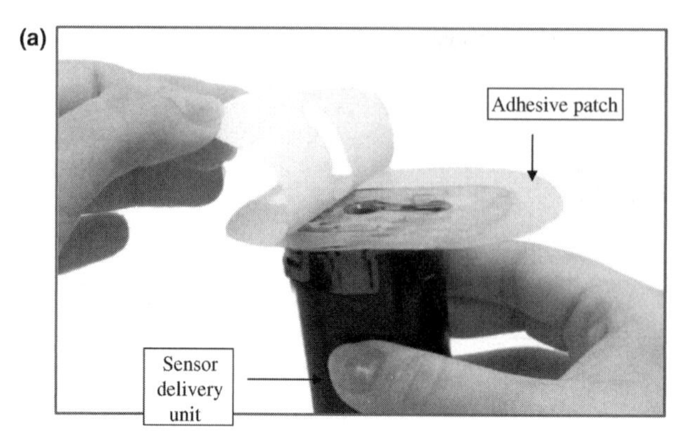

(b)

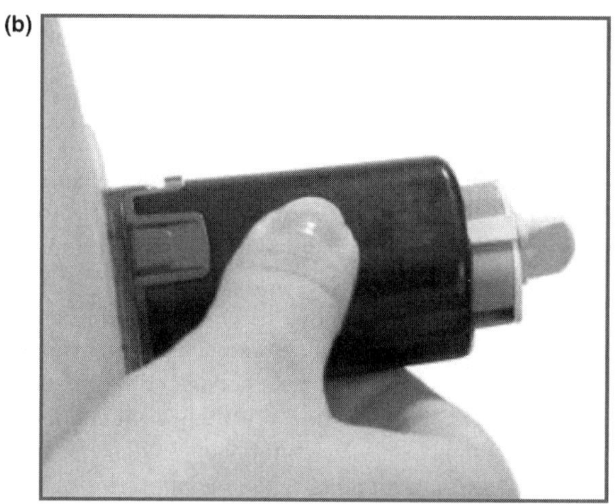

(c)

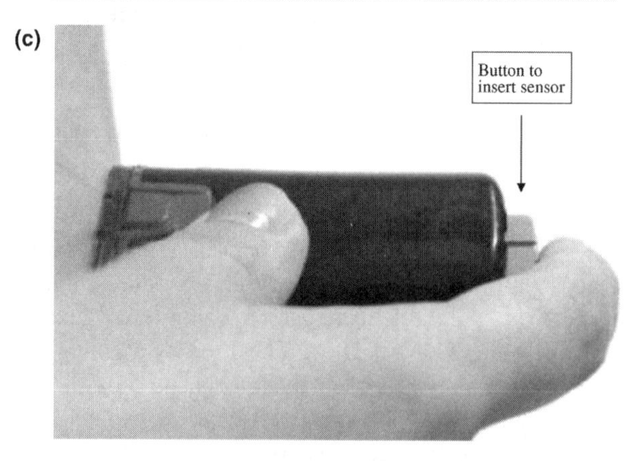

Figure 5.19 Application of the Abbott system to the body. (a) Removing adhesive from patch. (b) Applying patch and sensor delivery unit to the skin. (c) Pressing button to insert sensor. (d) Removing sensor delivery unit. (e) applying transmitter. Copyright 2008 Abbott. Used with permission.

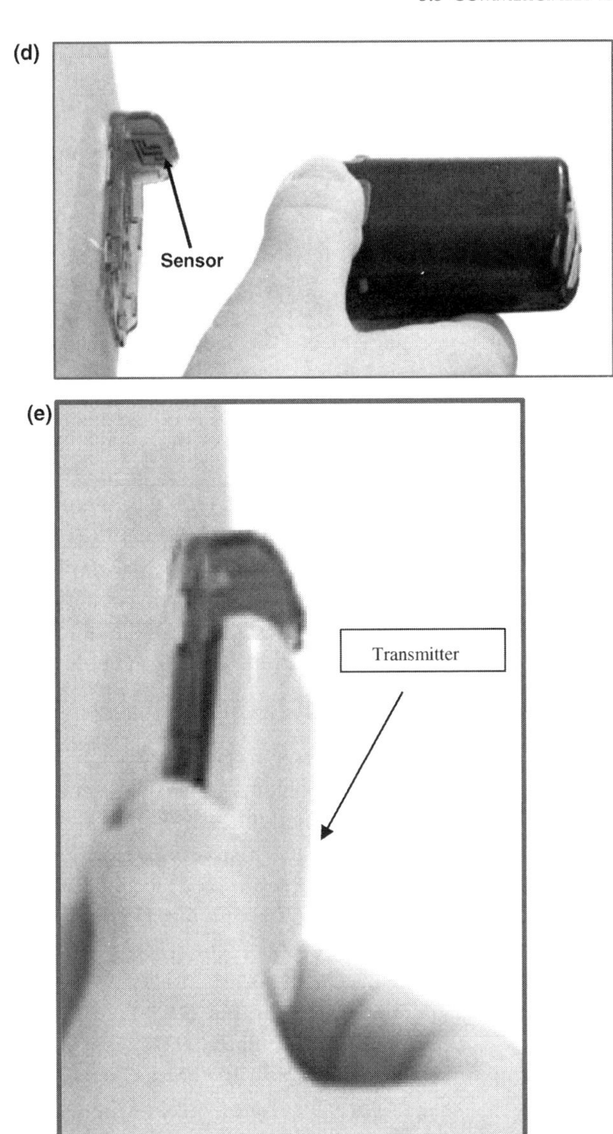

Figure 5.19 *(continued).*

adhesive bandage (Figure 5.19a and b). The user then removes a locking pin on the inserter and presses the insertion button (Figure 5.19c). The inserter is spring loaded and moves a needle with a sensor held with the needle into the skin. The needle is automatically withdrawn from the skin while the tip of the sensor remains in the skin.

A large portion of the sensor remains outside the skin and is used to provide the electrical contacts to the sensor. The inserter is removed from the patch and discarded (Figure 5.19d). A transmitter device is slid into the sensor support mount (Figure 5.19e). The transmitter contains electrical contacts that contact the conductive pads on the sensor. The transmitter contains a circuit that measures the electrochemical reaction occurring at the working electrode and transmits the signal information to a receiver. The transmitter measures the skin temperature, which is used in the calculations of interstitial glucose. All electrochemical sensor have some temperature dependence due to both the thermodynamics of the electrode processes and because diffusion rates are temperature dependent. The receiver displays the present glucose result and is updated every 1 min. The receiver stores a glucose value in its memory every 10 min.

The system does not display a glucose result for the first 10 h of operation to allow the equilibration of the sensor inside the body. The user is instructed to calibrate the sensor at hours 10, 12, 24, and 72, after insertion, by a blood glucose measurement. The blood glucose meter is built into the receiver, so the calibration is done automatically when a blood glucose measurement is made. After 5 days the user is instructed to remove and dispose of the sensor support mount. The transmitter is reusable and contains replaceable batteries. The transmitter and mount are water resistant and can be worn during showers. The receiver is not water resistant because the receiver contains the open port for insertion of a blood glucose test strip.

The tip of the sensor is inserted at an angle of approximately 90° to the skin to a depth of about 5 mm (Figure 5.20). The sensor contains electrodes that are patterned by screen printing. The electrode system functions as a three-electrode system with a working electrode, counter electrode, and a silver/silver chloride reference electrode. The working electrode contains glucose oxidase bound to a polymer that contains mediator molecules also bound to the polymer (Figure 5.2). Glucose oxidase reacts with the glucose and the reduced enzyme reacts with the mediator in the polymer. The reduced mediator reacts with other mediator sites in the polymer until eventually a reduced mediator near the metal surface of the electrode is oxidized by the metal surface poised at a low potential, 0.04 V, resulting in a current. The reaction uses an exogenous mediator, so the sensor is second generation. The use of a mediator lessens the requirement for oxygen, so the sensor can operte in low-oxygen environments better than first-generation sensor that depend on oxygen as a coreactant. Lessening the oxygen dependence also lessens the need to slow down the glucose diffusion because of the high concentration of glucose compared to oxygen. The current density of the Navigator sensor can be significantly higher than first-generation sensors because the diffusion of glucose does not need to be slowed down as much to accommodate oxygen. The low potential that is a function of the mediator potential means that common electrochemical interferences found in blood will not react at the working electrode. This eliminates the need for complex membranes to reduce or eliminate the diffusion of interferences to the metal surface that are required for first-generation sensor designs.

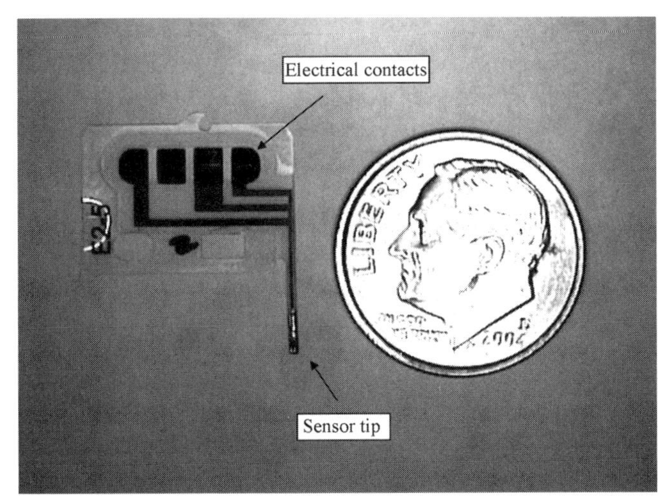

Figure 5.20 Abbott sensor showing size relative to U.S. dime. Copyright 2008 Abbott. Used with permission.

5.6 CLINICAL PERFORMANCE OF COMMERCIALLY AVAILABLE CGM PRODUCTS

5.6.1 Typical Performance

Blood glucose monitors have been commercially available for over 20 years. Specific international standards have been developed for the clinical performance of blood glucose monitors. The clinical testing required to get such a monitor through regulatory bodies and bring them to the market is fairly well understood. Continuous glucose monitors are much younger and international standards for clinical performance have not been established. The manufacturers of these new devices face a challenge in designing clinical trials to prove the efficacy of these new devices. CGM devices can generate hundreds to thousands of data points per day compared to the 2–4 blood glucose measurements that a typical person with diabetes generates daily. CGM devices also have the capability to set off alarms when abnormal glucose values are detected or even projected to happen. The accuracy of these alarms is another new area not found in blood glucose monitors and for which no standards exist. This implies that the field of clinical testing of CGM devices is evolving as the technology evolves. Making comparisons between the different manufacturers' clinical trials is challenging because the trials are run differently, but at present there is no better way of looking at relative performance. In the future we will probably see head-to-head comparisons of the different devices in a clinical trial, which will improve our understanding of the relative performance of the devices.

Two types of clinical trials are typically performed by CGM manufacturers for submission to regulatory bodies to obtain approval to sell their devices. The first type of trial is generally called the accuracy trial. Clinical trial subjects are placed in a clinic or hospital for a large portion of their day. The CGM device is calibrated by taking capillary blood and measuring the blood on a glucose meter. The times at which the calibration occurs are specified by the manufacturer. Blood samples are taken from the subjects on a regular basis and analyzed on a laboratory analyzer that is considered to be very accurate. The glucose analyzers made by YSI are considered to be the "gold standard" for whole-blood glucose analysis and are frequently used as reference instruments. Because the blood sampling must be made frequently and the amount of blood required for the YSI analyzer is not as small (few hundred microliters) as required by a blood glucose meter, a catheter is normally placed in a vein for blood sampling. The blood is sampled at the same time as the CGM device reports a glucose result, so the two glucose results are paired together. The clinical trial attempts to generate glucose values over as wide a range as is safely possible. During the several day trials, the subject may be asked to eat and delay their insulin shot to create a temporary high glucose reading as well as inject a larger amount of insulin than normal to create a temporary low glucose reading. The subjects are in a clinic at the time of the glucose extremes, so they are closely monitored by the study staff to prevent any adverse events. The YSI glucose values can be plotted versus the CGM glucose values resulting in a scatter plot that can be used to access the accuracy of the new device using a number of different metrics.

The second main type of clinical trial is the home use trial. Subjects use the CGM devices at home as they would expect a normal customer to use the device. They wear a device for the intended length of time and then start a new device. The subject may do this for 21 days to 6 months. They record their glucose values from capillary glucose measurements made on glucose meters. The capillary glucose results are paired with continuous glucose results made at the same time. The paired glucose values are plotted in a scatter plot and standard metrics are used to access the accuracy of the device. Home glucose meters do not have the accuracy and precision of a laboratory analyzer such as the YSI glucose analyzer, so more error is introduced into the home use trial by the reference instrument than in the accuracy trial. The home use trial is also conducted under normal living conditions and not in a well-controlled clinic situation, so the control over the trial is less leading to an expected lower accuracy.

The different manufacturers publish their own results in their user manuals. Mean absolute relative difference and bias results from the three manufacturers are shown in Table 5.2. The MARD measures indicates the average difference while the direction of the difference and the bias indicates if the differences are uniform or skewed to positive or negative values. The Clarke error grid analysis for the three manufacturers (Table 5.3) shows a wide difference in A zone results between the

TABLE 5.2 Accuracy Determined by Relative Difference and Bias

CGM system	Number of subjects	Number of points	Mean absolute relative difference	Bias
Guardian RT	16	3941	19.7%	−15.0 mg/dL
Dexcom STS	91	676	26%	~11%
FreeStyle Navigator	56	20362	12.8%	0.8 mg/dL

TABLE 5.3 Clarke Error Grid Analysis

CGM system	A	B	C	D	E
Guardian RT	61.7	34.4	0.2	3.5	0.2
Dexcom STS	49	41	4	7	0
FreeStyle Navigator	81.6	16.7	0.1	3.0	0.01

TABLE 5.4 Guardian RT Clarke Error Grid by Glucose Range

Glucose range (mg/dL)	Number of data points	A	B	C	D	E
40–80	356	60.1	16.0	0.6	22.5	0.8
81–120	769	60.2	39.7	0.1	N/A	N/A
121–240	2362	62.5	37.	0.2	N/A	0.2
> 240	454	61.0	25.8	0.2	3.5	0.2

N/A: This zone is not in the Clarke error grid for that glucose range.

different manufacturers. Tables 5.2 and 5.3 summarize the most common way of looking at the data, but the following examples show that the same raw data can be looked at in many different ways. The Clarke error grid results for the Guardian RT are shown by glucose range in Table 5.4. The table shows the challenge in the low glucose range with data points falling in the D zone. As can be seen from the grid (see Figure 1.7), the A zone becomes narrow around 75 mg/dL and there is no B zone between the A and D zone on the upper side of the A zone. In the low glucose range, D zone values are easier to reach than other portions of the grid. The FreeStyle Navigator system grid results by day of wear are shown in Table 5.5. The day-to-day performance is very consistent, indicating that performance does not deteriorate over the 5 days of wear. The sensor lifetime as a function of hours of wear is shown in Table 5.6 for the Dexcom STS device. The table shows how long a sensor is expected to last for that time window. For example 8% of the sensors lasted for 4–18 h. The sensor lifetime from the manufacturers, 3–7 days, is an expression of how long a typical sensor will last, but there is no guarantee that each sensor will last the full time. As can be seen from Table 5.6, it cannot be predicted as to how long a sensor will continue to function inside the body.

TABLE 5.5 FreeStyle Navigator System Clarke Error Grid by Day of Wear

Day of wear	A	B	C	D	E
Day 1	82.5	16.4	0.2	0.9	0.0
Day 2	82.4	16.6	0.1	0.9	0.1
Day 3	79.4	18.3	0.0	2.2	0.0
Day 4	84.0	14.2	0.0	1.8	0.0
Day 5	80.9	16.9	0.0	2.1	0.0

TABLE 5.6 Dexcom STS Estimated Sensor Lifetime

Hours of use	Percentage of sensors
4–18 h	8%
20–35 h	6%
45–64 h	11%
65–72 h	75%

TABLE 5.7 Precision of CGM Systems

CGM system	Number of data points	Coefficient of variation
Guardian RT	11,475	17.2%
Dexcom STS	N.A. (not available)	21%
FreeStyle Navigator	312,953	10%

Precision is normally one of the easiest performance parameters to measure for an *in vitro* medical diagnostic test. Multiple tests are run using the same sample and the standard deviation is calculated. Precision is not so easily tested for *in vivo* medical diagnostic tests. If we want to perform the normal precision test on an *in vivo* device, we would need to insert multiple devices in the same person. This would introduce the question of the glucose being the same at every spot in the body. The best that the CGM manufacturers have been able to do is to place two CGM devices in the same person and compare the results from the two sensors for precision. Because both devices are in the same person, they should read the same glucose level. This *in vivo* precision measurement is not the same as the *in vitro* measurement. The *in vivo* measurement can produce thousands of paired points from one subject, but are the data points independent because they are coming from the same person? The precision data from the three manufacturers are shown in Table 5.7.

5.6.2 Performance When Glucose is Changing Rapidly

Glucose passes fairly freely between the blood capillary walls and the interstitial space. In nonhospitalized subjects, glucose enters the blood from the digestive tract or is released from the liver. Glucose is consumed by the cells of the body as long as the insulin is present. A complex dynamics of glucose entering the blood stream and being used up by the cells exists. Blood glucose meters measure the glucose in capillary blood, while CGM devices measure glucose in interstitial fluid. When the blood glucose is not changing, it is expected that the interstitial glucose will be approximately the same as the blood glucose. When blood glucose is changing rapidly, the interstitial space may not be able to equilibrate fast enough to match the blood glucose. There may then exist conditions under which the blood glucose values measured using a blood glucose meter will not agree with the interstitial glucose

TABLE 5.8 Accuracy of FreeStyle Navigator System by Rate of Glucose Change

Rate of change (mg/dL/min)	Percentage of data points	Clarke error grid zone				
		A (%)	B (%)	C (%)	D (%)	E (%)
Less than −2	3.1	54.6	42.3	1.3	1.8	0.0
−2 to −1	8.8	71.7	26.2	0.3	1.8	0.0
−1 to 1	74.7	84.9	13.5	0.0	1.5	0.0
1–2	10.0	79.8	18.9	0.0	1.3	0.0
>2	3.5	63.5	34.7	0.0	1.7	0.0

measurement made with the CGM device. The situations are typically temporary and the two will soon agree. The important consideration is that calibration of the CGM with blood glucose concentrations may lead to a poor calibration of the CGM device when blood glucose is changing rapidly. Since it is known that blood glucose will rise rapidly after a meal, this is not a recommended time for calibration. The overall accuracy of the blood glucose and interstitial glucose values is also influenced by changing blood glucose values. The accuracy at various rates of change has been calculated and as expected the accuracy is best when the blood glucose is not changing rapidly (Table 5.8). Two manufacturers also display arrows that indicate the rate of glucose change of the sensor. These arrows will provide a helpful indicator to the user of when to calibrate.

5.7 FUTURE OF CGM

5.7.1 Long-Term Implanted Sensors

The CGM sensors described in this chapter have a lifetime of 3–7 days inside the body. The sensors will, therefore, need to be changed by the user on a regular basis if they want to know their glucose concentrations at all times. A sensor that could be implanted for a long period of time would eliminate the need to frequently change the sensor. The development of a long-term sensor is however very challenging. Sensors based on the reaction of glucose with glucose oxidase leading to the passage of current will likely require large batteries to be implanted with the sensor if they are expected to last for a year. The larger the device, the more the damage will be caused by the implantation procedure. Larger devices may also require surgical implantation that is not very appealing for the users or payers for the technology. A larger device may also be more prone during movement to break blood capillaries and cause internal bleeding that is believed to be bad for sensor performance. In addition, most enzymes such as glucose oxidase are not very stable at elevated temperatures, such as inside the body, and for long periods of time. Because of the preceding reasons, more effort has been put into developing small optical sensors that can be implanted under the skin. The proximity of the sensor to the surface of the skin means the light excitation and light transduction devices can be outside the skin, which minimizes the size of the device that needs to be implanted under the skin. A device that includes the measurement technology to also be implanted will present miniaturization challenges. Such a device would need telemetry capability to be able to transmit the signal outside the body to a receiver. The devices typically do not rely on an enzyme to transduce the glucose. The sensor technologies often involve a glucose binding event that changes an optical property of the sensor. Glucose binding sites, such as conconavalin A, are more stable than most enzymes. The optical change could be a fluorescent signal or an optical interference change, or possibly even a color change. All the systems, whether optical or electrochemical, will face long-term problems with the body capsule formed around the sensor. The body will initially try to destroy the implanted sensor but once the body gives up trying to destroy the sensor, it will start walling off the sensor with a thick fibrous layer. The thick fibrous layer will slow down the diffusion of glucose to the sensor and lead to long response times. Efforts have been made to promote the ingrowth of capillaries around an implant by using textured surfaces or by releasing substances that promote capillary growth around the implant.

5.7.2 Tight Glycemic Control in the Hospital

The body responds to serious illness or trauma by allowing the blood glucose level to rise significantly, hyperglycemia. This occurs in patients with diabetes and patients without diabetes. Hyperglycemia was thought to be a natural response of the body to try and heal itself and went untreated in the hospital unless it was severely high. Clinical studies were begun in the early 1990s in patients undergoing heart surgery where the glucose level was left high in some patients and other patients received treatment to lower their blood glucose.[39] Surprisingly, the group that received the glucose lowering treatment had a significant reduction in infection after the surgery. It was not known if the effect was specific to patients undergoing heart surgery or would benefit the general hospital population. Another study was done with hospitalized diabetic patients that involved glucose lowering treatment in the hospital and after their release from the hospital.[40] The mortality rate was nearly half for the treated group and again the number of infections in the treated group was reduced. In 2001, Dr. Greet van den Berghe published a controlled clinical trial studying the effect of reducing the glucose level in hospitalized ICU patients to near normal levels.[41] A statistically significant lowering of the mortality rate was found in those studies. Although the results are controversial, Dr. van den Berghe suggests that the nearer to normal the glucose can be maintained, the better off the patient is. In most of the published studies from the hospital intensive care units, the lowering of glucose levels seems to benefit the nondiabetic patients more than the diabetic ones.

Since 2001 most hospitals have decided to aggressively reduce glucose levels in their ICU patients but this is not easy for the hospital to do so and does carry with it health risks to the patient. Reducing the glucose level requires frequent blood glucose determinations and insulin adjustments. A commonly quoted statistic is that each patient under strict glucose control requires an additional 2 h of nursing time per day. Lowering the blood glucose level brings with it the risk that the glucose level can be lowered too far and the patient will become hypoglycemic. Hypoglycemia increases the risk of significant complication and can cause death. Two recent European clinical studies involving strict glycemic control have been suspended because of higher mortality rates in the strict glucose control arm of the studies. There is at present a significant need for a technology that can continuously monitor the patient and alarm the staff to any low or high glucose levels. Such a sensor would reduce staffing costs and lower the risk of hypoglycemia for strictly controlled patients.

Significant challenges exist for the implementation of current continuous glucose sensors in the ICU. Clinical studies are needed that demonstrate that the subcutaneous tissue where the current sensors reside in will accurately track blood glucose values in severely ill patients. Severely ill patients can experience edema that might affect subcutaneous glucose values by diluting the glucose. Microcirculation can be shut down in patients in shock as well as by various vasodilating drugs. The decrease in peripheral circulation could lead to less glucose transport to the subcutaneous space around the sensor and glucose values lower than the blood. One solution is to adapt the continuous sensors for use in blood. The sensor could be placed directly in an artery or blood could be sampled from an artery using a pump. Once such system was sold but was not commercially successful. Using the sensors directly in blood has its own problems associated with getting access to the blood and clotting. The need for a continuous glucose sensor in critical care is so great that companies will be driven to provide technical solutions in the near future.

5.7.3 Closed-Loop Control of Glucose

The ultimate use for continuous glucose monitoring would be to combine the glucose information with an insulin pump to provide a system that automatically maintains the user's glucose level within prescribed limits. The sensor and pump are combined in a "closed loop" to mimic the function of the pancreas and therefore form an "artificial pancreas." The performance of continuous glucose sensors has been steadily improving and should reach the state where they could be relied on to provide guidance for the pump. Current insulin pumps deliver insulin with the precision and accuracy necessary for the closed loop. A third vital element needed is the algorithm to translate the glucose values into the correct insulin infusion instructions for the pump.

Many factors contribute to the current glucose level in your body as well as what it will be in the future. How big a meal will you eat and how many carbohydrates will it contain? When do you plan on exercising and for how long and how strenuous? Are you under stress today at work? The glucose sensor has no way of knowing any of these things in advance, so it is always reacting to changes in your condition but cannot anticipate those changes. The glucose sensor and pump will always be chasing to control the glucose. One suggestion is that the user will need to input information, such an upcoming meal, into the control algorithm so that the system can react, pump insulin, ahead of time. This is referred to as "open-loop" control. Another possibility is that a second drug will be added to the pump that raises glucose. The current systems using only insulin can only lower glucose. Infusion of a drug such as glucagon causes an increase in glucose by releasing glucose that is stored in the liver. The closed-loop system would then have the power to both raise and lower glucose that would make closed-loop control easier. There is, however, a limited supply of glucose, enough for 1–2 days, stored in the liver, so there is some concern that a system that requires significant corrections using glucagon will end up depleting the glucose in the liver. If that happens, then the closed-loop system can only lower glucose levels because the glucagon will not be effective. The future looks very challenging and exciting for closed-loop control of glucose.

REFERENCES

1. Bode BW, Schwartz S, Stubbs HA, Block JE. Glycemic characteristics in continuously monitored patients with type 1 and type 2 diabetes: normative values. *Diabetes Care* 2005, 28, 2361–2366.
2. Henning TP, Cunningham DD. *Commercial Biosensors: Applications to Clinical, Bioprocess, and Environmental Samples*. John Wiley & Sons, New York, 1998.
3. Suri JT, Cordes DB, Cappucio FE, Wessling RA, Singaram B. Continuous glucose sensing with a fluorescent thin-film hydrogel. *Angewandte Chemie* 2003, 115, 6037–6039.
4. Russsell RJ, Pishko MV, Gefrides CC, McShane MJ, Cote GL. A fluorescence-based glucose biosensor using concanavalin A and dextran encapsulated in a poly(ethylene glycol) hydrogel. *Analytical Chemistry* 1999, 71, 3126–3132.
5. Nichols RJ, Cote GL. Optical glucose sensing in biological fluids: an overview. *Journal of Biomedical Optics* 2000, 5, 5–16.
6. Marvin JS, Hellinga HW. Engineering biosensors by introducing fluorescent allosteric signal transducers: construction of a novel glucose sensor. *Journal of the American Chemical Society* 1998, 120, 7–11.
7. Benson DE, Conrad DW, de Lorimer RM, Trammell SA, Hallinga HW. Design of bioelectronic interfaces by exploiting hinge-bending motion of proteins. *Science* 2001, 293, 1641–1644.
8. de Lorimier RM, Tian Y, Hellinga HW. Binding and signaling of surface-immobilized reagentless fluorescent biosensors derived from periplasmic binding proteins. *Protein Science* 2006, 15, 1936–1944.
9. Ballerstadt R, Evans C, Gowda A, McNichols R. *In vivo* performance evaluation of a transdermal near-infrared fluorescence resonance energy transfer affinity sensor for continuous glucose monitoring. *Diabetes Technology & Therapeutics* 2006, 8, 296–311.
10. Khalil O. Spectroscopic and clinical aspects of noninvasive glucose measurements. *Clinical Chemistry* 1999, 45, 165–177.
11. Khalil O. Non-invasive glucose measurement technologies: an update from 1999 to the dawn of the new millennium. *Diabetes Technology & Therapeutics* 2004, 6, 660–697.
12. Cass AE, Davis G, Francis DD, Hill HA, Aston WJ, Higgins IJ, Plotkin EV, Scott DL, Turner AP. Ferrocene-mediated enzyme electrode for amperometric determination of glucose. *Analytical Chemistry* 1984, 56, 667–671.
13. Bard AJ, Faulkner LR. *Electrochemical Methods*. John Wiley & Sons, New York, 1980.
14. Habermuller K, Mosbach M, Schuhmann W. Electron-transfer mechanisms in amperometric biosensors. *Fresenius Journal of Analytical Chemistry* 2000, 366, 560–568.
15. Chaubey A, Malhotra BD. Mediated biosensors. *Biosensors & Bioelectronics* 2002, 17, 441–456.
16. Pickup JC, Shaw GW, Claremont DJ. Implantable glucose sensors: choosing the appropriate sensing strategy. *Biosensors* 1987/1988, 3, 335–346.

17. Pickup JC, Shaw GW, Claremont DJ. *In vivo* molecular sensing in diabetes mellitus: an implantable glucose sensor with direct electron transfer. *Diabetologia* 1989, 32, 213–217.
18. Shaw GW, Claremont DJ, Pickup JC. *In vitro* testing of a simply constructed, highly stable glucose sensor suitable for implantation in diabetic patients. *Biosensors & Bioelectronics* 1991, 6, 401–406.
19. Heller A. Electrical connection of enzyme redox centers to electrodes. *Journal of Physical Chemistry* 1992, 96, 3579–3587.
20. Wilner I, Heleg-Shabtai V, Blonder R, Katz E, Tao G. Electrical wiring of glucose oxidase by reconstitution of FAD-modified monolayers assembled onto Au-electrodes. *Journal of the American Chemical Society* 1996, 118, 10321–10322.
21. Xiao Y, Patolsky F, Katz E, Hainfeld J, Willner I. "Plugging into enzymes": Nanowiring of redox enzymes by a gold nanoparticle. *Science* 2003, 299, 1877–1881.
22. Sotiropoulou S, Chaniotakis N. Carbon nanotube array-based biosensor. *Analytical and Bioanalytical Chemistry* 2003, 375, 103–105.
23. Liu S, Ju H. Reagentless glucose biosensor based on direct electron transfer of glucose oxidase immobilized on colloidal gold modified carbon paste electrode. *Biosensors & Bioelectronics* 2003, 19, 177–183.
24. Heller A. Plugging metal connectors into enzymes. *Nature Biotechnology* 2003, 21, 631–632.
25. Updike SJ, Hicks GP. The enzyme electrode. *Nature* 1967, 214, 986–988.
26. Updike SJ, Shults MC. Method for monitoring blood glucose levels and elements, U.S. Patent 4,240,438, 1980.
27. Updike SJ, Shults M, Ekman B. Implanting the glucose enzyme electrode: problems, progress, and alternative solutions. *Diabetes Care* 1982, 5, 207–212.
28. Shults MG, Capelli CC, Updike SJ. Biological fluid measuring device, U.S. Patent 4,994,167, 1991.
29. Updike SJ, Shults MC, Capelli CC, von Heimburg D, Rhodes RK, Joseph-Tipton N, Anderson B, Koch DD. Laboratory evaluation of new reusable blood glucose sensor. *Diabetes Care* 1988, 11, 801–807.
30. Johnson KW, Mastrototaro JJ, Howey DC, Brunelle RL, Purden-Brady PL, Bryan NA, Andrew CC, Rowe HW, Allen DJ, Noffke BW. *In vivo* evaluation of an electroenzymatic glucose sensor implanted in subcutaneous tissue. *Biosensors & Bioelectronics* 1992, 7, 709–714.
31. Gilligan BJ, Shults MC, Rhodes RK, Updike J. Evaluation of a subcutaneous glucose sensor out to 3 months in a dog model. *Diabetes Care* 1994, 17, 882–887.
32. Updike SJ, Shukts MC, Rhodes RK, Gilligan BJ, Luebow JO, von Heimburg D. Enzymatic glucose sensors improved long-term performance *in vitro* and *in vivo*. *ASAIO Journal* 1994, 40(2), 157–163.
33. Updike SJ, Shults MC, Gilligan BJ, Rhodes RK. A subcutaneous glucose sensor with improved longevity, dynamic range, and stability of calibration. *Diabetes Care* 2000, 23, 208–214.
34. Giligan BJ, Shults MC, Rhodes RK, Jacobs PG, Brauker JH, Pintar TJ, Updike SJ. Feasibility of continuous long-term glucose monitoring from a subcutaneous glucose sensor in humans. *Diabetes Technology & Therapeutics* 2004, 6, 378–386.
35. Csoregi E, Quinn CP, Schmidtke DW, Lindquist S, Pishko MV, Ye L, Katakis I, Hubbell JA, Heller A. Design, characterization, and one-point *in vivo* calibration of a subcutaneously implanted glucose electrode. *Analytical Chemistry* 1994, 66, 3131–3138.

36. The Diabetes Research in Children Network (DirecNet) Study Group. Accuracy of the GlucoWatch G2 Biographer and the continuous glucose monitoring system during hypoglycemia. *Diabetes Care* 2004, 27, 722–726.

37. Cox DJ, Clarke WL, Gonder-Frederick L, Pohl S, Hoover C, Snyder A, Zimbelman L, Carter WR, Bobbitt S, Pennebaker J. Accuracy of perceiving blood glucose in IDDM. *Diabetes Care* 1985, 8, 529–536.

38. Clarke WL, Cox D, Gonder-Frederick L, Carter W, Pohl S. Evaluating clinical accuracy of systems for self-monitoring of blood glucose. *Diabetes Care* 1987, 10, 622–628.

39. Furnary AP, Kerr KJ, Grunkemeier GL, Starr A. Continuous intravenous insulin infusion reduces the incidence of deep sternal wound infection in diabetic patients after cardiac surgical procedures. *Annals of Thoracic Surgery* 1999, 67, 352–360.

40. Diabetes Mellitus Insulin Glucose Infusion in Acute Myocardial Infarction Study Group. Prospective randomized study of intensive insulin treatment on long term survival after acute myocardial infarction in patients with diabetes mellitus. *British Medical Journal* 1997, 314, 1512.

41. van den Berghe G, Wouters P, Weekers F, Verwaest C, Bruyninckx F, Schetz M. Intensive insulin therapy in the critically ill patients. *New England Journal of Medicine* 2001, 345, 1359–1367.

CHAPTER 6

MEMBRANE-BASED SEPARATIONS APPLIED TO IN VIVO GLUCOSE SENSING—MICRODIALYSIS AND ULTRAFILTRATION SAMPLING

Julie A. Stenken

In Vivo Glucose Sensing, Edited by David D. Cunningham and Julie A. Stenken

6.1 MEMBRANE SEPARATIONS FOR *IN VIVO* ANALYTE COLLECTION

6.1.1 Historical Development

Cells within the human body chemically communicate with each other to maintain homeostasis. This chemical communication process ranges from regulation of pH to complex protein signaling networks including hormones and cytokine proteins. Many life science researchers have an interest in quantifying selected chemical targets involved with either metabolic or communication processes within living systems. From an analytical chemistry perspective, if specific sensors were created for each analyte of interest, this would pose an extremely challenging, time-consuming, and expensive task. An alternative to this approach is to use a device that will allow access to the site of interest to collect a representative sample containing the targeted analyte. With this approach, analysis development time and effort are significantly reduced since many different instruments are available that provide selective, rather than specific, detection capabilities. An example is with the catecholamine neurotransmitters, dopamine and norepinephrine, which differ in chemical structure by only a hydroxyl group, but can be separated using liquid chromatography (a selective analytical technology) and then detected using amperometric electrochemical detection.

To perform an *in vivo* chemical analysis, an appropriate sampling scheme to reach the target analyte must be considered. For living systems, the most desirable analytical detection method would be a noninvasive measurement using a device such as the fictional Star Trek tricorder that seems to have the capability to detect every possible analyte. At this time, the analytical transduction technology for such an extremely useful device has not been created. There are noninvasive methods including nuclear magnetic resonance (magnetic resonance spectroscopy) and infrared spectroscopy (see Chapter 13) that are used for measuring glucose under different conditions and for different clinical needs. Magnetic resonance spectroscopy works best with ^{13}C-labeled compounds and IR requires extensive calibration sets. The opposite of a noninvasive detection method would be to invasively remove tissue via a biopsy for a chemical analysis, which would not be appropriate on a repeated basis. What remains are procedures that are minimally invasive and are tolerated by many individuals. In such a procedure, a sensing device or probe that is small in size, for example, less than 1 mm o.d., might be implanted into a targeted tissue space.

A straightforward way to collect solutes from the interstitial fluid (ISF) space would be to have a semipermeable, hollow fiber, membrane-based device as originally described by Bito et al.[1] Two semipermeable membrane-based devices that have been used to collect different types of analytes from various mammalian tissues include microdialysis sampling probes (catheters) and ultrafiltration probes. The heart of each of these devices is the semipermeable polymeric membrane shown in Figure 6.1. The membranes allow for collection of analytes from the ISF that are below the membrane molecular weight cutoff (MWCO). Each of these devices provides a sample that has a significantly reduced amount of protein when compared to either blood or tissue

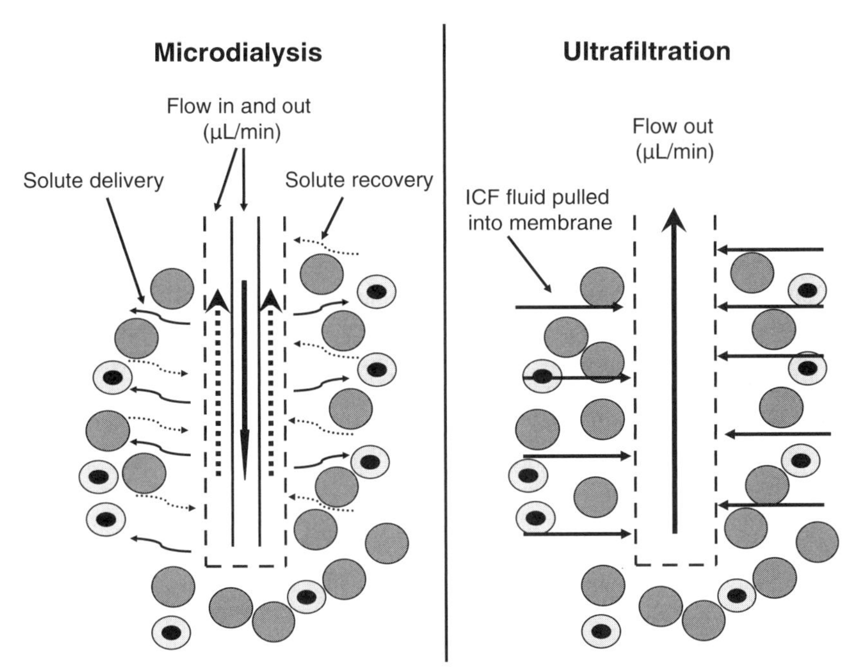

Figure 6.1 Schematic illustrating *in vivo* membrane separation processes. *Left*: microdialysis sampling probe with curved lines indicating analyte tortuous diffusion through the tissue into and out of the probe. *Right*: ultrafiltration representing interstitial fluid flow into the membrane device. Tissue is represented with cells and blood vessels (dark circles in light circles).

samples. The samples are then referred to as being analytically clean since no additional sample preparation is necessary prior to chemical analysis. During sampling with either microdialysis or ultrafiltration membranes (5 kDa or greater molecular weight cutoff), cells and the majority of tissue proteins would be excluded from the sample precluding the need to extract the low molecular weight solutes from the protein component, which is often necessary for a blood sample. Since both these devices allow for the collection of analytically clean samples that can then be subjected to a wide array of already developed analytical methods, it is not surprising that researchers in widely diverse fields have applied these conceptually simple membrane devices to solving important *in vivo* chemical analysis problems. The combined publications reporting use of microdialysis sampling or ultrafiltration techniques is presently over 11,000 and includes both basic and clinical research studies.

While both of these devices use hollow fiber membranes similar to the primary components of kidney dialyzer units, the difference between the two techniques lies in how the analyte undergoes mass transport into the device. Microdialysis sampling is a diffusion-based separation process that requires the analyte to freely diffuse from the tissue space into the membrane inner lumen in order to be collected by the perfusion fluid that passes through the inner lumen of the fiber. Ultrafiltration pulls sample fluid into the fiber lumen by applying a vacuum to the membrane (Figure 6.1).

Microdialysis sampling has been applied to numerous tissues, especially the brain since the brain is sensitive to alterations in volume and ionic composition. Ultrafiltration has been primarily used for peripheral tissue sampling from subcutaneous tissue since the removal of fluid from the brain is believed to cause alterations in brain chemistry.[2] For basic research use, microdialysis sampling devices are typically called microdialysis probes. For clinical studies, the device is called a microdialysis sampling catheter since in clinical medicine a catheter is defined as a small tube that can be implanted.

Extensive descriptions of microdialysis sampling and ultrafiltration techniques are available. Ultrafiltration techniques have been reviewed in the literature.[2,3] Additional information is available from the commercial vendor of ultrafiltration probes, BASi.[4] Three special journal volumes devoted to different aspects of microdialysis sampling have been published.[5–7] Two books are available that describe numerous aspects of microdialysis sampling techniques.[8,9] Microdialysis sampling has been widely applied to clinical studies and is frequently used in clinical settings for energy metabolism monitoring in head trauma patients. This application of using microdialysis sampling to monitor glucose and its metabolites in human brain has been recently reviewed by Ungerstedt and Rostami.[10] Additional information about clinical applications of microdialysis sampling is available on the CMA/Microdialysis web page.[11] There are two companies, CMA Microdialysis AB (Stockholm, Sweden) and Menarini (Italy), that sell microdialysis sampling devices that can be used in humans for monitoring glucose concentrations (Table 6.1). Catheters from CMA Microdialysis AB are available for implantation into many tissues including adipose and brain. Recently i.v. catheters for microdialysis sampling have become available. Menarini markets the GlucoDay device that incorporates microdialysis sampling collection with electrochemical biosensor detection of glucose. The use of the GlucoDay device has been reviewed.[12] Roche Diagnostics reported a microdialysis-based glucose sensing device (SCGM 1) as late

TABLE 6.1 Commercially Available Microdialysis Devices for Glucose Measurement

Device	Company[c]	Insertion site	Flow rates (μL/min)	Glucose detection method
GlucoDay	A. Menarini Diagnostics S.r.l.	Subcutaneous	~10	Wall-jet amperometric electrode
Enzyme Sensor[a]	Sycopel International	Typically brain	~0.5–2	Oxidase-based wire electrodes
CMA 60[b]	CMA Microdialysis AB	Subcutaneous or other tissues	~0.2–1.0	Off-line clinical analyzer

[a] Different enzyme electrodes are available including glutamate, glucose, glycerol, adenosine, acetyl-choline, and lactate.
[b] Multiple products are available for insertion into brain, liver, GI tract, and other peripheral tissue sites.
[c] Company websites are as follows: www.menarini.com, www.sycopelinternational.co.uk/invivo.htm, www.microdialysis.se.

as 2003. However, up to 2009, updated reports in the primary as well as the patent literature had not appeared regarding this technology.

This chapter will focus on describing briefly the fundamentals along with potential pitfalls that must be considered during membrane sampling processes with a primary focus on glucose measurements. Since both of these membrane-based separation techniques have been widely reviewed in the literature, this chapter will focus more on issues related to glucose sensing in humans. The reader may find the chapter to be lopsided with a greater emphasis on microdialysis sampling. The rationale for this decision is based on the much wider application and description in the literature for microdialysis sampling as compared to ultrafiltration sampling.

A primary reason for the high number of microdialysis sampling publications compared to ultrafiltration has been its wide acceptance among the neuroscience community. The majority of published reports of microdialysis sampling have been applications of brain dialysis focused on collection of different neurotransmitters such as dopamine, serotonin, and glutamate. Other reasons may include perceived difficulties with the performance of ultrafiltration (e.g., concerns about repeatable fluid volume collection, variable flow rates, or tubing collapse) and reluctance to try the ultrafiltration approach for certain chemical analysis problems due to greater expertise with existing microdialysis sampling techniques.

6.1.2 Membrane Basics and Theory

Both microdialysis sampling and ultrafiltration incorporate hollow fiber cylindrical semipermeable membranes with a variety of different polymeric compositions typically obtained from a kidney dialyzer unit to achieve separations. This variety is necessary because the membranes used for microdialysis and ultrafiltration sampling are hollow fiber membranes that have also been used in kidney dialyzer units. This extensive variety in the types and chemistry of the fibers exists because of the different clinical needs among hemodialysis patients as well as the sometimes unwanted immune responses of individuals to certain dialyzer chemistry that requires alternative polymers. The polymeric materials used for creating the hollow fibers range from homogeneous compositions including cellulose-based materials to designed copolymers such as the polycarbonate/polyether compositions. It is important to point out that because of the variety in different chemical preparations, membranes that are a blend of copolymers will be different despite incorporating the same name (e.g., polycarbonate/polyether).[13] This is an important point because it makes it very difficult to compare different microdialysis sampling and ultrafiltration membranes and their performance relative to published reports in the membrane science literature because the different companies (e.g., CMA Microdialysis AB and BASi) will not reveal the original source of the membranes.

Many classification schemes for hemodialysis membranes exist. Water permeability through the porous membranes is frequently used.[14] Water permeability for a dialyzer is defined by the ultrafiltration coefficient for the particular device (K_{UF}, mL/h/mmHg). The K_{UF} of any individual fiber will be related to the pore size and has an approximate correlation of $K_{UF} \cong r^4$, where r is the radius of the pore.[15] Additional parameters include overall membrane porosity as well as MWCO.

Based on the above-mentioned classifications, the necessary performance characteristics needed for a membrane used for microdialysis sampling is different from that for ultrafiltration sampling. Large values of K_{UF} can be detrimental during microdialysis sampling since under some conditions where the pressure across the membrane is not balanced (e.g., osmotic versus hydrostatic), fluid is lost across the membrane resulting in significantly decreased sample volumes. For ultrafiltration sampling, larger K_{UF} values may be desirable since lower vacuum pressure would be required to obtain fluids. Yet, membranes with larger K_{UF} may possess larger pore radii and thus have larger MWCO values allowing unwanted proteins to be obtained in the sample.

An important point to consider about hollow fiber membranes is their morphology. Hollow fiber membranes can be either symmetric or asymmetric.[16] Symmetric membranes have continuous pore structure throughout. Asymmetric membranes have a dense upper layer or skin layer that is then supported with a sublayer that is significantly more porous. Figure 6.2 shows SEM images of

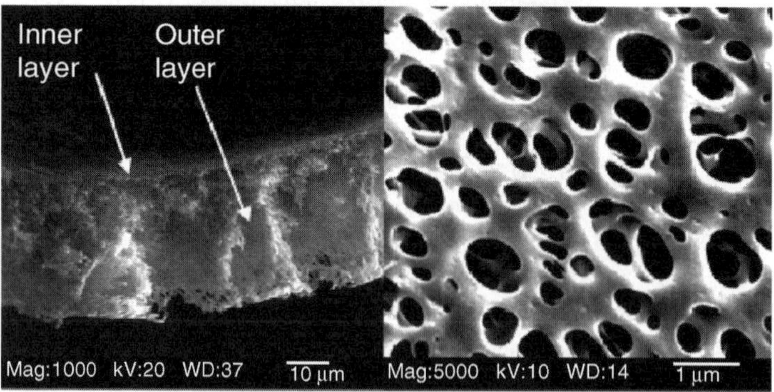

Figure 6.2 Scanning electron microscope images of symmetric and asymmetric membranes. *Top*: SEM image of a polycarbonate/polyether symmetric membrane (MWCO 20 kDa) from a CMA/20 microdialysis sampling probe. Picture is showing the outside of the hollow fiber membrane. *Bottom*: SEM image of a polyethersulfone asymmetric membrane (MWCO 100 kDa) from a CMA/20 microdialysis sampling probe. *Left*: SEM depicting the skin layer and the support layer. *Right*: SEM image of the outside of the membrane.

microdialysis membranes polycarbonate/polyether (symmetric) and polyethersulfone (asymmetric). The advantage of the asymmetric membranes is that mass transfer through the membrane is largely dictated by the skin layer rather than the entire membrane diameter. This allows for better control of membrane selectivity (e.g., MWCO) as well as higher mass transfer rates.

Both microdialysis and ultrafiltration collection obtain analytes from a sample in the reverse direction regardless of how a normal hemodialysis membrane is used. In hemodialysis, the blood is passed through the inner fiber lumen and filtrate is then collected on the outside of the hollow fiber. When these fibers are used as microdialysis or ultrafiltration devices for collection of samples, the outside of the fiber is interfaced with the sample and the analyte is collected into the inner fiber lumen of the hollow fiber. This is important particularly for the asymmetric membranes that have their large porous support layer on the outside facing the tissue sample.

6.2 ULTRAFILTRATION

Similar to microdialysis sampling, the use of ultrafiltration probes for collecting samples from an *in vivo* setting provides analytically clean samples (Figure 6.3). These probes can be used in nearly every tissue to which microdialysis sampling has been applied except for the brain.[2,3] As currently practiced, ultrafiltration sampling is not appropriate for sampling from the brain since there is a general consensus that neuronal processes are sensitive to significant alterations in fluid volume.

The main advantage of the use of ultrafiltration is that compared to microdialysis sampling device calibration is straightforward. Compared to microdialysis sampling, the calibration is considerably less dependent on parameters that influence the analyte diffusion and kinetic properties within the tissue. The ultrafiltrate is simply filtered interstitial fluid. A series of different analytes including glucose (MW 180) and cefazolin (MW 454) exhibited greater than 90% absolute recovery during calibration studies with *in vitro* ultrafiltration using polyacrylonitrile (PAN) membranes with 30,000 MWCO and 310 μm o.d.[17] Most of the analytes, including glucose, exhibited recovery values greater than 95%. The high recovery capacity of ultrafiltration probes has been additionally shown for proteins with PAN membranes (340 μm o.d.) and polysulfone membranes (1.1 mm o.d. with 100 kDa MWCO). Additional reports of proteins up to 68 kDa exhibiting greater than 90% recovery with a 100,000 Da MWCO polysulfone membrane during *in vitro* ultrafiltration have been described.[18]

In the same year, 1987, that microdialysis sampling was first described for use in humans for collecting glucose from subcutaneous tissue, the possibility of using *in vivo* ultrafiltration (sometimes called microfiltration) for glucose monitoring in humans with diabetes was reported by Janle-Swain et al.[19] As with microdialysis sampling, one of the potential clinical goals of applying ultrafiltration was to collect samples for glucose analysis in hospitalized patients who require frequent blood chemistry measurements. Removing the need for repeated blood draws is highly desirable in clinical settings and has been an important topic in chemical analysis.[20]

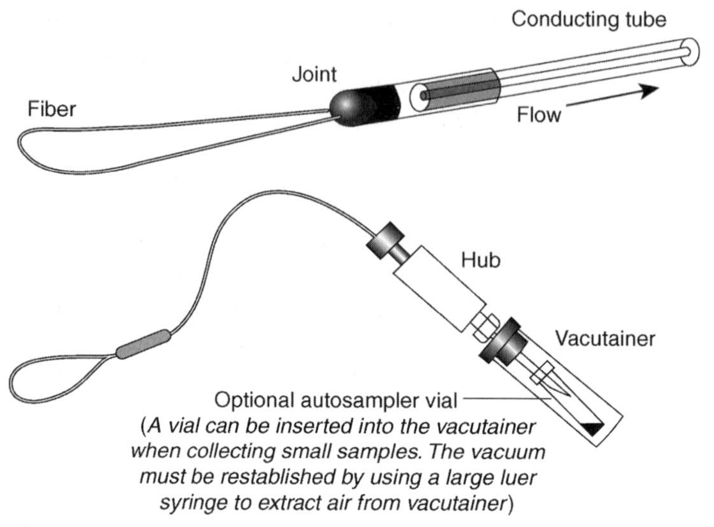

Figure 6.3 Schematic of an ultrafiltration probe manufactured by BASi. Figure reproduced with permission. Copyright BASi. www.bioanalytical.com.

At the present time, the only commercially available *in vivo* ultrafiltration devices can be obtained from BASi and are sold with a variety of membrane lengths from 2 to 12 cm.[21]

For glucose sensing that uses ultrafiltration as the collection method, only a few published reports are available. Ultrafiltration has been used to collect glucose subcutaneously in different mammalian species[19,22,23] including humans.[24–26] In the work of Tiessen et al., concerns were raised that the collected ultrafiltrate glucose concentrations sometimes did not track blood levels.[26] This group suggested that intravascular sampling would overcome this limitation with subcutaneous measures using the ultrafiltration method.

Despite fewer complications during calibration as compared to microdialysis sampling, ultrafiltration sampling seems to be a far more viable technique for acquiring representative samples from the extracellular fluid of peripheral tissues. Yet, far fewer reports of using ultrafiltration sampling compared to microdialysis sampling have appeared in the literature. It is not entirely clear why this discrepancy exists since miniaturized vacuum pumps based on microelectromechanical (MEMS) technology have been described.[27] In addition, ultrafiltration samples can be analyzed using the same methods as microdialysis sampling, including the possibility of integrating sensors.

6.3 MICRODIALYSIS SAMPLING

6.3.1 Historical Perspective

Microdialysis sampling was first developed to allow direct collection of chemicals involved with neurotransmission within the mammalian brain. Developing this technology for such studies was important as having direct access to the chemical transmission occurring within the brain created the possibility for observing how different types of treatments modulate neurotransmission. The success and ultimate research potential for the microdialysis sampling technique during the early 1980s led Professor Urban Ungerstedt, a pharmacology professor and medical doctor affiliated with the Karolinska Institute in Stockholm, Sweden, to form the company that has evolved into CMA Microdialysis AB. Knowing this history is important as Professor Ungerstedt has worked tirelessly over the last few decades to promote the use of microdialysis sampling to improve human health via both basic research studies and its use in the clinic to make chemically informed decisions about care and appropriate interventions.

Microdialysis sampling was used in humans as early as 1987.[28] The original study was performed in healthy volunteers and was designed for glucose measurements and developing appropriate *in vivo* calibration techniques. This focus in humans to develop calibration methods was important to microdialysis sampling since prior to this approach there was a more frequent reliance on simply comparing trends of dialysate absolute concentrations rather than correcting the obtained analyte concentration based on a validated *in vivo* calibration method. For a clinically important analyte such as glucose, it is essential to have an appropriate calibration for the implanted microdialysis sampling probe.

With this first report of applying microdialysis sampling in humans with a targeted analyte of glucose, an obvious use for the technique was to help elucidate glucose concentrations in persons with diabetes. CMA Microdialysis AB introduced catheters for human use in the mid-1990s. One of the primary aims for these original catheters was to provide physicians a way to monitor uncontrolled glucose concentrations. The patient had the catheter implanted into the subcutaneous space of the abdomen and used it for approximately 3 days outside of a hospital setting. With no attached sensor to the device, the patient collected discrete samples and was responsible for storing individual samples. Samples would then be transferred back to their physician for analysis of glucose and other metabolites (typically lactate and pyruvate).

The next logical step with microdialysis sampling, particularly with its application to glucose monitoring, was to connect a sensor to the device so that the patient did not need to collect discrete samples. Microdialysis samples are typically analytically clean. This provides an advantage to attaching a sensor since blood and other tissue components that could potentially foul the sensing device are excluded from the dialysis perfusion fluid. Soon initial *in vitro* descriptions of microdialysis sampling combined with glucose sensors began to appear in the literature.[29–31] The first descriptions of attaching a glucose sensor to the end of a microdialysis sampling probe used in a mammalian setting appeared in the early 1990s. The Albery group reported an incorporated glucose sensor with microdialysis sampling for the detection of glucose in dialysates collected from rat brain.[32] The Schoonen group reported microdialysis sampling coupled with a glucose sensor for subcutaneous monitoring of glucose in humans.[33,34] Several additional reports of glucose sensors attached to microdialysis probes with intent to detect or report detection of glucose in the subcutaneous space of humans appeared in 1992.[35–38] This led to many other descriptions of sensors attached to microdialysis sampling devices for human studies.[39] Many additional groups have published on the combination of sensors with microdialysis sampling probes since the early 1990s. The numerous advances made in the 1990s with respect to device miniaturization allowed for the creation of integrated microdialysis sampling–sensor devices. These initial reports along with the research efforts of many others have ultimately led to three companies aiming to market microdialysis sampling/glucose sensor devices (Table 6.1). Two of the three companies have approval to use their devices in humans.

6.3.2 Microdialysis Sampling Principles—The Basics of Mass Transport

To best understand microdialysis sampling at a conceptual level, it is important to consider that it is a diffusion-based separation technique that involves volumetric fluid flow through the inner membrane lumen. Some researchers have referred to the technique as an implanted artificial blood capillary. Any process or impediment to analyte diffusion in either the tissue space or within the membrane affects the amount of analyte that can be collected. The fluid flow rate that sweeps the inner part of the membrane fiber also affects collected analyte concentrations. Microdialysis sampling is a fairly straightforward process that most find easy to grasp. However, diffusion itself is also a concept that most individuals can easily comprehend, yet several textbooks are available that describe the rigorous mathematics associated with diffusion.[40,41] The most often cited and relevant research paper for understanding the underlying diffusive and mass transport processes that affect microdialysis sampling is the work of Bungay et al.[42]

Diffusion is the principal mass transport process that drives analytes from the tissue space into the microdialysis probe. Only analytes that can freely diffuse into the probe lumen will be collected into the microdialysis sampling device. If an analyte is confined within a cellular environment and cannot diffuse out of a cell, then it will not be sampled by the microdialysis device. Any process or procedure that serves to either increase or decrease the diffusive flux of analyte will have an impact on the analyte amount collected into the microdialysis probe. A small hydrophilic analyte such as glucose is well suited for *in vivo* microdialysis sampling, whereas larger analytes such as peptides and proteins can be difficult to collect with high efficiency during microdialysis sampling.[43] While there are many facets of the microdialysis sampling technique that can lead to time lags (e.g, length of tubing, dead volumes, etc), it would be expected that glucose rapidly diffuses across the membrane as has been demonstrated *in vitro* for dopamine, another small hydrophilic analyte.[44]

Calibration is necessary to allow correlation between collected dialysis concentrations to external sample concentrations surrounding the microdialysis probe. Extraction efficiency (EE) is used to relate the dialysis concentration to the sample concentration. The steady-state EE equation is shown in equation (6.1), where C_{outlet} is the analyte concentration exiting the microdialysis probe, C_{inlet} is the analyte concentration entering the microdialysis probe, $C_{tissue,\infty}$ is the analyte tissue concentration far away from the probe, Q_d is the perfusion fluid flow rate and R_d, R_m, R_e, and R_t are a series of mass transport resistances for the dialysate, membrane, external

Resistance terms

$$R_d = \frac{13(r_i - r_\alpha)}{70\pi L r_i D_d};\ R_m = \frac{\ln(r_o / r_i)}{2\pi L D_m \phi_m};\ R_e = \frac{\Gamma[K_o(r_o/\Gamma)/K_1(r_o/\Gamma)]}{2\pi r_o L D_s \phi_s}$$

$$\Gamma = \sqrt{\frac{D_s}{(k_{ep}(r) + k_m(r) + k_c(r))}}$$

Scheme 6.1 The multiple mass transport equations used to describe microdialysis sampling. D is the diffusion coefficient through the dialysate, D_d, membrane, D_m, and sample, D_s. L is the membrane length. Γ (cm) is a composite function; k_{ep}(r), $k_m(r)$, and $k_c(r)$ are kinetic rate constants as a function of radial position (r) from the microdialysis probe. Additional term definitions can be found in Ref. 42.

sample, and a trauma layer that exists at the interface of the probe membrane and the tissue.[42,45] The resistance term for a trauma layer, R_t, cannot be completely described with analytical parameters (e.g., exact diffusion coefficient, kinetics, etc.), but it does provide mass transport resistance that can be incorporated into finite element models.[45]Scheme 6.1 shows the parameters that can be used to model steady-state microdialysis sampling EE for the resistance terms with the exception of R_t.

$$\mathrm{EE} = \frac{C_{\mathrm{inlet}} - C_{\mathrm{outlet}}}{C_{\mathrm{inlet}} - C_{\mathrm{tissue},\infty}} = 1 - \exp\frac{-1}{Q_d(R_d + R_m + R_e + R_t)} \tag{6.1}$$

If C_{inlet} has a value of zero, then equation (6.1) reduces to what is commonly called relative recovery and denoted in the literature as RR. The terms EE and RR are frequently interchanged and sometimes the general word, recovery, is used to describe the sampling efficiency of the microdialysis probe. It is important to note that the equivalence of EE versus RR under certain conditions such as with highly regulated concentrations of neurotransmitters has been debated in the literature.[46]

EE is highly dependent upon numerous physiochemical parameters described above, including analyte diffusion coefficient, perfusion fluid volumetric flow rate, membrane pore size, membrane porosity, membrane surface area, analyte kinetic uptake into cells[47] and transport to and from microvasculature,[48] as well as the overall ISF volume fraction. The ISF volume fraction has been measured to a value of 0.2 in the brain[49] and has a generally accepted value of 0.2 in other tissues.[50] Adipose tissue has a reported ISF volume fraction of approximately 0.1.[51] Using equation (6.1) combined with Scheme 6.1, it is possible to develop theoretical curves for different parameters as shown in Figure 6.4. Figure 6.4 shows that flow rate is a predominant factor influencing the EE value for any analyte. Increasing the surface area of the microdialysis probe by increasing the length also increases EE. Decade differences in the analyte diffusion coefficient also significantly affect EE values. Finally, EE is not as sensitive to changes in the kinetics, but is affected by changes in the ISF volume fraction.[52] This change in EE as a function of ISF volume fraction is important to note since after initial implantation a localized edema would be expected that would affect analyte EE surrounding the probe.

A maximum value of EE can be estimated for any microdialysis sampling probe by immersing it into a well-stirred fluid at the same temperature at which sampling will occur (typically ~37°C). The stirring minimizes the influence of the solution external

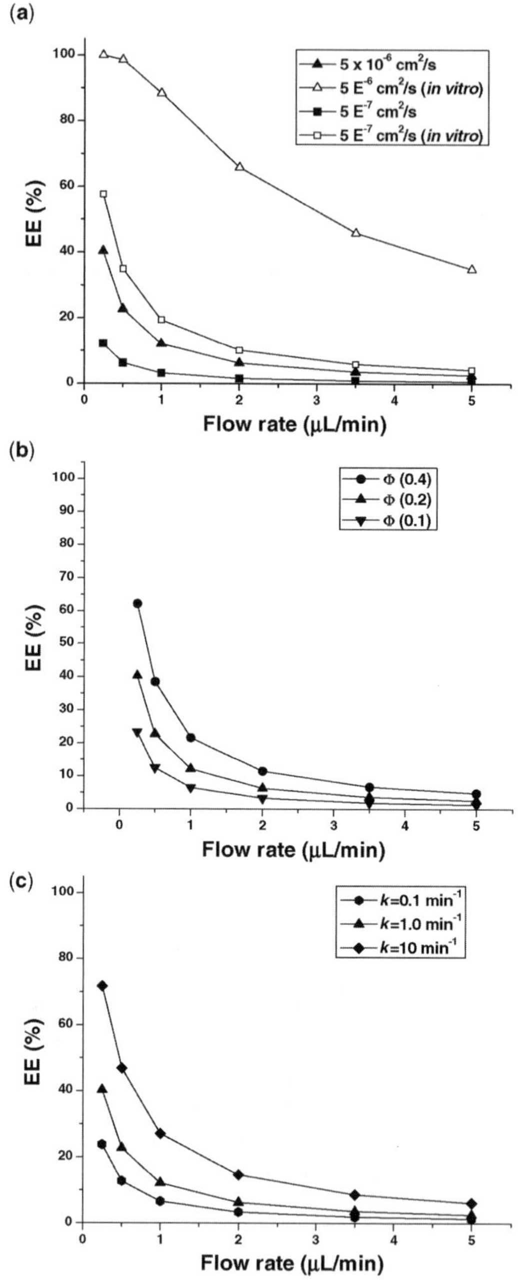

Figure 6.4 Tissue parameter effects on microdialysis sampling extraction efficiency(EE) using the published Bungay et al. model.[42] (a) Parameters in graph (a) denote changes with diffusion coefficients combined with the following parameters: probe length, L, 0.3 cm; inner radius, r_i, 0.021 cm; outer radius, r_O, 0.025 cm; cannula radius, $r_{cannula}$, 0.0175 cm; tissue sample volume fraction, ϕ_s, 0.2; membrane diffusion coefficient, D_m, $0.2D_d$; tissue sample diffusion coefficient, $D_s = D_d/2.25$; K_{ep} 1.0 min^{-1}, K_m 0.1 min^{-1}. *In vitro* simulation sets the external tissue resistance, R_e to a value of zero. (b) Variation in tissue sample volume fraction, ϕ_s, with all parameters used. $D_d - 5 \times 10^{-6}$ cm^2/s. (c) Variation in sum of external rate constants. All other parameters are as in (a) with $D_d - 5 \times 10^{-6}$ cm^2/s.

resistance (R_e) and a trauma layer would not be present (R_t) in a beaker. This approach allows the user to identify parameters that would need to be altered if a 100% EE is desired such as the need for a longer membrane or lower volumetric flow rate.

Once the microdialysis probe is implanted into a tissue space, the mass transport resistance imposed by the tissue is believed to be significantly greater than that for the combined value of the membrane and dialysate (e.g., $R_e \gg (R_d + R_m)$). The analyte diffusive and kinetic properties within the tissue will dictate how reduced the *in vivo* EE will be from the maximum possible *in vitro* EE value. Tissue tortuosity alters the analyte diffusion properties within the tissue by causing the diffusive path to be longer than under *in vitro* conditions. This decrease in the diffusion coefficient caused by the increased analyte path length can be approximated using $D_{isf} = D_d/\lambda^2$, where λ has a value of approximately 1.6 for tissues such as those of healthy brain.[53] Solute diffusion and biophysical properties in the brain have been widely studied and the methods for determining diffusion and tortuosity parameters have been recently described by Hrabetova and Nicholson.[54] Biophysical parameters such as tortuosity have not been well described for other tissues and these types of studies deal with complicated solute diffusion through various water compartments such as the intracellular fluid.[55] In addition to the alteration in diffusive characteristics, tissues are vascularized and have active cellular components. Depending on the analyte, removal or delivery kinetics will affect the overall microdialysis EE. It is difficult to obtain *in vivo* diffusion and kinetic properties for analytes. Thus, it is quite challenging to predict expected EE for a specified analyte in any given tissue.

6.3.3 Implantation and Tissue Effects on Probe Performance for Glucose Measurements

Implantation of microdialysis probes into skin tissue requires the use a guide needle. This initial trauma of the tissue space will disrupt or break cells and can damage capillaries resulting in the release of different components. Most microdialysis reports allow for a set period of time to pass before initial sampling procedures are initiated. Only a few reports have actually described measured compounds associated with the insertion trauma immediately after probe insertion. These measurements and the temporal response of the released cellular components that serve as trauma biomarkers give a more detailed understanding of the trauma response to insertion and the waiting time necessary for the tissue to return to homeostasis. For example, Groth et al. measured histamine release after microdialysis probe insertion into rat skin and determined that after 20 min the histamine concentrations decreased from approximately 350 ng/mL to less than 25 ng/mL.[56]

Of particular importance to glucose collection at the site of a microdialysis probe would be changes in localized blood flow. Anderson et al. used laser Doppler perfusion imaging to determine how localized blood flow is altered after microdialysis probes insertion into the forearm of human volunteers and determined that after 60 min blood flows were equivalent to those at 24 h.[57] However, Wientjes et al. found for longer term implants that it appeared that 2–3 days were necessary to achieve a stable glucose concentration or baseline.[58] The GlucoDay system has been reported to be acceptable at 60 min after insertion and improved accuracy at 120 min.[12] A review of the literature suggests that an acceptable waiting period prior to first sample collection appears to be approximately 2 h after device insertion.

After insertion, there will be an immediate immune response initiated by the wound caused during insertion. Signaling molecules including leukotrienes and cytokines will be released due to the inflammatory response. Thus, after device insertion, it is critical to realize that tissue surrounding the probe is in a dynamic and not static state. Inflammatory cells are migrating into and out of the area of implantation. Studies by Mou and Stenken using a marker of glucose uptake, 2-deoxyglucose, showed that any potential alterations in uptake due to increased metabolic activity of the incoming cells did not affect microdialysis sampling extraction efficiency.[59] However, it is important to point out that significantly large changes in localized kinetics (approximately 10-fold) are necessary for observable changes in the microdialysis EE for any given analyte (Figure 6.4).

For any implanted device that is intended for long-term use, collagen encapsulation of the device will occur due to the immune response to the device.[60] Collagen deposition typically begins at roughly 5–7 days postimplantation and takes up to a month to reach completion. In terms of calibration of the microdialysis probe, this layer of material will provide additional mass transport resistance and could be denoted as a trauma layer.

Wisniewski et al. have described how mass transport for glucose changes for implanted microdialysis probes in rats[61] and have noted important differences in glucose flux between rats and humans for microdialysis sampling devices during a long-term implantation.[62] For microdialysis sampling probes implanted into the rat subcutaneous space, the glucose concentrations increased between the implant day and 24 h postimplantation most likely due to initial edema upon implantation. Then between postimplant days 2–7 the glucose concentrations obtained in the implanted dialysis probe decreased for the rat. For the human studies, the glucose concentration slowly rose after the initial implantation. The rationale for the differences between the two is the lack of vascularization between human subcutaneous space that has more adipose tissue and rat subcutaneous space. The trend of increased glucose collection over a week long period postimplantation has also been observed by Wientjes et al.[63] They have reported that this increase in glucose concentrations collected into the microdialysis probe over a period of 4–6 days is due to capillary repair since capillaries are damaged by the necessary needle-based insertion procedure.[64]

Adipose tissue poses additional calibration differences for microdialysis sampling devices. The thickness of the tissue (e.g., lean versus obese individuals) and thus the capillary density will affect interindividual microdialysis sampling recovery values. This has been shown by Lutgers et al., who demonstrated decreases in glucose recovery of up to 50% between human volunteers with a skin fold thickness of 20 versus 45 mm.[65] This points to how microdialysis sampling recovery is dependent upon analyte supply since glucose will be supplied better to the microdialysis probe in lean individuals with a higher density of capillaries per unit tissue than obese subjects with a lower density of capillaries and thus an increased mass transfer resistance to the probe. Additional reports have also shown less interindividual differences between microdialysis probes implanted in the forearm versus in the subcutaneous tissue.[66]

Glucose recovery in the probes relative to the blood glucose concentration was $91.1 \pm 4.1\%$ in the forearm (leaner) versus $82.7 \pm 18.0\%$ in the abdomen.

6.3.4 Basic Versus Clinical Research Designed Microdialysis Sampling Devices

There are many different microdialysis sampling probes available for both basic and clinical research studies shown in Figure 6.5 and these differences are highlighted in Table 6.2. Microdialysis sampling probes that are used for basic research studies using laboratory animals have membrane lengths between 1 and 10 mm. Perfusion fluid flow rates that are typically used with these probes range between 0.5 and 2.0 μL/min. Using basic research microdialysis probes with low flow rates and longer membranes (10 mm), it is possible to reach 100% EE for glucose. However, with shorter membranes it is difficult to reach 100% EE *in vivo* unless flow rates below 0.5 μL/min are used.[67] During most animal studies involving microdialysis sampling, the sampling conditions of higher flow rate combined with shorter membrane lengths cause a situation where equilibrium concentrations between the dialysate and the sample are rarely obtained. That is, the EE is not near 100% and therefore knowing the EE value becomes necessary to elucidate analyte concentration within the vicinity of the sampling device. In other words, *in vivo* calibration is essential for these situations if absolute concentrations of analyte surrounding the probe are desired.

Compared to basic research microdialysis sampling devices, those used for clinical studies are much longer with typical membrane lengths between 10 and 30 mm. Flow rates of the perfusion fluid through these devices are also much lower (0.3 μL/min) than typically applied in basic research studies. This combination results in high EE% values for low molecular weight and hydrophilic analytes such as glucose.

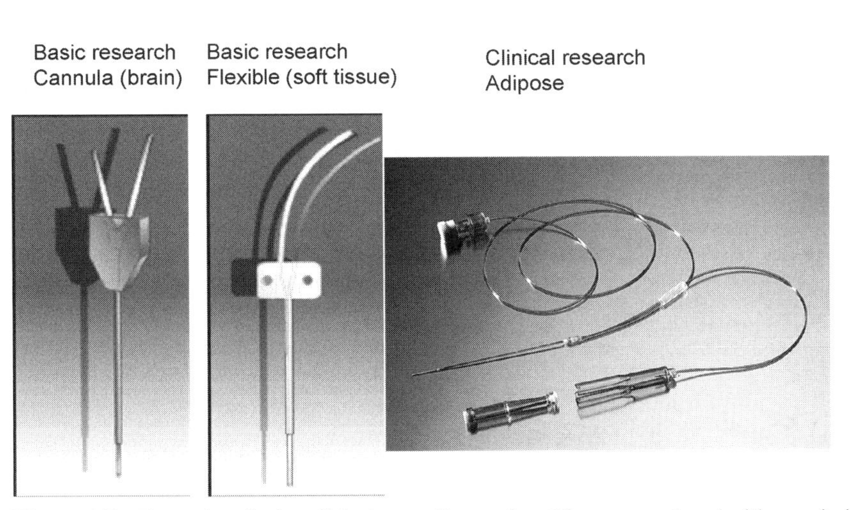

Figure 6.5 Examples of microdialysis sampling probes. Figure reproduced with permission. Copyright CMA Microdialysis AB. www.microdialysis.se.

TABLE 6.2 Comparison of Basic Research (Animal) Versus Human Microdialysis Sampling Probes

	Laboratory Animal	Human
Available Vendors	BASi	CMA Microdialysis AB (CMA)
	CMA Microdialysis AB (CMA)	
	Harvard Apparatus	
	SciPro	
Available Geometry	Cannula style for brain applications—stainless steel and nonmetallic	Linear cannula for peripheral tissues
	Cannula style with internal fluid line for simultaneous dialysis and localized infusion	Cannula style for insertion into brain
	Linear probes	
	Soft style flexible cannula probes (blood vessels or peripheral tissues)	
Membrane type and molecular weight cutoff in kilodaltons (kDa)	Polyarylethersulfone (PAES) (20 kDa)	Polyimide
	Polyethersulfone (PES) (55 and 100 kDa)[a]	PES
	Polyacrylonitrile (PAN) (30 kDa)	
	Cuprophane (CUP) (6 kDa)	
Typical flow rates	0.5–2.0 μL/min	0.3 μL/min
Sterile in package	No	Yes
Maximum reported implantation time	Variable[b]	21 days
	Note: typical use is acute (1-day use)	

[a] 55 kDa PES probes are only available on a linear probe from CMA Microdialysis AB.

[b] (1) CMA/20 polycarbonate/polyether (PC) probes have been implanted subcutaneously in rats in our laboratory up to 7–10 days. This membrane chemistry has been discontinued. Failure after 7 days is typical and appears to be due to the foreign body reaction as localized anti-inflammatory administration (dexamethasone) increased the lifetime. (2) CMA/20 polyarylethersulfone (PAES) probes have been implanted subcutaneously for up to 7 days in our laboratory. Failure rates have not been determined. (3) CMA/20 polyethersulfone (PES) probes have been implanted subcutaneously for up to 14 days in our laboratory. Failure rates have not been determined.

6.3.5 Calibration of Microdialysis Sampling Devices

A frequent debate among chemical analysts who perform measurements *in vivo* is the issue of how to perform appropriate device calibration. A mammalian system is different from a beaker and thus the true concentration outside the device is unknown, may not be in a steady-state, and may not be homogeneous in the vicinity of the implanted device. Since the inception of microdialysis sampling, the issue of device calibration has been a widely published and frequently discussed aspect of this sampling approach. For the majority of microdialysis sampling applications, only relative changes with respect to a baseline value are typically reported. This data interpretation approach is most commonly observed with neurotransmitter collection where some modulating stimulus of either pharmacological or physical origin is applied and the direction of neurotransmitter concentration change within the dialysate (positive or negative) is then reported. In many of the neurochemistry studies, the pharmacological intervention may affect one of the many parameters (kinetics, metabolism, blood flow, capillary permeability, and even tissue volume fraction) that affect EE during sampling. Yet, this possibility of changes in the EE due to either insertion trauma or some type of pharmacological application changing different tissue parameters known to affect EE is typically not considered.

Over 200 review articles describing various aspects of microdialysis sampling have been published. Few specific reviews exist with titles or searchable terms that suggest that the review is primarily focused on calibration of microdialysis sampling devices.[68,69] Nearly all microdialysis sampling calibration techniques require long sampling times or lengthy procedures. Table 6.3 contrasts the different calibration methods, which will be described briefly below. Many of these techniques require a steady-state concentration surrounding the microdialysis sampling device throughout the duration of the calibration procedure. In individuals with diabetes, the requirement for a steady-state glucose concentration during a calibration period that can last several hours would be difficult to achieve.

***6.3.5.1 In Vitro* Calibration** While frequently derided as an inappropriate calibration method, an *in vitro* calibration can at least give an idea of the approximate range of concentrations for the targeted analyte. Among the many publications citing the use of microdialysis sampling, few researchers actually report device performance metrics for new analytes with an *in vitro* calibration. Since the validity of *in vivo* calibration methods continues to be debated and discussed, the approximation obtained with an *in vitro* calibration at least provides a starting point with respect to asking questions about how the localized biology may affect the membrane performance.

6.3.5.2 Approach to Zero Flow As mentioned above, microdialysis sampling EE depends on numerous factors. Depending on the application, membrane length and the volumetric flow rate through the dialysis probe can be externally controlled. EE increases with decreasing flow rate since there is a greater time period during which the concentrations inside the probe lumen can equilibrate with the external solution. If the flow rate is decreased such that it begins to approach a zero value, then EE should approach 100% (Figure 6.4) as the internal solution concentration should match the external concentration. For the long microdialysis sampling membranes combined with such low flow rates used for clinical studies, most of these typically reach a high EE value. This approach requires a steady-state concentration of analyte and therefore in a person with diabetes would not be an appropriate calibration method.

TABLE 6.3 Microdialysis Sampling Calibration Methods

Calibration method	Brief description	Concerns/requirements	Glucose monitoring applicability
In vitro calibration	Use an *in vitro* calibration and relate to *in vivo* concentrations	A *stirred in vitro* calibration at an appropriate temperature will give the maximum EE% for any flow rate. Using this procedure, all *in vivo* values would be lower than the *in vitro* EE%	Possible, but cannot be relied upon to give accurate *in vivo* glucose concentrations
Approach to zero flow	Vary the perfusion flow rate (*Q*) and create a calibration graph	Requires steady-state concentrations at all times	Possible, if it can be confirmed that low flow rates always produce nearly 100 EE%
		Fluid evaporation at low flow rates (<0.5 μL/min) is a concern	EE% has to approach 100%
Internal standards	Loss of internal standards (e.g., 2-deoxyglucose or antipyrine) to calibrate for changes in extraction efficiency caused by either metabolic or tissue variations.	Performed *in situ*	Possible
		Useful under dynamic conditions	
No net flux	Vary concentrations through the probe to straddle the anticipated tissue concentrations	Requires a steady-state external concentration	Not practical
		Long experimental collection times requiring a steady state that lasts at least >4 h	Glucose concentrations in diabetics are typically not at steady state
Fingerstick	Ratio of [glucose](dialysate)/[glucose](fingerstick)	Clinically preferred	Clinically preferred
		Used for implanted biosensors	

6.3.5.3 *Internal Standards and Retrodialysis* The internal standards used for analytical measurements that involve separations are widely used to correct the analyte loss during complicated sample preparation procedures, as is typically seen with either biological or environmental samples. Judiciously chosen internal standards help correct for potential losses of analyte on materials or correct for analyte mass transport-related changes (e.g., metabolism, edema, etc.) that may occur within the sampling matrix. For microdialysis sampling, a good internal standard would possess similar solubility or $\log P$ (octanol–water partition coefficient) characteristics of the targeted analyte and similar mass transport through the tissue space including similar metabolic and uptake characteristics. The advantage of using internal standards is that external analyte (e.g., glucose) concentrations do not need to be in a steady state. Thus, changes in extraction efficiency of the internal standard would be contributed to alterations in diffusive (potential localized edema) or other tissue properties (e.g., kinetic properties such as uptake into capillaries or onto receptors as well as metabolism). Given the desired characteristics for internal standards, the best possible choices would typically be either labeled or close structural analogues of the analytical target. Typically, either radioisotopes (^{14}C, ^{3}H) or stable isotopes (^{13}C or ^{2}H) are used as labeled analogues of glucose.[70–72] These few reported studies were focused on measuring the localized glucose pool or dynamics within a targeted tissue rather than a specific calibration procedure. The structural analogue, 2-deoxyglucose, has long been applied to various studies of glucose metabolism and has also been used as an internal standard during microdialysis studies.[59] Finally, different fluorinated labels such as [^{18}F]-2-fluoro-2-deoxy-D-glucose have been used as standards to allow microdialysis sampling to be combined with positron emission tomography.[73]

The term retrodialysis has been applied to the calibration procedure that uses an internal standard to infuse through the microdialysis sampling probe.[74,75] During the retrodialysis approach, the standard is locally delivered to the tissue site. Alterations in the EE of the internal standard during sampling would then be attributable to alterations in the analyte mass transport resistance and all the encompassing terms associated with that variable denoted in Scheme 6.1. The underlying assumption with the retrodialysis approach is that the EE values for the loss of the internal standard and the gain of the targeted across the microdialysis probe are identical.

Additionally, there has been considerable interest in identifying endogenous compounds that might serve as useful recovery markers for *in situ* microdialysis probe calibration. In other words, are there endogenous compounds that exist at a steady-state concentration in the tissue that would serve as appropriate EE markers? Glucose has actually been proposed by some researchers as a useful endogenous standard, but it appears that the use of urea has become more accepted as an endogenous standard.[76–78] There are other researchers who have not been able to validate the use of urea as a suitable recovery marker for glucose.[79]

6.3.5.4 *No Net Flux/Zero Net Flux Calibration* The most time intensive of all the *in vivo* microdialysis calibration methods is the method of no net flux (NNF), which is sometimes referred to as the method of zero net flux (ZNF). The NNF calibration method requires the analyte to be in a steady state. So, for healthy

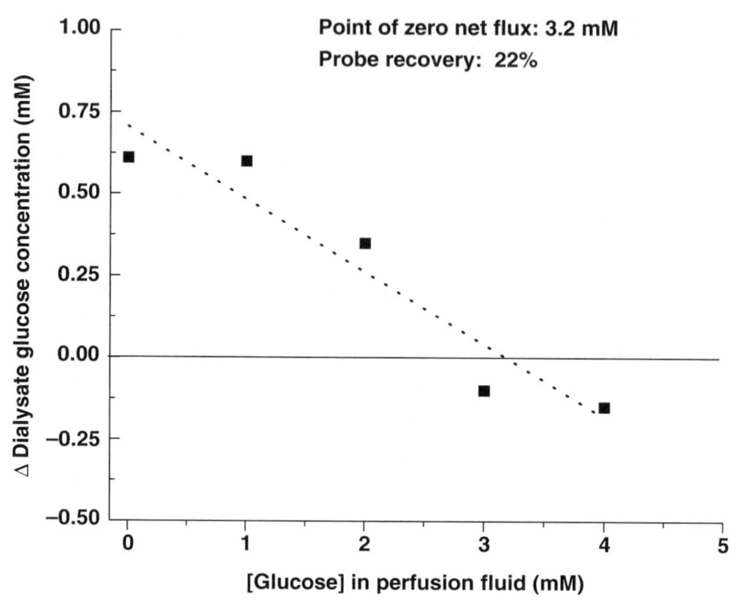

Figure 6.6 Example of a zero net flux graph.

individuals where glucose is the targeted analyte, this requirement can be maintained in a situation where no glucose challenges via food or drink occur. Steady-state glucose levels are obviously difficult to achieve in individuals with diabetes.

The aim of this approach is to straddle the external analyte concentration. If the assumption is that loss of analyte occurs at the same rate as collection of analyte, then a series of different analyte concentrations can be passed through the probe and the net gain or loss of analyte would define the probe EE value and the intercept would define the local concentration surrounding the probe (Figure 6.6).

6.3.5.5 ***Fingerstick/One-Point (Two-Point) Calibrations*** All glucose sensor devices are compared against blood glucose values. For microdialysis sampling applications seeking to compare dialysate glucose with blood glucose values, the commonly applied practice of collecting a baseline measurement versus a glucose challenge measurement used for electrochemical-based sensing devices has become standard. These calibration methods for electrochemical-based sensors have been described in Chapter 1 and have also been reviewed by Wilson and Reach.[80]

6.4 MEASUREMENT OF GLUCOSE FROM MEMBRANE DEVICES

6.4.1 Off-Line Dialysis Measurements

Microdialysis and ultrafiltration techniques are both collection methods. Neither of these two techniques by themselves directly detects analyte within a sample. An early problem with analyte measurements in samples collected with microdialysis sampling was dealing with the low microliter per minute volumetric flow rates when most analytical methods required sample volumes in the mid-to-high microliter range. When collecting samples off-line, it is important to keep in mind dead volume and thus delay time between the probe membrane and the tubing outlet. Additionally, it is important to remember that microdialysis samples represent a weighted average of the concentration over the collection time. If concentrations are rapidly changing as in a pharmacokinetics type of study, the dialysate does not represent an instantaneous time point and appropriate averaging techniques must be considered in order to properly fit data.[81] Even with these averaging techniques, an appropriate kinetic model is necessary. The excursions in glucose concentrations within a diabetic human are not as predictable as drug pharmacokinetics and thus could pose potential errors if related to blood glucose values.

The first report of microdialysis sampling in humans in which glucose was a targeted analyte was in 1987 by Lonnroth et al.[28] In that report, glucose was measured spectrophotometrically from the collected dialysate using a glucose oxidase colorimetric assay. The following year, the same group of authors extended their study and also performed an off-line dialysis measurement using a colorimetric glucose dehydrogenase method.[82] A few years later, an adapted liquid chromatography–electrochemistry methodology using glucose oxidase-based electrodes was reported and later reviewed by the electrochemistry group at BASi.[83] This same group has also reported flow injection analysis methods combined with glucose sensing electrodes to measure glucose in microdialysis and ultrafiltration samples.

At the present time, the vast majority of clinical studies incorporating microdialysis sampling measure the contents in the dialysate using colorimetric assays that are available with the different clinical analyzers available from CMA Microdialysis AB. The CMA 600 instrument allows for off-line measurement in 2 μL of sample for four of the following six analytes: glucose, glutamate, pyruvate, glyercol, lactate, and urea. There are other oxidase enzymes including cholesterol oxidase that if incorporated into the instrumental capability may have some importance with management of different aspects of metabolic syndrome. However, these instruments are off-line analyzers that require someone to perform a discrete sample collection (i.e., someone has to fill the sample collection vial, time the sample collection period, remove the sample vial, and finally place the sample into the instrument for chemical analysis). Collecting dialysate samples and delivering these samples to a clinical lab is highly labor intensive and does not allow for rapid information about chemical concentrations. Incorporating appropriate sensing technologies onto the end of microdialysis probes can overcome these limitations.

6.4.2 In-Line Dialysis Measurements

6.4.2.1 Electrochemical-Based Sensors The ability of the microdialysis sampling probe to provide an analytically clean sample has led to numerous possibilities for the coupling of different detectors at the end of the dialysis probe. Numerous research groups have described in-line measurements of the glucose concentrations within the flowing dialysate fluid.[84,85] Much of this early work with integrating sensors to the end of dialysis probes focused on glucose measurements in the rodent brain. The advantage of this approach, shown schematically in Figure 6.7, is the reduced fouling incurred with the sensing device, particularly enzyme-based biosensing electrodes. Over a decade ago, a company named Sycopel was formed that marketed combination microdialysis sampling devices with various biosensors based on oxidase-enzyme technology. In particular, the initial biosensors offered were for glucose, glutamate, and glycerol, all of which have commercially available FAD oxidase enzymes that allow coupling to electrodes for biosensing of their targeted analyte.[86] Many other commercially available FAD oxidase enzymes are available for including ethanol, D-amino acids, lactate, L-amino acids, and pyruvate.

6.4.2.2 *Other Measurement Technologies*

The inclusion of optical-based sensors to the outlet flow of microdialysis sampling devices has also been reported in the literature, particularly from the sensors group under the direction of Professor Heise at Dortmund. This group has been interested in using infrared spectroscopy to detect targeted analytes in the microdialysis sampling fluid.[87] Again, the main advantage of the microdialysis sampling approach for these studies is the analytically clean sample obtained. Unlike direct infrared spectroscopic measurements of glucose and other critical metabolites in more complex media such as a bioreactor,[88] the microdialysis samples will contain little to no protein that can require additional steps to ensure appropriate sensor calibration. However, the concern with this type of sensing procedure combined with microdialysis sampling devices is the issue of potential interferents that are unique to individuals.

6.5 ALTERNATIVE USES OF MICRODIALYSIS SAMPLING

One primary advantage of microdialysis sampling, when compared to traditional electrochemical sensors, is that components can be delivered through the probe

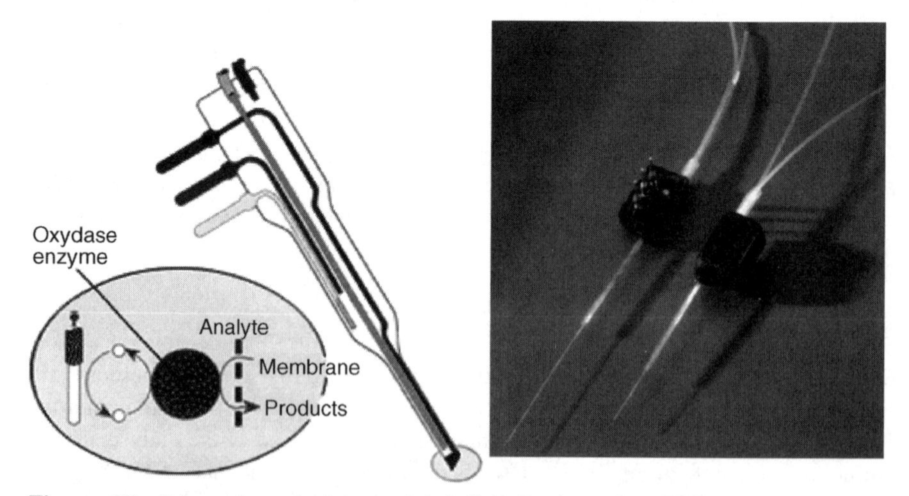

Figure 6.7 Schematic and photograph of dialysis electrodes (Figure reproduced with permission. Copyright Sycopel International.)

TABLE 6.4 Modulators and Internal Standards Used in Microdialysis Sampling Related to Glucose and Biosensor Biocompatibility

Modulator	Research question asked	Output collected or obtained/measured	Reference
2-Deoxyglucose	Macrophage activation	EE% change (would expect EE% to increase with increased macrophage activation)	59,89
Antipyrine	Capillary permeability or density change	EE% change (would expect EE% to decrease with decreased capillary permeability or density around the probe)	59,89
Gd-DTPA	Capillary permeability change	Changes in MRI contrast intensity	88
Monocyte chemoattractant protein-1	Increase monocytes to the microdialysis probe	[Glucose] collected	89
		Changes in EE% for internal standards (2-DG and antipyrine)	
	Question asked: How does this affect glucose recovery and internal standard delivery?	Changes in size and histology of capsule formed during the foreign body reaction	
Isotope-labeled glucose (e.g., ^{13}C, ^{14}C)	Metabolism changes	Measure glucose converted to labeled lactate and other metabolites.	70–72
F-labeled glucose analogues	Uptake differences	PET imaging	73

membrane to achieve a localized delivery to the tissue space. This procedure allows for collection of solutes that are released in response to the locally delivered substance (Table 6.4). Locally delivered substances could include internal standards, markers for alterations in metabolism or capillary permeability, imaging agents,[89] or immune modulators.

Different reports using either radiolabeled or stable isotope labels for glucose have been reported in the literature.[70–73] Of particular interest are the stable isotope labels (SIL) as these allow the observation of changes in metabolic flux by following the uptake and metabolism of the SIL compound. With the increased use and familiarity of mass spectrometry among biomedical researchers along with great interest in metabolomics, it is likely that SIL glucose (^{13}C label) will be used in the future with microdialysis sampling.

We and others have shown that during chronic microdialysis sampling probe implantation, there are discrepancies between observations with a localized delivery and recovered amounts.[59,61] Internal standards such as antipyrine (a marker for capillary permeability) and 2-deoxyglucose exhibited no statistically significant changes in extraction efficiency (EE) over a 7-day implantation. However, glucose concentrations were significantly decreased, suggesting that a localized supply problem existed. Pharmacokinetic studies of antipyrine were performed and significant decreases in the amount of antipyrine recovered into the microdialysis probes were observed.[90]

There is significant interest in the field of implantable biosensors to create controlled release materials that locally deliver different modulating agents as described in Chapters 3 and 9. With microdialysis sampling, the probe itself could serve as a rapid prototype for drug releasing strategies. While some have actually coated microdialysis probes with vascular endothelial growth factor (VEGF) and dexamethasone releasing hydrogels,[91] these materials can be locally infused through the dialysis probe. Preliminary studies in rats with infusing monocyte chemoattractant protein-1 (MCP-1) that recruits monocytes to the probe showed no alterations in the EE of antipyrine or 2-DG, yet a significantly larger capsule surrounding the probe was clearly evident and was measured to range between 2 and 2.5 times larger than the control (Figure 6.8). The fibrous capsule that surrounded these 500 μm o.d. microdialysis sampling probes was approximately 200–225 μm (radial coordinate) for the controls versus 500–600 μm for the MCP-1-infused probes. Localized delivery through the dialysis probe of dexamethasone via its phosphate form greatly reduced the size of the capsule surrounding the probes, resulting in partially formed capsules that were difficult to quantify using histological methods. Interestingly, the infusion of dexamethasone did not completely remedy the supply problem as shown in Figure 6.9. It should be noted that in these studies, the local animal ethics committee required animals to be returned to their cages overnight. Thus, continuous infusion was not possible and only occurred during sampling periods that lasted for 3 h. A possible way to overcome this problem would be to implant osmotic pumps to provide continuous flow through the implanted microdialysis probes.[92]

6.6 IMPLICATIONS OF FLUX DIFFERENCES BETWEEN MICRODIALYSIS SAMPLING VERSUS GLUCOSE SENSORS

There is great potential for individual variation in calibration properties of an implanted microdialysis sampling probe. The main cause of this variation is that dialysis membranes are porous and are not diffusion limiting. Glucose sensors contain a coating on the outside that serves to restrict glucose diffusion allowing the diffusion step through the membrane rather than some other mass transport step (e.g., tissue diffusion) to be rate limiting.

Both microdialysis sampling and electrochemical-based glucose sensing are flux-based procedures. In other words, both these methods remove glucose from the sample space. The amount of glucose removed per minute from the tissue site can be easily determined with microdialysis sampling by knowing the glucose concentration obtained in the dialysate and the volumetric flow rate. As an example, if 5 mM glucose is obtained in the dialysate and the flow rate is set to 0.3 μL/min (typical flow rate for microdialysis catheters in humans), then the glucose removal rate from the tissue at the probe site is 1.5 nmol/min. It should be noted that typically the flow

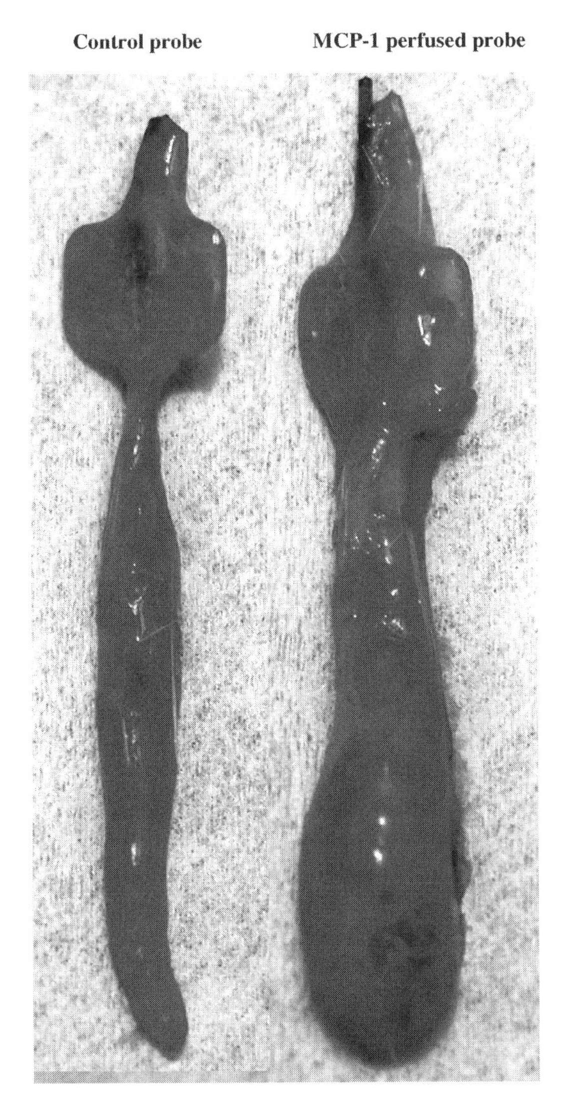

Figure 6.8 Capsule differences between probes infused with or without MCP-1. (See the color version of this figure in Color Plates section.)

rate used for human studies with microdialysis sampling probes is less than flow rates used in animal studies (~1.0 μL/min). Mass collected with microdialysis sampling increases with increasing flow rates. Thus, a 2.5 mM glucose sample collected through a microdialysis probe perfused at 2 μL/min (assuming 50% EE from a 5 mM sample) yields a removal rate of 5 nmol/min. Even though the sample concentration is far lower with higher perfusion flow rates, the total amount of material removed per unit time from the tissue is greater. This is a critically important point to consider between microdialysis sampling devices and glucose sensors since glucose sensors remove far less material per minute. A glucose sensor bathed in 5 mM glucose with a glucose consumption rate of 5×10^{-13} mol/s would

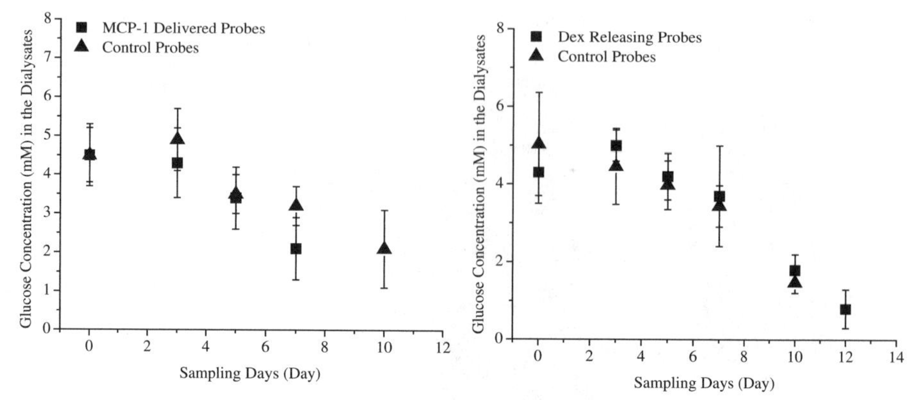

Figure 6.9 Infusion of modulators and long-term collected glucose concentration from microdialysis sampling probes implanted in rat subcutaneous tissue. *Left*: MCP-1 (200 ng/mL) was infused through the microdialysis probes to recruit monocytes. *Right*: dexamethasone phosphate (200 μg/mL) was infused through the probe each day during the 3 h collection period.

yield a removal glucose rate of 0.03 nmol/min.[93] While this may not be a grave concern for measurements of glucose within human adipose tissue, it could cause alternative interpretations to be made between a sensor-based versus microdialysis-based approach for detection of localized glucose concentrations in more energy-sensitive tissues such as the brain.

This potential concern of the measurement device inducing physiological changes due to alteration of normal flux for the targeted analyte has been raised by and debated among many scientists. Many equilibrium-based sensing technologies for glucose have been cited in the literature and discussed in Chapter 10. Well-designed experiments comparing flux- versus equilibrium-based measurements combined with a greater understanding of the analyte flux within a tissue that has a sensor or microdialysis probe implanted are necessary to more fully understand if the measurement technique itself affects measured analyte concentrations.

6.6.1 Effects of Fluid Composition and Probe Type/Geometry for Glucose Collection in Brain

As suggested above, there is considerable interest in collection of glucose in mammalian brain. For collection of glucose in the brain, it has been shown by McNay and Sherwin that minor alterations in the microdialysis sampling perfusion fluid ionic composition causes significant alterations in collected glucose concentrations using a zero net flux-based calibration from probes implanted into the hippocampus.[94] Depending on the perfusion fluid used through the microdialysis probes, the external glucose concentration ranged from 0.5 to 2.2 mM. Perfusion fluid composition that contained low Mg^{2+} resulted in low glucose concentrations (0.5 mM), whereas perfusion fluids containing low K^+ resulted in higher glucose concentrations (2.2 mM). In addition, the choice of dialysis probe used (CMA-11 versus CMA-12) also affected the glucose concentrations with CMA-11 probes exhibiting higher glucose concentrations than the CMA-12. CMA-11 probes have a smaller external diameter and different polymeric membrane composition compared to the CMA-12.

A consensus value from this carefully designed study from McNay and Sherwin is that 1.25 mM glucose be included in microdialysis sampling perfusion fluids used for brain dialysis. This consensus matches previously reported values obtained using NMR techniques.[95]

A more recent study by Robert Kennedy's group where they used in-house created probes along with the capability to have segmented flow as a means to ultimately obtain information about the dynamic concentration changes in the extracellular fluid space surrounding the microdialysis sampling probe. In this work, they found a 1.5 mM glucose concentration in rat brain (nucleus accumbens) and a significant reduction 60% in glucose concentration with a high infusion K^+ spike.[96]

A review of all the different basal concentrations found for glucose in rat brain is well beyond the scope of this chapter. However, these two brief examples highlight one of the constant concerns with microdialysis sampling in sensitive tissues such as the brain. The main concern is how the measurement event itself is actually affecting the analyte that is being measured.

6.7 MICRODIALYSIS FOR CONTROL OF BLOOD GLUCOSE AND/OR DIABETES IN HUMANS

After microdialysis sampling was first reported for applications in clinical studies of glucose in subcutaneous tissue in 1987, various investigations of glucose and overall carbohydrate metabolism began to appear in the literature. The significant clinical questions that could be answered using the more nonselective nature of this sampling method (as compared to a specific method such as an electrochemical glucose sensor) became recognized early on by many researchers. In 1991, Arner and Bolinder pointed out and reviewed the usefulness of microdialysis sampling for both glucose and lipoloysis (glycercol) studies for individuals with various forms of metabolic syndrome.[97]

Many interesting and important clinical problems related to glucose have been addressed using microdialysis sampling over the past two decades. A significant number of these reports have focused on the use of the device for collection of glucose in the brain after head trauma, which has recently been reviewed.[98] Microdialysis has been used to determine the level of glucose changes as well as lipolysis under periods of stress.[99] An important contribution has been the use of microdialysis sampling with a glucose sensor for use in understanding how prevalent nocturnal hypoglycemia is among type 1 diabetics.[100] The technique has also been used in different pediatric settings.[101] Recently, the GlucoDay microdialysis system has been compared with three electrochemical-based sensors for subcutaneous glucose monitoring including the Guardian, DexCom, and Navigator systems. The accuracy of the GlucoDay microdialysis system was similar to the Guardian and Navigator electrochemical-based sensor systems.[102]

6.8 CONCLUSIONS AND FUTURE PERSPECTIVES

Microdialysis and ultrafiltration sampling, combined with electrochemical glucose detection, have been successfully used for continuous glucose monitoring in persons with diabetes. The advantage of microdialysis as well as ultrafiltration sampling in humans with respect to glucose detection is that additional solutes can be collected and detected provided suitable analysis methods exist. For example, glucose and other metabolites such as lactate and glycerol could be detected in the dialysate along with different endocrine peptides such as insulin or leptin under a wide variety of conditions. With the wide availability of microliter volume (5–25 μL) bead-based immunoassays for different endocrine hormones and cytokines,[103] the possibility for measures of analytes associated with metabolism and protein signaling is feasible with both membrane-based sampling techniques. This is the area (e.g., detection of multiple analytes within the dialysate) where microdialysis sampling will likely find its greatest utility in the future with respect to glucose collection in clinical settings. Additional uses of the microdialysis approach will be to use it as a rapid prototype drug delivery device to see how the foreign body reaction or other targeted outcomes change with different pharmaceutical agents.

There has been interest in how localized infusion of insulin through the microdialysis sampling probe affects glucose uptake in the tissue surrounding the microdialysis probe. For skeletal tissue, the local delivery of insulin through the probe caused a decrease in the amount of glucose collected.[104,105] However, for an implanted Menarini microdialysis-based glucose sensor implanted into subcutaneous tissue next to an insulin infusion catheter, it has been reported that localized insulin infusion did not affect glucose measurements from an implanted microdialysis sampling probe.[106]

ACKNOWLEDGMENTS

Microdialysis sampling research in the Stenken laboratory has been funded by the NIH through different grants including NIH EB001441. Many helpful discussions over the years with Professors Bruce Klitzman, Monty Reichert, Ken Ward, and George Wilson, as well as Dr. Natalie Wisniewski and my coeditor, Dr. David Cunningham, are gratefully acknowledged.

REFERENCES

1. Bito L, Davson H, Levin E, Murray M, Snider N. The concentrations of free amino acids and other electrolytes in cerebrospinal fluid, *in vivo* dialysate of brain, and blood plasma of the dog. *Journal of Neurochemistry* 1966, 13, 1057–1067.
2. Leegsma-Vogt G, Janle E, Ash SR, Venema K, Korf J. Utilization of *in vivo* ultrafiltration in biomedical research and clinical applications. *Life Science* 2003, 73, 2005–2018.
3. Huinink KD, Korf J. Ultraslow microfiltration and microdialysis for *in vivo* sampling: principle, techniques, and applications. In: Westerink BHC, Cremers TIFH (Eds), *Handbook of Microdialysis: Methods, Applications and Clinical Aspects*. Elsevier, Amsterdam, 2007.

4. http://www.bioanalytical.com/products/iv/ivultsmp.php.
5. Lunte CE.(ed.), *Analytica Chimica Acta* 1999, 379, 227–369.
6. Elmquist WF, Sawchuk RJ. Microdialysis sampling in drug delivery. *Advanced Drug Delivery Reviews* 2000, 45, 123–307.
7. Stanford C. Microdialysis: recent developments. *Pharmacology Biochemistry and Behavior* 2008, 90, 1–173.
8. Robinson T, Justice JB. *Microdialysis in the Neurosciences*. Elsevier, Amsterdam (The Netherlands), 1991.
9. Westerink BHC, Cremers TIFH. *Handbook of Microdialysis: Methods, Applications and Clinical Aspects*, Elsevier, Amsterdam, 2007.
10. Ungerstedt U, Rostami E. Microdialysis in the human brain: clinical applications. In: Westerink BHC, Cremers TIFH (Eds), *Handbook of Microdialysis: Methods, Applications and Clinical Aspects*. Elsevier, Amsterdam, 2007, pp. 675–686.
11. http://www.microdialysis.se/.
12. Ricci F, Moscone D, Palleschi G. *Ex vivo* continuous glucose monitoring with microdialysis technique: the example of GlucoDay. *IEEE Sensors Journal* 2008, 8, 63–70.
13. Vienken J. Membranes in hemodialysis. In: Peinemann K-V, Nunes SP (Eds), *Membranes for the Life Sciences*, Wiley-VCH, Weinheim, Germany, Vol 1, 2008 pp. 1–48.
14. Clark WR, Hamburger RJ, Lysaght MJ. Effect of membrane composition and structure on solute removal and biocompatibility in hemodialysis. *Kidney International* 1999, 56, 2005–2015.
15. Lysaght MJ. Hemodialysis membranes in transition. *Contributions to Nephrology* 1988, 61, 1–17.
16. Mulder M. *Basic Principles of Membrane Technology*, 2nd edn. Kluwer Academic Publishers, Dordrecht, The Netherlands, 1996.
17. Linhares MC, Kissinger PT. Capillary ultrafiltration: *in vivo* sampling probes for small molecules. *Analytical Chemistry* 1992, 64, 2831–2835.
18. Schneiderheinze JM, Hogan BL. Selective *in vivo* and *in vitro* sampling of proteins using miniature ultrafiltration sampling probes. *Analytical Chemistry* 1996, 68, 3758–3762.
19. Janle-Swain E, Van Vleet JF, Ash SR. Use of a capillary filtrate collector for monitoring glucose in diabetics. *American Society for Artificial Internal Organs* 1987, 33, 336–340.
20. Collison ME, Meyerhoff ME. Chemical sensors for bedside monitoring of critically ill patients. *Analytical Chemistry* 1990, 62, 425A–437A.
21. http://www.bioanalytical.com/products/iv/ivultsmp.php.
22. Janle EM, Clark T, Ash SR. Use of an ultrafiltrate sampling probe to control glucose levels in a diabeteic cat: case study. *Current Separations* 1992, 11, 3–6.
23. Janle EM, Kissinger PT. Monitoring physiological variables with membrane probes. *Acta Astronautica* 1998, 43, 87–99.
24. Schmidt FJ, Sluiter WJ, Schoonen AJM. Glucose concentration in subcutaneous extracellular space. *Diabetes Care* 1993, 16, 695–700.
25. Ash SR, Rainier JB, Zopp WE, Truitt RB, Janle EM, Kissinger PT, Poulos JT. A subcutaneous capillary filtrate collector for measurement of blood chemistries. *American Society for Artificial Internal Organs* 1993, 39, M699–M705.
26. Tiessen RG, Kaptein WA, Venema K, Korf J. Slow ultrafiltration for continuous *in vivo* sampling: applications for glucose and lactate in man. *Analytica Chimica Acta* 1999, 379, 327–335.

27. Doms M, Muller J. Design, fabrication, and characterization of a micro vapor-jet vacuum pump. *Journal of Fluids Engineering* 2007, 129, 1339–1345.
28. Lonnroth P, Jansson PA, Smith U. A microdialysis method allowing characterization of intercellular water space in humans. *American Journal of Physiology* 1987, 253, E228–E231.
29. Schoonen AJ, Schmidt FJ, Hasper H, Verbrugge DA, Tiessen RG, Lerk CF. Development of a potentially wearable glucose sensor for patients with diabetes mellitus: design and *in-vitro* evaluation. *Biosensors & Bioelectronics* 1990, 5, 37–46.
30. Van der Kuil JHF, Korf J. On-line monitoring of extracellular brain glucose using microdialysis and a NADPH-linked enzymic assay. *Journal of Neurochemistry* 1991, 57, 648–654.
31. Keck FS, Kerner W, Meyerhoff C, Zier H, Pfeiffer EF. Combination of microdialysis and glucosensor permits continuous (on line) s.c. glucose monitoring in a patient operated device: I. *In vitro* evaluation. *Hormone and Metabolic Research* 1991, 23, 617–618.
32. Albery WJ, Boutelle MG, Galley PT. The dialysis electrode—a new method for *in vivo* monitoring. *Journal of the Chemical Society, Chemical Communications* 1992, 12, 900–901.
33. Aalders AL, Schmidt FJ, Schoonen AJ, Broek IR, Maessen AG, Doorenbos H. Development of a wearable glucose sensor; studies in healthy volunteers and in diabetic patients. *The International Journal of Artificial Organs* 1991, 14, 102–108.
34. Schmidt FJ, Aalders AL, Schoonen AJM, Doorenbos H. Calibration of a wearable glucose sensor. *International Journal of Artificial Organs* 1992, 15, 55–61.
35. Moscone D, Pasini M, Mascini M. Subcutaneous microdialysis probe coupled with glucose biosensor for *in vivo* continuous monitoring. *Talanta* 1992, 39, 1039–1044.
36. Moscone D, Mascini M. Microdialysis and glucose biosensor for *in vivo* monitoring. *Annales de Biologie Clinique* 1992, 50, 323–327.
37. Keck FS, Meyerhoff C, Kerner W, Siegmund T, Zier H, Pfeiffer EF. Combination of microdialysis and glucosensor permits continuous (online) s.c. glucose monitoring in a patient-operated device. II. Evaluation in animals. *Hormone and Metabolic Research* 1992, 24, 492–493.
38. Meyerhoff C, Bischof F, Sternberg F, Zier H, Pfeiffer EF. Online continuous monitoring of subcutaneous tissue glucose in men by combining portable glucosensor with microdialysis. *Diabetologia* 1992, 35, 1087–1092.
39. Meyerhoff C, Mennel FJ, Bischof F, Sternberg F, Pfeiffer EF. Combination of microdialysis and glucose sensor for continuous online measurement of subcutaneous glucose concentration: theory and practical application. *Hormone and Metabolic Research* 1994, 26, 538–543.
40. Crank J. *The Mathematics of Diffusion*, 2nd edn. Clarendon Press, Oxford, 1975.
41. Cussler EL. *Diffusion: Mass Transfer in Fluid Systems*, Cambridge University Press, Cambridge, 1984.
42. Bungay PM, Morrison PF, Dedrick RL. Steady-state theory for quantitative microdialysis of solutes and water *in vivo* and *in vitro*. *Life Sciences* 1990, 46, 105–119.
43. Clough GF. Microdialysis of large molecules. *AAPS Journal* 2005, 7, E686–E692.
44. Newton AP, Justice JB. Temporal response of microdialysis probes to local perfusion of dopamine and cocaine followed with one-minute sampling. *Analytical Chemistry* 1994, 66, 1468–1472.

45. Bungay PM, Newton-Vinson P, Isele W, Garris PA JusticeJrJB. Microdialysis of dopamine interpreted with quantitative model incorporating probe implantation trauma. *Journal of Neurochemistry* 2003, 86, 932–946.
46. Peters JL, Yang H, Michael AC. Quantitative aspects of brain microdialysis. *Analytica Chimica Acta* 2000, 412, 1–12.
47. Justice JB. Quantitative microdialysis of neurotransmitters. *Journal of Neuroscience Methods* 1993, 48, 263–276.
48. Clough GF, Boutsiouki P, Church MK, Michel CC. Effects of blood flow on the *in vivo* recovery of a small diffusible molecule by microdialysis in human skin. *Journal of Pharmacology and Experimental Therapeutics* 2002, 302, 681–686.
49. Nicholson C. Diffusion and related transport mechanisms in brain tissue, *Reports on Progress in Physics* 2001, 64, 815–884.
50. Guyton AC, Hall JE. *Textbook of Medical Physiology*, 9th edn. W.B Saunders Company, Toronto, 1996, pp. 305–311.
51. Kety SS. The theory and applications of the exchange of inert gas at the lungs and tissues. *Pharmacological Reviews* 1951, 3, 1–41.
52. Dykstra KH, Hsiao JK, Morrison PF, Bungay PM, Mefford IN, Scully MM, Dedrick RL. Quantitative examination of tissue concentration profiles associated with microdialysis. *Journal of Neurochemistry* 1992, 58, 931–940.
53. Lehmenkühler A, Syková E, Svoboda J, Zilles K, Nicholson C. Extracellular space parameters in the rat neocortex and subcortical white matter during postnatal development determined by diffusion analysis. *Neuroscience* 1993, 55, 339–351.
54. Hrabetova S, Nicholson C. Biophysical properties of brain extracellular space explored with ion-selective microelectrodes, integrative optical imaging and related techniques. In: Michael AC, Borland LM (Eds), *Electrochemical Methods for Neuroscience*. CRC Press/ Taylor & Francis, Boca Raton, FL, 2007 167–204.
55. Stanisz GJ. Diffusion MR in biological systems: tissue compartments and exchange. *Israel Journal of Chemistry* 2003, 43, 33–44.
56. Groth L, Jorgensen A, Serup J. Cutaneous microdialysis in the rat: insertion trauma and effect of anaesthesia studied by laser Doppler perfusion imaging and histamine release. *Skin Pharmacology and Applied Skin Physiology* 1998, 11, 125–132.
57. Anderson C, Andersson T, Wardell K. Changes in skin circulation after insertion of a microdialysis probe visualized by laser Doppler perfusion imaging. *Journal of Investigative Dermatology* 1994, 102, 807–811.
58. Wientjes KJC, Grob U, Hattemer A, Hoogenberg K, Jungheim K, Kapitza C, Schoonen AJM. Effects of microdialysis catheter insertion into the subcutaneous adipose tissue assessed by the SCGM1 system. *Diabetes Technology & Therapeutics* 2003, 5, 615–620.
59. Mou X, Stenken JA. Microdialysis sampling extraction efficiency of 2-deoxyglucose: role of macrophages *in vitro* and *in vivo*. *Analytical Chemistry* 2006, 78, 7778–7784.
60. Anderson JM, Rodriguez A, Chang DT. Foreign body reaction to biomaterials. *Seminars in Immunology* 2008, 20, 86–100.
61. Wisniewski N, Klitzman B, Miller B, Reichert WM. Decreased analyte transport through implanted membranes: differentiation of biofouling from tissue effects. *Journal of Biomedical Materials Research* 2001, 57, 513–521.
62. Wisniewski N, Rajamand N, Adamsson U, Lins PE, Reichert WM, Klitzman B, Ungerstedt U. Analyte flux through chronically implanted subcutaneous polyamide

membranes differs in humans and rats. *American Journal of Physiology* 2002, 282, E1316–E1323.

63. Wientjes KJ, Vonk P, Vonk-van KY, Schoonen AJ, Kossen NW. Microdialysis of glucose in subcutaneous adipose tissue up to 3 weeks in healthy volunteers. *Diabetes Care* 1998, 21, 1481–1488.
64. Wientjes KJC, Hoogenberg K, Schoonen AJM. Transport of glucose to a probe in adipose tissue. In: Westerink, Cremers (Eds), *Handbook of Microdialysis*. Academic Press, 2007.
65. Lutgers HL, Hullegie LM, Hoogenberg K, Sluiter WJ, Dullaart RPF, Wientjes KJ, Schoonen AJM. Microdialysis measurement of glucose in subcutaneous adipose tissue up to three weeks in type 1 diabetic patients. *Netherlands Journal of Medicine* 2000, 57, 7–12.
66. Thomas K, Kiwit M, Kerner W. Glucose concentration in human subcutaneous adipose tissue: comparison between forearm and abdomen. *Experimental and Clinical Endocrinology & Diabetes* 1998, 106, 465–469.
67. Kaptein WA, Zwaagstra JJ, Venema K, Korf J. Continuous ultraslow microdialysis and ultrafiltration for subcutaneous sampling as demonstrated by glucose and lactate measurements in rats. *Analytical Chemistry* 1998, 70, 4696–4700.
68. Stenken JA. Methods and issues in microdialysis calibration. *Analytica Chimica Acta* 1999, 379, 337–357.
69. Klonoff DC. Microdialysis of interstitial fluid for continuous glucose measurement. *Diabetes Technology & Therapeutics* 2003, 5, 539–543.
70. MacLean DA, Ettinger SM, Sinoway LI, Lanoue KF. Determination of muscle-specific glucose flux using radioactive stereoisomers and microdialysis. *American Journal of Physiology* 2001, 280, E187–E192.
71. Newman JMB, Ross RM, Richards SM, Clark MG, Rattigan S. Insulin and contraction increase nutritive blood flow in rat muscle *in vivo* determined by microdialysis of L-[14C] glucose. *Journal of Physiology* 2007, 585, 217–229.
72. Gustafsson J, Eriksson J, Marcus C. Glucose metabolism in human adipose tissue studied by 13C-glucose and microdialysis. *Scandinavian Journal of Clinical and Laboratory Investigation* 2007, 67, 155–164.
73. Peltoniemi P, Lonnroth P, Laine H, Oikonen V, Tolvanen T, Gronroos T, Strindberg L, Knuuti J, Nuutila P. Lumped constant for [18F]fluorodeoxyglucose in skeletal muscles of obese and nonobese humans. *American Journal of Physiology* 2000, 279, E1122–E1130.
74. Wang Y, Wong SL, Sawchuk RJ. Microdialysis calibration using retrodialysis and zero-net flux: application to a study of the distribution of zidovudine to rabbit cerebrospinal fluid and thalamus. *Pharmaceutical Research* 1993, 10, 1411–1419.
75. Bouw MR, Hammarlund-Udenaes M. Methodological aspects of the use of a calibrator in *in vivo* microdialysis—further development of the retrodialysis method. *Pharmaceutical Research* 1998, 15, 1673–1679.
76. Eisenberg EJ, Eickhoff WM. A method for estimation of extracellular concentration of compounds by microdialysis using urea as an endogenous recovery marker *in vitro* validation. *Journal of Pharmacology and Toxicology* 1993, 30, 27–31.
77. Strindberg L, Lonnroth P. Validation of an endogenous reference technique for the calibration of microdialysis catheters. *Scandinavian Journal of Clinical and Laboratory Investigation* 2000, 60, 205–212.
78. Sorg BS, Peltz CD, Klitzman B, Dewhirst MW. Method for improved accuracy in endogenous urea recovery marker calibrations for microdialysis in tumors. *Journal of Pharmacological and Toxicological Methods* 2005, 52, 341–349.

79. Brunner M, Joukhadar C, Schmid R, Erovic B, Eichler HG, Muller M. Validation of urea as an endogenous reference compound for the *in vivo* calibration of microdialysis probes. *Life Sciences* 2000, 67, 977–984.

80. Reach G, Wilson GS. Can continuous glucose monitoring be used for the treatment of diabetes. *Analytical Chemistry* 1992, 64, 381A–386A.

81. Ståhle L. Pharmacokinetic estimations from microdialysis data. *European Journal of Clinical Pharmacology* 1992, 43, 289–294.

82. Jansson PA, Fowelin J, Smith U, Lonnroth P. Characterization by microdialysis of intracellular glucose level in subcutaneous tissue in humans. *American Journal of Physiology* 1988, 255, E218–E220.

83. Kissinger PT, Shoup RE. Optimization of LC apparatus for determinations in neurochemistry with an emphasis on microdialysis samples. *Journal of Neuroscience Methods* 1990, 34, 3–10.

84. Albery WJ, Boutelle MG, Galley PT. The dialysis electrode: a new method for *in vivo* monitoring. *Journal of the Chemical Society Chemical Communications* 1992, 12, 900–901.

85. Van der Kuil JHF, Korf J. On-line monitoring of extracellular brain glucose using microdialysis and a NADPH-linked enzymic assay. *Journal of Neurochemistry* 1991, 57, 648–654.

86. http://www.biotechproducts.com/syco-mdb.html (accessed on June 23, 2008).

87. Kondepati VR, Heise HM. Recent progress in analytical instrumentation for glycemic control in diabetic and critically ill patients. *Analytical and Bioanalytical Chemistry* 2007, 388, 545–563.

88. Riley MR, Arnold MA, Murhammer DW, Walls EL, DelaCruz N. Adaptive calibration scheme for quantification of nutrients and byproducts in insect cell bioreactors by near-infrared spectroscopy. *Biotechnology Progress* 1998, 14, 527–533.

89. Stenken JA, Reichert WM, Klitzman B. Magnetic resonance imaging of a tissue/implanted device biointerface using *in vivo* microdialysis sampling. *Analytical Chemistry* 2002, 74, 4849–4854.

90. Mou X. Modulation of foreign body response towards implanted microdialysis sampling probes. Ph.D. Dissertation. Rensselaer Polytechnic Institute, 2007.

91. Norton LW, Koschwanez HE, Wisniewski NA, Klitzman B, Reichert WM. Vascular endothelial growth factor and dexamethasone release from nonfouling sensor coatings affect the foreign body response. *Journal of Biomedical Materials Research, Part A* 2007, 81A, 858–869.

92. Cooper JD, Heppert KE, Davies MI, Lunte SM. Evaluation of an osmotic pump for microdialysis sampling in an awake and untethered rat. *Journal of Neuroscience Methods* 2007, 160, 269–275.

93. Wilson GS. Personal Communication. May 2008.

94. McNay EC, Sherwin RS. From artificial cerebro-spinal fluid (aCSF) to artificial extracellular fluid (aECF): microdialysis perfusate composition effects on *in vivo* brain ECF glucose measurements. *Journal of Neuroscience Methods* 2004, 132, 35–43.

95. Mason GF, Behar KL, Rothman DL, Shulman RG. NMR determination of intracerebral glucose concentration and transport kinetics in rat brain. *Journal of Cerebral Blood Flow and Metabolism* 1992, 12, 448–455.

96. Wang M, Roman GT, Schultz K, Jennings C, Kennedy RT. Improved temporal resolution for *in vivo* microdialysis by using segmented flow. *Analytical Chemistry* 2008, 80, 5607–5615.

97. Arner P, Bolinder J. Microdialysis of adipose tissue. *Journal of Internal Medicine* 1991, 230, 381–386.
98. Bellander B-M, Cantais E, Enblad P, Hutchinson P, Nordstrom C-H, Robertson C, Sahuquillo J, Smith M, Stocchetti N, Ungerstedt U, Unterberg A, Olsen NV. Consensus meeting on microdialysis in neurointensive care. *Intensive Care Medicine* 2004, 30, 2166–2169.
99. Hagstroem-Toft E, Arner P, Wahrenberg H, Wennlund A, Ungerstedt U, Bolinder J. Adrenergic regulation of human adipose tissue metabolism *in situ* during mental stress. *Journal of Clinical Endocrinology and Metabolism* 1993, 76, 392–398.
100. Wentholt IM, Maran A, Masurel N, Heine RJ, Hoekstra JB, DeVries JH. Nocturnal hypoglycaemia in Type 1 diabetic patients, assessed with continuous glucose monitoring: frequency, duration and associations. *Diabetic Medicine* 2007, 24, 527–532.
101. Hack A, Busch V, Gempel K, Baumeister FAM. Subcutaneous microdialysis for children—safe biochemical tissue monitoring based on a minimal traumatizing no touch insertion technique. *European Journal of Medical Research* 2005, 10, 419–425.
102. Kovatchev B, Anderson S, Heinemann L, Clarke W. Comparison of the numerical and clinical accuracy of four continuous glucose monitors. *Diabetes Care* 2008, 31, 1160–1164.
103. Duo J, Fletcher H, Stenken JA. Natural and synthetic affinity agents as microdialysis sampling mass transport enhancers: current progress and future perspectives. *Biosensors & Bioelectronics* 2006, 22, 449–457.
104. Rosdahl H, Hamrin K, Ungerstedt U, Henriksson J. A microdialysis method for the *in situ* investigation of the action of large peptide molecules in human skeletal muscle: detection of local metabolic effects of insulin. *International Journal of Biological Macromolecules* 2000, 28, 69–73.
105. Hamrin K, Henriksson J. Interstitial glucose concentration in insulin-resistant human skeletal muscle: influence of one bout of exercise and of local perfusion with insulin or vanadate. *European Journal of Applied Physiology* 2008, 103, 595–603.
106. Hermanides J, Wentholt IM, Hart AA, Hoekstra JB, DeVries JH. No apparent local effect of insulin on microdialysis continuous glucose-monitoring measurements. *Diabetes Care* 2008, 31, 1120–1122.

CHAPTER 7

TRANSDERMAL MICROFLUIDIC CONTINUOUS MONITORING SYSTEMS

David D. Cunningham

7.1 INTRODUCTION

7.1.1 Miniaturization and Transdermal Microfluidics

A number of new medical devices have grown out of recent fundamental scientific and engineering programs aimed at control of small volumes of fluid and microfabrication of sensors and electronics. In a general context, these fundamental efforts fall under the umbrellas of microelectromechanical systems (MEMS) and micro-total analytical systems (μTAS). In the last decade, microfluidic systems based on a range of fundamental principles, and microfabrication technology developed in specialized foundries, have progressed to the stage where reliable devices can be constructed. Glucose monitoring is one of the most attractive commercial opportunities for MEMS

In Vivo Glucose Sensing, Edited by David D. Cunningham and Julie A. Stenken

and μTAS technologies, so several companies have invested heavily in product development. The general goal of these commercial projects is to produce an untethered device adhering to the skin that extracts and senses small amounts of physiological fluid, producing a continuous glucose reading for up to a week. The classification of transdermal microfluidic systems is used to denote that the physiological fluid is extracted prior to the measurement of glucose rather than inserting a sensor into tissue where a capsule may form around the sensor. In a transdermal microfluidic system, development of the glucose sensor component may be easier since calibration can be performed *ex vivo* and the sensor is not in direct contact with body tissue so the foreign body reaction is avoided. However, balancing these potential advantages is the very small amount of extracted physiological fluid available for measurement.

A clear trend toward increasing complexity and miniaturization is evident in the transdermal microfluidic devices that have reached research-stage clinical trials and the market. The remaining sections of this chapter are organized according to the method used to permeabilize the skin and extract physiological fluid, specifically (1) direct current (DC) electrical current to produce reverse iontophoresis, (2) low-frequency ultrasound, (3) ablation to disrupt the top layer of skin, and (4) surface penetration of the skin by microneedles. Reverse iontophoresis utilizes DC electrical current to form small pores in the skin and iontophoresis (or electrophoresis) to extract interstitial fluid. Cygnus developed and marketed the GlucoWatch™ based on extraction with reverse iontophoresis and measurement with an electrochemical glucose oxidase enzyme electrode.[1–8] Sontra Medical, now Echo Therapeutics, led development of the SonoPrep low-frequency ultrasound device that permeabilizes the skin and allows osmotic extraction and continuous measurement.[9] SpectRx (now Guided Therapeutics, Inc.) developed a handheld laser that porates the non-viable, superficial portion of the epidermis allowing extraction of interstitial fluid with vacuum for several days.[10–12] So far, two companies have reported results for integrated single-use microneedle systems[13,14] and many academic groups are beginning to report clinical results with components of a continuous microneedle-based system.[15–17] The glucose measuring systems entering the market have increased in complexity with some devices incorporating a vacuum pump and micromechanical movement of the glucose sensor to the sample.[18] Given the scientific, technical, and manufacturing infrastructure that is now in place, development of additional miniaturized, transdermal microfluidic systems seems likely. At some point, the line between microneedle systems (defined as having penetration depths less than 1 mm) and transdermal sensors (see Chapter 5) may intersect if sensors become smaller and shorter.

7.1.2 Properties of the Skin and Interstitial Fluid (ISF)

Interstitial fluid bathes the cells of the body and is protected from the external environment by the top layer of skin termed as the epidermis. The epidermis has an irregular thickness of 40–100 μm and is not vascularized, so viable cells receive nutrients by diffusion. The outer, dead, nonviable portion of the epidermis, termed as the stratum corneum, is 6–22 μm thick and about one-cell layer per day flakes off.[19,20]

The glucose content of interstitial fluid is close to that of capillary blood; however, a small time lag may exist between a change in blood glucose concentration and a change in ISF glucose concentration.[21–23] Numerous physiological and environmental factors may influence the time lag, but blood flow in the skin, as affected by exercise and temperature, is likely a principal driver.[24]

Perhaps the oldest method for accessing interstitial fluid for scientific study is by puncturing the skin with a needle and placing a short fiber wick through the hole. Only a small amount of fluid can be sampled and some of the ISF evaporates. An early observation was that ISF is at a slightly negative pressure compared to the atmosphere, increasing the difficulty of obtaining fluid. Application of mild vacuum over a period of time results in a well-defined separation of the epidermis from the dermis and a suction blister filled with ISF.[22,25] Sequential applications of adhesive tape can be used to remove the stratum corneum layer by layer; however, removal becomes less effective as more hydrated cell layers are exposed. After 30 or more applications of tape, a thin basement membrane remains, in some cases, limiting access to ISF. As discussed further in Section 7.4, thermal and laser ablation have been used to make micropores on the order of 100 μm diameter through the epidermis. Application of a DC voltage across the skin during iontophoresis results in the formation of negatively charged meso- to nanoscale pores through areas where the initial electrical resistance is lowest, which is near hair follicles and sweat glands.[26] Low-frequency ultrasound increases the permeability of the skin through implosion of hot gaseous bubbles in the liquid coupling media, termed cavitation, resulting in discrete, permeabilized regions over the treated area.[27] Ultrafiltration and microdialysis are generally used to sample ISF from deeper in the skin and subcutaneous tissue as described in Chapter 6. Interstitial fluid is generally clear with a yellow tinge, while any pink or red color denotes the presence of red blood cells. Unlike blood, interstitial fluid does not easily clot, so extraction for continuous glucose testing is possible.

The glucose concentration in small amounts of extracted ISF is susceptible to contamination by glucose in the stratum corneum and dilution by sweat that has a low glucose concentration. Water or simple buffer solutions placed on the skin dissolve measurable amounts of glucose (i.e., 60 ng/cm^2), while the total glucose content of the stratum corneum is much higher (i.e., 360 ng/cm^2).[28] Normal, invisible water loss through the skin is $\sim$2 μL/cm^2/h, while the nominal level of visible perspiration from light work is much higher, 12 μL/cm^2/h. Obviously, a small ISF sample can be significantly diluted by sweat. Sweat can also build up, hydrating the skin and leading to a failure of adhesion between the skin and a device. To avoid hydration of the skin and delay loss of adhesion, moisture vapor transport channels can be designed into the structure of the device to improve the long-term attachment of the device to the body.[29]

Iontophoresis produces an unusual tingling sensation, and mild erythema and edema skin reactions in most cases. A more severe skin reaction is encountered 10–15% of the time, but preapplication of a corticosteroid may mitigate the effects.[30] The cavitation of low-frequency ultrasound also produces an unusual sensation that is generally well tolerated. Tape stripping of the stratum corneum is tedious and relatively unpleasant. Healing after tape stripping generally takes about a month, but the skin color can remain darker, in some cases, for a much longer time. During vacuum blister, formation a tingling sensation is often experienced prior to separation of the skin layers and healing is commonly a little faster than after that tape stripping. The experience with microneedles is expected to be similar to that of standard lancets. Lancet sticks on body sites other than the finger are most often painless; however, about 5% of the time the lancet may hit a nerve and be as painful as a fingerstick.[31] Erythema or red spots from lancet sticks generally resolve in 3 days but are more visible than lancet sticks on the finger since the thicker stratum corneum on the finger obscures the damage.[32]

Sample collection efficiencies are expected to vary from person to person and from test to test within a person due to variations in the device as well as variations in the skin and the underlying tissue. Unfortunately, data sets reported in the literature for sampling of interstitial fluid are generally not detailed enough to calculate this variability. As an indication of the degree of variability that may be encountered, a relative standard deviation was 70–88%, depending on the study design, was observed in the case of blood extraction from lancet wounds using vacuum.[33] Thus, care should be taken to design clinical studies with adequate statistical power to discern changes due to the processes under study or prototype design from normal anatomical and physiological variability.

The interstitial fluid content of the skin is higher than in the subcutaneous fat layer and normal fluid movement is intrinsically linked to lymphatic drainage as governed by mechanical stresses of the tissue. A model of temporal profiles of pressure, stress, and convective ISF velocity has been developed based on hydraulic conductivity, overall fluid drainage (lymphatic function and capillary absorption), and elasticity of the tissue.[34] Measurements on excised tissue and *in vivo* measurement on the one-dimensional rat tail have defined bulk average values for key parameters of the model and the hydration dependence of the hydraulic flow conductivity. Numerous *in vivo* characterization studies with nanoparticles and vaccines are currently underway, so a more detailed understanding of the interstitial/lymphatic system will likely be forthcoming.

7.1.3 Glucose Sensing at Low Flow Rates

Many physical and chemical methods have been applied to the measurement of glucose, but only enzymatic methods have found commercial success, so far, in the form of single-use test strips. In these test strips, optical detection of color formation predominated a decade ago, but electrochemical detection is now more common. The drivers for this change are the lower volume required for electrochemical strips and the lower cost and complexity of the components needed for detection. Large-scale manufacturing now produces electrochemical glucose test strips requiring only 300 nL of sample with a retail price of less than 1 U.S. dollar (USD) per test strip.[35] The most capable low-flow sensors reported to date have been fabricated from silicon wafers with photopolymerized coatings.[36,37] In one format, the sensor also contains an integrated spike to open an ISF-filled blister formed by vacuum, and the sample volume is on the order of 2.5 μL, as shown in Figure 7.1. Lower cost manufacturing options such as screen printing, metal sputtering on plastic, lamination of layers, and laser cutting would likely be implemented in commercial products.

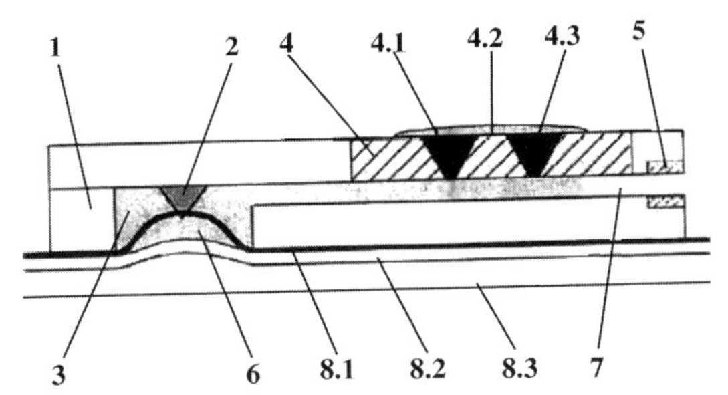

Figure 7.1 Electrochemical sensor with (1) plastic spacer, (2) spike for opening ISF-filled blister, (3) ISF, (4.1) working electrode, (4.2) encapsulation of electrical leads, (4.3) reference electrode, (5) connection to vacuum, (6) ISF-filled blister, (7) fluidic pathway, (8.1) stratum corneum, (8.2) epidermis, and (8.3) dermis. Reprinted with permission from Ref. 36. Copyright 2002 Verlag Walter de Gruyter GmbH & Co. KG.

Sensing undiluted interstitial fluid flowing at a slow rate presents several challenges. Extraction rates achieved so far typically run less than half a microliter per minute. Soluble reagents are generally used in single-use test strips but are not amenable to continuous sensing. Fortunately, technology for immobilization of the sensing reagents is well developed as described in Section 5.3. The glucose oxidase reaction consumes one mole of oxygen for each mole of glucose; however, the oxygen concentration (ca. 0.3 mM) is much lower than the ISF glucose concentration (3–35 mM). In single-use test strips, high concentrations of mediator, that take the place of oxygen in the enzyme reaction, are used to allow the reaction to go to completion. However, in a continuous sensor, completion of the reaction would lead to consumption of glucose in the sample and the sensor would be very dependent on the flow rate. To address both the oxygen limitation and glucose depletion issues, coatings that significantly slow down the diffusion of glucose and are more permeable to oxygen than glucose are applied. Although most coatings are proprietary, silicons are known to have a very high permeabilty to oxygen, and urethanes are relatively easy to apply from nonaqueous solutions. At very low flow rates, hydrogen peroxide generated in the sensing layer may diffuse out and accumulate on the top of the sensor leading to a higher signal at low flow rates. Inclusion of catalase enzyme in the outermost layer leads to the formation of oxygen and water from hydrogen peroxide diffusing out of the sensing layer and eliminates any increase in hydrogen peroxide signal at low flow rates. Some sensors also include a layer near the working electrode to exclude electroactive interferences such as ascorbic acid and acetaminophen. One sensor of note has a layered composition platinum/electropolymerized diaminobenzene/glucose oxidase in a photopolymerized 2-hydroxyethyl methacrylate (HEMA) gel/diffusion limiting catalyase layer.[37] The sensor response was constant at 0.4 nA/mM down to a flow rate of 0.1 μL/min with a total time lag of ~3.5 min (1.5 min due to the sensor and 3 min due to the 300 nL fluidic dead space).[37] Measurement of ISF flow through the sensor would be useful to assure that there was adequate flow to produce an accurate reading. However, there is no obvious flow sensing technology available for an inexpensive wearable device at present.

7.2 REVERSE IONTOPHORESIS-BASED SYSTEMS

The GlucoWatch developed by Cygnus is based on reverse iontophoresis that extracts small amounts of glucose using a DC electrical current through the skin. The original development and clinical performance of the system are exceptionally well documented in the literature (see selected Refs 1–8, 30, and citations within). Reverse iontophoresis was originally proposed by Richard Guy for sampling glucose based on his previous experience with iontophoresis-based transdermal drug delivery.[38] Under typical iontophoresis conditions, application of a 0.5 mA/cm^2 DC current across the skin leads to formation of pores near the hair follicles and sweat glands, and the initial 150–400 kΩ electrical resistance of the skin decreases to a negligible value after a few seconds[39] Battery-operated iontophoretic drug delivery devices apply up to 70 V across two patches applied to the skin to generate the desired DC current. As noted, the stratum corneum layer of the skin has a high electrical resistance compared to the interstitial fluid, so the path of least electrical resistance is down through the stratum corneum, across the conductive interstitial fluid, and up through the stratum corneum to the second electrode. Sodium ions carry most of the ionic current since sodium is abundant ($\sim$140 mM) and components of the skin impart a net fixed negative charge ($pK_a \sim 4$). A plug-like osmotic flow is generated through the pores in the skin, so electrically neutral solutes including glucose are carried toward the cathode, similar to the electro-osmotic flow observed in glass capillary electrophoresis. The volume of ISF extracted is directly proportional to the amount of DC electrical current passed through the skin.

A photograph of the Glucowatch and an exploded view of the sensor are shown in Figures 7.2 and 7.3. To use the system, a disposable hydrogel/sensor portion is placed securely into the reusable electronic portion of the watch using a special clamp device to assure uniform contact between the gel and electrodes. The liner covering the gel is removed and the system placed on the wrist. After 3 h, a whole-blood glucose measurement is taken with a standard glucose meter and the reading is entered into the

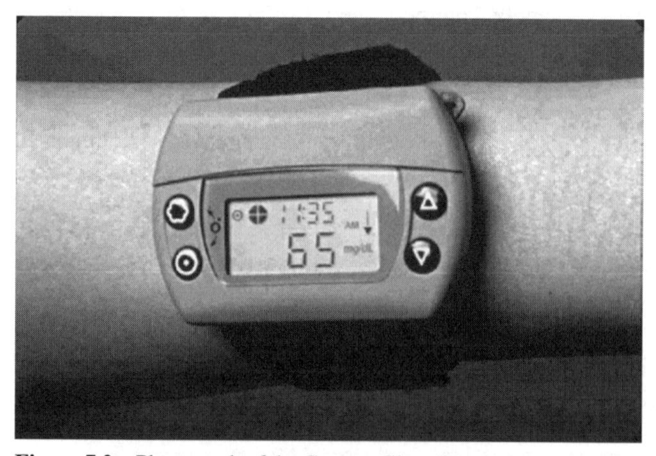

Figure 7.2 Photograph of the Cygnus GlucoWatch showing the time of day, glucose reading, and glucose trend arrow. Reprinted with permission from Ref. 6. Copyright 2001 Elsevier.

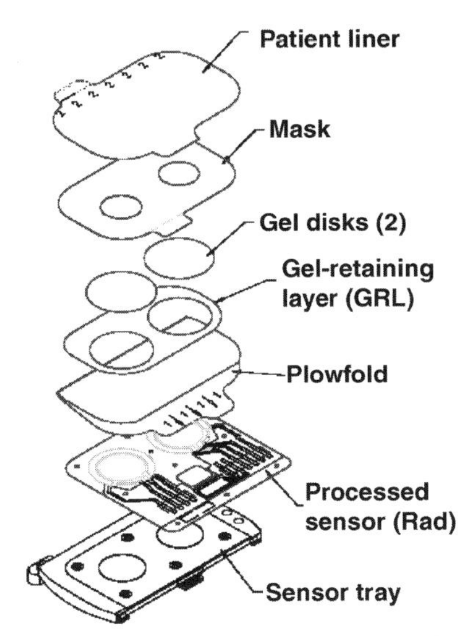

Figure 7.3 Exploded view of the Cygnus GlucoWatch showing the components of the disposable hydrogel electrode and the underlying reusable electronic leads. Reprinted with permission from Ref. 6. Copyright 2001 Elsevier.

GlucoWatch to standardize the sensor. Afterward, the device provides up to three glucose readings per hour for as long as 12 h.

Each sensor contains two identical sets of iontophoresis electrodes and glucose sensing electrodes. For each measurement, 0.3 mA of current is passed for 3 min, and then the extracted glucose is measured for 7 min. The polarity of the system is then reversed and the other electrodes are used for extraction and measurement. The iontophoresis electrodes are screen printed Ag and Ag/AgCl layers patterned as a 1.6 cm diameter, 1 cm^2 ring around the glucose sensor disk. The silver and silver chloride are active components since when used as an anode, silver is oxidized and precipitates as silver chloride, while when used as a cathode, the silver chloride layer is reduced to silver. An electrochemically inert iontophoresis electrode, such as gold or platinum, is less desirable since hydrolysis of water would occur to generate acid at the anode and hydroxide ion at the cathode leading to extremes in pH that could inactivate the sensitive sensor components and irritate the skin.

The glucose sensing electrodes consist of 0.65 cm diameter disks of screen printed platinum–carbon ink, and the iontophoresis electrodes are used as the reference/counter electrode during measurement. Both sets of electrodes are covered with a 127 μm thick glucose oxidase containing hydrogel. The 3 min application of 0.3 mA current during iontophoresis typically draws 15–150 nL of interstitial fluid through the skin. The extracted glucose reacts with glucose oxidase in the hydrogel forming hydrogen peroxide. The hydrogen peroxide then diffuses to the platinum–carbon electrode and is oxidized at 0.42 V versus a reference electrode Ag/AgCl. As an additional complication, interstitial fluid contains both anomeric forms of glucose (37% α-form and 63% β-form) but only the β-form reacts with glucose oxidase. As the β-form is depleted, the α-form mutarotates (converts) to the β-form to reestablish the equilibrium. The sensor is designed to measure all of the glucose extracted without allowing buildup of glucose from cycle to cycle. Thus, sufficient measurement time is required for diffusion, enzyme reaction, and mutarotation. An impermeable collimating mask is used at the skin surface, so glucose can only enter the hydrogel from directly under the glucose electrode. The reaction–diffusion parameters have been modeled to optimize the collection and measurement times.[2]

Considerable effort was made to develop signal checking and processing algorithms for the GlucoWatch.[4] Conductivity and temperature sensors were incorporated into the system to flag measurements that could be inaccurate due to sweating or temperature influences on diffusion. Limits were also set for the average glucose electrode signal and iontophoresis electrode voltage. Typically, about 20% of the time the measurements were flagged or failed to meet software signal quality checks and the glucose readout was "skipped." The integrated sensor current was generally linear with glucose concentration, with a 2000 nC intercept; however, better accuracy was achieved by inclusion of other factors: (1) elapsed time since calibration, (2) blood glucose value at calibration, and (3) signal value at calibration. A number of algorithms were evaluated for converting the raw signal data into a glucose result. Simple linear regression gave a mean absolute relative difference of 23%, while the values for the other methods were generally lower: adaptive nearest neighbor (23%), multivariate adaptive regression splines (15%), neural networks (19%), and mixture of experts (14%). The mixture of experts method is computationally less demanding than the other methods and was incorporated into the GlucoWatch software after optimization and weighting of the linear summation of input parameters.

In vitro testing showed that the sensor gave precise results across the expected glucose range with coefficients of variation in the 5–7% range. However, the clinical trial results on 92 subjects showed a mean absolute relative error compared to standard measurements on fingerstick samples of 16% with a time lag of 18 min.[5] Continuous monitoring potentially should allow detection of hypoglycemia and triggering of a warning alarm. The receiver operator characteristic (ROC) curves for the system were analyzed at various alert levels and ~5.6 mM suggested as optimal.[7] However, at this level the alarm would go off almost every day with about 10% false alarms. The performance of a second-generation system that reduces the warm-up time before fingerstick calibration to 2 h and reports out values every 10 min was evaluated in 89 children and adolescents with similar results to the study above; mean absolute relative error of was 22%.[8]

The GlucoWatch was approved by the FDA in 2000 and marketed in the United States, but relatively few were sold. The devices were expensive relative to current meters, required a fingerstick, left visible irritation after use, in many cases, and did not seem to work well in hot weather. Cygnus was purchased by Animas in 2004 and Animas was purchased by Johnson & Johnson in 2005. More recently, an internal standard approach has been suggested to eliminate the need for fingerstick calibration.[40] The physiological sodium ion concentration does not vary significantly, so measurement of sodium ion in the extracted ISF might be used to measure the amount of sample extracted.

7.3 ULTRASOUND-BASED SYSTEMS

Low-frequency ultrasound was originally identified as a method for increasing the permeability of the skin for transdermal drug delivery and has since been applied to glucose monitoring (see Refs 9 and 38–43 and citations within). Ultrasound has been divided into high frequency (above 3 MHz), therapeutic (0.75–3 MHz), and low frequency (less than 750 KHz). In 1995, a remarkable increase in the permeability of human skin was reported with modest intensities of 20 kHz ultrasound.[41] Factors associated with the enhancement were thought to include cavitation, microstreaming fluid flow caused by cavitation, and an increase in the temperature. At 20 kHz, the surface of the ultrasonic device moves about 11 μm, causing oscillating pressures in the liquid medium. Nucleation of the pressure leads to transient cavitation (formation, growth, and collapse of gas bubbles) and stable cavitation (slow oscillatory motion of the bubble). Cavitation is inversely proportional to frequency since at higher frequencies the time between the positive and negative acoustic pressures becomes short compared to the time required for dissolved gas within the liquid to diffuse into the cavitation nuclei. Besides frequency and intensity, the temperature and surface tension of the liquid affect the amount of cavitation.

Implosion of the gaseous bubble close to the skin during transient cavitation leads to the formation of micropores through the stratum corneum, as originally measured by a large decrease in electrical resistance.[41,42] Additional work demonstrated that 20 kHz ultrasound produced localized areas about 3–5 mm across that exhibit increased permeability.[27,47,48] Higher intensities produced smaller, more disperse areas of increased permeability, as shown in Figure 7.4, and the number of areas increased with the duration of ultrasound treatment. Based on these results, frequencies near 60 kHz are preferred for achieving good dispersity of transport regions using a reasonable energy density and application time. Additional insight into the tortuosity, hindrance factor, and transcellular component of the transdermal pathways has recently been uncovered using fluorescent two-photon imaging and radiolabeled diffusion masking experiments.[47,48] Other factors such as the distance between the ultrasonic horn and skin have been studied; however, a complete

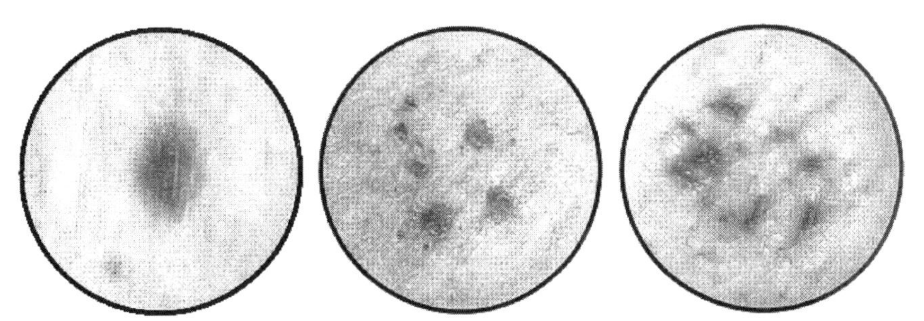

Figure 7.4 Optical images of transport regions created in full thickness pig skin using 0.84 W/cm^2 low-frequency ultrasound as visualized with sulforhodamine B. 20 kHz for 6 min (left), 37 kHz for 9 min (middle), and 59 kHz for 14 min (right). Image area is 1.8 cm^{-2}. Reprinted with kind permission from Springer Science + Business Media, from Ref. 27. Copyright 2001 Springer.

understanding of key factors and methods to control formation of local areas of increased permeability is not yet available.

A relatively simple quantitative model describing transport of small molecules (<500 MW) by low-frequency ultrasound was described as follows:

$$\text{Permeability} = \frac{K\phi D}{N\ell}\ (\text{cm/h}) \tag{7.1}$$

where K is the permeant partition coefficient in aqueous channels (assumed to be 1 for glucose), ϕ is the average fractional area occupied by the aqueous channels generated by ultrasound, D is the diffusion coefficient, N is the number of intercellular lipid regions that a molecule has to cross, and ℓ is the thickness of each lipid region.[42] The permeability calculated using this model was very high (0.057) due to the small denominator (15 lipid regions × 0.05 μm = 0.75 μm). A more appropriate value for the denominator might be a greater percentage of the full thickness of the stratum cornium (~13 ± 5 μm).[19] The extraction rate or flux is directly related to the permeability (extraction rate or flux = permeability × difference in concentration), so the volume of extracted fluid is readily calculated from permeability results or glucose concentrations reported in the literature. A much more complex mathematical model including 21 variables has been used to describe vacuum-induced transport of ISF through ultrasonically permeabilized skin.[44] Under vacuum extraction condition, 90% of the glucose is extracted by vacuum-induced convection and 10% by diffusion. This comprehensive model was also used to calculate that the lag time between changes in whole-blood glucose concentration and the concentration in extracted ISF would be 2 min.

A variety of ultrasonic treatment and extraction conditions have been evaluated as summarized in Table 7.1. Two early studies showed that a much higher permeability was observed for sucrose than for insulin upon continuous, pulsed application of ultrasound.[41,42] Later studies focused on extraction of glucose, demonstrating that while ultrasonically produced pores remained open for many hours, continuous application of pulsed ultrasound produced greater fluxes, albeit with significant subject-to-subject variability as shown in Figure 7.5. Application of modest vacuum levels (−0.6 atm) after a short ultrasonic treatment gave higher fluxes from Sprague Dawley rats (34 μL/cm^2/h) than from human subjects (12 μL/cm^2/h) under similar conditions.[43] The average 1.8 μL ISF sample extracted in the 5 min vacuum sampling period was collected from a 1.8 cm^2 area of skin. Allowing access to vacuum and then sensing this relatively small amount of ISF over such a large area appears challenging. The possible evaporation of small amounts of ISF was not discussed. Very large differences in extraction rates (coefficient of variation ~75%) were observed from human subject to human subject, as determined from the calibration factors reported in this study.[43] The extracted ISF samples were shown to track blood glucose values over the few hours of the study with a mean absolute relative error of 23%.

Application of a high ionic strength media on the skin after ultrasonic treatment was also found to be effective in extracting glucose.[9] In the SonoPrep device, the electrical impedance of the skin is continuously monitored, so application of the ultrasound can continue until a fixed permeability is achieved (see Figure 7.6). Generally, about 10 s is required to reduce the impedance below 10 kΩ. The ultrasonic

TABLE 7.1 Summary of Low-Frequency Ultrasound Studies

Skin type/condition	Ultrasound (US) application/extraction conditions	US frequency (kHz)	Power (W/cm^2)	Duty cycle (%)	Permeabilty	Flux ($\mu L/cm^2/h$)	Reference
Human *in vitro*	4 h continuous US extraction	20	0.225	10	0.0033	3[a]	41
Human *in vitro*	3 h continuous US extraction	20	0.125	10	0.026	26[b]	43
Rat *in vivo*	2 min US, and then extraction with PBS buffer	20	7	50	0.0034	3	45
Rat *in vivo*	2 min US pretreatment, and then 3 h continuous US extraction	20	7 (pretreatment), 1.6 (extraction)	50	0.026	26	43
Rat *in vivo*	2 min US, and then −0.6 atm vacuum extraction	20	7	50	0.034	34	43
Human *in vivo*	2 min US, and then −0.6 atm vacuum extraction	20	10	50	0.012	12	43
Human *in vivo*	US until skin resistance drops, and then osmotic buffer extraction	55	15	Not available	0.003	3	9

[a] Reported permeability for insulin used to calculate flux.

[b] Reported permeability for sucrose used to calculate flux.

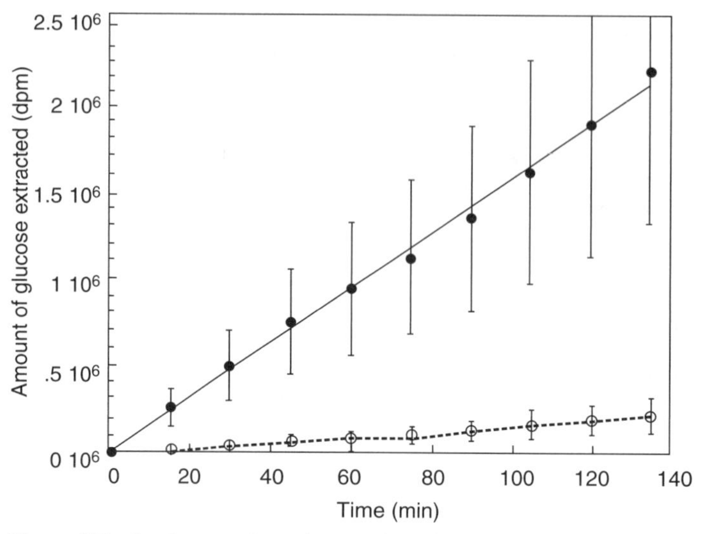

Figure 7.5 *In vivo* transdermal extraction of glucose from rats after a single low-frequency ultrasound treatment (open circles) and during continuous ultrasound treatment (closed circles). Reprinted with permission from Ref. 45. Copyright 2000 American Chemical Society.

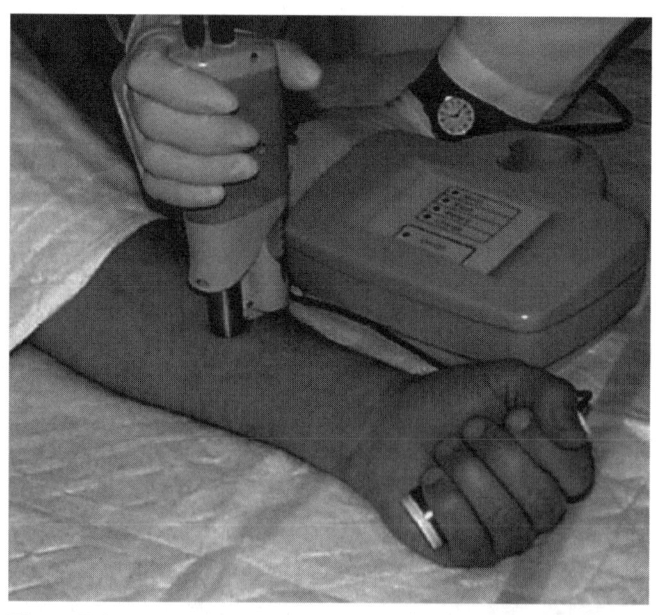

Figure 7.6 Clinical prototype for treatment with low-frequency ultrasound. 12 W of 55 kHz ultrasound is applied to a skin area of 0.8 cm^{-2} until the impedance is below 10 KΩ. The hand grip serves as the return electrode for the impedance measurement. The coupling media and ultrasonic horn are within the handpiece housing. Reprinted with permission from Ref. 9. Copyright 2004 Mary Ann Liebert, Inc. publishers.

horn is coupled to the skin using a phosphate buffered saline solution containing 1% sodium dodecyl sulfate. The sodium dodecyl sulfate reduces the surface tension of the solution, enhancing cavitation, and likely extracts lipids from the stratum corneum. After ultrasound treatment, a 0.25 M buffered lactic acid solution is applied in the form of a hydrogel over the site. The hydrogel has an osmotic pressure near 500 mOsm/kg while the osmotic pressure of ISF is ~280 mOsm/kg. This difference in osmotic pressures provides a large driving force for extraction, as may be calculated using the conversion factor between mOsm/kg and mmHg pressure (difference of $210\ \text{mOsm/kg} \times 19.3\ \text{mmHg (kg/mOsm)} = 4053\ \text{mmHg} = 5.3\ \text{atm}$). Glucose extraction data were generated at two anatomical sites on each of 10 human subjects over an 8 h period. The average glucose sample extraction rate was about 3 $\mu L/cm^2/h$. The mean absolute relative error in extraction rates between the two sites on a subject was 23%, while the mean absolute relative error from person to person was 55%. In a few cases, very good glucose tracking data and good correlations between the glucose concentration in extracted ISF samples and the whole-blood glucose values were observed. However, the correlation was greater than 0.9 for only 4 of 20 sites and greater than 0.7 for only 12 of 20 sites. All but one of the skin sites closed up a day after the study, as measured by an increase in electrical impedance. The mild-to-modest erythema and edema that occurred was generally well tolerated and usually resolved after a couple of days.

Advancement of ultrasound technology appears to hinge on overcoming the unique challenges inherent in design of a reliable glucose sensor for measuring the small amounts of extracted ISF in a high ionic strength hydrogel environment. Various attempts to increase the transdermal flux such as the use of cavitation enhancers, topical pretreatments, and combination with iontophoresis appear to have met with limited success. On the other hand, an attractive path toward miniaturization of the ultrasonic transducer has been identified based on inexpensive 1 mm thick lead-zirconate-titanate (PZT) ceramic disk resonators. Attachment of 3 mm high tapered titanium caps on either side of the 13 mm diameter disk amplifies the piezoelectric motion so that powers up to ~100 mW/cm^{-2} at 20 kHz can be achieved.[46] Many academic research groups are actively investigating transdermal ultrasound technology for both drug delivery and diagnostic purposes, so additional insight into methods for improvement may be forthcoming. Sontra Medical (now Echo Therapeutics) received FDA approval in 2004 to market a handheld low-frequency ultrasound-based device for delivery of lidocaine, so safety and regulatory concerns surrounding the technology are minimal. A 2008 press release from Echo Therapeutics reported that Clarke error grid analysis of glucose monitoring in adult patients undergoing cardiac surgery gave 70% A-zone and 27% B-zone results, with a mean absolute relative error of 16%.

7.4 ABLATION-BASED SYSTEMS

Ablation occurs when a large amount of energy is applied to a small amount of material in a short time. At least a portion of the material heats up above the boiling point and rapid expansion of vapor removes the vapor along with some pieces of surrounding material. Pulsed laser systems are able to supply a significant amount of energy in a short time and ablation of human tissue occurs if the energy deposited in the tissue, Q, is above a threshold of about 600 J/cm[3,49] The energy deposited in the tissue is a product of the radiant exposure, ϕ, and the absorption coefficient, μ, as shown in equation(7.2).

$$Q = \mu\phi \tag{7.2}$$

The tissue must strongly absorb the laser radiation, so only the top surface heats up and thermal damage to deeper tissue is limited. Both the ArF excimer laser at 193 nm, which is absorbed by components of the tissue, and the Er:YAG laser at 2040 nm, which is strongly absorbed by water in the tissue, have proven suitable for the controlled removal of the stratum corneum.[49] The penetration depth of the 193 nm radiation increases from about 0.4 μm in the stratum corneum to 1.5 μm in the epidermis since the epidermis contains more nonabsorbing water and a lower concentration of protein. On the other hand, the penetration depth of the 2040 nm radiation decreases from 4 μm during removal of the stratum corneum to 1 μm in the epidermis due to the increase in water content. When several pulses are used, dehydration of the stratum corneum during removal will also affect the penetration depths. Conditions found to remove the stratum corneum are summarized in Table 7.2. Both lasers can ablate the stratum corneum in a fairly reproducible number of pulses, but the total radiant power density needed using the Er:YAG laser is larger. Attempts to commercialize direct ablation of the skin have focused on use of higher powered Er: YAG lasers to obtain blood samples for testing with conventional glucose meters. Transmedica developed the Laser Lancet that cut a 0.25 mm × 1.1 mm slit through the thick stratum corneum on the finger. The Cell Robotics Lasette was brought to the market with an ~$1000 price tag. Neither system has attained significant commercial success so far.

Ablation can be carried out with lower cost laser diodes when a broadband energy absorber is applied to the skin. For example, shallow pores are created by application of twenty to thirty 20 ms pulses of 810 nm radiation at a power of 250 mW to a 50 μm thick carbon containing black tape on the skin.[50] A pattern of these shallow pores is easily generated on the skin by movement of the laser. Application of about one-half atmosphere of vacuum extracts about 0.2 μL ISF per minute with a relative standard deviation of ~65%, which includes person-to-person and site-to-site variabilities. Larger volumes of ISF (~0.5 μL/min) can be exuded from the ablated site by application of positive pressure to the tissue around the site. In general, the volume exuded increases as the number of micropores at the site increases from one to

TABLE 7.2 Conditions Used to Ablate the Stratum Corneum

Laser system/ wavelength	Irradiated diameter (μm)	Pulse (ns)	Power per pulse (mJ)	Number of pulses to remove stratum corneum	Total radiant exposure (J/cm^2)
ArF 193 nm	476	14	48	8.3 ± 1.5	27
Er:YAG 2940 nm	210	170	9	73 ± 2	52

three and as the applied pressure increases. The geometry of the harvesting head applied to skin for extraction of ISF is also important. In one study, a harvesting head with an open aperture of ~2.5 mm over a 1 mm pattern of micropores was found to be optimal.[50] The harvesting head may be placed in a holder and the force needed for extraction is supplied by a vacuum piston mechanism or a pressure cuff-like wrap.[50]

SpectRx (now Guided Therapeutics, Inc.) has a strong intellectual property position in the field of laser ablation and has published clinical study results for extraction of ISF using continuous vacuum pressure.[10–12] In these studies, an energy adsorbing dye film in direct contact with the skin is irradiated with a low-cost handheld laser system and four <100 μm pores are formed. A vacuum pump is connected to the ablated site and clear, straw-colored interstitial fluid collected at a flow rate of 5–15 μL/h for the duration of the study (up to 56 h). The laser system, alignment ring, harvesting head, and vacuum pump are shown in Figure 7.7. Ablation removes the stratum corneum and does not penetrate the epidermis or dermis as shown in Figure 7.8. The mechanism of pore formation may be complex, involving ablation of the black dye and thermal transfer of energy to the upper layers of the skin. Vacuum blisters exhibit the same clean separation of the stratum corneum from the epidermis as observed with the SpectRx system (Figure 7.8). In a clinical study of 56 adults, several examples of good glucose tracking were reported and Clarke error grid analysis showed that 82% of the ISF samples were in the A-zone and 16% were in the B-zone. Slightly poorer performance was found in a study of 110 children with diabetes, where 69% of the results were in the A-zone and 21% in the B-zone. Loss of vacuum due to an inadequate seal between the harvesting head and the skin and mechanical failure of the vacuum pump were the main reasons for inadequate sample collection. After removal of the harvesting head at the end of the study, a red area was observed where the skin had been ablated and the redness persisted for at least 1 day. The closest correlation between fingerstick whole-blood glucose readings and the

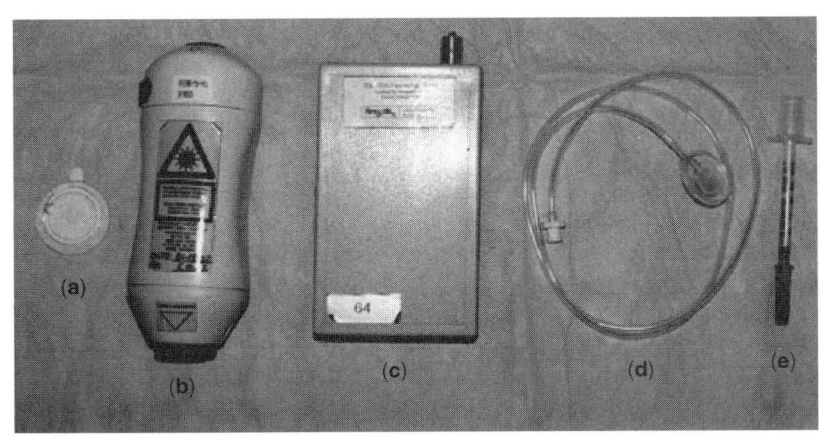

Figure 7.7 Research clinical prototype for laser ablation and vacuum extraction of interstitial fluid. (a) Alignment ring with energy-absorbing dye, (b) laser porator, (c) low-pressure, continuous vacuum unit, (d) harvesting head, and (e) 1 mL insulin syringe. Reprinted with permission from Ref. 12. Copyright 2006 Mary Ann Liebert, Inc. publishers.

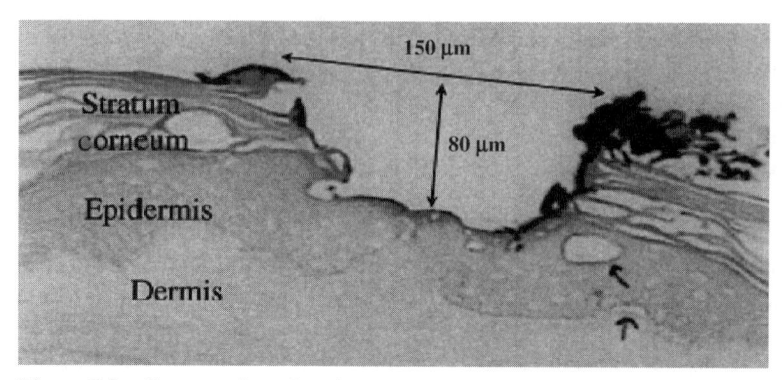

Figure 7.8 Cross-section of a micropore created by laser ablation using an energy absorbing dye on the skin. The stratum corneum is removed without greatly affecting the viable epidermis or dermis. Reprinted with permission from Ref. 12. Copyright 2006 Mary Ann Liebert, Inc. publishers.

continuous glucose sensor readings of the ISF indicated a short 15 min lag between the SpectRx system values and the whole-blood values. The lag time includes any physiological lag in glucose changes between whole blood and ISF, as well as the time required for ISF collection and sensor measurements. Miniaturization and improvements in the design to minimize the physical space between the fluid collection portion of the device and the sensor would decrease the lag time.

Radiofrequency thermal ablation can also be used to create micopores of high permeability through the skin. Application of high-voltage radiofrequency (100–500 kHz) to an electrode on the skin surface causes ions within the tissue to attempt to follow the change in the direction of the alternating current. This movement of ions results in frictional heating of the tissue, producing local heating, liquid evaporation, and ablation of the cells in a local area. Arrays of stainless steel wires of length 100 μm and 40 μm diameter on 1 mm centers have been used to produce 100 micopores per square centimeter in rat and pig skin. A micropore across the stratum corneum and into the viable epidermis is typically formed using five 1 ms bursts of 200 V at 100 kHz with 15 ms between bursts.[51] The resulting dimensions of the micropore are similar to the dimensions of the wire. *In vivo* studies in rats gave permeability coefficients of $2–4 \times 10^{-3}$ cm/h that suggest that ISF collection rates on the order of 3 μL/cm^2/h could be achieved, similar to the fluxes observed after low-frequency ultrasound treatment.

Thermal ablation with an array of electrically resistive filaments on the skin surface has also been used to create an array of micropores. Application of a short controlled pulse of DC electrical current to an array of 80 μm wide and 300 μm long wires produced 72 micropores per square centimeter. A peak wire temperature of 750°C was set using an optical pyrometer. Control of the electrical current, pulse width, number of pulses, pulse spacing, and tip contact pressure gave elliptical-shaped pores approximately 80 μm wide, 300-μm long, and 40-μm deep with a variability of 20–30% (%CV) as measured using light microscopy. Iontophoresis enhanced the transdermal flux through the micropores about twofold compared to passive diffusion.[52]

Initial development work has also been reported on an integrated patch-like device containing 25 individually addressable DC microablation elements, microfluidic connectors, and colorimetric glucose membranes on a 1 cm^2 size footprint.[53] Preliminary studies using surrogate human graft skin (Apligraf, Organogenesis, and Novartis Pharmaceuticals) demonstrated that a 50 mJ pulse of thermal energy ablated a 50 μm diameter pore through the stratum corneum and multiple pulses produced a pore about 40 μm deep. The patch included a fluid reservoir for bathing the ablated region on the skin and transporting sample to the detection membrane. While ISF fluid was shown to wet the micropore area after ablation of the Apligraf and the colorimetric glucose membrane was responsive over a reasonable concentration range, complete operation of the integrated device has not yet been demonstrated.

7.5 MICRONEEDLE-BASED SYSTEMS

The female mosquito's blood sampling ability has often been cited as an inspiration for development of microneedle-based systems. A few features of the mosquito anatomy and extraction ability are noteworthy. The mosquito's labium is about 3.5 mm long and narrows to an inner diameter of about 30 μm. The labium is applied to the skin with a hammer-like motion at the rate of 6–7 Hz for penetration. A muscle valve and mouth pump move in concert to create about 7 kPa of negative pressure that is sufficient to extract 1.9 μL of blood in 2 min.[54] There is some indication that the mosquito can sense when a source of blood is reached and can change the direction of insertion while partway in the skin to achieve a greater rate of success. While the mosquito has served as inspiration for design and function parameters, man-made devices to date have relatively rudimentary functionality in comparison.

Man-made microneedles have been fabricated from a wide range of materials including metals, alloys, ceramics and polymers. Traditionally, single needles with an outside diameter down to about ~125 μm have been commercially manufactured from stainless steel stock with an internal diameter of about 50 μm. A final electropolishing step and lubricant coating are often applied to increase the sharpness of the point and reduce the force required for penetration. More recently, metal vapor deposition has been used to fabricate needles from the surface up. Chemical etching across the crystal lattice of chemically masked silicon wafers has also been used to produce arrays of very strong, sharp microneedles as shown in Figure 7.9. Other pattern, etch, and release methods developed in the microelectronics industry have been adopted to fabricate laminated structures with microfluidic channels and fine points. Alza Corporation developed a very simple approach for fabricating an array of microblades. A thin, flat sheet of titanium is patterned with a slanted leading edge and open area surrounding the blade, the area around the blade pattern is etched away with acid, and then the blades are pushed downward to make a 90° angle with the original plane of the titanium sheet.[55] Typical arrays contain about 330 μm long blades at a density of 200 cm^{-2} with a large percentage of open area that can be used to collect sample from the skin. Patterning barbs on the blades may also be used to help keep the blades in the skin after application. In the field of polymeric microneedles, major industrial companies such as Procter & Gamble and 3M have established strong

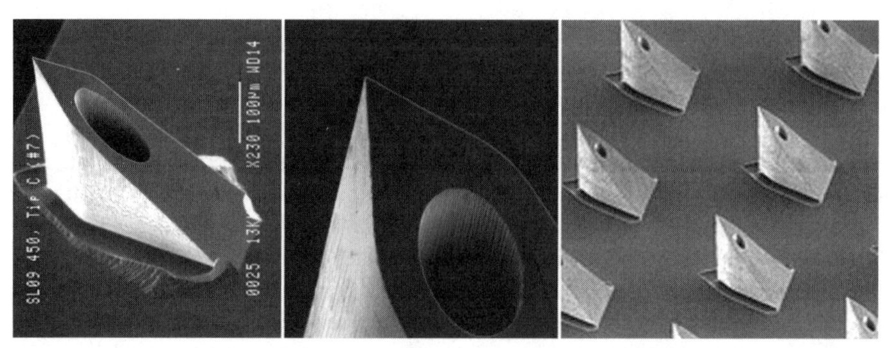

Figure 7.9 An extremely sharp hollow microneedle with an off-set hole manufactured as an array. The sharp point and edge are produced using a mask and chemical etching process that is faster along one crystal plane of the silica wafer. Reprinted with permission of Nanopass Technologies.

intellectual property positions surrounding their manufacture and use. Over the past 10 years, numerous innovations have resulted in about 1800 U.S. patents, and scientific publications referencing the use of microneedles have increased more than fivefold.

Common modes of failure reported for microneedles include fracture, buckling during insertion, and clogging of the orifices from punched-out tissue during insertion. Design of the tip geometry is a trade-off between reducing the tip radius or taper angle to the tip, which minimizes the penetration force, and maintaining sufficient material strength to avoid breakage. The optimal taper angle must be empirically determined for each material but values bracketing $\sim 25°$ can be recommended for study. The obvious increase in force necessary with larger diameter needles has been demonstrated: 100 μm diameter requires 0.05 N, 200 μm diameter requires 0.45 N, and 900 μm diameter requires 0.8 N.[54] Concurrent vibration of a single microneedle at ultrasonic frequency was shown to reduce the force required for insertion by 70%.[56] Short microneedles may deform the skin without penetration, so various spring-actuated application fixtures have been designed. Better penetration may be achieved if the skin is held taunt. The depth of penetration may be limited by irregularities of the skin (wrinkles, hair, etc), but even with the most advanced prototype systems, penetration is often much more shallow than the target depth. For example, application of an array of 330 μm long microblades to taunt skin with an impact applicator showed only 50% penetrated more than 100 μm and less than 10% penetrated more than 200 μm.[57] Indentation of the skin also appears to have the undesirable effect of pressing liquid out of the tissue, which dramatically lowers the flow conductivity of interstitial fluid in the skin beneath the tip. Retraction of the needle after penetration may be used to reverse the compaction of the tissue as demonstrated in a study where a flow rate of 15 μL/h was achieved from a microneedle inserted about 1 mm, while retraction of the needle to a depth of 300 μm gave an increased flow rate of nearly 200 μL/h.[59] Fairly sophisticated programmable electromechanical devices based on rotary voice coil drives have been designed to control the entire velocity profile during penetration over the range of 2–10 m/s with penetration depths as shallow as 400 μm.[58] Another electromechanical mechanism involves electrically heating a Ti-Ni-Cu shape memory alloy to supply the force for penetration.[54]

The first generation of hollow MEMS microneedles had a volcano tip shape that acted as a punch during insertion and easily became clogged. Later generations had an offset through-hole and were more resistant to clogging. More recent designs have tended toward geometries with slots in the side to provide a larger area for extraction than a single through-hole. Masking and metal vapor deposition techniques commonly used in MEMS fabrication allow integration of microfluidic channels and electrical traces to sensing elements in a wide variety of geometries. The fluidic path of one recent design incorporates a microdialysis membrane over the flow channel in the microneedle to protect the sensor from high molecular weight protein.[15] The membrane was formed by deposition of a thin layer of permeable polysilicon or etching of a 2 μm thin sacrificial oxide layer to produce 30 nm pores. The flow channel in the needle has a 2.6 nL volume (height 10 μm, width 65 μm, and length 4 mm), so there is a very little time lag associated with flow rates, as low as a few nanoliters per minute. With a 100 μL reservoir of dialysis fluid, the device could operate for an extended period of time. Piezoelectric devices have often been incorporated into MEMS and μTAS systems to provide pressure for controlled fluid transfer.

An interesting nonmechanical physical–chemical approach has been used to generate negative pressures in a continuous microneedle glucose monitoring system.[16] Certain polymers are known to shrink to less than 80% of their original volume at elevated temperatures or with a change in the pH. In particular, copolymerized gels of *N*-isopropylacrylamide and acrylic acid exhibit a large decrease in volume, generally in less than an hour, when going from 25 to 37°C. Placing the gel near the skin would naturally raise the temperature and shrink the gel, creating vacuum pressure for fluid extraction. However, in a continuous sampling device a more gradual shrinkage is desired to maintain a relatively constant extraction rate and to perhaps automatically recover if the vacuum pressure should be lost at some point due to a leak. Longer shrinkage times were achieved by incorporating glucose oxidase into the gel and allowing the extracted fluid to enter the gel. The reaction of glucose with glucose oxidase produces gluconic acid, which lowers the pH causing additional shrinkage of the gel as sample is extracted. *In vitro* testing of an integrated prototype device showed that the volume change of a small portion of gel (11 mm $\times$ 11 mm 4 mm) placed behind a silicon rubber diaphragm shrank $\sim$30% in a fairly linear fashion over an 8 h period to extract a total of 40 μL of glucose solution.[16] At these modest flow rates, traditional coated glucose oxidase-based sensing systems may consume a significant amount of glucose in the sample and respond to a buildup of hydrogen peroxide, so the sensor signal exhibits an undesirable dependence on the flow rate. Excessive consumption of glucose by the sensor has been addressed by decreasing the glucose permeability with a more limiting coating, and buildup of hydrogen peroxide has been minimized by adding peroxidase to the sensor overcoat.

In a human clinical study, glass microneedles were inserted 700–1500 μm deep into the skin to produce 7–10 openings, a modest level of vacuum applied for 2–10 min, and the extracted interstitial fluid (1–10 μL) measured with a conventional glucose test strip.[17] Good glucose tracking was observed for several hours. Since the ISF was collected by vacuum over a large area, the glucose concentration was higher due to evaporation. Minimization of the vacuum headspace can reduce evaporative loss, but outgassing of dissolved gas can lead to the formation of bubbles in the ISF stream, which confound glucose detectors. Leaks may also produce bubbles in the ISF stream. Several microfluidic bubble traps have been described but not yet applied to microneedle systems.

In another human clinical study, ISF was extracted by applying external pressure around a single 30-gauge stainless steel microneedle inserted 2 mm into the skin.[14] ISF was extracted from different sites on the forearm by applying 15 N force on a 5.5 mm diameter pressure ring and ISF was collected continuously from the same site using a pressure ring. When the ISF was collected from different sites, 1 μL of ISF was collected in an average of 3 s and the ISF glucose values showed good tracking over 6 h with a mean absolute relative error of 22%. When ISF was collected repeatedly from the same site using modulated pressure application for continuous measurement, 320 nL was collected in a median time of 85 s and the tracking was even better with a mean absolute relative error of only 11%. In both studies, the glucose was measured with a modified electrochemical test strip. The repeated pressure modulation at the same site reduced the time lag between ISF glucose concentration and whole-blood glucose concentration from 38 to 2.5 min, clearly showing the huge effect that stimulation of local blood flow at the sampling site can have. The patent literature contains numerous examples of milking-type devices composed of fingers or edges that are movable between a relaxed position and a contracted position with respect to the surface of the skin to draw up and pinch a portion of the skin for expression of fluid. In general, there are few clinical results in the patents, so it is difficult to determine which approach might be most effective. Given the current level of research in the areas of microneedles and microfluidics, application of new developments to glucose sensing should continue for some time.

7.6 OTHER METHODS OF ACCESSING INTERSTITIAL FLUID

Many other methods have been evaluated for removing the stratum corneum and accessing interstitial fluid; however, they generally suffer from difficulties in control and greater damage to the skin. Small openings can be made with high-powered jets of liquid or powder, but it is difficult to consistently form a pore while limiting the depth of penetration into the tissue. As briefly mentioned earlier in the chapter, repeated application of tape removes the stratum corneum layer by layer but the process is tedious. Glue can also be applied but it is difficult to limit the area of skin removed and hydration levels on the surface greatly affect the penetration depth and number of layers removed. Blisters can be formed by chemical treatment, friction, light, or vacuum. Vacuum is the most easily controlled, but the minimum size is still significant and healing lengthy. It should be noted that the interstitial fluid collected from vacuum blisters has nearly the same glucose concentration as whole blood and closely tracks changes in blood glucose values.[22,25] Many methods of abrasion have been tested with various types of lapidary agents and motion. Rotary microabrasion may hold some promise since relatively small areas can be treated. The depth of long blade-type systems has been relatively difficult to control, but a dual-blade geometry has the advantage of introducing a capillary space and capillary force that might be utilized for ISF extraction.

ISF can also be accessed without the need to remove the stratum corneum at mucus membranes. The bucca mucosa of the inside of the mouth is an example. A small amount of interstitial fluid traverses the membrane, but the large concentration of active enzymes in the mouth can quickly consume glucose and saliva can dilute the sample. A number of concepts in the form of contact lens have received attention. The inner eyelid may supply a significant amount of the glucose to the surface liquid on the eye; however, concentration due to evaporation and dilution due to tearing present practical challenges.

7.7 CONCLUSIONS AND FUTURE OUTLOOK

Interstitial fluid can be extracted continuously over a period of several days using a variety of methods allowing continuous monitoring of glucose. The glucose concentration in ISF closely tracks changes in blood glucose values with a modest time lag. Increasing the local blood perfusion at the extraction site essentially eliminates any difference in glucose concentration between ISF and blood. Some transdermal microfluidic systems, such as iontophoresis and ultrasound-based systems, dilute extracted ISF during extraction and measurement, so they must be calibrated with a fingerstick measurement. Ablation and microneedle-based systems allow extraction and measurement of undiluted ISF, so there is no need for a fingerstick. Damage to the skin caused by accessing and extracting ISF generally resolves in a few days but cosmetic issues may significantly impact the acceptance of a commercial product. The precision and accuracy achieved in many of the research clinical trials needs to be improved to achieve levels acceptable in a commercial product. Further miniaturization of the systems and incorporation of smaller sensors could drive the field forward. A significant MEMS and μTAS infrastructure exists that will bring new innovation forward. These advancements coupled with the clinical experience summarized in this chapter suggest that more capable transdermal microfluidic devices for continuous glucose monitoring will be developed over the next several years.

REFERENCES

1. Tamada JA, Bohannon NJV, Potts RO. Measurement of glucose in diabetic subjects using noninvasive transdermal extraction. *Nature Medicine* 1995, 1, 1198–1201.
2. Kurnik RT, Berner B, Tamada J, Potts RO. Design and simulation of a reverse iontophoretic glucose monitoring device. *Journal of the Electrochemical Society* 1998, 145, 4119–4125.
3. Tierney MJ, Jayalakshmi Y, Parris NA, Reidy MP, Uhegbu C, Vijayakumar P. Design of a biosensor for continual, transdermal glucose monitoring. *Clinical Chemistry* 1999, 45, 1681–1683.
4. Kurnik RT, Oliver JJ, Waterhouse SR, Dunn T, Jayalakshmi Y, Lesho M, Lopatin M, Tamada J, Wei C, Potts RO. Application of the mixtures of experts algorithm for signal

processing in a noninvasive glucose monitoring system. *Sensors and Actuators B* 1999, 60, 19–26.

5. Tamada JA, Garg S, Jovanovic L, Pitzer KR, Fermi S, Potts RO. Noninvasive glucose monitoring comprehensive clinical results. *Journal of the American Medical Association* 1999, 282, 1839–1844.
6. Tierney MJ, Tamada Ja Potts RO, Jovanovic L, Garg S. Clinical evaluation of the GlucoWatch® biographer: a continual, non-invasive glucose monitor for patients with diabetes. *Biosensors & Bioelectronics* 2001, 16, 621–629.
7. Pitzer KR, Desai S, Dunn T, Edelman S, Jayalakshmi Y, Kennedy J, Tamada JA, Potts RO. Detection of hypoglycemia with the GlucoWatch™ biographer. *Diabetes Care* 2001, 24, 881–885.
8. The Diabetes Research in Children Network (Directnet) Study Group. The accuracy of the GlucoWatch™ G2 Biographer in children with type 1 diabetes. *Diabetes Technology & Therapeutics* 2003, 5, 791–800.
9. Chuang H, Taylor E, Davison TW. Clinical evaluation of a continuous minimally invasive glucose flux sensor placed over ultrasonically permeated skin. *Diabetes Technology & Therapeutics* 2004, 6, 21–30.
10. Gebhart S, Faupel M, Fowler R, Kapsner C, Lincoln D, McGee V, Pasqua J, Steed L, Wangsness M, Xu F, Vanstory M. Glucose sensing in transdermal body fluid collected under continuous vacuum pressure via micropores in the stratum corneum. *Diabetes Technology & Therapeutics* 2003, 5, 159–166.
11. Burdick J, Chase P, Faupel ML, Schultz B, Gebhart S. Real-time glucose sensing using transdermal fluid under continuous vacuum pressure in children with type 1 diabetes. *Diabetes Technology & Therapeutics* 2005, 7, 488–455.
12. Nindl BC, Tuckow AP, Alemany JA, Harman EA, Rarick KR, Staab JS, Faupel ML, Khosravi MJ. Minimally invasive sampling of transdermal body fluid for the purpose of measuring insulin-like growth factor-I during exercise training. *Diabetes Technology & Therapeutics* 2006, 8, 244–252.
13. Smart WH, Subramanian K. The use of silicon microfabrication technology in painless blood glucose monitoring. *Diabetes Technology & Therapeutics* 2000, 2, 549–559.
14. Stout PJ, Racchini JR, Hilgers ME. A novel approach to mitigating the physiological lag between blood and interstitial fluid glucose measurements. *Diabetes Technology & Therapeutics* 2004, 6, 635–644.
15. Zahn JD, Trebotich D, Liepmann D. Microdialysis microneedles for continuous medical monitoring. *Biomedical Microdevices* 2005, 7, 59–69.
16. Suzuki H, Tokuda T, Miyagishi T, Yoshida H, Honda N. A disposable on-line microsystem for continuous sampling and monitoring of glucose. *Sensors and Actuators B* 2004, 97, 90–97.
17. Wang PM, Cornwell M, Prausnitz MR. Minimally invasive extraction of dermal interstitial fluid for glucose monitoring using microneedles. *Diabetes Technology & Therapeutics* 2005, 7, 131–141.
18. Fineberg SE, Bergenstal RM, Bernstein RM, Laffel LM, Schwartz SL. Use of an automated device for alternative site blood glucose monitoring. *Diabetes Care* 2001, 24, 1217–1220.
19. Pirot F, Berardesca E, Kakia YN, Singh M, Maibach HI, Guy RH. Stratum corneum thickness and apparent water diffusivity: facile and noninvasive quantitation *in vivo*. *Pharmaceutical Research* 1998, 15, 492–494.

20. Goldsmith LA. *Physiology, Biochemistry, and Molecular Biology of the Skin*, 2nd edn. Oxford University Press, New York, 2001.
21. Service FJ, O'Brian PC, Wise SD, Ness S, LeBlanc SM. Dermal interstitial glucose as an indicator of ambient glycemia. *Diabetes Care* 1997, 20, 1426–1429.
22. Thennadil SN, Rennert JL, Wenzel BJ, Hazen KH, Ruchti TL, Block MB. Comparison of glucose concentration in interstitial fluid, and capillary and venous blood during rapid changes in blood glucose levels. *Diabetes Technology & Therapeutics* 2001, 3, 357–365.
23. Koschinsky T, Jungheim K, Heinemann L. Glucose sensors and the alternate site testing-like phenomenon: relationship between rapid blood glucose changes and glucose sensor signals. *Diabetes Technology & Therapeutics* 2003, 5, 829–842.
24. Haupt A, Berg B, Paschen P, Dreyer M, Häring HU, Smedegaard J, Matthaei S. The effects of skin temperature and testing site on blood glucose measurements taken by a modern blood glucose monitoring device. *Diabetes Technology & Therapeutics* 2005, 7, 597–601.
25. Svedman C, Samra JS, Clark ML, Levy JC, Frayn KN. Skin mini-erosion technique for monitoring metabolites in interstitial fluid: its feasibility demonstrated by OGTT results in diabetic and non-diabetic subjects. *Scandinavian Journal of Clinical and Laboratory Investigation* 1999, 59, 115–123.
26. Uitto OD, White HS. Electroosmotic pore transport in human skin. *Pharmaceutical Research* 2003, 20, 646–652.
27. Tezel A, Sens A, Tuchscherer J, Mitragotri S. Frequency dependence of sonophoresis. *Pharmaceutical Research* 2001, 18, 1694–1700.
28. Cunningham D, Young D. Measurements of glucose on the skin surface, in stratum corneum and in transcutaneous extracts: implications for physiological sampling. *Clinical Chemistry and Laboratory Medicine* 2003, 41, 1224–1228.
29. Cunningham D, Lowery M. Moisture vapor transport channels for the improved attachment of medical devices to the human body. *Biomedical Microdevices* 2004, 6, 149–154.
30. Tamada JA, Davis TL, Leptien AD, Lee J, Wang B, Lopatin M, Wei C, Wilson D, Comyns K, Eastman RC. The effect of preapplication of corticosteroids on skin irritation and performance of the GlucoWatch G2™ biographer. *Diabetes Technology & Therapeutics* 2004, 6, 357–367.
31. Cunningham D, Henning T, Shain E, Young D, Elstrom T, Taylor E, Schroder S, Gatcomb P, Tamborlane W. Vacuum-assisted lancing of the forearm: an effective and less painful approach to blood glucose monitoring. *Diabetes Technology & Therapeutics* 2000, 2, 541–548.
32. Cunningham D, Henning T, Young D, Berardesca E. Wound healing after lancing the skin. *Wounds* 2000, 12, 131–137.
33. Cunningham D, Henning T, Shain E, Hanning J, Barua E, Lee R. Blood extraction from lancet wounds using vacuum combined with skin stretching. *Journal of Applied Physiology* 2002, 92, 1089–1096.
34. Swartz MA, Kaipainen A, Netti PA, Brekken C, Boucher Y, Grodzinsky AJ, Jain RK. Mechanics of interstitial-lymphatic fluid transport: theoretical foundation and experimental validation. *Journal of Biomechanics* 1999, 32, 1297–1307.
35. Feldman B, McGarraugh G, Heller A, Bohannon N, Skyler J, DeLeeuw E, Clarke D. FreeStyle: a small-volume electrochemical glucose sensor for home blood glucose testing. *Diabetes Technology & Therapeutics* 2000, 2, 221–229.

36. Scheipers A, Hinkers H, Wassmus O, Sundermeier C, Ross B. Chemical and biochemical microsensors in silicon for the clinical on-line monitoring. *Biomedizinische Technik (Berlin)* 2002, 47, 209–212.

37. Petrou PS, Moser I, Jobst G. Microdevice with integrated dialysis probe and biosensor array for continuous multi-analyte monitoring. *Biosensors & Bioelectronics* 2003, 18, 613–619.

38. Glikfeld P, Hinz RS, Guy RH. Noninvasive sampling of biological fluids by iontophoresis. *Pharmaceutical Research* 1989, 6, 988–990.

39. Oh SY, Guy RH. Effects of iontophoresis on the electrical properties of human skin *in vivo*. *International Journal of Pharmaceutics* 1995, 124, 137–142.

40. Sieg A, Guy RH, Delgado-Charro MB. Noninvasive glucose monitoring by reverse iontophoresis *in vivo*: application of the internal standard concept. *Clinical Chemistry* 2004, 50, 1383–1390.

41. Mitragotri S, Blankschtein D, Langer RS. Ultrasound-mediated transdermal protein delivery. *Science* 1995, 269, 850–853.

42. Mitragotri S, Blankschtein D, Langer RS. Transdermal drug delivery using low-frequency sonophoresis. *Pharmaceutical Research* 1996, 13, 411–420.

43. Kost J, Mitragotri S, Gabbay RA, Pishko M, Langer RS. Transdermal monitoring of glucose and other analytes using ultrasound. *Nature Medicine* 2000, 6, 347–350.

44. Mitragotri S, Coleman M, Kost J, Langer RS. Analysis of ultrasonically extracted interstitial fluid as a predictor of blood glucose levels. *Journal of Applied Physiology* 2000, 89, 961–966.

45. Mitragotri S, Kost J. Low-frequency sonophoresis: a noninvasive method of drug delivery and diagnostics. *Biotechnology Progress* 2000, 16, 488–492.

46. Lee S, Nayak V, Dodds J, Pishko M, Smith NB. Glucose measurements with sensors and ultrasound. *Ultrasound in Medicine and Biology* 2005, 7, 971–977.

47. Kushner J, Kim D, So PTS, Blankschtein D, Langer RS. Dual-channel two-photon microscopy study of transdermal transport in skin treated with low-frequency ultrasound and a chemical enhancer. *Journal of Investigative Dermatology* 2007, 127, 2832–2846.

48. Kushner J, Blankschtein D, Langer RS. Evaluation of hydrophilic permeant transport parameters in the localized and non-localized transport regions of skin treated simultaneously with low-frequency ultrasound and sodium lauryl sulfate. *Journal of Pharmaceutical Sciences* 2008, 97, 894–906.

49. Jacques SL, Ejeckam F, Tittel F. How micro is microdissection? Laser removal of stratum corneum of skin to expose the epidermal battery. *SPIE Laser–Tissue Interaction IV* 1993, 1882, 23–33.

50. Grace J, Loomis N, Schapira T, Wong S, Noonan K, Lowery M, Bojan P, Schmidt D, Huang T, Hiltibran R, Pope M, Kotlarik J, Tarkowski V, Cunningham D. Apparatus and method for the collection of interstitial fluids. U.S. Patent 6,786,874, 2004.

51. Sintov AC, Krymberk I, Daniel D, Hannan T, Sohn Z, Levin G. Radiofrequency-driven skin microchanneling as a new way for electrically assisted transdermal delivery of hydrophilic drugs. *Journal of Controlled Release* 2003, 89, 311–320.

52. Badkar AV, Smith AM, Eppstein JA, Banga AK. Transdermal delivery of interferon alpha-2B using microporation and iontophoresis in hairless rats. *Pharmaceutical Research* 2007, 24, 1389–1395.

53. Paranjape M, Garra J, Brida S, Schneider T, White R, Currie J. A PDMS dermal patch for non-intrusive transdermal glucose sensing. *Sensors and Actuators A* 2003, 104, 195–204.

54. Tsuchiya K, Nakanishi N, Uetsuji Y, Nakamachi E. Development of blood extraction system for health monitoring system. *Biomedical Microdevices* 2005, 7, 347–353.

55. Cormier M.J.N, Neukermans A.P, Block B, Theeuwes F.T, Amkraut A.A. *Device and method for enhancing transdermal flux of agents being delivered or sampled.* U.S. Patent 7,184,826, 2007.

56. Yang M, Zahn JD. Microneedle insertion force reduction using vibratory actuation. *Biomedical Microdevices* 2004, 6, 177–182.

57. Matriano JA, Cormier M, Johnson J, Young WA, Buttery M, Nyam K, Daddona PE. Macroflux™ microprojection array patch technology: a new and efficient approach for intracutaneous immunization. *Pharmaceutical Research* 2002, 19, 63–70.

58. Martanto W, Moore JS, Couse T, Prausnitz M. Mechanism of fluid infusion during microneedle insertion and retraction. *Journal of Controlled Release* 2006, 112, 357–361.

59. Freeman D.M. Boecker D. *Method and Apparatus for Measuring Analytes.* U.S. Patent Application Publication 2006/0200044 A1, 2006.

CHAPTER 8

REDUNDANT ARRAYS AND NEXT-GENERATION SENSORS

Becky L. Clark and Michael V. Pishko

8.1 INTRODUCTION AND NEED FOR REDUNDANCY

For use in diabetes management, intensive research and commercial activity have focused on the development of simple painless techniques of monitoring glucose with increased frequency. Glucose sensor development for neonatal care has not been a primary research focus. Techniques under investigation can be categorized as noninvasive, minimally invasive, and invasive (i.e., implantable sensors). Noninvasive and the so-called minimally invasive monitoring methods have been areas of intense research for the past decade. The techniques fall into two general categories: transdermal spectroscopy[1–4] and interstitial fluid sampling.[5–8] Transdermal spectroscopic methods of glucose determination use visible, near-infrared, or infrared light transmitted through or backscattered from tissue (i.e., skin). The transmitted or scattered light is then collected and glucose concentration determined from a change in the polarization of the light, or a change in absorbance, or the Raman scattering as described in more detail in Chapters 12–15. These techniques have the advantage that

In Vivo Glucose Sensing, Edited by David D. Cunningham and Julie A. Stenken

they are truly noninvasive and thus painless. However, a major challenge of these nonspecific spectroscopic techniques is the need to measure glucose in a highly complex matrix of water, proteins, polysaccharides, and lipids. As a result, the acquired signal is very complex and the glucose measurement must be extracted from this signal by chemometric methods such as principal component regression or partial least squares (PLS) analysis. Because these techniques fail to provide a direct measure of blood or interstitial fluid glucose, there is considerable debate as to if glucose is actually measured or is rather a phenomenon correlated to glucose concentration under the experimental conditions.

The primary technique for interstitial fluid sampling researched thus far has been reverse iontophoresis.[7,8] In brief, when an electric field is applied between two electrodes placed on the skin, anions in the skin will move to the cathode and cations move toward the anode. This results in an electro-osmotic flow of interstitial fluid toward these electrodes. The interstitial fluid is collected in and diluted by the electrolyte that electrically couples the electrodes to the skin. Glucose can then be measured using a glucose sensor in contact with the electrolyte. A significant advantage of this method is that a physiologically relevant fluid sample is collected. Interstitial fluid has been shown in a number of studies to have a glucose concentration identical to or closely correlated to that in plasma, though there are significant variations from study to study as a result of differing collection methods.[9–16] The volume of interstitial fluid collected using reverse iontophoresis is small, primarily because the fluid is transported through hair follicles and sweat ducts. Thus, the entire skin surface area exposed to the electric field is not used for transport. In addition, the interstitial fluid sample is diluted in the electrolyte used to couple the electrode to the skin. As a result, glucose concentrations in the receiver are low and difficult to measure using existing sensor technology. The concept, however, of interstitial fluid sampling and subsequent measurement is sound and will be feasible if the volume of interstitial fluid extracted can be increased, can be extracted without dilution, and the extraction system integrated with a reliable sensor. Similar results to those demonstrated for reverse iontophoresis were also demonstrated using reverse sonophoresis, that is, an acoustic field was used to extract interstitial fluid transdermally.[17] Minimally invasive monitoring systems are more comprehensively reviewed in Chapter 7 and future incorporation of redundant arrays holds the promise of increasing the performance of these systems.

Implantable glucose sensors have been under investigation for nearly three decades and commercial systems are now reaching the market, as discussed in greater detail in Chapter 5. These devices are implanted either in subcutaneous tissue, where they measure glucose associated with interstitial fluid,[18–24] or are implanted intravascular.[25] In comparison to noninvasive and minimally invasive techniques, implanted sensors are in direct contact with undiluted, physiological relevant fluids (blood or interstitial fluid). The expectation is that a continuous glucose monitoring system will result in increased patient compliance with intensive treatment and also decrease the frequency of hypoglycemic episodes because of the increased awareness of blood glucose levels. One potential method of achieving tighter metabolic control is a closed-loop insulin delivery system, incorporating a microprocessor-controlled insulin pump and a glucose sensor. Glucose sensors, which use an enzyme (glucose oxidase) to achieve specificity, are at present not stable or sensitive enough to meet the demands of a closed-loop delivery system. As a result, the application of glucose biosensors has been primarily limited to home glucose test meter and blood–gas instruments containing sensors for glucose. There are a number of reasons for the slow commercial progress, both technical and economic. Technically, many proposed biosensors for glucose simply do not have the accuracy and stability (operational or storage) to meet the desired need. Inaccuracy and imprecision in sensor performance are frequently due to imprecision in sensor manufacturing, for example, immobilized biomolecules cannot be deposited on transducer surfaces at the same density and with the same mass transfer limitations. Instability is often a problem inherent in the biomolecule, the result of poor immobilization methods resulting in leaching, or inactivation of the biomolecule by species present in the sensing environment. For implantable glucose sensors to be successful, the issues of reproducibility and instability must be addressed.

8.2 SENSOR REDUNDANCY

The reliability of the glucose sensors is extremely important since diabetic patients rely on glucose sensors to monitor and control their glucose levels. Glucose sensors can fail in many ways when placed in an *in vivo* environment, as summarized in Figure 8.1. One way to increase reliability of the glucose readings is to use redundant sensors that generate several signals for the glucose concentration. Methods for fabrication of redundant sensors are described and the techniques used to process the signals obtained are discussed.

8.2.1 Sources of Error and Failure of *In Vivo* Glucose Sensors

8.2.1.1 Loss of Enzyme Activity and Leaching In most electrochemical glucose sensors, the enzyme glucose oxidase is used for its specificity to glucose.

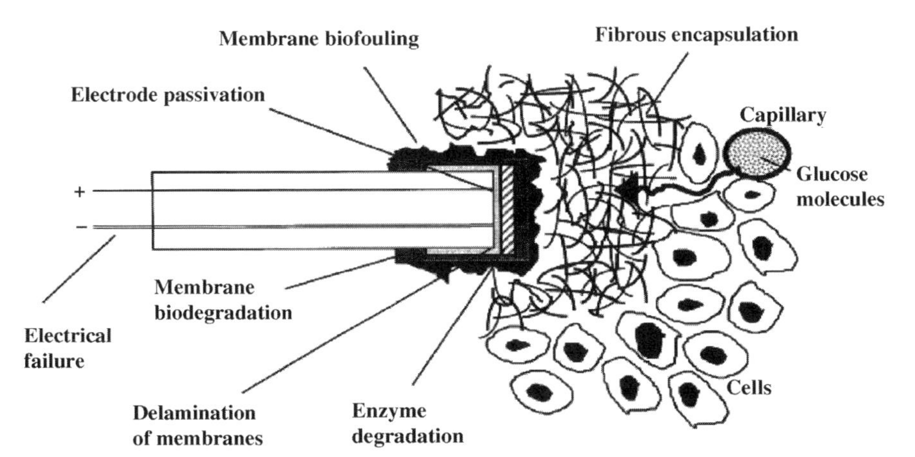

Figure 8.1 Diagram of the sources of error and failure in *in vivo* glucose sensors. Reprinted with permission from Ref. 31. Copyright 2000 Elsevier.

Enzymes can lose their activity because of protein denaturation and loss of active site.[26] Glucose oxidase can also lose its activity when exposed to excessive amounts of hydrogen peroxide. Most enzymes utilized in the fabrication of biosensors for implantation are derived from nonmammalian proteins, which can lead to an allergic response if the enzymes leach (or leak out of the sensor) into the body. Enzyme leaching can result when the attachment or entrapment of the enzyme is not robust. For example, if a method of entrapment via cross-linking polymers is utilized and the reaction does not go to completion during entrapment, then the enzyme can leach out and cause a loss of signal.[27]

8.2.1.2 Loss of Mediator Activity and Leaching Many *in vivo* electrochemical glucose sensors utilize a mediator to shuttle the electron to the electrode and decrease the interference by electro-oxidizing species.[28] When small mediators are used, the mediators tend to diffuse out of the membrane and into the body. If the mediator enters the body, it may result in interference with biological reactions.[28–30] The method of attachment for the mediator is very important because if the mediator leaches out or loses activity, the sensor may fail. The mediator may lose activity if it undergoes a detrimental reaction or denatures, causing the potential to shift or the electron mediator ability to be lost.

8.2.1.3 Film Delamination Most sensors have a film or membrane used as a sensing layer and/or a biocompatible layer. When this film is not attached properly or degrades, it can fail and delaminated from the electrode surface, causing the sensor to lose function through increased mass transfer resistance, diminished electron transfer, and/or its ability to deter the foreign body reaction.[31]

8.2.1.4 Corrosion and Failure of Electrodes, Leads, and Insulation Some of the component-based failures of sensors that can occur include failure of the different parts of the electrode. If the metal used is prone to corrosion in the presence of water and is not properly protected, the electrode can fail through metal corrosion. Other methods for failure also include the failure of the insulation or leads. The leads can fail if their connection is lost and the signal cannot be obtained, this may be the case if the leads are attached poorly or are also prone to corrosion. Another failure mode of the electrodes is electrode fouling, which occurs when small molecules come into contact with the surface of the electrode after penetration of the sensor.[31]

8.2.1.5 Biofouling and Foreign Body Reaction Biofouling is described when a sensor is implanted in the body and there is a buildup of cells, proteins, and other biological substances on the surface of the sensor, as discussed in more detail in Chapters 2–4. Briefly, when the body detects a foreign object, a foreign body reaction is begun and acts to dispel the foreign body through the wound-healing stages. The stages for wound healing are homeostasis, inflammation, repair, and scar formation and are affected by the material on the outer surface of the membrane. With the fibrous capsule surrounding the implanted sensor, membrane biofouling has occurred and an even longer diffusive path is created for glucose to reach the sensor for detection, which leads to a smaller signal and eventually can lead to complete failure of the sensor.[31,32]

8.2.2 Statistics of Reliability and Redundancy

Microelectrode arrays where each array element detects the same analyte may have superior properties compared to a single large electrode. Because of radial mass transfer effects, the flux of analyte to microelectrodes is higher than that for a large planar electrode where semi-infinite linear diffusion dominates. If the current at the electrode is mass transfer limited, then the increased rate of mass transfer resulting from the geometry of a microelectrode will result in a higher current density.[33] As a starting approximation, higher signal-to-noise ratios could be observed starting dimensions near 100 μm and the increase might approach 10-fold for electrode dimensions near 10 μm.[34,35] Independently addressable array elements for the same analyte also have the advantage of allowing advanced signal processing techniques to be used to reduce noise and improve the accuracy of the overall sensor. This is particularly useful in medical applications such as an implantable glucose sensor where erroneous results may harm the patient. An early study using two implanted glucose electrodes monitoring simultaneously in the subcutaneous tissue of a rat combined with a signal processing algorithm demonstrated that the overall glucose measurement accuracy could be improved over that of a single sensor.[36,37] Likewise, a sensor array with four coated platinum electrodes (each 0.25×10 mm in exposed surface) with an algorithm designed to reject outlying signals also demonstrated improved accuracy in studies with diabetic rats.[38,39]

8.2.2.1 ***Reliability*** Large numbers of redundant sensors allow signal averaging to improve accuracy and the use of fault detection algorithms to detect the failure of individual array elements.[36] For example, the variance of a measurement based on the average of N identical sensors is

$$\sigma = \frac{\sigma_N}{\sqrt{N}} \tag{8.1}$$

where σ is the variance of the measurement, σ_N is the variance of each individual sensor, and N is the number of sensors. In addition, the reliability of the overall device will increase because of redundancy. In the most basic sense, reliability is defined as the probability of a component surviving for some period of time t. If $R_m(t)$ is the average sensor reliability among a group of N sensors (i.e., the number of sensors functioning correctly at time t divided by the total number of sensors), then the reliability of an array of these sensors operating in parallel is

$$R_s(t) = 1-[1-R_m(t)]^N \tag{8.2}$$

As shown in Table 8.1, the reliability of the system where the average reliability of the individual components is 0.75 is greatly increased by increasing the number of elements in the array. Thus, for an array of four sensors each with a reliability of 0.75, the reliability of the array is $1 - (1 - 0.75)^4$ or 0.996, a large increase as compared to a single sensor.[37,40,41]

Since the implanted glucose sensor readings have a direct impact on the patient, the reliability of the sensor's reading is vital and can be influenced by developing a system that requires more functioning elements to produce a reading. The reliability

TABLE 8.1 Increase in Reliability with the Number of Sensors in an Array

Number in array	Reliability of system
1	0.75[a]
2	0.938
3	0.984
4	0.996
5	0.999

[a] By definition.

of the system where K-out-of-N parallel elements must be functional for the system to be functional is described with respect to redundancy by the following equation:

$$R_s(t) = \sum_{r=K}^{N} \binom{N}{r} [R(t)]^r [1-R(t)]^{N-r} \tag{8.3}$$

$R_s(t)$ is the reliability of the system, $R(t)$ is the reliability of an individual component, N is the total number of components,

$$\binom{N}{r} = \frac{N!}{r!(N-r)!},$$

and K is the minimum number of functioning components for a functioning array.[37,42,43] As shown in Table 8.2, if the K-out-of-N system of sensors is used, the reliability for each required number of functioning components is seen for five components with $R(t)$ equal to 0.75.

As seen from the table, the reliability of the system when all five elements must be functional is very low, which means that to increase the reliability of the sensor functioning in this system, the number of functioning elements required needs to be examined and adjusted for optimal performance of the sensor.

8.2.2.2 *Mean-Time-to-Failure* The most common metric for reliability is the mean-time-to-failure or MTTF where

$$\text{MTTF} = \int_0^{\infty} R(t)\,\mathrm{d}t \tag{8.4}$$

TABLE 8.2 Decrease in Reliability as the Required Number of Functioning Sensors Increases

Number of functioning electrodes (out of five), K	Reliability of system ($R_s(t)$)
1	0.999
2	0.984
3	0.897
4	0.633
5	0.237

TABLE 8.3 Increase in Mean-Time-to-Failure with an Increasing Number of Electrodes

	$MTTF_s$	
Number of electrodes, N	MTTF = 5	MTTF = 15
1	5	15
2	7.5	22.5
3	9.2	27.5
4	10.4	31.3
5	11.4	34.3

For many systems, component lifetimes are distributed exponentially, thus

$$R(t) = e^{-\frac{t}{\lambda}} \tag{8.5}$$

where λ is the component's mean lifetime. Thus, for a single component,

$$\text{MTTF} = \frac{1}{\lambda} \quad \text{and} \quad \text{MTTF}_s = \text{MTTF}\left(1 + \frac{1}{2} + \cdots + \frac{1}{N}\right) \tag{8.6}$$

for a redundant array of identical sensors. As is apparent in Table 8.3, the MTTF of the array ($MTTF_s$) increases as the number of components increase. However, there is a diminishing return, that is, each additional component contributes less to the MTTF. Thus, an optimum number of components (or in our case sensors) exists to maximize reliability and minimize cost of the array.

8.2.3 Fabrication of Redundant Sensor Arrays

8.2.3.1 Fabrication of Base Sensors For fabrication of redundant sensor arrays on metal or flexible insulating substrates, processes used in the semiconductor industry can be utilized as shown in Figure 8.2.[44,45] Photolithography, wet etch, dry etch, sputtering, and evaporation are a few processes that have been used.[46,47] The techniques described in this section can be used many times on the same sample to lead to numerous metals (or different electrodes) being fabricated on the same substrate, which can then be implanted *in vivo*.

Photolithography Photolithography is defined as a process that uses light to develop a pattern. Photolithography uses photoresists to develop desired patterns and go through a multistep process to obtain the final desired product, as shown in Figure 8.3. There are two main types of photoresists used for photolithography: positive and negative. A negative resist will result in the exposed regions remaining and the nonexposed regions being removed. The positive resist results in the regions that are exposed to UV light through a mask and are removed and the nonexposed regions remain. When choosing a photoresist and patterning method, a number of process issues need to be determined: the pattern of the mask, if a metal should be deposited first and removed through an etch step or a metal should be deposited after photolithography and excess removed through a lift-off procedure, the desired

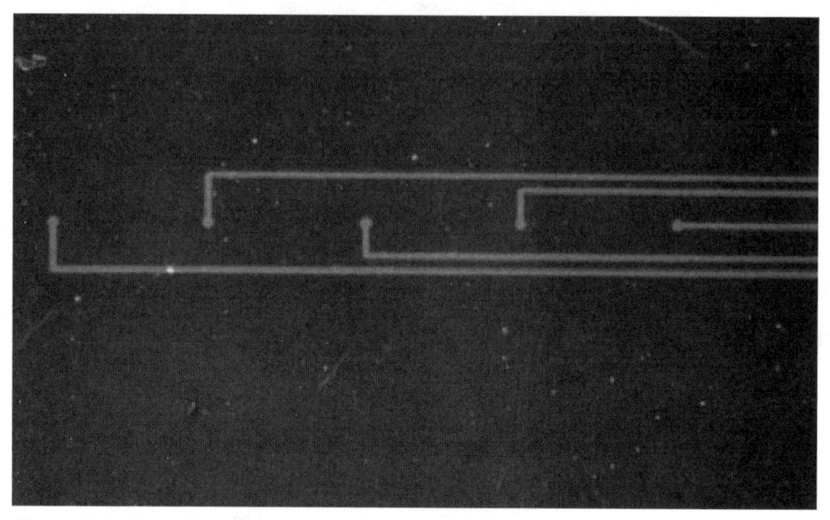

Figure 8.2 An array of redundant gold electrodes fabricated by photolithographic methods, where the electrode diameters are 50 μm and the leads are 10 μm. Reprinted with permission from Ref. 2. Copyright 2000 Elsevier.

thickness of the photoresist, and the properties of the substrate and resist/developers to determine cross-reactivity.

When using photoresist, the first step is to obtain a clean substrate, which could be an insulating layer or an insulating layer with a metal deposited, and then deposit the photoresist on the surface through spin coating (the spin coating settings can be determined through the specifications of the desired thickness and given conditions in the photoresist product information). When there is an even layer of photoresist on the surface, a hard bake step is needed to remove the solvent, this is done by placing the substrate on a hot plate set to a designated temperature for a given amount of time (the time and temperature depend on the photoresist used and will be in the specifications of the photoresist). Once the solvent is evaporated and the sample is allowed to cool, the sample is brought into close contact with a mask with the desired pattern. When the mask and sample are exposed to UV light, only the openings in the pattern allow the sample exposure to UV light; therefore, the exposed photoresist undergoes a chemical reaction leaving the unexposed regions unchanged.

A developer solution is utilized to remove the excess photoresist and obtain the desired pattern. Once the pattern is obtained, the next step can be an etching or a metal deposition step, depending on the process chosen, and the remaining photoresist can be removed through etching or lift-off procedures.[44,48]

Metal Deposition Metal deposition can be accomplished by the sputtering or evaporation of a metal onto the surface. Both methods require clean substrates to be inserted into the machines used for deposition. Sputtering takes place when the target metal is in an argon plasma and bombarded by the positive ions of argon. The process then uses the momentum transfer from the bombardment of positive ions to sputter the metal away from the target and onto the desired substrate. A negative charge is

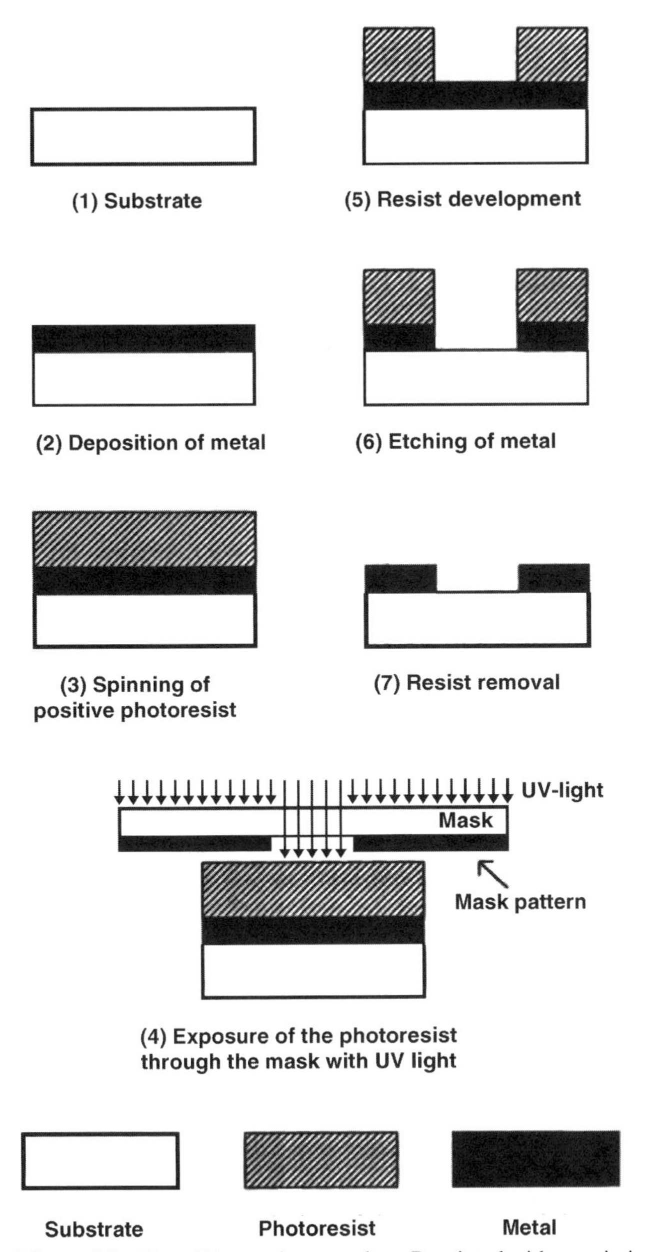

Figure 8.3 Photolithography procedure. Reprinted with permission from Ref. 44. Copyright 1992 IOP Publishing.

produced on the target by its connection to a negative RF or DC power source, leading to the positive ions being directed to the target.[44]

Evaporation is a process in which the target metal is heated in a high-vacuum environment and deposited onto the substrate. There are two main ways to heat the target metal: resistive heating, where a high current is passed through a boat or filament, and using an electron gun, which is focused on the target metal at a high energy and makes the metal melt and evaporate. The substrate is affixed to holders inside the evaporator, which are rotated around the target metal to increase uniformity of the deposited layer.[44] Sputtering is often preferred because of difficulties in the evaporation of alloys. Sputtering generally gives a higher deposition rate, and results in better step coverage, increased adhesion, and higher purity than evaporation.[44]

Lift-Off When utilizing the lift-off technique, the photoresist is deposited and developed onto the substrate and then the metal is deposited on top of the photoresist, as shown in Figure 8.4. The excess photoresist and metal are then removed by a lift-off technique, in which the sample is submersed in a solution in which the photoresist is soluble. For example, when using the photoresist ma-P 100, submersion of the sample into acetone can complete the lift-off procedure.

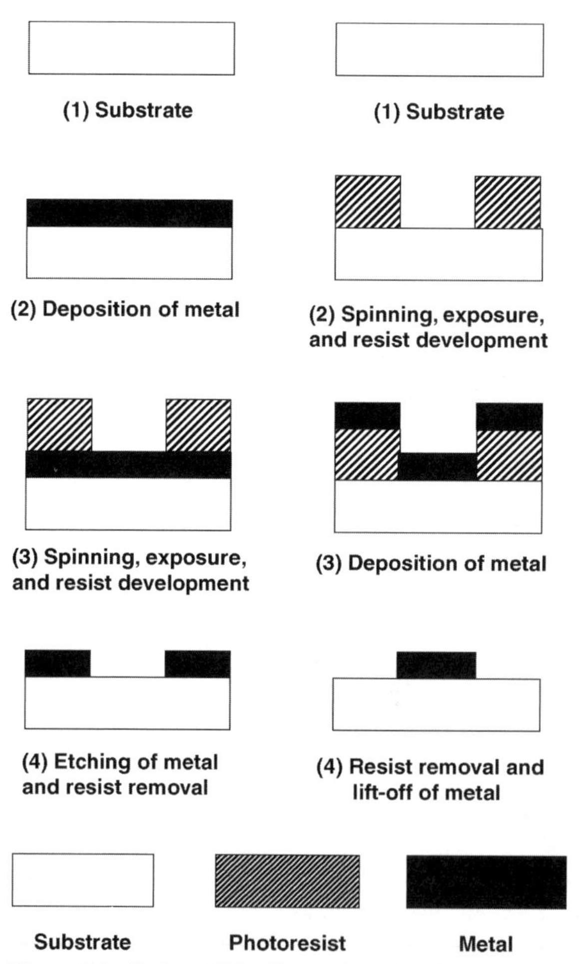

Figure 8.4 Etch and lift-off procedure. Reprinted with permission from Ref. 44. Copyright 1992 IOP Publishing.

Etching In the silicon industry, etching is known as the process of removing layers of metal, mostly for removing oxide layers from the metal. Etching can be done in a dry, wet, or plasma environment. For electrode fabrication, photolithography is performed on an insulating substrate with a metal layer deposited and then an etching step is used to remove the excess metal not covered by the photoresist. Then the photoresist is removed and the desired pattern is obtained.

8.2.3.2 Deposition of Sensing Chemistries Deposition of sensing chemistries can be accomplished by many means: covalent attachment, electrostatic attachment, entrapment, or a combination of these methods. Covalent attachment is when the sensing elements are chemically bonded to the surface through chemical reactions creating covalent bonds.[49,50] In an interesting example, after a self-assembled monolayer of aminoethane thiol is attached to the gold surface, ferrocene-modified glucose oxidase is covalently attached to the aminoethane thiol by glutaraldehyde.[50]

Electrostatic attachment can be accomplished when the elements used for sensing are charged oppositely to the surface or a layer attached to the surface. The sensing elements have been attached to the electrode via electrostatic attachment, utilizing the positive charge of the mediator and the negative charge of glucose oxidase.[51] One of the benefits of electrostatic attachment is the ease of fabrication when the charged surface is exposed to a liquid containing the oppositely charged element; the oppositely charged element then forms a layer on the surface. This layering approach can be repeated many times and the time needed per layer is less than 1 h.[51]

Entrapment is another method to incorporate the sensing elements into the sensor and involves entrapping the elements behind a membrane or inside a polymeric material.[52] This method is useful when the effective pore size of the membrane or polymer is smaller than the sensing protein but larger than the analyte molecule. Since glucose oxidase is much bigger than glucose, the entrapment method can be utilized.[26] The glucose oxidase and mediator can be trapped in the sensor, while the glucose is free to diffuse in and out of the sensor through the membrane leading to the reaction and detection of glucose. The two methods of entrapment may be combined to develop the sensor. In one example, the sensing layer is fabricated by entrapping the mediator and glucose oxidase into a polymeric layer formed by cross-linking with poly(ethylene glycol) diglycidyl ether (PEGDGE). Another membrane was used for biocompatibility and the rest of the sensor components was trapped behind a membrane developed by photo-cross-linking tetraacrylated poly(ethylene oxide) in solution with a photoinitiator via exposure to UV light.[22]

Membranes can also be used as a protective, diffusional, or biocompatible barrier. When implanting the sensor in the body, an outer membrane can be utilized as a size exclusion or protective barrier to deter larger molecules from interfering and also from sensing elements leaking out. A diffusional barrier is utilized to alter the enzyme kinetics and extend the linear range of the sensor by increasing the diffusion-limited regime with respect to Michaelis–Menten kinetics.

The fabrication procedures for membranes vary, but many are relatively straightforward and simply involve casting or dip coating. In the example above with poly(ethylene glycol) diglycidyl ether, a precursor solution is made up of glucose oxidase, mediator, and PEGDGE. The solution is mixed, applied to the surface of the electrode, and then allowed to dry to cross-link the solution.[22] Other methods, such as photopolymerization or sol–gel formation, require specific instruments or more complicated procedures for fabrication, but have benefits, as discussed in the following sections.

8.2.3.3 *Improving Biocompatibility* Significant research has focused on improving biocompatibility of different materials and methods. The following sections are devoted to review the approaches that may find utility in the next generation of sensors and redundant sensor arrays. The general issues of biocompatibility and other specific strategies for improvement are covered elsewhere in the book.

Hydrogels Hydrogels consist of a cross-linked polymer system or a mesh-like structure that is flexible and can swell in the presence of water, producing a hydrophilic surface and masking the components below. The most widely used hydrogels consist of poly(hydroxyethyl methacrylate) (PHEMA), poly(ethylene glycol) (PEG), or poly(vinyl alcohol) (PVA) and tend to have reduced interaction with tissue.[22,31,44,53–55]

Hydrogels can be formed by applying UV light to, for example, poly(ethylene glycol) diacrylate (PEGDA) in the presence of a photoinitiator inducing a gelation process that forms the hydrogel.[44,56] PEGDA hydrogels are used to encapsulate sensing elements and allow the analyte to diffuse into the hydrogel for detection with a controllable mass transfer rate. The mass transfer into the PEG hydrogels depends on the mesh size, which can be thought of as the size of the hole in the 3D structure. The mesh size of the structure can be controlled by altering the molecular weight (length) or branching of the precursor PEG. When the molecular weight of PEG is increased, the mesh size and diffusivity increases.[57–62]

Some problems that can occur with the fabrication of hydrogels are difficulty in adherence to the surface, mechanical stability problems when implanted, some precursors (monomer, solvent, cross-linking agent, UV light) can harm the enzyme or other sensor components, and some hydrogels can limit the diffusion of the desired analyte to the sensor.[31]

Phospholipid-Based Biomimicry Utilizing phospholipids, components that make up the cellular membrane, some biosensors are constructed to mimic cells on the outside.[31,49,55] Phospholipid membranes are difficult to fabricate and fragile, leading to polymers modified with phospholipids being utilized to increase stability. So in one study, a Langmuir–Blodgett technique was utilized to apply the glucose oxidase and phospholipid analogous vinyl polymer (PCP).[63] In another study, the phospholipid polymer 2-methacryloyloxyethyl phosphorylcholine (MPC) was utilized in a polymer form (poly(MPC-co-DMA), where DMA is dodecyl methacrylate) to entrap glucose oxidase in the sensor through solution casting.[65]

Polymer Membranes Polymer membranes can be formed from naturally derived materials and synthetic materials to improve biocompatibility. Membranes made from naturally derived materials such as silk fibroin,[65,66] cellulose,[67] and chitosan[68] have been used to increase biocompatibility. Eisele et al. showed that bacterial cellulose membranes increased the stability of glucose sensors in whole and diluted blood compared to cuprophran (wood cellulose) membranes.[67] Chemical modification by acylation of the C-2 position of cellulose acetate increases biocompatibility of the membranes versus unmodified.[69] Cellulose acetate can be applied through precast membranes and spin coating or dip coating procedures; the method of application depends on the size of the sensor.[52]

The synthetic polymer Nafion is an anionic, inert polymer that has hydrophilic and hydrophobic properties. Nafion is composed of perfluorosulfonic acid and has shown to increase the lifetime of sensors and to decrease the inflammatory response in the short term and decrease interferents (believed to be due to the anionic property). Nafion can be applied as an outer coating to sensors through a dip coating or spin coating procedure to entrap the enzyme and improve biocompatibility of the sensor.[31,50,70,71]

Surfactants Surfactants are molecules made up of a polar hydrophilic head group and a hydrophobic hydrocarbon tail, which leads these molecules to favor as being at the interface between two phases.[31] Membranes that were previously used, such as solvent-cast cellulose acetate[72] or dip-coated polyurethane membranes,[73] can be modified by surfactants to decrease biofouling. Lindner et al. showed that membranes with lower surfactant plasticizer ratios exhibited lower anion interference and better biocompatibility.[74]

Flow-Based Systems Needle-type sensors with a fluid flowing over the sensor tip seem to resist biofouling and extend sensor lifetime.[31] There are numerous methods that have been investigated for flow-based sensors, such as microperfusion systems,[75] microdialysis,[76,77] and ultrafiltration.[78] Reduced fouling was found with an open microflow system where slow flow of protein-free fluid over the sensor surface at the implant site is effected.[73] Different from the other flow-based sensors, the open microflow is controlled by the subcutaneous tissue hydrostatic pressure and does not require a pump.

Covalent Attachments Covalent attachment of antifouling agents to the sensor surface can increase sensor longevity; some covalent modifications that have been shown to improve biocompatibility are poly(*o*-phenylenediamine),[79] diols,[80] glycols,[80] and silanes.[81]

8.2.4 Operation of Redundant Sensor Arrays

8.2.4.1 *Addressing Sensors and Multiplexing* Wires will be attached to the developed redundant sensors separately and connected to a potentiostat or biopotentiostat for signal gathering. A multiplexer can be utilized to obtain separate signals using one potentiostat. After implantation of the sensor, blood sampling is not started until the sensor is allowed at least 1 h to reach a basal signal level.[22]

8.2.4.2 *Signal Processing and Fault Detection* A "one-point" *in vivo* calibration method is desired for implantable sensors, but the signal must be accurate enough for the patient's life to be dependent on the sensor. The "one-point" calibration method relies on a zero point taken from the sensor in buffer with zero glucose present and then measuring one blood glucose concentration.[22]

In Ref. 22, the "one-point" calibration method is tested on fabricated sensors and then a pair of redundant sensors is utilized to improve the reliability of the system. When using the pair of redundant sensors, the first step is to normalize the readings of the separate electrodes to each other by multiplying each reading of electrode A by the average output from electrode B divided by the average output from electrode A. The next step is to determine the standard deviation of the electrodes by obtaining 24 sets of readings of electrode B and the normalized electrode A and calculating the standard deviation. Then, the sets of readings that are outside of one standard deviation are deemed unreliable and not used. When using this method, the sensors reliability to lie within the clinically correct zones of the Clarke error grid increased from 94% to 99%.[22]

Another method to process and detect faulty responses of the redundant sensor signal is called the likelihood test. The likelihood test used is borrowed from nuclear reactor safety for pressure sensors deciding shutdown or no shutdown. Instead of shutdown or no shutdown for this case, the sensor reading would be rejected or accepted. For accuracy of the readings, the test chosen needs to reject readings that have even the smallest likelihood of being false and leading to the patient's inappropriate response.[36]

When comparing the readings of the pair of implanted sensors, a hypothesis must be tested. The null hypothesis is when the sensors should be equal, meaning the hypothesis is that is there is no statistical difference between the two readings. To determine the validity of the null hypothesis, there can be two errors, type I and type II. Type I error is caused by rejecting the null hypothesis when it is correct, which is often termed a false positive. The probability of type I error occurring is expressed as α. Type II error accepts the null hypothesis when it is false; the probability of this error occurring is termed β. The likelihood test described here takes into account the two types of errors above that can occur and skews the data in favor of type I errors versus type II to favor the safer of the two errors and decrease the number of incorrect treatments given by the patient.[36]

To utilize the likelihood test, the readings from two redundant sensors, g_1 and g_2, are based on the "one-point" calibration previously described.[22] Then, a probability ratio test can be utilized to determine if the difference ($y = |g_1 - g_2|$) between the two readings at time t was significant statistically. Next, an algorithm is applied to two hypotheses to reject risky points. Hypothesis 1 (H_1) inspects the set for a statistical difference between the pairs, which is greater than $|M|$, the system disturbance. A clinically significant disturbance is deemed when the signal is above $+M$ or below $-M$. Hypothesis 2 (H_2) inspects the set for the probability that there is no difference between the sensor signals.

H_1: y is drawn from a Gaussian probability density function (pdf) with mean M and variance σ^2.

H_2: y is drawn from a Gaussian pdf with mean 0 and variance σ^2.

Next, the ratio of the likelihood of hypothesis 1 versus hypothesis 2 is examined and defined as R.

$$R = \frac{f(y|H_1)}{f(y|H_2)} \tag{8.7}$$

where $f(y|H_i)$ is the probability that H_i is true. Then error probabilities are defined to determine if H_1 or H_2 is true:

H_1 is decided with probability $1-\beta$.

H_2 is decided with probability $1-\alpha$

where α is the probability of accepting H_1 when H_2 is true, which is the probability of a false alarm or rejecting an accurate glucose reading. Also, β is the probability of a missed alarm, or accepting H_2 when H_1 is true and accepting a glucose reading when it should have been accepted.

For any value of R, there are three possible scenarios: H_1 is true, H_2 is true, or H_1 and H_2 are false. For the case at hand, the only scenario that is a concern is if the sensor readings are too different and need to be rejected, or when H_1 is true.

Next, the thresholds must be established for when to accept the different hypotheses, this is done through the acceptance criteria below that are related to the error probabilities.

accept H_1 if $R \geq \frac{1-\beta}{\alpha}$.

accept H_2 if $R \leq \frac{\beta}{1-\alpha}$.

The value of α is set fairly high at 0.5, because there is no penalty in rejecting a valid point. The value of β is then set very low at 0.05, because there could be a very dangerous result if a false point is thought valid when it should be rejected.

With the assumption that y is normally distributed, the likelihood (L) that H_1 is true is

$$L(y|H_1) = \frac{1}{(2\pi)^{1/2}\sigma} \exp\left[\frac{-1}{2\sigma^2}(y^2 - 2yM + M^2)\right] \tag{8.8}$$

and the likelihood that H_2 is true is

$$L(y|H_2) = \frac{1}{(2\pi)^{1/2}\sigma} \exp\left[\frac{-1}{2\sigma^2}(y^2)\right] \tag{8.9}$$

which leads to the likelihood ratio being defined as

$$R = \exp\left[\frac{-1}{2\sigma^2}(M(M-2y))\right] \tag{8.10}$$

In Ref. 36, the likelihood test described above is applied to five pairs of implanted sensors and examined on the Clarke error grid for clinical relevance. The percentage of readings that fell into the clinically correct regions (Zones A and B) increased from 92.4% to 98.8% when utilizing the likelihood test.[36]

Finally, a median-based exclusion algorithm has been applied to data obtained with a fully implanted array of four microelectrochemical sensors.[38,39] The *Z*-score was calculated from the values of individual sensors and the median absolute deviation, as shown in equation (8.11), and sensor readings were removed from the data set if the value was greater than one.

$$Z\text{-score} = \frac{|\text{medianofallsensorvalues} - \text{sensor} \times \text{value}|}{|\text{medianofallsensordeviations}| \times 1.483} \tag{8.11}$$

Clarke error grid analysis of a study of 15 diabetic rats showed the percentage of readings that fell into the clinically correct regions (Zones A and B) increased from 92% to 96% when applying the *Z*-score rejection criteria.[38] During the long-term implantation (25 ± 4 days), *Z*-score calculations removed 32% of the individual sensor data from six fully implanted four-sensor arrays.[39]

8.3 NEXT-GENERATION ELECTROCHEMICAL SENSORS

8.3.1 New Chemistries

In previous sensors, a mediator is required for the electron transfer from the glucose oxidase to the electrode surface, but new techniques seem promising to allow direct electron transfer from glucose oxidase to the electrode surface for detection.[46,82] In Ref. 83, the distance separating the enzyme and the electrode is greatly reduced by direct communication of the enzyme and electrode through immobilization of a low molecular weight biocatalyst immobilized on the thiol-monolayer electrode, which in turn increases the signal produced.[83] Khan used a stable charge transfer complex (CTC) electrode to achieve direct electron transfer between glucose oxidase and the CTC electrode resulting in improved sensor performance.[84] More recent work has focused on connecting the active site of glucose oxidase, flavin adenine dinucleotide (FAD), to the electrode and then applying reconstituted glucose oxidase (Apo-GOX) for glucose specificity. Various methods have been utilized when electrically connecting FAD to the electrode surfaces, some examples are pyrroloquinoline quinone (PQQ),[85,86] carbon nanotubes,[87] and gold nanoparticles.[88]

Nanomaterials can also be applied to glucose biosensors to enhance the properties of the sensors and, therefore, can lead to smaller sensors with higher signal outputs. Carbon nanotubes have been incorporated in previously developed sensors and seen to increase the peak currents observed by threefold.[89] Platinum nanoparticles and single-wall carbon nanotubes have been used in combination to increase sensitivity and stability of the sensor.[90,91] CdS quantum dots have also been shown to improve electron transfer from glucose oxidase to the electrode.[92,93] Yamato et al. dispersed palladium particles in a polypyrrole/sulfated poly(beta-hydroxyethers) and obtained an electrode response at 400 mV, compared to 650 mV, at a conventional platinum electrode.[94]

Other work has been focused on lowering the operating potential to minimize the interference of electroactive species through "wiring" of the mediator and enzyme[95] or varying the mediator.[96,97]

8.3.2 Fabrications Technologies

To further miniaturize the sensors, nanotechniques such as surface probe microscopy and lithography, lateral force microscopy, atomic force microscopy (AFM), and AFM lithography can be utilized.[46] Nanoimprint lithography is a low-cost technique that has been shown to produce patterns on the nanometer scale.[98,99]

Electrodeposition is one method that has promise to be utilized in micro-, nano-, and biotechnologies due to its use in water-based electrolytes, ability to grow complex 3D structures, and ability to develop intricate patterns on the nanoscale or larger scales (an example can be seen in Figure 8.5).[100–103] One method using electroless deposition was the fabrication of arrays of gold nanotubes, which were then used to attach glucose oxidase and a mediator and shown to have high sensitivity.[104]

Devices such as microelectromechanical systems (MEMS) have promise to be able to "sense, think, act, and communicate." Some advantages with using MEMS include low cost, small size, and reliability. More work must be done on advancing MEMS to biomedical sensor applications.[105]

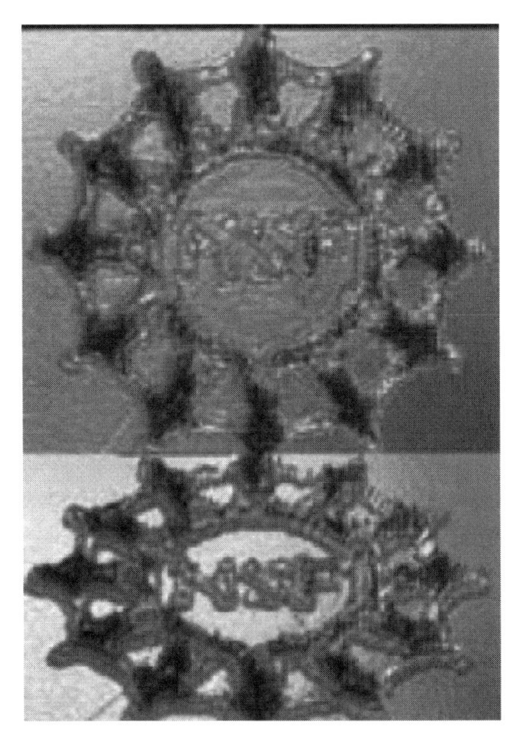

Figure 8.5 Example of electrochemical printing and the intricate patterns that can be obtained with no mask. The diameter of the entire pattern is 1.8 mm and the thickness is 700 nm. Reproduced from Ref. 102 by permission of The Electrochemical Society.

8.3.3 Signal Processing

For glucose sensors, the method of calibration helps to determine the accuracy of the sensor. The theory of mixtures of experts has previously been applied to the Glucowatch system, where the inputs are elapsed time, blood glucose value at calibration point, integrated current, and calibrated signal. The algorithm used multiple linear regressions and a switching algorithm to calculate outcomes.[106] Another calibration method used presteady-state kinetic data to determine glucose measurements.[107]

Many numerical techniques are available for developing signal processing algorithms with applications in biosensor arrays. These are typically divided into two categories: univariate statistical techniques and multivariate methods such as neural networks (NN), partial least squares, and principal component analysis (PCA). Univariate statistical methods take advantage of upper and lower bounds; if the data do not conform to these bounds, the sensor is considered faulty. Unfortunately, this type of analysis often results in numerous indicators for the root cause of failure, making identification of the primary problem difficult. In addition, all types of failure cannot be diagnosed with univariate methods, for example, consider the case of a sensor functioning improperly and yet displaying measurements within the predetermined upper and lower bound limits.[108,109] A univariate method would not be able to identify this failed sensor and thus more sophisticated multivariate techniques become necessary.

PLS, NN, and PCA are frequently employed multivariate statistical techniques for analyzing large data sets. PLS, like PCA, is a dimension reduction method with the goal of maximizing the correlation between input and output matrices. This permits focusing fault detection efforts on the variables that most specifically influence another set of variables.[110] Neural networks are the most sophisticated of the three multivariate techniques mentioned above. They have been employed in numerous studies, including the determination of blood glucose by near-infrared spectroscopy,[111] the calibration of an array of voltammetric microelectrodes, and for online monitoring with an electronic nose.[112] Despite the sophistication of the neural networks employed in these and other studies, various drawbacks exist with respect to their application for *in vivo* biosensing. The necessity for training sets, which may be quite large in some instances, is a considerable complication associated with neural networks. The generation of such training sets *in vivo* may be difficult or impractical. This drawback is not inherent with PCA, as a PCA model may be applied to analyze a large data set without prior training, minimizing the requirements for *a priori* model development. Because of these advantages, PCA has been used for many applications, albeit unrelated to biosensor arrays, including the analysis of data generated by an electronic tongue[113] and disturbance detection in wastewater treatment plants.[110]

8.4 FUTURE OUTLOOK

As commercial continuous glucose sensors come into use for the treatment of diabetes, the real-world performance required of the glucose sensing system will become more apparent. Initial attempts to use sensing data to control insulin pump medication may be less than optimal due to limitations in the accuracy and reliability of the sensors. Medical providers will also be exploring the use of commercial systems in delicate treatment situations such as neonatal care. The combination of redundant sensor arrays and the next-generation electrochemical chemistries, fabrication technologies, and signal processing strategies that have been presented here may serve as a guide to overcoming the expected limitations of the first commercial continuous glucose sensing systems.

REFERENCES

1. Coté G, Fox M, Northup R. Noninvasive optical polarimetric glucose sensing using a true phase measurement technique. *IEEE Transactions on Biomedical Engineering* 1992, 39, 752–756.
2. Heise HM, Marbach R, Koschinsky T, Gries FA. Noninvasive blood glucose sensors based on near-infrared spectroscopy. *Artificial Organs* 1994, 18, 439–447.
3. Coté G. Noninvasive optical glucose sensing: an overview. *Journal of Clinical Engineering* 1997, 22, 253–259.
4. Klonoff DC. Noninvasive blood glucose monitoring. *Clinical Diabetes* 1998, 16, 43–45.
5. Glikfeld P, Hinz RS, Guy RH. Noninvasive sampling of biological fluids by iontophoresis. *Pharmaceutical Research* 1989, 6, 988–990.
6. Kimura J. Noninvasive blood glucose concentration monitoring method with suction effusion fluid by ISFET biosensor. *Applied Biochemistry and Biotechnology* 1993, 41, 55–58.
7. Rao G, Glikfeld P, Guy RH. Reverse iontophoresis: development of a noninvasive approach for glucose monitoring. *Pharmaceutical Research* 1993, 10, 1751–1755.
8. Rao G, Guy R, Glikfeld P, LaCourse W, Leung L, Tamada J, Potts R, Azimi N. Reverse iontophoresis: noninvasive glucose monitoring *in vivo* in humans. *Pharmaceutical Research* 1995, 12, 1869–1873.
9. Fischer U, Ertle R, Abel P, Rebrin K, Brunstein E, Dorsche HHv, Freyse EJ. Assessment of subcutaneous glucose concentration: validation of the wick technique as a reference for implanted electrochemical sensors in normal and diabetic dogs. *Diabetologia* 1987, 30, 940–945.
10. Jansson P-A, Fowelin J, Smith U, Lonnroth P. Characterization by microdialysis of intercellular glucose level in subcutaneous tissue in humans. *American Journal of Physics* 1988, 255, E218–E220.
11. Bruckel J, Kerner W, Zier H, Steinbach G, Pfeiffer EF. *In vivo* measurement of subcutaneous glucose concentrations with an enzymatic glucose sensor and a wick method. *Wiener Klinische Wochenschrift* 1989, 67, 491–495.
12. Hagstrom E, Arner P, Engfeldt P, Rossner S, Bolinder J. *In vivo* subcutaneous adipose tissue glucose kinetics after glucose ingestion in obesity and fasting. *Scandinavian Journal of Clinical and Laboratory Investigation* 1990, 50, 129–136.
13. Meyerhoff C, Sternberg F, Bischof F, Mennel FJ, Pfeiffer EF. Measurement of subcutaneous glucose concentration. *Diabetes Care* 1993, 16, 1626–1627.

14. Poitout V, Moattisirat D, Reach G, Zhang Y, Wilson GS, Lemonnier F, Klein JC. A glucose monitoring-system for on line estimation in man of blood-glucose concentration using a miniaturized glucose sensor implanted in the subcutaneous tissue and a wearable control unit. *Diabetologia* 1993, 36, 658–663.
15. Schmidt FJ. Measurement of subcutaneous glucose concentration. *Diabetes Care* 1993, 16, 1627–1628.
16. Quinn CP, Pishko MV, Schmidtke DW, Ishikawa M, Wagner JG, Raskin P, Hubbell JA, Heller A. Kinetics of glucose delivery to subcutaneous tissue in rats measured with 0.3-mm amperometric microsensors. *American Journal of Physiology: Endocrinology and Metabolism* 1995, 32, E155–E161.
17. Kost Y, Mitragotri S, Gabbay R, Pishko M, Langer R. Non-invasive measurement of glucose and other analytes. *Nature Medicine* 2000, 6, 347–350.
18. Bindra DS, Zhang Y, Wilson GS. Design and *in vitro* studies of a needle-type glucose sensor for subcutaneous monitoring. *Analytical Chemistry* 1991, 63, 1692–1696.
19. Poitout V, Moatti D, Velho G, Reach G, Sternberg R, Thevenot D, Bindra D, Zhang Y, Wilson GS. *In vitro* and *in vivo* evaluation in dogs of a miniaturized glucose sensor. *American Society for Artificial Internal Organs* 1991, 37, M298–M300.
20. Wilson GS, Zhang Y, Reach G, Moatti-Sirat D, Poitout V, Thevenot DR, Lemmonnier F, Klein J. Progress toward the development of an implantable sensor for glucose. *Clinical Chemistry* 1992, 38, 1613–1617.
21. Kerner W, Lindquist S-E, Pishko MV, Heller A. Amperometric glucose sensor containing glucose oxidase cross-linked in redox gels. In: Turner APF, Alcock SJ (Eds), *In Vivo Chemical Sensors: Recent Developments*. Cranfield Press, Cranfield UK, 1993.
22. Csöregi E, Quinn CP, Schmidtke DW, Lindquist S-E, Pishko MV, Ye L, Katakis I, Heller A. Design, characterization, and one-point *in vivo* calibration of a subcutaneously implanted glucose electrode. *Analytical Chemistry* 1994, 66, 3131–3138.
23. Linke B, Kerner W, Kiwit M, Pishko M, Heller A. Amperometric biosensor for *in vivo* glucose sensing based on glucose oxidase immobilized in a redox hydrogel. *Biosensors & Bioelectronics* 1994, 9, 151–158.
24. Csöregi E, Schmidtke D, Heller A. Design and optimization of a selective subcutaneously implantable glucose electrode based on "wired" glucose oxidase. *Analytical Chemistry* 1995, 67, 1240–1244.
25. Armour JC. Long-term intravascular glucose sensors with telemetry. *Artificial Organs* 1989, 13, 171.
26. Aehle W. *Enzymes in Industry: Production and Applications*, 2nd edn. Wiley-VCH, Weinheim, 2004.
27. House J, Anderson E, Ward W. Immobilization techniques to avoid enzyme loss from oxidase-based biosensors: a one-year study. *Journal of Diabetes Science and Technology* 2007, 1, 18–27.
28. Heller A. Electrical wiring of redox enzymes. *Accounts of Chemical Research* 1990, 23, 128.
29. Gregg BA, Heller A. Cross-linked redox gels containing glucose oxidase for amperometric biosensor applications. *Analytical Chemistry* 1990, 62, 258.
30. Heller A. Electrical connection of enzyme redox centers to electrodes. *Journal of Physical Chemistry* 1992, 96, 3579–3587.
31. Wisniewski N, Reichert M. Methods for reducing biosensor membrane biofouling. *Colloids and Surfaces B* 2000, 18, 197–219.

32. Wilson GS, Hu Y. Enzyme-based biosensors for *in vivo* measurements. *Chemical Reviews* 2000, 100, 2693–2704.
33. Pishko MV, Michael AC, Heller A. Amperometric glucose microelectrodes prepared through immobilization of glucose-oxidase in redox hydrogels. *Analytical Chemistry* 1991, 63, 2268–2272.
34. Weber S. Signal-to-noise ratio in microelectrode-array-based electrochemical detectors. *Analytical Chemistry* 1989, 61, 295–302.
35. Aoki K. Theory of ultramicroelectrodes. *Electroanalysis* 1993, 5, 627–639.
36. Schmidtke DW, Pishko MV, Quinn CP, Heller A. Statistics for critical clinical decision making based on readings of pairs of implanted sensors. *Analytical Chemistry* 1996, 68, 2845–2849.
37. Revzin A, Sirkar K, Pishko M. Glucose, lactate, and pyruvate biosensor arrays based on redox polymer/oxidoreductase nanocomposite thin films deposited on photolithographically patterned gold electrodes. *Sensors and Actuators B* 2002, 81, 359–368.
38. Ward WK, Casey HM, Quinn MJ, Federiuk IF, Wood MD. A fully implantable subcutaneous glucose sensor array: enhanced accuracy from multiple sensing units and a median-based algorithm. *Diabetes Technology & Therapeutics* 2003, 5, 943–952.
39. Ward WK, Wood MD, Casey HM, Quinn MJ, Federiuk IF. An implantable subcutaneous glucose sensor array in ketosis-prone rats: closed loop glycemic control. *Artificial Organs* 2005, 29, 131–143.
40. Modarres M, Kaminsky M, Krivtsov V. *Reliability Engineering and Risk Analysis: A Practical Guide*. Marcel Dekker, New York, 1999.
41. ReliaSoft, Simple Parallel Systems. http://www.weibull.com/SystemRelWeb/simple_parallel_systems.htm (accessed on September 5, 2008).
42. Freund JE, Simon GA. *Statistics: A First Course*, 5th edn. Prentice Hall, Englewoods Cliff, 1991.
43. ReliaSoft, *k*-out-of-*n* Parallel Configuration. http://www.weibull.com/SystemRelWeb/k-out-of-n_parallel_configuration.htm (accessed on September 5, 2008).
44. Lambrecht M, Sansen W. *Biosensors: Microelectrochemical Devices*. Institute of Physics Publishing, Bristol, Philadelphia, and New York, 1992.
45. Revzin A, Russell RJ, Yadavalli VK, Koh WG, Deister C, Hile DD, Mellott MB, Pishko MV. Fabrication of poly(ethylene glycol) hydrogel microstructures using photolithography. *Langmuir* 2001, 17, 5440–5447.
46. Plummer JD, Deal MD, Griffin PB. *Silicon VLSI Technology: Fundamentals, Practice and Modeling*. Prentice Hall, Upper Saddle River, NJ, 2000.
47. Zhang S, Wright G, Yang Y. Materials and techniques for electrochemical biosensor design and construction. *Biosensors & Bioelectronics* 2000, 15, 273–282.
48. MicroChem, MicroChem Resist Datasheets. www.microchem.com (accessed on September 5, 2008).
49. Chen D, Li J. Interfacial design and functionization on metal electrodes through self-assembled monolayers. *Surface Science Reports* 2006, 61, 445–463.
50. Imamura M, Haruyama T, Kobatake E, Ikariyama Y, Aizawa M. Self-assembly of mediator-modified enzyme in porous gold–black electrode for biosensing. *Sensors and Actuators B* 1995, 24, 113–116.

51. Sirkar K, Pishko M. Amperometric biosensors based on oxidoreductases immobilised in photopolymerised poly(ethylene glycol) redox polymer hydrogels. *Analytical Chemistry* 1998, 70, 2888–2894.
52. Thevenot D, Toth K, Durst R, Wilson G. Electrochemical biosensors: recommended definitions and classification. *Biosensors & Bioelectronics* 2001, 16, 121–131.
53. Quinn C, Connor R, Heller A. Biocompatible, glucose-permeable hydrogel for *in situ* coating of implantable biosensors. *Biomaterials* 1997, 18, 1665–1670.
54. Quinn CP, Pathak CP, Heller A, Hubbell JA. Photo-crosslinked copolymers of 2-hydroxyethyl methacrylate, poly(ethylene glycol) tetra-acrylate and ethylene dimethacrylate for improving biocompatibility of biosensors. *Biomaterials* 1995, 16, 389–396.
55. Xu Z, Chen X, Dong S. Electrochemical biosensors based on advanced bioimmobilization matrices. *Trends in Analytical Chemistry* 2006, 25, 899–908.
56. Yadavalli VK, Koh WG, Lazur GJ, Pishko MV. Microfabricated protein-containing poly (ethylene glycol) hydrogel arrays for biosensing. *Sensors and Actuators B* 2004, 97, 290–297.
57. Peppas N, Merrill E. Poly(vinyl alcohol) hydrogels: reinforcement of radiation-crosslinked networks by crystallization. *Journal of Polymer Science* 1976, 14, 441–457.
58. Peppas N, Merrill E. Determination of interaction parameter for poly(vinyl alcohol) in gels crosslinked from solution. *Journal of Polymer Science* 1976, 14, 459–464.
59. Peppas NA, Huang Y, Torres-Lugo M, Ward JH, Zhang J. Physicochemical foundations and structural design of hydrogels in medicine and biology. *Annual Review of Biomedical Engineering* 2000, 2, 9–29.
60. Kim B, Peppas N. Synthesis and characterization of pH-sensitive glycopolymers for oral drug delivery systems. *Journal of Biomaterials Science* 2002, 13, 1271–1281.
61. Berger J, Reist M, Mayer JM, Felt O, Peppas NA, Gurny R. Structure and interactions in covalently and ionically crosslinked chitosan hydrogels for biomedical applications. *European Journal of Pharmaceutics and Biopharmaceutics* 2004, 57, 19–34.
62. Russell RJ, Axel AC, Shields KL, Pishko MV. Mass transfer in rapidly photopolymerized poly(ethylene glycol) hydrogels used for chemical sensing. *Polymer* 2001, 42, 4893–4901.
63. Yasuzawa M, Hashimoto M, Fujii S, Kunugi A, Nakaya T. Preparation of glucose sensors using the Langmuir–Blodgett technique. *Sensors and Actuators B* 2000, 65, 241–243.
64. Kudo H, Sawada T, Kazawa E, Yoshida H, Iwasaki Y, Mitsubayashi K. A flexible and wearable glucose sensor based on functional polymers with Soft-MEMS techniques. *Biosensors & Bioelectronics* 2006, 22, 558–562.
65. Demura M, Asakura T, Kuroo T. Immobilization of biocatalysts with *bombyx mori* silk fibroin by several kinds of physical treatment and its application to glucose sensors. *Biosensors* 1989, 4, 361–372.
66. Liu H, Liu Y, Qian J, Yu T, Deng J. Feature of entrapment of glucose oxidase in regenerated silk fibroin membranes and fabrication of a 1,1′-dimethylferrocene-mediating glucose sensor. *Microchemical Journal* 1996, 53, 241–252.
67. Eisele S, Ammon H, Kindervater R, Grobe A, Gopel W. Optimized biosensor for whole blood measurements using a new cellulose based membrane. *Biosensors & Bioelectronics* 1994, 9, 119–124.

68. Yang M, Yang Y, Liu B, Shen G, Yu R. Amperometric glucose biosensor based on chitosan with improved selectivity and stability. *Sensors and Actuators B* 2004, 101, 269–276.
69. Gunasingham H, Teo P, Lai Y, Tan S. Chemically modified cellulose acetate membrane for biosensor applications. *Biosensors* 1989, 4, 349–359.
70. Moussy F, Hassison D, Rajotte R. A miniaturized Nafion-based glucose sensor: *in vitro* and *in vivo* evaluation in dogs. *International Journal of Artificial Organs* 1994, 17, 88–94.
71. Wilson GS, Gifford R. Biosensors for real-time *in vivo* measurements. *Biosensors & Bioelectronics* 2005, 20, 2388–2403.
72. Maines A, Ashworth D, Vadgama P. Diffusion restricting outer membranes for greatly extended linearity measurements with glucose oxidase enzyme electrodes. *Analytica Chimica Acta* 1996, 333, 223–231.
73. Ahmed S, Dack C, Farace G, Rigby G, Vadgama P. Tissue implanted glucose needle electrodes: early sensor stabilisation and achievement of tissue–blood correlation during the run in period. *Analytica Chimica Acta* 2005, 537, 153–161.
74. Lindner E, Cosofret V, Ufer S, Buck R, Kao W, Neuman M, Anderson J. Ion-selective membranes with low plasticizer content: electroanalytical characterization and biocompatibility studies. *Journal of Biomedical Materials Research* 1994, 28, 591–601.
75. Rigby GP, Ahmed S, Horseman G, Vadgama P. *In vivo* glucose monitoring with open microflow: influences of fluid composition and preliminary evaluation in man. *Analytica Chimica Acta* 1999, 385, 23–32.
76. Meyerhoff C, Bischof F, Sternberg F, Zier H, Pfeiffer EF. On line continuous monitoring of subcutaneous tissue glucose in men by combining portable glucosensor with microdialysis. *Diabetologia* 1992, 35, 1087–1092.
77. Bolinder J, Ungerstedt U, Arner P. Long-term continuous glucose monitoring with microdialysis in ambulatory insulin-dependent diabetic patients. *Lancet* 1993, 342, 1080–1085.
78. Ash SR, Rainier JB, Zopp WE, Truitt RB, Janle EM, Kissinger PT, Poulos JT. A subcutaneous capillary filtrate collector for measurement of blood chemistries. *American Society for Artificial Internal Organs* 1993, 39, M699–M705.
79. Myler S, Eaton S, Higson S. Poly(*o*-phenylenediamine ultra-thin polymer-film composite membranes for enzyme electrodes. *Analytica Chimica Acta* 1997, 357, 55–61.
80. Guo B, Anzai J, Osa T. Modification of a glassy carbon electrode with diols for the suppression of electrode fouling in biological fluids. *Chemical and Pharmaceutical Bulletin (Tokyo)* 1996, 44, 860–862.
81. Keedy F, Vadgama P. Determination of urate in undiluted whole blood by enzyme electrode. *Biosensors & Bioelectronics* 1991, 6, 491–499.
82. Murphy L. Biosensors and bioelectrochemistry. *Current Opinion in Chemical Biology* 2006, 10, 177–184.
83. Lotzbeyer T, Schuhmann W, Schmidt H. Electron transfer principles in amperometric biosensors: direct electron transfer between enzymes and electrode surface. *Sensors and Actuators B* 1996, 33, 50–54.
84. Khan G. Construction of SEC/CTC electrodes for direct electron transferring biosensors. *Sensors and Actuators B* 1996, 36, 484–490.
85. Willner I, Heleg-Shabtai R, Blonder R, Katz E, Tao G, Buckmann A, Heller A. Electrical wiring of glucose oxidase by reconstitution of fad-modified monolayers assembled onto Au-electrodes. *Journal of the American Chemical Society* 1996, 118, 10321–10322.

86. Zayats M, Katz E, Willner I. Electrical contacting of glucose oxidase by surface-reconstitution of the apo-protein on a relay-boronic acid–FAD cofactor. *Journal of the American Chemical Society* 2002, 124, 2120–2121.
87. Liu J, Chou A, Rahmat W, Paddon-Row M, Gooding JJ. Achieving direct electrical connection to glucose oxidase using aligned single walled carbon nanotube arrays. *Electroanalysis* 2005, 17, 38–46.
88. Xiao Y, Patolsky F, Katz E, Hainfield J, Willner I. Plugging into enzymes: nanowiring of redox enzymes by a gold nanoparticle. *Science* 2003, 299, 1877–1881.
89. Joshi P, Merchant S, Wang Y, Schmidtke D. Amperometric biosensors based on redox polymer–carbon nanotube–enzyme composites. *Analytical Chemistry* 2005, 77, 3183–3188.
90. Hrapovic S, Liu Y, Male K, Luong J. Electrochemical biosensing platforms using platinum nanoparticles and carbon nanotubes. *Analytical Chemistry* 2004, 76, 1083–1088.
91. Xu T, Zhang N, Nichols H, Shi D, Wen X. Modification of nanostructured materials for biomedical applications. *Materials Science and Engineering* 2007, 27, 579–594.
92. Huang Y, Zhang W, Xiao H, Li G. An electrochemical investigation of glucose oxidase at a CdS nanoparticle modified electrode. *Biosensors & Bioelectronics* 2005, 21, 817–821.
93. Pumera M, Sanchez S, Ichinose I, Tang J. Electrochemical nanobionsensors. *Sensors and Actuators B* 2007, 132, 1195–1205.
94. Yamato H, Koshiba T, Ohwa M, Wemet W, Matsumura M. A new method for dispersing palladium microparticles in conducting polymer films and its application to biosensors. *Synthetic Metals* 1997, 87, 231–236.
95. Kenausis G, Chen Q, Heller A. Electrochemical glucose and lactate sensors based on "wired" thermostable soybean peroxidase operating continuously and stably at 37°C. *Analytical Chemistry* 1997, 69, 1054–1060.
96. Mano N, Heller A. Detection of glucose at 2 fM concentration. *Analytical Chemistry* 2005, 77, 729–732.
97. Mano N, Mao F, Heller A. Electro-oxidation of glucose at an increased current density at a reducing potential. *Chemical Communications* 2004, 18, 2116–2117.
98. Jeong J, Sim Y, Sohn H, Lee E. UV-nanoimprint lithography using an elementwise patterned stamp. *Microelectronic Engineering* 2004, 75, 165–171.
99. Dauksher W, Le N, Ainley E, Nordquist K, Gehoski K, Young S, Baker J, Convey D, Mangat P. Nano-imprint lithography: templates, imprinting and wafer pattern transfer. *Microelectronic Engineering* 2006, 83, 929–932.
100. Deligianni H. Electrodeposition and microelectronics. *Interface* 2006, Spring, 33–35.
101. Miscoria S, Barrera G, Rivas G. Glucose biosensors based on immobilization of glucose oxidase and polytyramine on rodhinized glassy carbon and screen printed electrodes. *Sensors and Actuators B* 2006, 115, 205–211.
102. Schwartz D. Electrodeposition and nanobiosystems. *Interface* 2006, Spring, 34–35.
103. Schwarzacher W. Electrodeposition: a technology for the future. *Interface* 2006, Spring, 32–33.
104. Delvaux M, Demoustier-Champagne S. Immobilisation of glucose oxidase within metallic nanotubes arrays for application to enzyme biosensors. *Biosensors & Bioelectronics* 2003, 18, 943–951.

105. Romig Jr A, Dugger M, WcWhorter P. Materials issues in microelectromechanical devices: science, engineering, manufacturability and reliability. *Acta Materialia* 2003, 51, 5837–5866.

106. Kurnik R, Olier J, Waterhouse S, Dunn T, Jayalakshmi Y, Lesho M, Lopatin M, Tamada J, Wei C, Potts R. Application of the mixtures of experts algorithm for signal processing in a noninvasive glucose monitoring system. *Sensors and Actuators B* 1999, 60, 19–26.

107. Rinken T, Rinken A, Tenno T, Jarv J. Calibration of glucose biosensors by using pre-steady state kinetic data. *Biosensors & Bioelectronics* 1998, 13, 801–807.

108. Dunia R, Qin S. A unified geometric approach to process and sensor fault identification and reconstruction: the unidimensional fault case. *Computers and Chemical Engineering* 1998, 22, 927–943.

109. Dunia R, Qin SJ. Joint diagnosis of process and sensor faults using principal component analysis. *Control Engineering Practice* 1998, 6, 457–469.

110. Rosen C, Olsson G. Disturbance detection in wastewater plants. *Water Science and Technology* 1998, 37, 197–205.

111. Jagemann K-U, Fischbacher C, Danzer K, Muller UA, Mertes B. Application of near-infrared spectroscopy for non-invasive determination of blood/tissue glucose using neural networks. *Zeitschrift fur Physikalische Chemie* 1995, 191, 179–190.

112. Wehrens R, van der Linden WE. Calibration of an array of voltammetric microelectrodes. *Analytica Chimica Acta* 1996, 334, 93–101.

113. Winquist F, Wide P, Lundstrom I. An electronic tongue based on voltammetry. *Analytica Chimica Acta* 1997, 357, 21–31.

CHAPTER 9

NITRIC OXIDE-RELEASING SUBCUTANEOUS GLUCOSE SENSORS

Heather S. Paul and Mark H. Schoenfisch

9.1 INTRODUCTION

The clinical utility of electrochemical sensors for continuous glucose monitoring in subcutaneous tissue has been limited by numerous challenges related to sensor component and biocompatibility-based failures.[1,2] Sensor component failures include electrical failure, loss of enzyme activity, and membrane degradation,[3,4] while examples of biocompatibility-based failures include infection, membrane biofouling (e.g., adsorption of small molecules and proteins to the sensor surface), and fibrous

In Vivo Glucose Sensing, Edited by David D. Cunningham and Julie A. Stenken

encapsulation as a result of the foreign body reaction (FBR). A truly biocompatible sensor should maintain analytical performance (i.e., sensitivity, selectivity, response time, drift, and mechanical stability) while minimally perturbing the local environment of the host tissue.[3] As a result, most approaches for enhancing glucose sensor biocompatibility have focused on polymeric sensor coatings that trigger a less aggressive foreign body reaction. Unfortunately, the clinical utility of such "passive" strategies has been rather limited due to poor tissue integration. A more recent approach aims to address specific aspects of the foreign body reaction through the release of factors known to influence wound healing such as vascular endothelial growth factor or nitric oxide (NO), for example. To date, research devoted to the application of "active" biomaterials (i.e., materials or coatings that actively release therapeutic mediators) to glucose sensors has primarily involved NO release. This chapter will further introduce the biocompatibility challenges at the sensor/tissue interface, the current strategies to address these challenges, and the current and future benefits of NO release for enhancing sensor biocompatibility.

9.1.1 General Biocompatibility Concerns for Subcutaneous Sensors

The FBR drastically affects the analytical performance (i.e., sensitivity, selectivity, response time, drift, and mechanical stability) of subcutaneous glucose sensors. The primary effect of the FBR on sensor performance is the reduction of glucose and oxygen diffusion to the sensor's surface due to biofouling (i.e., the adsorption of small molecules, proteins, and cells to the sensor surface), collagen capsule formation, and the hypovascular nature of the tissue surrounding an implanted sensor. Biofouling, the topic of two recent reviews,[1,2] disturbs analyte diffusion around the sensor via two mechanisms. First, adsorbed proteins and inflammatory cells (i.e., monocytes, neutrophils, and macrophages) physically limit analyte diffusion to the sensor surface. Second, active inflammatory cells perturb local analyte concentrations due to their consumption of oxygen and glucose, as well as increased production of reactive oxygen species.[4] The FBR also involves the formation of multinucleated foreign body giant cells (FBGCs) that secrete exopolysaccharides and lead to the formation of fibrous, nonvascular tissue, further decreasing analyte diffusion. Persistent inflammatory cells attempt to destroy and eventually isolate the sensor, which leads to degradation of the sensor membrane and a subsequent loss of membrane stability. Ultimately, the FBR ends with encapsulation of the sensor within a collagen capsule that further limits analyte diffusion to the sensor. Decreased analyte diffusion increases sensor response time, decreases sensitivity, and often results in signal drift necessitating frequent recalibration.[5] Similar to the initial biofouling layer, the capsule continues to affect local analyte concentrations due to the presence of inflammatory cells and its nonvascular nature. Indeed, the interstitial glucose concentration surrounding the implant incorrectly reflects the blood glucose concentration in the healthy vasculature.

The FBR is increased in the presence of bacteria, a secondary challenge to sensor biocompatibility.[3,6,7] Initial bacterial association with an implant surface is quickly followed by more permanent cell attachment through cell surface adhesion compounds. Some bacterial strains, particularly *Staphylococci* species (e.g., *S. epidermidis* and *S. aureus*), are capable of further aggregation and secretion of an exopolysaccharide matrix to form a mature biofilm.[8,9] Biofilm formation often results in persistent and dangerous infections that necessitate the removal of the implanted device.[7] Notably, bacterial infection leads to an increased immune response and the recruitment of additional inflammatory cells to the implant, further limiting sensor performance.

In principle, the effects of the FBR and decreased analyte diffusion on sensor performance (e.g., sensitivity, selectivity, response time, drift, and mechanical stability) can be accounted for by recalibrating the sensor after a stable wound environment has been reached. The time frame for reaching a steady-state FBR varies, however, depending on an individual's immune response and the sensor (i.e., membrane) composition in contact with the tissue. In most cases, capsule formation does not begin until 2–3 weeks after sensor implantation.[10] Obviously an enduring inflammatory response is not ideal for designing clinically useful sensors. Next-generation glucose sensors are now focused on minimizing (or eliminating) the FBR.

9.1.2 Current Approaches to Biocompatibility

Passive coatings (i.e., chemical modification of the outer sensor coating) and active biomaterials that release antibacterial or anti-inflammatory agents represent two approaches for enhancing the tissue biocompatibility of subcutaneous sensors. Many of the current strategies for improving sensor biocompatibility are based on the passive approach. Recent publications have highlighted passive strategies to increase sensor biocompatibility.[1,11,12] Strategies include hydrophilic surface modifications and the use of polyurethane, silicone elastomers, biomimicry, hydrogels, Nafion, surfactant-derived membranes, and diamond-like carbon coatings as outer membranes.[1,3] Of these approaches, polyethylene glycol (PEG)-derived hydrogels,[13] Nafion,[14] and polyurethane[15] have been applied to subcutaneous glucose sensors most successfully and will be introduced briefly to illustrate the limited success of passive biomaterials.

The use of hydrophilic polymers to modulate protein adsorption has been widely studied.[10,16] Sensor modifications with PEG or polyethylene oxide (PEO) have been shown to prevent nonspecific protein adsorption to the sensor surface,[16] implying that sensor coatings functionalized with PEG/PEO may present a more biocompatible interface with respect to biofouling since surface proteins mediate cell adhesion. Quinn and colleagues studied the utility of glucose sensors coated with hydrogels composed of cross-linked PEG derivatives.[13] Fewer immune cells were adhered to glucose sensors coated with a PEG-derived hydrogel than controls (*in vivo* canine model). However, the study was limited to 7 days and did not address analytical performance characteristics.

A second coating material that has been studied extensively in the design of glucose sensors is Nafion, a perfluorinated polymer. Moussy et al. implanted Nafion-coated glucose sensors in the subcutaneous tissue of canines and observed limited tissue encapsulation and a stable sensor response up to 10 days.[14] However, fluctuating permeability and cracking of the sensor membranes proved to be a particular challenge for the long-term (i.e., weeks) stability of such devices.[1]

Polyurethanes have also been employed as outer sensor membranes. Yu et al. evaluated the biocompatibility and analytical performance of a subcutaneous glucose sensor with an epoxy-enhanced polyurethane outer membrane.[15] The membrane was mechanically durable and the resulting sensors were functional for up to 56 days when implanted in the subcutaneous tissue of rats. Despite the improved sensor lifetime, all of the polyurethane-coated sensors were surrounded by a fibrous capsule, indicating an enduring inflammatory response that is undesirable due to the aforementioned effects on analytical sensor performance. To date, the clinical success of most passive approaches has been rather limited. It is doubtful that one passive material alone will be capable of imparting long-term (i.e., weeks to months) biocompatibility for *in vivo* use due to the extremely dynamic nature of the wound environment.

As an alternative to passive sensor coatings, research concerning biomaterials that actively release antibacterial and anti-inflammatory agents has been initiated. Initial reports on such "active" biomaterials have demonstrated reduced initial bacterial adhesion and biofilm formation, and diminished immune response.[7,17] Examples of general active release strategies include controlled release of antibiotics,[18] silver ions,[19,20] IgG antibodies,[21,22] angiogenic factors,[23] and NO.[17,24] Of these approaches, only biomaterials that release vascular endothelial growth factor (VEGF) and NO have been explored for improving the biocompatibility of glucose sensors.

The active release of the angiogenic factor VEGF has been shown to promote the formation of new blood vessels, enhancing delivery of analytes such as glucose to the sensor surface. As such, recent research has focused on examining the release of VEGF. Studies performed by Ward and colleagues have demonstrated improved sensor biocompatibility via VEGF release.[23] VEGF release from electrodes implanted in canines up to 40 days resulted in a 200–300% increase in the vascularization of surrounding tissue. A more detailed description of VEGF release and other strategies to overcome biological responses is presented in Chapter 3.

Both passive strategies and the release of VEGF rely on targeting a single aspect of the FBR. While passive polymers such as Nafion offer the advantages of simple fabrication (e.g., dip coating), commercial availability, and increased selectivity due to the chemical nature of the coating,[1] their biocompatibility is limited to reducing protein-related biofouling. The dynamic *in vivo* environment surrounding the implanted sensor may still lead to membrane degradation and a loss of the physiochemical properties that rendered the sensor coating antifouling. Active release strategies show distinct benefits over passive approaches. However, the active release of VEGF, for example, relies on increasing the vascular nature surrounding the implant, but does not address initial biofouling, the activity of inflammatory cells, or capsule formation. Extensive research is still needed to develop active release polymers as sensor membranes. Due to beneficial roles of NO in wound healing and its short half-life, NO release may offer distinct advantages over VEGF with respect to mediating the FBR.

9.1.3 Chapter Objectives

The motivation for harnessing the power of NO to modulate the FBR and potentially improve the analytical performance of subcutaneous glucose sensors is the focus of this chapter. A brief discussion of strategies for designing NO-releasing polymers and the characterization of NO release from such coatings (i.e., flux, amount, and duration of release) is provided. The challenges of fabricating NO-releasing amperometric glucose sensors will then be reviewed. The ability of NO release to influence bacterial adhesion and factors associated with the host immune response will be discussed in detail. Finally, *in vivo* studies demonstrating the utility of NO release in a working glucose sensor will be presented.

9.2 NITRIC OXIDE RELEASE

Nitric oxide, a diatomic free radical, is a seemingly simple molecule that actually plays a complex role in physiology despite its short half-life (on the order of seconds) in biological milieu.[25] Nitric oxide is produced endogenously by endothelial cells, macrophages, and neutrophils. Such cells produce NO via a class of enzymes termed nitric oxide synthases (NOSs) that catalyze the conversion of L-arginine to L-citrulline with oxygen, flavine adenine dinucleotide (FAD), flavine mononucleotide (FMN), nicotinamide adenine dinucleotide phosphate (NADPH), and tetrahydrobiopterin as cofactors (Figure 9.1).[26] Three types of NOSs exist: endothelial NOS (eNOS) and neuronal NOS (nNOS) that release low-level constitutive amounts of NO, and inducible NOS (iNOS) that releases NO at a flux regulated by cytokines, growth factors, and inflammatory stimuli.[26] Of note, the flux of NO from iNOS is significantly greater than from eNOS or nNOS, indicating an increased level of NO during inflammation and the FBR. Nitric oxide produced endogenously has been reported to regulate aspects of the cardiovascular, respiratory, gastrointestinal, genitourinary, and nervous systems. As well, NO is known to inhibit thrombosis, battle microbial infection, mediate the immune response, and promote angiogenesis (Figure 9.2),[17] all of which make the concept of NO release particularly attractive for subcutaneous sensor applications.

During thrombosis platelets adhere to blood vessel walls and aggregate to form a blood clot or thrombus. Platelet aggregation is dependent on the intraplatelet Ca^{2+} concentration. Nitric oxide is able to inhibit thrombosis by indirectly decreasing

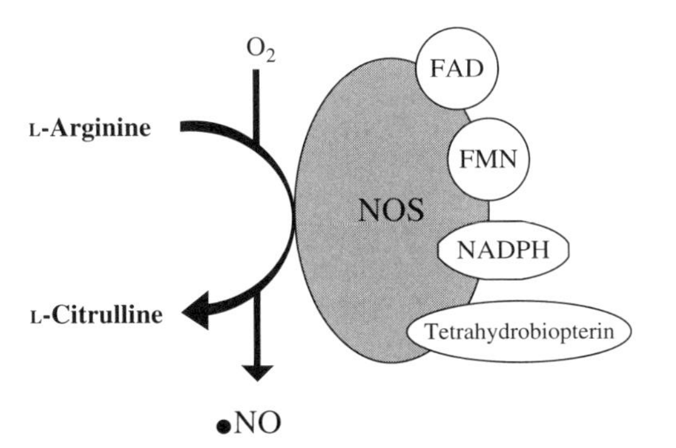

Figure 9.1 Endogenous production of NO catalyzed by NOS with known cofactors.

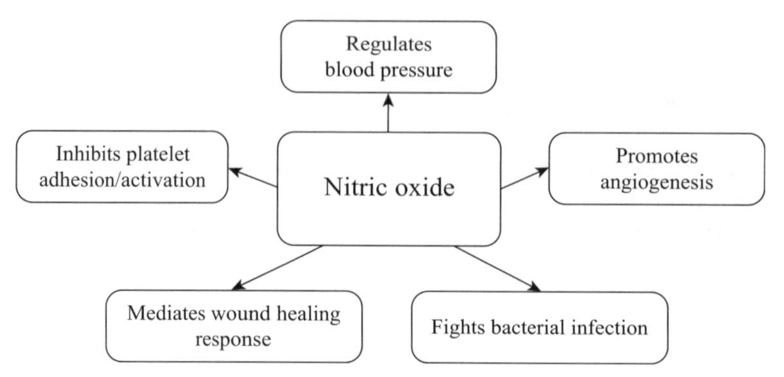

Figure 9.2 Areas of NO influence important for biomaterials.

Ca^{2+} concentration.[27] Nitric oxide and its reactive intermediates (e.g., NO_2, N_2O_3) also play a significant role in fighting bacterial infection. Production of NO by macrophages is stimulated when bacteria are present, and the oxidizing species produced from NO destroy bacteria through oxidation of membrane lipids[28] and/or DNA damage.[29] In addition to the role of NO in fighting bacteria, the inducible production of NO by macrophages is important in mediating broader aspects of wound healing including collagen formation and cell proliferation.[26,30] Cytokines and growth factors prevalent in the wound area induce increased synthesis and activity of iNOS initially, but the details of NO signaling during sustained wound healing are not well understood.[26] Nitric oxide also acts as an autocrine regulator of angiogenesis by promoting the growth and proliferation of the capillary endothelium that is necessary for the formation of new blood vessels.[31] The expression of the most prevalent angiogenesis factor, VEGF, is also dependent on NO.[30]

The influence of NO in thrombogenesis, bacterial infection, angiogenesis, and the immune response suggest that its active release into the tissue surrounding a sensor may minimize the FBR. Sensor coatings that release NO could reduce the occurrence and severity of bacterial infection, minimize inflammation and collagen capsule formation, and promote the formation of new blood vessels, all of which would create a more favorable implant environment. Since NO is reactive (i.e., has a short half-life), the effects of NO would remain localized to the area from which it is released.

9.2.1 Enhanced Biocompatibility at Nitric Oxide-Releasing Implants

The influence of NO release on the FBR at a subcutaneous implant has been demonstrated experimentally using model NO-releasing implants (i.e., substrates).[32] Hetrick et al. investigated the tissue biocompatibility of NO-releasing implants with respect to capsule thickness, collagen density, angiogenesis, and the extent of the inflammatory response.[32] Uncoated silicone rubber, silicone rubber with a coating not modified for NO release (i.e., control implants), and silicone rubber implants with a coating modified to release NO were implanted in the subcutaneous tissue of a rat for 1, 3, and 6 weeks. The NO-releasing coatings released a total of approximately 1.35 μmol NO per square centimeter of implant surface area with half the release occurring within the first 5 h. At predetermined times, the implants were removed and the surrounding tissue excised and evaluated using histological methods to determine the effect of NO release on capsule thickness, collagen density, angiogenesis, and the degree of inflammatory response.

The collagen capsule thickness and collagen density were examined in images of tissue slices taken from the area surrounding the implants. The capsule was defined as the dense collagen layer adjacent and parallel to the implant surface.[32] As expected, capsule formation was not observed for any of the implants after 1 week, since capsule formation does not begin until 2–3 weeks after implantation.[10] By week 3, a well-defined collagen capsule was noted at all implants. At both 3 and 6 weeks, the NO-releasing implants exhibited a capsule thickness significantly less than that observed at uncoated and control implants (Figure 9.3). The capsule also appeared to be fully developed after 3 weeks at the uncoated and control implants, while capsule development appeared to be delayed at the NO-releasing implants. Additionally, the collagen density measured within 100 μm of the NO-releasing implants was significantly less than uncoated and control implants at all time points.[32]

Hetrick et al. also treated tissue samples with CD31 immunohistochemical stain to reveal blood vessels in the sample area extending approximately 330 μm from the implant surface. The degree of angiogenesis in the tissue surrounding an implanted sensor is important for maintaining blood and analyte diffusion to the sensor surface and relevant for both wound healing and glucose sensing. The NO-releasing implants showed a greater number of blood vessels near the sensor surface than coated control implants at all three time points. However, the difference was not significant at 6 weeks. This behavior was somewhat expected as most of the NO in these experiments had been released in the first 10 h after implantation. At the earliest time point (1 week), the NO-releasing implants showed 77% more blood vessels than coated controls, indicating that the NO stimulated angiogenesis early on but the effect subsided by week 6.[32]

The inflammatory response was investigated by staining and quantifying the nuclei of immune cells in the collected tissue samples using hematoxylin and eosin (H&E). The inflammatory response at NO-releasing implants was not significantly decreased at 1 week, but was reduced by more than 30% at 3 and 6 weeks when compared to uncoated and control implants. The similar inflammatory response seen for each experimental group after only 1 week may be due to the surgical trauma common to all implants. The significant reduction in inflammatory cells at NO-releasing implants at extended periods may have resulted from the downregulation of proinflammatory cytokines by NO.[32]

The results by Hetrick et al.[32] support the use of NO-release coatings for developing more tissue-compatible sensors. However, the impact of NO on the biocompatibility at a NO-releasing implant is a multifaceted question that is still not fully understood. Further study into the mechanisms by which NO decreases tissue encapsulation and chronic immune response while increasing angiogenesis will aid in optimization of the NO release properties (e.g., flux, concentration, and duration) of an implant coating for sensor applications.

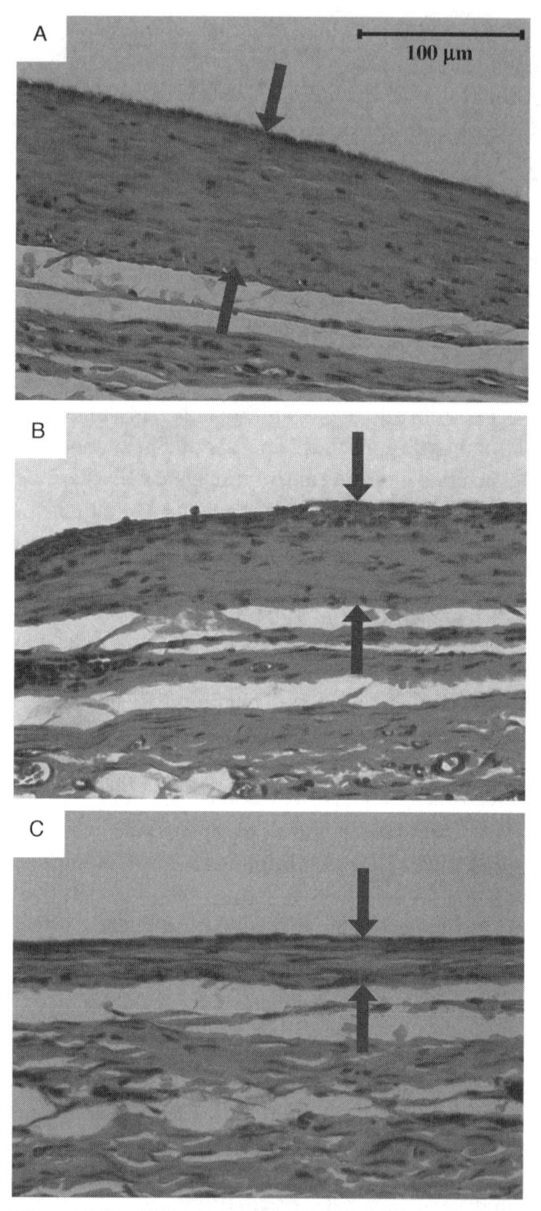

Figure 9.3 Optical micrographs of subcutaneous tissue showing the foreign body capsule (denoted by arrows) formed after 6 weeks at (A) bare silicone rubber elastomer, (B) xerogel-coated control, and (C) NO-releasing xerogel-coated implants. Reprinted with permission from Ref. 32. Copyright 2007 Elsevier. (See the color version of this figure in Color Plates section.)

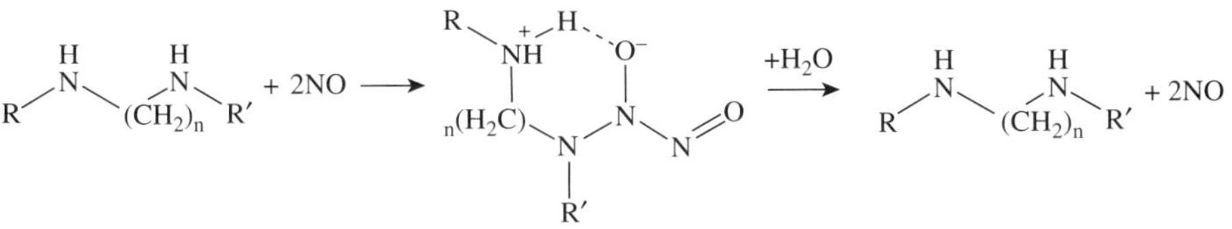

Scheme 9.1 *N*-Diazeniumdiolate synthesis and decomposition.

9.2.2 Nitric Oxide Donors

Due to the potential of NO as a therapeutic, the synthesis and biological applications of synthetic NO donors have been the focus of intense study.[33,34] Despite the enormous variety among NO donors, diazeniumdiolates and nitrosothiols have emerged as the most widely studied systems. With respect to glucose sensor membranes, diazeniumdiolates have been used exclusively as NO release sources.[7,17,24]

Nitrogen-bound diazeniumdiolates (*N*-diazeniumdiolates or 1-amino-substituted diazen-1-ium-1,2-diolates) spontaneously produce NO under physiological conditions.[34] This class of NO donors is synthesized by exposing polyamine precursors to high pressures of NO (Scheme 9.1).

The direct reaction of the polyamine with two equivalents of NO under appropriate conditions results in the formation of a stable zwitterionic diazeniumdiolate moiety. Upon exposure to aqueous conditions, the diazeniumdiolate spontaneously decomposes into the polyamine precursor and two equivalents of NO. The kinetics of NO release are easily controlled by varying the structure of the amine precursor and/or environment (e.g., pH, temperature, solvent, etc.) In hydrophobic polymers, the NO release is slowed down as diazeniumdiolate decomposition is dependent on water uptake.

Initial efforts to prepare NO-releasing polymers using diazeniumdiolates were based on dispersing the diazeniumdiolate-modified small molecules within the polymer. A wide array of NO-releasing polymers have been developed as sensor membranes for electrochemical ion and gas sensors for the purpose of improving their thromboresistivity in blood.[24] Unfortunately, the diazeniumdiolate NO donors and their decomposition by-products were found to leach from the polymer membranes,[35] an undesirable outcome due to the carcinogenic nature of certain polyamines. As a result of NO donor precursor leaching, alternative strategies for permanent incorporation of the diazeniumdiolate within polymer matrices were explored. For example, Batchelor et al. synthesized NO donor precursors with longer alkyl chains to increase the lipophilic character of the NO donor and enhance its retention within the polymer.[36] As expected, sensors fabricated using more lipophilic NO donors resulted in reduced NO donor precursor leaching. However, the NO release capacity was significantly reduced.[37] An alternative approach was to covalently attach the NO donor to the polymer backbone. Frost et al. reported on the performance of NO-releasing oxygen catheter sensors fabricated with silicone rubber covalently modified with diazeniumdiolate NO donors.[37] Initially, the NO-releasing sensors showed an increased thromboresistivity relative to controls. However, the NO release was depleted after 20 h, at which point an increase in platelet adhesion was observed with a concomitant decrease in sensor performance.[37] Thicker silicone rubber films would be required to extend NO release beyond 20 h. Unfortunately, thicker coatings impeded analyte diffusion to the sensor.

9.2.3 Sol–Gel Chemistry for Designing NO-Releasing Sensor Membranes

Sol–gel derived materials (i.e., xerogels) are well suited for sensing applications due to their chemical inertness, stability, porosity/permeability, and ability to encapsulate key sensor components such as enzymes, dyes, and electron mediators.[38] Furthermore, the sol–gel process occurs under mild conditions (aqueous, low temperature, etc.), thus preserving the utility of incorporated biomolecules, for example.[39] Encapsulation of glucose oxidase (GOx) in a xerogel membrane is accomplished by simply adding GOx to the sol (solution) and allowing hydrolysis and condensation reactions to occur around the enzyme. As the siloxane network condenses, the enzyme is encapsulated within the resulting xerogel. The porosity of the xerogel permits diffusion of enzyme substrates (i.e., oxygen and glucose) and products (i.e., hydrogen peroxide) necessary for glucose sensing. In some cases, GOx encapsulation imparts additional stability, allowing the enzyme to maintain biological function under nonphysiological conditions.[39] The increased enzyme stability upon encapsulation has been attributed to motion restriction that may impede irreversible structural changes.[39]

Xerogel films that release NO have been synthesized using classical sol–gel chemistry (Scheme 9.2).[40] Briefly, alkylalkoxysilanes and aminoalkoxysilanes are mixed in the presence of water, a cosolvent (e.g., methanol), and either an acid or a base catalyst. Hydrolysis of the silane precursors results in the formation of reactive silanol groups that undergo condensation reactions to form a highly branched siloxane network. The amines within the xerogel can be converted to the diazeniumdiolate form that subsequently is capable of decomposition and NO release in aqueous solution (Scheme 9.1).

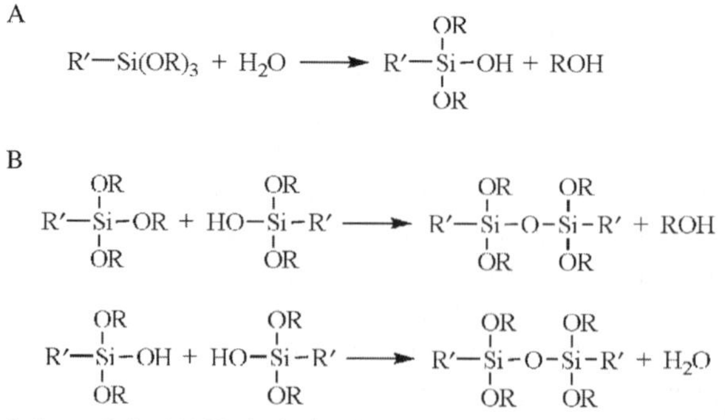

A

$$R'\text{—}Si(OR)_3 + H_2O \longrightarrow R'\text{—}Si(OR)_2\text{—}OH + ROH$$

B

$$R'\text{—}Si(OR)_2\text{—}OR + HO\text{—}Si(OR)_2\text{—}R' \longrightarrow R'\text{—}Si(OR)_2\text{—}O\text{—}Si(OR)_2\text{—}R' + ROH$$

$$R'\text{—}Si(OR)_2\text{—}OH + HO\text{—}Si(OR)_2\text{—}R' \longrightarrow R'\text{—}Si(OR)_2\text{—}O\text{—}Si(OR)_2\text{—}R' + H_2O$$

Scheme 9.2 (A) Hydrolysis of silane precursors and (B) condensation to form silane network, where R = alkyl group and R′ = amine-containing group.

9.2.4 Properties of NO-Releasing Xerogels

Specific characteristics of NO-releasing xerogels including stability, permeability, NO release, and biocompatibility may be tailored by varying a range of synthetic conditions including the solution pH, the catalyst, the ratio of alkyl- to aminoalkoxysilanes, the water content, and the time and the temperature of drying/conditioning.[17] Several alkyl- and aminoalkoxysilanes (Table 9.1) may be employed to modify the properties of the resulting xerogels. For use as subcutaneous sensor coatings, xerogels must be physically robust to survive the dynamic wound environment. Stability is especially important considering the potentially hazardous nature of the diazeniumdiolate by-products upon NO release. Leaching of the NO donor is assumed to be avoided due to covalent attachment to the xerogel backbone. However, fragmentation of the xerogel might present toxicity concerns if the material swells or breaks apart in solution. The stability of NO-releasing xerogels in solution has been evaluated by Marxer and coworkers.[40] Briefly, xerogel films were soaked in phosphate buffered saline (PBS; 0.01 M, pH 7.4) for up to 14 days. The amount of silicon in solution, a measure of the amount of xerogel fragmentation, was measured at different soak times using direct current plasma-optical emission spectrometry (DCP-OES). The percent fragmentation (as moles of silicon) was calculated for various xerogel compositions. The stability of the NO-releasing xerogels tested varied

TABLE 9.1 Structures of Alkylalkoxysilane and Aminoalkoxysilane Precursors Used in the Synthesis of NO-Releasing Xerogels

	Alkylalkoxysilane precursors
Methyltrimethoxysilane (MTMOS)	—$Si(OCH_3)_3$
Ethyltrimethoxysilane (ETMOS)	$CH_3CH_2Si(OCH_3)_3$
Isobutyltrimethoxysilane (BTMOS)	$(CH_3)_2CHCH_2Si(OCH_3)_3$
	Aminoalkoxysilane precursors
(Aminoethylaminomethyl) phenethyltrimethoxysilane (AEMP3)	$H_2N(CH_2)_2NHCH_2C_6H_4(CH_2)_2Si(OCH_3)_3$
N-(6-Aminohexyl) aminopropyltrimethoxysilane (AHAP3)	$H_2N(CH_2)_6NH(CH_2)_3Si(OCH_3)_3$
N-[3-(Trimethoxysilyl) propyl]diethylenetriamine (DET3)	$H_2N(CH_2)_2NH(CH_2)_2NH(CH_2)_3Si(OCH_3)_3$
N-(2-Aminoethyl)-3-aminopropyl-trimethoxysilane (AEAP3)	$H_2N(CH_2)_2NH(CH_2)_3Si(OCH_3)_3$

greatly depending on the amount and the type of aminoalkoxysilane precursor. There was an observed decrease in stability with increasing amine content (either increasing the mole percent of the aminoalkoxysilane precursor or using a precursor with multiple free amines.) This may be a consequence of hindered or incomplete xerogel condensation due to the increased hydrogen bonding that is possible with higher amine content. For example, significant fragmentation was measured for 40% (v/v) *N*-[3-(trimethoxysilyl)propyl]diethylenetriamine (DET3)/isobutyltrimethoxysilane (BTMOS) xerogels, which exhibited 4% (Si mol%) fragmentation after only 1 day in buffer.[40] In contrast, *N*-(6-aminohexyl)aminopropyltrimethoxysilane (AHAP3)/BTMOS and (aminoethylaminomethyl)phenethyltrimethoxysilane (AEMP3)/BTMOS xerogels showed less than 0.5% fragmentation after 14 days for all concentrations less than 40% AHAP3.[40] The negligible fragmentation of these systems in solution indicates that xerogels are a promising avenue for designing highly stable sensor coatings.

As glucose sensor membranes, xerogel coatings must facilitate glucose and hydrogen peroxide diffusion to enable sensor response. Shin et al. measured the permeability of hydrogen peroxide through aminoalkoxysilane-modified xerogels.[41] Platinum electrodes were coated with methyltrimethoxysilane (MTMOS) and AEMP3/MTMOS xerogel films. The response of the xerogel-modified electrodes to 0.1 M hydrogen peroxide was monitored at 0.7 V (versus Ag/AgCl). The permeability of hydrogen peroxide through the films was determined by comparing the peak current measured at the xerogel-coated electrode to the peak current measured at bare platinum prior to xerogel modification. As determined by this unitless ratio, the MTMOS xerogel was characterized by a greater permeability than the AEMP3/MTMOS xerogel (0.059 and 0.034, respectively). An even greater decrease in permeability occurred upon exposing the aminoalkoxysilane-based xerogels to high pressures of NO (necessary for NO donor synthesis). After formation of the diazeniumdiolate NO donor moiety, the permeability of the AEMP3/MTMOS films decreased from 0.034 to less than 0.0001. The observed decline in permeability was attributed to enhanced condensation reactions between residual silanol groups in the xerogel catalyzed by NO. The decreased permeability to hydrogen peroxide resulted in significantly decreased sensor response to changing glucose concentrations.[41] Multiple sensor configurations were thus pursued to successfully couple NO release with glucose sensing (to be discussed in Section 9.3).

The NO release properties of xerogels are best described by the amount, duration, and rate of NO release. Marxer et al. measured the NO release from AEMP3/BTMOS, AHAP3/BTMOS, and DET3/BTMOS xerogel films using chemiluminescence.[40] As expected, the amount of NO generated varied for the type of aminoalkoxysilane precursors tested and the amino- to alkylalkoxysilane ratio. The duration of NO release increased with an increasing amino- to alkylalkoxysilane ratio. Measurable release was observed up to 20 days for 40% (v/v) AHAP3/BTMOS xerogels; however, the most stable xerogels only released antibacterial levels of NO for 6 days. Future studies may require the combination of multiple aminoalkoxysilanes to achieve greater NO release over extended periods (weeks to months).

Nablo et al. were the first to evaluate the antibacterial properties of NO-releasing xerogels using *Pseudomonas aeruginosa* and *Staphylococcus aureus* as model bacteria. Both *P. aeruginosa* and *S. aureus* readily form biofilms and cause implant-related infections. In general, bacterial adhesion on NO-releasing xerogels was 2–4 times less than on xerogel controls.[42] The efficacy of NO-releasing xerogels at reducing bacterial adhesion was also evaluated *in vivo*.[43] Uncoated silicone rubber implants and implants coated with an NO-releasing 40% (v/v) AHAP3/BTMOS xerogel were implanted into the subcutaneous tissue of adult rats. An aggressive *S. aureus* infection (10^8 cfu/mL) was then injected at each implant site. Tissue samples from each site were tested after 8 days. Of note, the 40% AHAP3/BTMOS xerogels released NO for up to 7 days with 90% of the total release occurring within the first 24 h. Microbiological examination after 8 days showed that only 13.3% of the NO-releasing implants were infected with *S. aureus*, as compared to 73.3% of uncoated control samples.[43] Histological examination of the tissue surrounding uncoated and xerogel-coated implants also supported effectiveness of NO at reducing infection. Uncoated silicone rubber controls showed the formation of a biofilm-like layer of bacterial cells at the implant surface that was absent in all NO-releasing implants. Additionally, the capsule observed around the uncoated controls was roughly four times thicker than that observed around the NO-releasing implants. The tissue around NO-releasing implants also showed an increase in blood vessel density, possibly illustrating the effect of NO on angiogenesis.[43]

Collectively, NO's endogenous production, short half-life, and integral role in many areas of the FBR make it an ideal candidate for active release in the design of more biocompatible subcutaneous glucose sensors. The influence of NO on thrombogenesis, infection, immune response, and angiogenesis is well documented in the literature and has been demonstrated *in vivo*.[32] Diazeniumdiolates represent useful NO release agents due to their ability to spontaneously release NO in a controlled manner upon exposure to physiological conditions. Sol–gel chemistry provides a method for both covalently immobilizing the diazeniumdiolate NO donor precursor and creating stable sensor films capable of NO release. Extensive investigation into the properties of NO-releasing xerogel films by Marxer et al.,[40] Shin et al.,[41] and Nablo et al.[42] suggest promise for application of xerogels as thin sensor coatings that impart biocompatibility while maintaining sensor analytical performance.

9.3 SENSOR DESIGN AND EVALUATION

The unique NO release and antibacterial properties of xerogel materials led to their use as scaffolds for designing more biocompatible glucose sensor membranes. Shin et al. were the first to report on the fabrication of a functional electrochemical NO-releasing glucose sensor.[41] Of note, the coupling of NO release with enzyme-based glucose sensing chemistry proved significantly more challenging than anticipated.

9.3.1 Nitric Oxide-Releasing Xerogel Particle/Polyurethane Glucose Sensor

Due to the low analyte permeability of the aminoalkoxysilane-based xerogel films, Shin et al. determined that NO release was best accomplished by doping NO-releasing

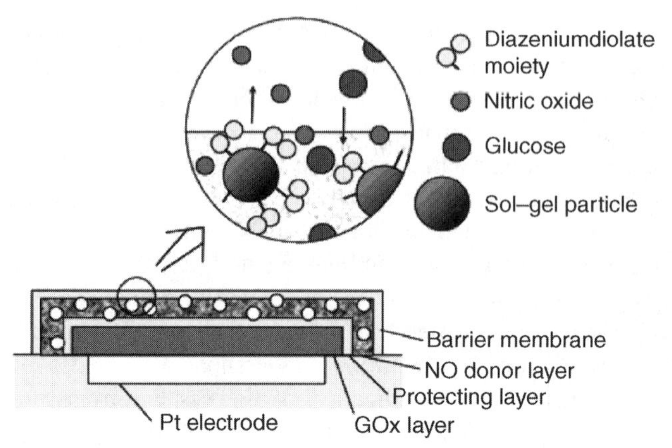

Figure 9.4 NO-releasing xerogel particle/polyurethane glucose sensor. Reprinted with permission from Ref. 41. Copyright 2004 American Chemical Society.

xerogel particles into the outermost polyurethane membrane used to fabricate the sensor. Particles were prepared by simply grinding xerogel films. As shown in Figure 9.4, the sensor design had four distinct layers. First, GOx was encapsulated within a xerogel and cast directly on a platinum electrode. Next, a thin layer of polyurethane was added to protect the enzyme layer from the subsequent NO-releasing layer. To prepare the NO-releasing layer, 10–200 μm (diameter) xerogel particles (AHAP3/MTMOS) were charged with NO to form diazeniumdiolate NO donors. The xerogel particles were then dispersed in a polyurethane solution and cast as the third layer. Finally, a thin polyurethane coating was cast to limit glucose diffusion (relative to oxygen) and minimize particle leaching.[41] The sensor response to glucose concentrations was evaluated on the benchtop using chronoamperometry for three membrane configurations: (1) two-layer design with GOx and protective polymer, (2) four-layer design containing xerogel particles not modified for NO release, and (3) four-layer design with diazeniumdiolate-modified particles capable of NO release (Figure 9.5). The NO-releasing xerogel particle/polyurethane sensor showed a linear response over all glucose concentrations tested (0–60 mM). The sensitivity of this sensor was only slightly lower than both the four-layer and two-layer membrane configurations without NO release (-3.8×10^{-2}, -5.7×10^{-2}, and -4.8 10^{-2} μA/mM, respectively). The response of the sensor was both rapid (<20 s) and stable for at least 18 days. The physical stability of the four-layer design was evaluated after storage in PBS (via DCP-OES) and revealed minimal particle leaching. The extent of leaching at 24 h was $1.1 \pm 0.4\%$ and nearly tripled to $3.0 \pm 0.2\%$ at 48 h.[41]

Despite successful proof of principle that NO-releasing materials can be employed to fabricate a functional glucose sensor, the toxicity of the particles that leached from the polymer membrane remained a concern. Additionally, the amount and the duration of NO release were limited by the mass of the particles in the polymer film. Upon device miniaturization, the NO release may not prove sufficient to sustain biocompatibility. Two alternative strategies were explored to address these

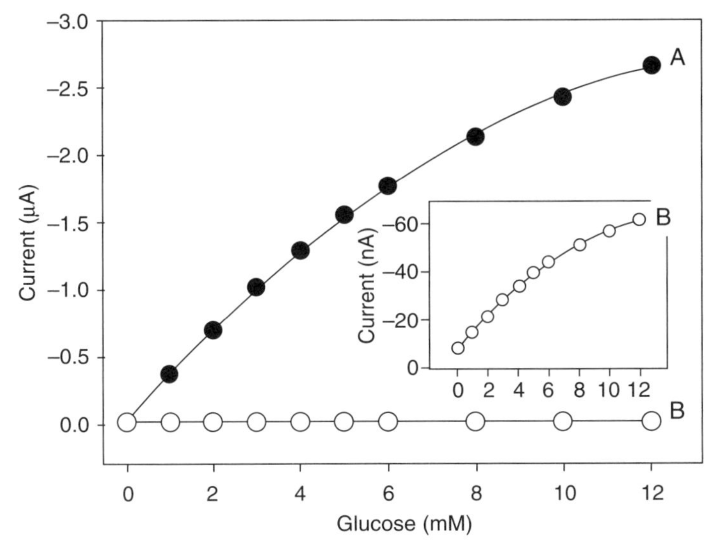

Figure 9.5 Amperometric response to changing glucose concentrations for (A) four-layer design without NO release, and (B) four-layer design with NO release. The response of the NO-releasing sensor is expanded in the (inset).

limitations: a patterned NO-releasing xerogel/polyurethane sensor[44] and a hydrophilic polymer-doped NO-releasing xerogel sensor.[45]

9.3.2 Patterned Nitric Oxide-Releasing Xerogel Arrays

Micropatterning of diazeniumdiolate-modified xerogels allowed for NO release without complete modification of the sensor surface. Robbins et al. were the first to report the synthesis of xerogel microarrays that released biologically relevant levels of NO.[46] Such interfaces showed similar blood compatibility compared to cohesive xerogel films. When applied to oxygen sensors, the xerogel micropattern improved the blood compatibility of the sensor without compromising oxygen diffusion or analytical sensitivity.[46] Likewise, Oh et al.[44] reported the fabrication of a NO-releasing glucose microsensor via xerogel microarrays to avoid the challenges Shin et al. encountered. Briefly, the microsensor consisted of a platinum electrode and a Ag/AgCl reference electrode assembled in a dual-barrel glass capillary with a total sensor geometry measuring 1 mm in diameter (Figure 9.6). After formation of the reference electrode by immersion in an aqueous $FeCl_3$ solution and platinization of the working electrode in chloroplatinic acid, the entire sensing tip was coated with a GOx-containing xerogel layer. A protective polyurethane layer was also added to minimize enzyme leaching and increase selectivity. Nitric oxide release ability was imparted to the sensor by depositing an AEMP3/MTMOS sol onto a PDMS microarray template. The sensing tip was then brought into conformal contact with the PDMS surface. After allowing the xerogel to cure for 24 h, the diazeniumdiolate

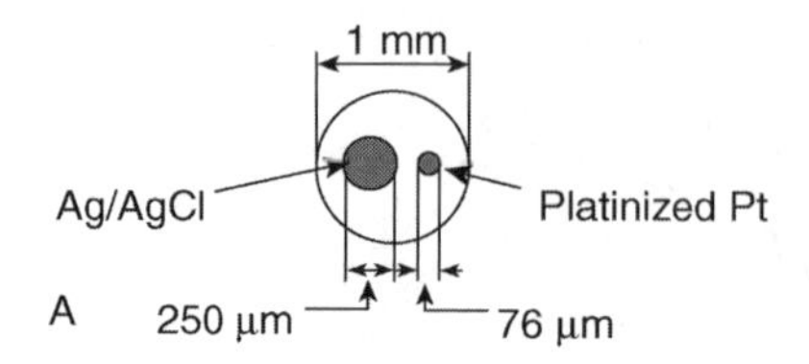

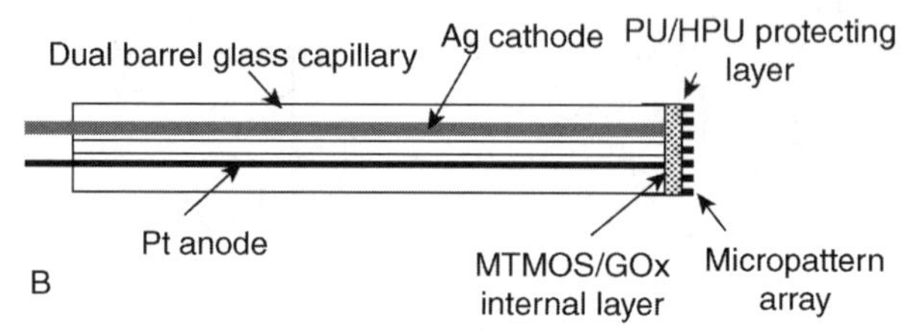

Figure 9.6 (A) Sensor tip and dimensions prior to modification with enzyme and xerogel pattern; (B) sensor after surface modification. Reprinted with permission from Ref. 44. Copyright 2005 Elsevier.

moiety was created by exposure to high pressures of NO as previously described, resulting in an NO-releasing xerogel microarray.[44]

The patterned NO-releasing xerogel microsensor was characterized by faster response times and greater sensitivity to changing glucose concentrations relative to sensors coated with uniform NO-releasing xerogel films.[44] Indeed, the sensor response to glucose was linear over 0–20 mM and selective over common interferences including ascorbic acid, uric acid, and 4-aminoacetophenol. DCP-OES stability analysis confirmed minimal fragmentation of the xerogel microarray. Furthermore, the performance of the sensor was robust, remaining stable for up to 7 days. As measured via *in vitro* bacterial adhesion assays, the surface coverage of *P. aeruginosa* on NO-releasing microarrays was 70–80% that of control xerogel microarrays not capable of NO release. Platelet adhesion was also reduced by more than 40% on NO-releasing microarrays relative to microarray controls. Of note, platelet and bacterial adhesion at the NO-releasing microarrays increased as the flux of NO subsided, indicating the need for sustained NO release with respect to minimizing biofouling.[44]

9.3.3 Hydrophilic Polymer-Doped Nitric Oxide-Releasing Xerogel Membranes

In comparison to the particle and micropattern approaches, cohesive xerogel coatings have the potential to simplify microsensor fabrication and extend NO release due to the significantly greater volume of NO donor. Unfortunately, the dense xerogel coating investigated by Shin et al. drastically compromised the sensor's analytical performance.[41] To improve the xerogel's permeability, NO-releasing xerogels were doped with a hydrophilic, water-soluble polymer.[45,47] The incorporation of hydrophilic polymers into xerogels was recently reported as a strategy for altering the chemical and physical properties of xerogel membranes. Xerogels doped with poly (vinylpyrrolidone) (PVP) were shown to improve membrane permeability while maintaining membrane stability.[45]

The integration of a hydrophilic polymer during xerogel synthesis reduces the number of condensation reactions that occur, thus creating a more fluid-like environment. By doping aminoalkoxysilane-modified xerogels with up to 20% (w/w) PVP, membranes were modified to be more permeable to hydrogen peroxide, even after NO exposure.[45] As expected, *N*-(2-aminoethyl)-3-aminopropyltrimethoxysilane (AEAP3)/BTMOS xerogels doped with 20% PVP and exposed to NO for 48 h were stable in solution for up to 17 days. When these xerogels were applied as coatings for amperometric glucose sensors, the response to glucose was linear over 1–30 mM glucose with a sensitivity of 4.6 ± 0.5 nA/mM, representing a 30-fold increase in sensitivity compared to sensor membranes prepared without PVP. After an initial burst of NO from the PVP-doped xerogels, the NO release properties of the doped xerogels and nondoped xerogels were similar. The enhanced NO release was attributed to increased water uptake by the xerogel as a result of the film's increased hydrophilic nature upon PVP incorporation.[45] The PVP-doped xerogels also showed similar bacterial adhesion characteristics as compared to nondoped "blank" substrates when incubated with *P. aeruginosa*, indicating that the presence of the hydrophilic polymer did not significantly alter the antibacterial properties of the NO-releasing xerogel. The NO-releasing PVP-doped xerogel exhibited a 50–60% reduction in bacterial surface coverage relative to the nondoped blanks and PVP-doped controls without NO release, respectively.[45]

Schoenfisch et al. reported that the incorporation of PVP within an NO-releasing xerogel (film)-based sensor membrane enhanced sensor performance without sacrificing xerogel stability or biocompatibility.[45] The design also avoided the concerns of previously reported sensor approaches including poor analytical response and particle leaching for the xerogel particle/polyurethane glucose sensor as discussed in Section 9.3.1,[41] and the complicated fabrication and limited NO release upon miniaturization that were challenges with the patterned xerogel array discussed in Section 9.3.2.[44] These three approaches to sensor fabrication were reported with different aminoalkoxysilane precursors (AHAP3, AEMP3, and AEAP3). Of note, each aminoalkoxysilane could be used for any of the fabrication strategies to tune the NO release characteristics for the particular application. The PVP-doped NO-releasing xerogel membrane sensor showed good sensitivity, linear range, response time, stability, and reduced *in vitro* bacterial adhesion, indicating that the design might be practical for *in vivo* glucose sensing applications. However, further exploration of methods to extend the duration of NO release, miniaturize the sensor, and evaluate both sensor performance and biocompatibility *in vivo* are required.

9.4 *IN VIVO* TESTING OF NO-RELEASING GLUCOSE BIOSENSORS

The ultimate goal of fabricating a functional NO-releasing glucose sensor for clinical use requires extensive *in vivo* investigation. Frost et al.[37] and Gifford et al.[48] recently

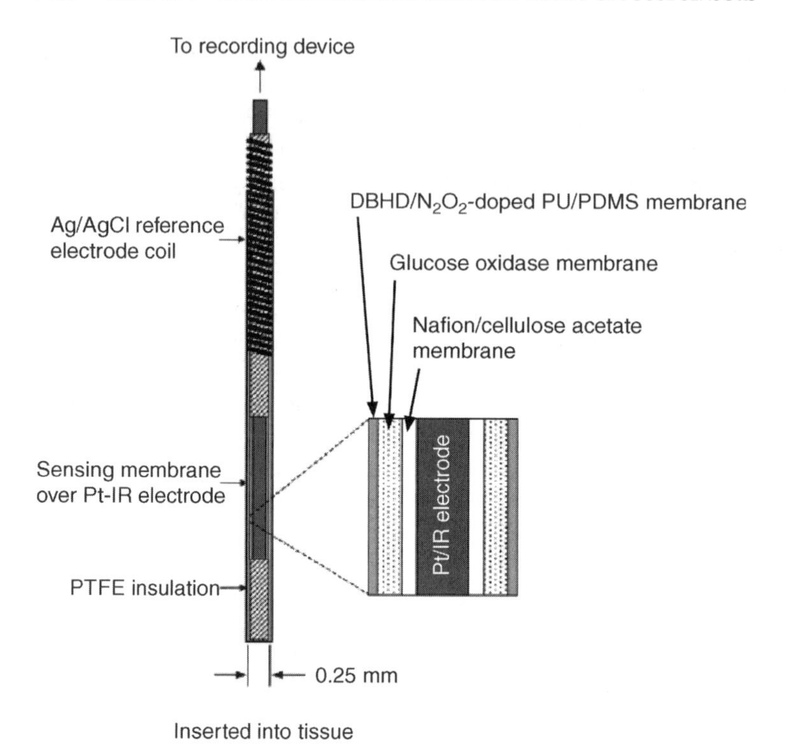

Figure 9.7 Schematic of NO-releasing subcutaneous glucose sensor. Reprinted with permission from Ref. 37. Copyright 2003 Elsevier.

evaluated the acute performance and biocompatibility of a functional NO-releasing glucose sensor in a rat model. To start, a conventional needle-type amperometric sensor[37,49] was modified to release NO (Figure 9.7). Specifically, the outer polyurethane(PU)/polydimethylsiloxane(PDMS) layer of the working electrode was doped with a small molecule diazeniumdiolate compound (*Z*)-1-[*N*-methyl-*N*-[6-(*N*-butylammoniohexyl)amino]]-diazen-1-ium-1,2-diolate (DBHD/N_2O_2) to confer NO release upon exposure to physiological conditions (Scheme 9.3). Control sensors were prepared without the diazeniumdiolate NO donor in the PU/PDMS layer. Both control and NO-releasing sensors were prepared with and without GOx for a total of four sensor configurations. When evaluated *in vitro*, the analytical performance of the NO-releasing sensors was comparable to controls with respect to sensitivity, stability, linearity, response time, and selectivity.[48]

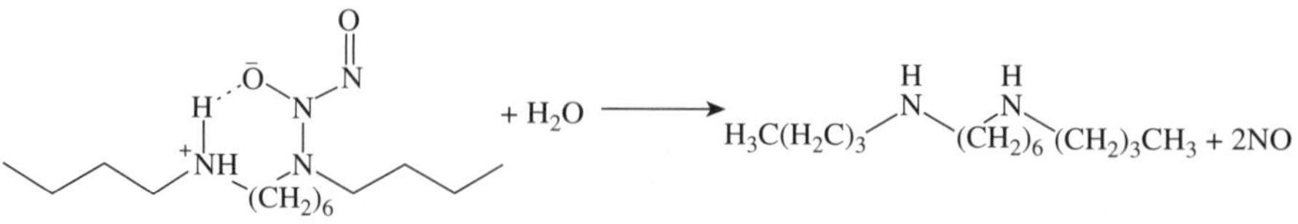

$$\text{DBHD/N}_2\text{O}_2 + H_2O \longrightarrow H_3C(H_2C)_3\text{–NH–}(CH_2)_6\text{–NH–}(CH_2)_3CH_3 + 2NO$$

Scheme 9.3 Release of NO from DBHD/N_2O_2.

The *in vivo* performance of the NO-releasing sensors was evaluated on the day of implantation and 48 h after implantation.[48] Generally, after 48 h the sensors showed a decreased sensitivity (using both one-point and two-point calibration) and an increased background current relative to sensor performance at $t = 0$ h. The sensor sensitivity from initial implant to 48 h was more consistent using a one-point calibration, but for both calibration methods, the sensitivity of NO-releasing sensors was more stable than controls. The linearity of sensor response was also evaluated and 33% of NO-releasing sensors displayed an increase in linearity from implant to 48 h postimplant, compared to 12% of controls. Gifford et al.[48] evaluated the sensor's accuracy at predicting glycemic levels based on a Clarke error grid analysis.[50] The NO-releasing sensors performed better than control sensors, with 99.7% of points occurring within the acceptable zones compared to 96.3% for controls at the time of implant. This value dropped slightly after 48 h, but 98.9% of values were still within the acceptable zones for NO-releasing sensors.[48] Of note, the study did not include glucose concentrations lower than 4 mM where the conditions for an accurate response are more demanding.

The NO-releasing sensors exhibited a shorter response time than would be beneficial for clinical use.[48] The majority of NO-releasing sensors reached a stable baseline response after 30 min postimplantation and polarization, which was 8–20 times faster than control sensors. Nitric oxide-releasing sensors also displayed a small peak in glucose response within the first few hours that was not observed with control sensors. *In vitro* experiments were performed to investigate the hypothesis that the NO release was interfering with the H_2O_2 oxidation signal. Differential pulse voltammetry confirmed that the oxidation peak of NO overlapped the oxidation peak of H_2O_2, and thus the large amount of NO released from the sensor coating within the first few hours of implantation contributed an interfering signal at early time points. This interference became inconsequential after most of the NO had been released, and the glucose response of NO-releasing sensors then matched that of control sensors.[48]

Histological evaluation of the tissue surrounding the implanted sensor was performed to monitor the inflammatory response at each implant site.[48] Each slice was then scored qualitatively from 0 to 4 to indicate the relative degree of inflammation, with 0 indicating no inflammation and 4 indicating a dense concentration of neutrophils and tissue necrosis surrounding the implant. An average histological score was calculated for each type of implant after 24 and 48 h of implantation. A significant reduction in the inflammatory response was observed after 24 h for NO-releasing sensors compared to controls (Figure 9.8), and the inflammation scores were calculated to be 1.2 ± 0.3 and 2.2 ± 0.7, respectively. As shown in Figure 9.8, healthy tissue appears pink, while inflammatory cells (neutrophils) appear purple, and necrosis as dark black clumps surrounding the control implant. The inflammatory response around NO-releasing implants was less than controls. After 48 h, the difference was no longer significant (2.4 ± 1.0 and 3.0 ± 0.5, respectively), indicating that NO served to delay the immune response but not prevent it.[48] It is important to note that the vast majority of NO release occurred within the first 20 h of implantation. Such NO release is characteristic for DBHD/N_2O_2. Thus, the level of NO release after 20 h may no longer be sufficient to modulate the immune response. The obvious

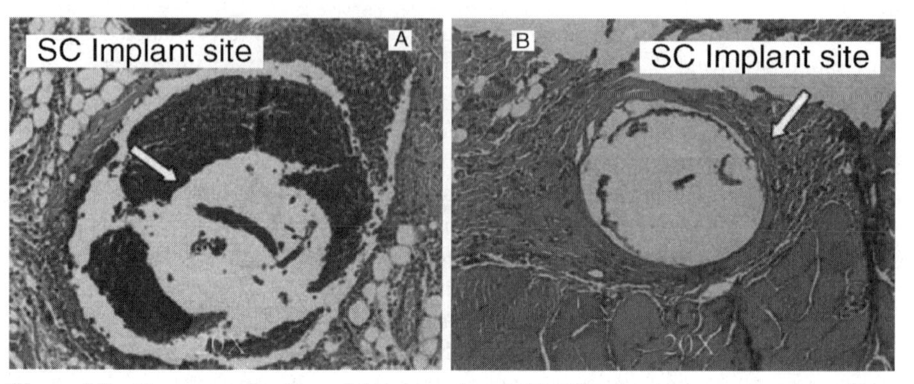

Figure 9.8 Tissue cross-section of (A) control and (B) NO-releasing sensor implant site. Reprinted from Ref. 48 with permission of John Wiley & Sons, Inc. Copyright 2005 John Wiley & Sons, Inc. (See the color version of this figure in Color Plates section.)

effectiveness of NO to reduce the immune response in the first 24 h postimplant, when the flux of NO from the sensor coating was greatest, support the use of NO-releasing membranes that release larger fluxes of NO for extended periods. Both efforts to optimize NO release and additional studies to observe sensor function and biocompatibility beyond 48 h are necessary to expand on the results of Gifford et al.[48]

9.5 FUTURE DIRECTIONS

The future of NO-releasing glucose sensors for long-term subcutaneous use is promising but continues to face certain challenges. Perhaps the greatest limitation in developing a functional and biocompatible glucose sensor is the foreign body reaction that is part of the host response.[3] Approaches for improving tissue biocompatibility should focus on reducing or eliminating the inflammatory host response to the implanted biosensor. The limited success of passive biomaterials to date has left the field poised for the development of active biomaterials. Nitric oxide may prove useful as an active release strategy due to its influence in numerous areas relevant to wound healing. The work of the Schoenfisch, Meyerhoff, and Wilson groups that has been reviewed herein points to a promising future for NO-releasing glucose sensors. Preliminary *in vivo* studies indicate a decreased inflammatory response at NO-releasing sensors after 48 h of implantation.[48] Hetrick et al. demonstrated that the effect of NO release on the inflammatory response is likely extended to several weeks.[32] Together, these studies demonstrate the need to understand whether extended NO release is necessary and at what levels.

One method for increasing the duration of NO release is to develop better NO release polymers. Studies are currently underway to design xerogel coatings consisting of multiple diazeniumdiolates (e.g., AHAP3 and DET3). Multiple NO donor systems should expand the ability to tune the NO flux, NO release longevity, and biocompatibility. The use of different NO donors and/or immobilization methods may also prove useful for increasing the stability of the NO-releasing sensor membrane. Recently, sol–gel chemistry has been expanded from the synthesis of xerogel polymers to NO-releasing silica nanoparticles of variable size ($d = 20$–500 nm) and NO release properties.[51,52] Acting as NO storage scaffolds, these nanoparticles may be doped into polymers to impart NO release. Both the NO release and membrane stability may be varied via nanoparticle composition and concentration, and the polymer.[51] Studies are currently underway to fabricate glucose sensor membranes using NO-releasing silica nanoparticles doped into polyurethane.

Another alternative NO donor strategy includes *S*-nitrosothiols, which also can be immobilized in xerogels or doped into polymer films as particles. Nitrosothiol NO donors offer multiple mechanisms for initiating and sustaining NO release.[53] In contrast to diazeniumdiolates, nitrosothiols are endogenous suggesting that the leaching of nitrosothiols into the surrounding tissue would not raise toxicity concerns. Strategies to generate NO donor species (i.e., nitrosothiols) endogenously via catalytic surfaces have been reported.[54–58] The use of immobilized organoselenium,[56] organoditelluride,[57] or copper[54,55,58] catalysts may allow long-term generation of NO at the sensor/tissue interface through the decomposition of endogenous nitrosothiols and thus avoid the limitation of finite NO donor reservoirs via immobilized diazeniumdiolate NO precursors.

Perhaps the most promising approach to designing more biocompatible glucose sensors may be to combine multiple passive and active biocompatibility strategies. For example, many of the aforementioned strategies are based on NO-releasing hydrophilic polymers. Additionally, the combination of multiple active release agents may prove useful for targeting different aspects of biocompatibility (i.e., immune response, wound healing, and bacterial infection).

Future efforts in designing biocompatible glucose sensors must also allow for straightforward fabrication and miniaturization to merit consideration for clinical use. Although a smaller implant causes less damage to the surrounding tissue, miniaturization of a functional NO-releasing glucose sensor may prove challenging due to the concomitant decrease in coating volume and NO donor loading. Miniaturized sensors must release sufficient levels of NO to impart biocompatibility. Indeed, sensor fabrication must balance the desire for a small wound area and need for NO.

To warrant consideration for use as a subcutaneous glucose sensor, extensive *in vivo* evaluation of analytical performance and biocompatibility is necessary for prospective sensors. Indeed, an ideal NO-releasing sensor must demonstrate a response time, sensitivity, selectivity, and linear range similar or better than existing devices. Additionally, the sensor must predict blood glucose levels and glycemia without frequent calibration, and function subcutaneously for several weeks post-implant. Nitric oxide release should increase the tissue biocompatibility and decrease the extent of bacterial infection in the implant area. Ultimately, future research efforts in NO-releasing subcutaneous glucose sensors must combine reliable sensor performance with potential of NO to facilitate sensor encapsulation into healthy, vascularized tissue.

REFERENCES

1. Wisniewski N, Reichert M. Methods for reducing biosensor membrane biofouling. *Colloids and Surfaces B: Biointerfaces* 2000, 18, 197–219.
2. Wisniewski N, Moussy F, Reichert WM. Characterization of implantable biosensor membrane biofouling. *Fresenius Journal of Analytical Chemistry* 2000, 366, 611–621.
3. Wilson GS, Gifford R. Biosensors for real-time *in vivo* measurements. *Biosensors & Bioelectronics* 2005, 20, 2388–2403.
4. Wilson GS, Hu Y. Enzyme-based biosensors for *in vivo* measurements. *Chemical Reviews* 2000, 100, 2693–2704.
5. Baker DA, Gough DA. Dynamic delay and maximal dynamic error in continuous biosensors. *Analytical Chemistry* 1996, 68, 1292–1297.
6. Gristina AG. Biomaterial-centered infection: microbial adhesion versus tissue integration. *Science* 1987, 237, 1588–1595.
7. Hetrick EM, Schoenfisch MH. Reducing implant-related infections: active release strategies. *Chemical Society Reviews* 2006, 35, 780–789.
8. An YH, Dickinson RB, Doyle RJ. *Handbook of Bacterial Adhesion: Principles, Methods, and Applications*, Humana Press, Totowa, 2000, pp. 1–27.
9. Habash M, Reid G. Microbial biofilms: their development and significance for medical device-related infections. *Journal of Clinical Pharmacology* 1999, 39, 887–898.
10. Ratner BD, Bryant SJ. Biomaterials: where we have been and where we are going. *Annual Review of Biomedical Engineering* 2004, 6, 41–75.
11. Schlosser M, Ziegler M. *Biosensors in the Body*. John Wiley & Sons, New York, 1997, pp. 138–170.
12. Reichert WM, Saavedra S. *Medical and Dental Materials*, VCH Publishers, Inc., New York, 1992, pp. 303–343.
13. Quinn CAP, Connor RE, Heller A. Biocompatible, glucose-permeable hydrogel for *in situ* coating of implantable biosensors. *Biomaterials* 1997, 18, 1665–1670.
14. Moussy F, Harrison DJ, O'Brien DW, Rajette RV. Performance of subcutaneously implanted needle-type glucose sensors employing a novel trilayer coating. *Analytical Chemistry* 1993, 65, 2072–2077.
15. Yu B, Long N, Moussy Y, Moussy F. A long-term flexible minimally-invasive implantable glucose biosensor based on an epoxy-enhanced polyurethane membrane. *Biosensors & Bioelectronics* 2006, 21, 2275–2282.
16. Chapman RG, Ostuni E, Takayama S, Holmlin RE, Yan L, Whitesides GM. Surveying for surfaces that resist the adsorption of proteins. *Journal of the American Chemical Society* 2000, 122, 8303–8304.
17. Shin JH, Schoenfisch MH. Improving the biocompatibility of *in vivo* sensors via nitric oxide release. *The Analyst* 2006, 131, 609–615.
18. Wu P, Grainger DW. Drug/device combinations for local drug therapies and infection prophylaxis. *Biomaterials* 2005, 27, 2450–2467.
19. Kumar R, Munstedt H. Silver ion release from antimicrobial polyamide/silver composites. *Biomaterials* 2004, 26, 2081–2088.
20. Furno F, Morley KS, Wong B, Sharp BL, Arnold PL, Howdle SM, Bayston R, Brown PD, Winship PD, Reid HJ. Silver nanoparticles and polymeric medical devices: a new approach to prevention of infection? *Journal of Antimicrobial Chemotherapy* 2004, 54, 1019–1024.

21. Rojas IA, Slunt JB, Grainger DW. Polyurethane coatings release bioactive antibodies to reduce bacterial adhesion. *Journal of Controlled Release* 2000, 63, 175–189.
22. Poelstra KA, Barekzi NA, Rediske AM, Felts AG, Slunt JB, Grainger DW. Prophylactic treatment of Gram-positive and Gram-negative abdominal implant infections using locally delivered polyclonal antibodies. *Journal of Biomedical Materials Research* 2002, 60, 206–215.
23. Ward WK, Quinn MJ, Wood MD, Tiekotter KL, Pidikiti S, Gallagher JA. Vascularizing the tissue surrounding a model biosensor: how localized is the effect of a subcutaneous infusion of vascular endothelial growth factor (VEGF)? *Biosensors & Bioelectronics* 2003, 19, 155–163.
24. Frost M, Reynolds MM, Meyerhoff ME. Polymers incorporating nitric oxide releasing/generating substances for improved biocompatibility of blood-contacting medical devices. *Biomaterials* 2005, 26, 1685–1693.
25. Ignarro LJ. *Nitric Oxide: Biology and Pathobiology*. Academic Press, San Diego, CA, 2000.
26. Witte MB, Barbul A. Role of nitric oxide in wound repair. *The American Journal of Surgery* 2002, 183, 406–412.
27. Gewaltig MT, Kojda G. Vasoprotection by nitric oxide: mechanisms and therapeutic potential. *Cardiovascular Research* 2002, 55, 250–260.
28. Rubbo H, Radi R, Trujillo M, Tellari R, Kalyanaraman B, Barnes S, Kirk M, Freeman, BA. Nitric oxide regulation of superoxide and peroxynitrite-dependent lipid peroxidation. *The Journal of Biological Chemistry*, 1994, 269, 26066–26075.
29. Miranda KM, Espey MG, Jourd'heuil D, Grisham MB, Fukuto J, Freelisch M, Wink DA. *Nitric Oxide: Biology and Pathology*, Academic Press, San Diego, CA, 2000, pp. 41–56.
30. Broughton GI, Janis JE, Attinger CE. The basic science of wound healing. *Plastic and Reconstructive Surgery* 2006, 117(7 Suppl.), 12S–34S.
31. Ziche M, Morbidelli L, Masini E, Amerini S, Granger HJ, Maggi CA, Geppetti P, Ledda F. Nitric oxide mediates angiogenesis *in vivo* and endothelial cell growth and migration *in vitro* promoted by substance P. *The Journal of Clinical Investigation* 1994, 94, 2036–2044.
32. Hetrick EM, Prichard HL, Klitzman B, Schoenfisch MH. Reduced foreign body response at nitric oxide-releasing subcutaneous implants. *Biomaterials* 2007, 28, 4571–4580.
33. Wang PG, Xian M, Tang X, Wu X, Wen Z, Cai T, Janczuk AJ. Nitric oxide donors: chemical activities and biological applications. *Chemical Reviews* 2002, 102, 1091–1134.
34. Hrabie JA, Keefer LK. Chemistry of the nitric oxide-releasing diazeniumdiolate ("nitrosohydroxylamine") function group and its oxygen-substituted derivatives. *Chemical Reviews* 2002, 102, 1135–1154.
35. Mowery KA, Schoenfisch MH, Saavedra JE, Keefer LK, Meyerhoff ME. Preparation and characterization of hydrophobic polymeric films that are thromboresistant via nitric oxide release. *Biomaterials* 2000, 21, 9–21.
36. Batchelor MM, Reoma SL, Fleser PS, Nauthakki VK, Callahan RE, Shanley CJ, Politis JK, Elmore J, Merz SI, Meyerhoff ME. More lipophilic dialkyldiamine-based diazeniumdiolates: synthesis, characterization, and application in preparing thromboresistant nitric oxide release polymeric coatings. *Journal of Medicinal Chemistry* 2003, 46, 5153–5161.
37. Frost MC, Batchelor MM, Lee Y, Zhang H, Kang Y, Oh BK, Wilson GS, Gifford R, Rudich SM, Meyerhoff ME. Preparation and characterization of implantable sensors with nitric oxide release coatings. *Microchemical Journal* 2003, 74, 277–288.

38. Kandimalla VB, Tripathi VS, Ju H. Immobilization of biomolecules in sol–gels: biological and analytical applications. *Critical Reviews in Analytical Chemistry* 2006, 36, 73–106.
39. Chen Q, Kenausis GL, Heller A. Stability of oxidases immobilized in silica gels. *Journal of the American Chemical Society* 1998, 120, 4582–4585.
40. Marxer SM, Rothrick AR, Nablo BJ, Robbins ME, Schoenfisch MH. Preparation of nitric oxide (NO)-releasing sol–gels for biomaterial applications. *Chemistry of Materials* 2003, 15, 4193–4199.
41. Shin JH, Marxer SM, Schoenfisch MH. Nitric oxide-releasing sol–gel particle/polyurethane glucose biosensors. *Analytical Chemistry* 2004, 76, 4543–4549.
42. Nablo BJ, Chen T-Y, Schoenfisch MH. Sol–gel derived nitric-oxide releasing materials that reduce bacterial adhesion. *Journal of the American Chemical Society*, 2001, 123, 9712–9713.
43. Nablo BJ, Rothrock AR, Schoenfisch MH. Nitric oxide-releasing sol–gels as antibacterial coatings for orthopedic implants. *Biomaterials* 2005, 26, 917–924.
44. Oh BK, Robbins ME, Nablo BJ, Schoenfisch MH. Miniaturized glucose biosensor modified with a nitric oxide-releasing xerogel microarray. *Biosensors & Bioelectronics* 2005, 21, 749–757.
45. Schoenfisch MH, Rothrick AR, Shin JH, Polizzi MA, Brinkley MF, Dobmeier KP. Poly(vinylpyrrolidone)-doped nitric oxide-releasing xerogels as glucose biosensor membranes. *Biosensors & Bioelectronics* 2006, 22, 306–312.
46. Robbins ME, Hopper ED, Schoenfisch MH. Synthesis and characterization of nitric oxide-releasing sol–gel microarrays. *Langmuir* 2004, 20, 10296–10302.
47. Marxer SM, Robbins ME, Schoenfisch MH. Sol–gel derived nitric oxide-releasing oxygen sensors. *The Analyst* 2005, 130, 206–212.
48. Gifford R, Batchelor MM, Lee Y, Gokulrangan G, Meyerhoff ME, Wilson GS. Mediation of *in vivo* glucose sensor inflammatory response via nitric oxide release. *Journal of Biomedical Materials Research* 2005, 75A, 755–766.
49. Bindra DS, Zhang Y, Wilson GS, Sternberg R, Thevenot DR, Moatti D, Reach G. Design and *in vitro* studies of a needle-type glucose sensor for subcutaneous monitoring. *Analytical Chemistry* 1991, 63, 1692–1696.
50. Clarke WL, Cox D, Gonder-Frederick LA, Carter W, Pohl SL. Evaluating clinical accuracy of systems for self-monitoring of blood glucose. *Diabetes Care* 1987, 10, 622–628.
51. Shin JH, Metzger SK, Schoenfisch MH. Synthesis of nitric oxide-releasing silica nanoparticles. *Journal of the American Chemical Society* 2007, 129, 4612–4619.
52. Shin JH, Schoenfisch MH. Inorganic/organic hybrid silica nanoparticles as a nitric oxide delivery scaffold. *Chemical Materials* 2008, 20, 239–249.
53. Hogg N. Biological chemistry and clinical potential of *S*-nitrosothiols. *Free Radical Biology & Medicine* 2000, 28, 1478–1486.
54. Oh BK, Meyerhoff ME. Spontaneous catalytic generation of nitric oxide from *S*-nitrosothiols at the surface of polymer films doped with lipophilic copper(II) complex. *Journal of the American Chemical Society* 2003, 125, 9552–9553.
55. Oh BK, Meyerhoff ME. Catalytic generation of nitric oxide from nitrite at the interface of polymeric films doped with lipophilic Cu(II)-complex: a potential route to the preparation of thromboresistant coatings. *Biomaterials* 2004, 25, 283–293.
56. Cha W, Meyerhoff ME. Catalytic generation of nitric oxide from *S*-nitrosothiols using immobilized organoselenium species. *Biomaterials* 2007, 28, 19–27.

57. Hwang S, Meyerhoff ME. Organoditelluride-mediated catalytic *S*-nitrosothiol decomposition. *Journal of Materials Chemistry* 2007, 17, 1462–1465.
58. Wu Y, Rojas AP, Griffith GW, Skrzypchak AM, Lafayette N, Bartlett RH, Meyerhoff ME. Improving blood compatibility of intravascular oxygen sensors via catalytic decomposition of *S*-nitrosothiols to generate nitric oxide *in situ*. *Sensors and Actuators B* 2007, 121, 36–46.

CHAPTER 10

FLUORESCENCE-BASED GLUCOSE SENSORS

Mike McShane and Erich Stein

10.1 INTRODUCTION

Fluorescence is a phenomenon in which a molecule is excited via energy absorption from a photon and subsequently releases energy in the form of a lower energy photon.

In Vivo Glucose Sensing, Edited by David D. Cunningham and Julie A. Stenken

The emitted light is shifted in wavelength relative to the excitation light, a unique feature of fluorescence that enables highly sensitive measurements, as the emission from the indicator can be isolated from excitation light and endogenous fluorophores by optical filters. For molecular detection, fluorescence-based techniques have proved to be highly sensitive, even affording *single*-molecule detection, due in part to low background signals.[1] Fluorescence is characteristic of certain chemical groups, including benzene rings and double and triple bonds. In addition, both energetics and kinetics of the process depend upon the environment of the fluorophores at the time of excitation. Thus, the time- and spectral-domain features of fluorescence carry information regarding the environment of these groups, making them extremely attractive candidates as optical transducers of chemical information. As a result, a large body of work has been dedicated to exploiting various fluorescence phenomena to "readout" interactions between glucose and molecular recognition elements. This chapter describes the general requirements for using fluorescence for *in vivo* glucose monitoring and provides a survey of reported work in the areas of glucose receptors, assays for glucose using these receptors, and various fluorescence phenomena, and the materials and methods for interfacing measurement instrumentation with receptors for construction of biosensors.

10.1.1 Requirements for *In Vivo* Sensing Via Fluorescence

This discussion begins with a general assumption that fluorescence-based *in vivo* sensing of glucose cannot be accomplished with endogenous materials. Glucose is not fluorescent, and no native fluorophores generate signals proportional to glucose concentration; therefore, exogenous reagents are required to produce a change in fluorescence properties proportional to glucose concentration. These reagents must be packaged and placed *in contact* with a biological fluid—blood, interstitial fluid (ISF), sweat, lacrimal fluid, saliva, or other—for purposes of *in vivo* measurement; this may take the form of an externally applied system, such as a contact lens or a patch placed on the skin, or an implant intended for longer term use. These then demand that materials used must be sufficiently *biocompatible* to avoid severe acute or sustained long-term host reaction to the foreign object. While contact lenses or other superficially located devices have less stringent requirements, an object inserted into tissue will elicit at least an acute inflammatory response due to local tissue trauma during the procedure and possibly a stronger and more prolonged reaction due to recognition of the foreign material. The host reaction is primarily, though not completely, determined by the surface properties of the implant.[2] Therefore, if the implant has suitable surface properties (smooth, anionic or neutral, hydrophilic, mechanically matched to environment, etc.) and so long as the contents are not released or degraded over time, a stable relationship will typically be formed between the implant surface and the host. Hence, the composition and purity of the materials used in sensor fabrication can be a key factor in determining biocompatibility. The reader is referred to Chapters 2 and 3 of this book for additional details on biocompatibility.

For short-term applications, the host reaction to the foreign material should reach steady state rapidly and resist fouling during the duration of use. For long-term implantations, the steady-state immune response must be limited to minimal fibrous tissue formation. The extent of fibrotic response, seen as the thickness and density of the collagen matrix deposited at the implant interface, may affect transport and optical properties of the sensor; dense collagen may inhibit glucose diffusion to the sensing assay, resulting in response delays or even altered sensitivity (if sensor relies on glucose flux). Moreover, fibrous encapsulation of the implants interfere with light propagation to and from the implant (most likely in the form of increased scattering), resulting in decreased light escaping skin, corresponding to a decreased signal-to-noise ratio (SNR) of measurements.

A second key requirement for fluorescence-based sensing systems is a *strong, fully reversible, and sufficiently rapid response over the range of interest.* For clinically viable systems, the measured signal must be strong enough to be detected with low-cost instrumentation, sensitive over the clinical range of interest, resistant to errors arising from noise or system perturbations, and must accurately reflect *current* glucose levels. Specifically, benchmark data have shown that fluctuations in blood glucose usually occur over a period of 30 min, typically fall between ~41–414 mg/dL, and change at a maximum rate of 3.6 mg/dL min.[3,4] Therefore, sensors should have <5 min response time and linear response over the range of 0–400 mg/dL; some sensitivity must be retained up to 600 mg/dL.[5] For affinity and competitive binding assays, coverage of this range with maximum sensitivity mandates that the effective K_d of the sensing system is within ~40–400 mg/dL (~2–20 mM); K_d values outside this range can still be useful, but the binding capabilities of the receptors are limited and sensitivity will be reduced.

Neglecting the influence of the optical instrumentation employed, this requires a combination of properties: adequate amounts of fluorophore with high quantum yield (typically, QY > 0.1; the trade-offs in indicator properties for use in implants are discussed in detail in a later section), proper selections of wavelengths to maximize light delivery and detection to the implants, and a glucose-sensitive assay that produces large percentage change from baseline measurements over the range of interest. Ideally, a linear response that produces a 2–10 times change in the measurand (e.g., intensity ratio, wavelength, or lifetime) over the 0–600 mg/dL range within a few seconds of change in glucose concentration could be achieved. However, even nonlinear profiles and response times on the order of a few minutes, sufficiently short to accurately reflect details of blood glucose excursions, would be sufficient if glucose could be measured with error <10%.[6]

A third desired property of fluorescent glucose sensors is a *stable assay response over the lifetime of the sensor.* Temporal changes in sensitivity and signal levels due to denaturation, relaxation, or poisoning of molecular recognition elements or photoinduced oxidation can be compensated, to a degree, by calibration. However, any "minimally invasive" measurement approach loses appeal if frequent readjustments using blood samples are required to obtain accurate glucose values. Thus, it is desirable to use maximally photostable elements and an environment that stabilizes proteins or imprinted polymers against irreversible conformational changes or irreversible binding to nonglucose interferents. In this case, synthetic systems such as boronic acid (BA)[7,8] or rigid molecularly imprinted polymers may be advantageous over protein receptors,[9–11] though they often exhibit drift in sensitivity to fluctuations in ionic strength and pH, as well as suffer from lack of specificity.[12,13]

Specificity is a key aspect of all sensing systems, and this is a prime reason for great attention given to enzymatic transduction schemes for electrochemical and other sensor types. However, it can be generally stated that specificity need only to be defined in terms of what will be encountered in extremes of normal operation. For example, specificity to glucose is desired in a glucose sensor, but if a sensor is to be implemented for monitoring interstitial fluid, the response to glucose need only to be such that the glucose sensitivity is minimally affected by the presence of potential interferents at their maximum expected physiological levels in the ISF. Thus, while sensors that demonstrate high specificity for glucose over other sugars such as mannose or dextrose or potentially interfering species are attractive, sensors exhibiting low specificity should not be excluded from consideration in systems on that basis alone. This fact has encouraged continual development and modification of polymeric and boronic acid receptors.

Another attribute desired in implants is *minimal or zero consumption of analytes and minimal or zero by-product formation*. Enzymatic systems, although attractive because of the specificity and potential for high sensitivity, generally consume glucose and cosubstrates, decreasing the local levels, while producing other species that may have deleterious effects on the sensors or the surrounding tissue. For example, when glucose oxidase (GOx) is used in enzyme-based glucose detection systems, molecular oxygen in consumed and gluconic acids and hydrogen peroxide are produced.[14] These cosubstrates and by-products can affect sensor performance through several means, including (1) changing enzymatic activity due to pH-induced conformational alterations; (2) changing fluorescence signals from slightly pH-sensitive fluorophores; (3) irreversible degradation of enzyme structure due to peptide bond cleavage from peroxide; and possibly other mechanisms. Furthermore, damage to surrounding tissue caused by removal of key nutrients (oxygen, glucose) or production of toxic materials (acid, peroxide, reactive oxygen species, etc.) must also be avoided. Thus, careful consideration must be given to reactants and by-products and appropriate design features used to reduce or eliminate their influence on the sensor and host, considerations that are often nontrivial. Thus, fluorescent glucose assays based on nonconsuming glucose binding reactions (e.g., concanavalin A (Con A),[15,16] boronic acid,[8,17–20] glucose/galactose binding protein, molecularly imprinted polymers,[9,11] and apo-enzymes[21]) are advantageous in this regard.

Fiber-optic probes enable deployment of fluorescence glucose assays in a manner similar to electrodes (hence, the term optodes), with the assay chemistry appropriately immobilized on the tip or on other optically accessible section of the fiber. This approach may enable use of low-cost disposable indwelling probes; however, fluorescent measurements from implanted assay reagents located in the skin or subcutaneous space may also be performed by transmitting excitation light and collecting emission light right through superficial tissue (see Figure 10.1). In such cases, the fluorescent glucose assay must have *optical properties that enable interrogation with reasonable signal-to-noise ratio*. SNR > 3 is a typical minimum target, whereas SNR > 10 is considered excellent. The achievement of a reasonable SNR essentially requires the implanted materials to have excitation and emission in spectral regions that can be probed with low-cost instrumentation, while still

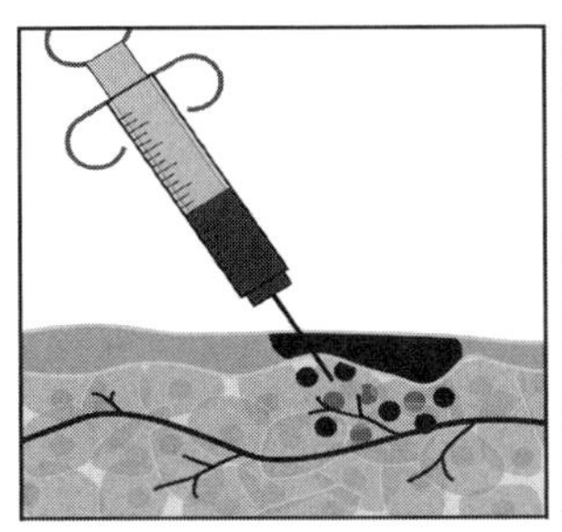

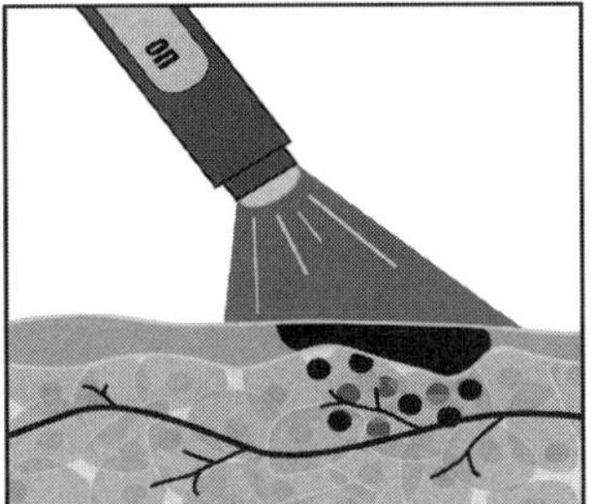

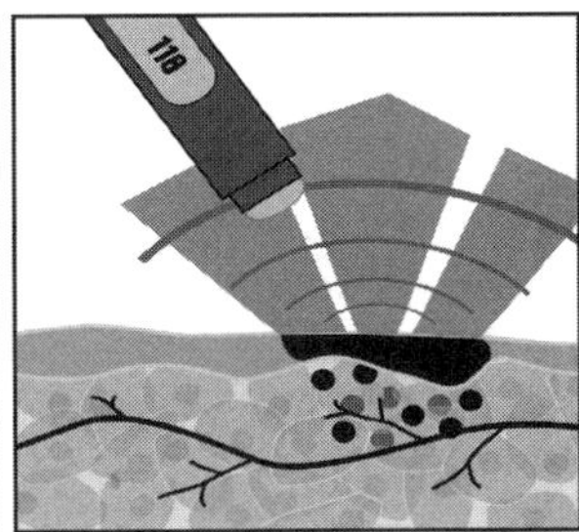

Figure 10.1 Schematic of the disconnected sensor concept: implantation, interrogation, and readout (microspheres, as illustrated). Reprinted with kind permission from Ref. 138. Copyright 2006 Springer.

allowing sufficient propagation of excitation light to elicit strong signals and collection of emitted photons at the skin surface. Given the reasonably well-understood optical properties of skin,[22,23] the ideal range for optical communication with implanted fluorescent sensors is in the long wavelength visible and near-infrared (NIR) region, for example, 600–1000 nm, due both to decreased scattering and the minimal absorption by tissue chromophores and water. However, it has been shown that even shorter wavelengths can be feasibly used with an appropriately designed system, even for steady-state measurements;[24,25] additional improvement in SNR can be expected when modulated light or phase lifetime methods are employed.

Figure 10.1 contains an illustration of an implanted fluorescence-based sensor; assuming the implanted materials and monitoring system can be designed to have each of the properties described above, this configuration is ideal, as it does not require any transdermal connections or implanted electronics.

10.1.2 Advantages and Current Limitations

Fluorescence-based techniques have proved to be highly sensitive, affording even *single*-molecule detection.[1] In addition, since fluorescence-sensing techniques rely on specific reporter molecules with unique optical properties to transduce concentration information, readings are highly specific, unlike many other spectroscopic techniques that rely on the intrinsic absorbing or scattering properties of glucose. Also, a variety of techniques that target different fluorescence phenomena can be used to measure analyte concentrations, providing researchers a means to develop an array of sensing schemes for a particular analyte of interest. Furthermore, given that fluorescence techniques are optically based, implantable devices could be developed to allow minimally invasive monitoring of analytes. Thus, it is evident why an increasing amount of work has been focused on developing various methods to quantify glucose levels using fluorescence techniques.[26]

However, translation of these quantitative analytical methods into useful devices for long-term *in vivo* sensing remains a significant challenge. Overcoming the intrinsic limitations of photostability and loss of recognition capability represents one challenging aspect of this problem, while materials and methods for packaging to create indwelling sensors or systems add another level of complexity. Options for meeting these needs are being aggressively explored individually and in combination by many groups worldwide, giving hope for practical devices to be available in the near future.

10.2 OVERVIEW OF FLUORESCENCE-BASED GLUCOSE ASSAYS

Assays for glucose employing fluorescence transduction have been developed using a number of different molecular recognition strategies, from naturally occurring glucose binding proteins (GBPs) to genetically engineered mutants and even synthetic materials, and encompassing a broad region of the spectrum, from the ultraviolet (UV) to the near-infrared. Optical transduction has been accomplished using intrinsic fluorescence, dual-label systems that transduce binding or conformational changes as alterations in energy transfer or solvent effects, or complex enzymatic systems that monitor consumption of coreactants or formation of reaction products with specific indicators. The glucose-dependent changes have been monitored using both intensity and fluorescence lifetime measurements. These assays have been developed first as solution-phase reagents that exhibit optical changes in proportion to glucose concentration, and some of these have been further demonstrated in many sensor configurations use in sample analysis, from waveguide-based systems to hydrogel materials used for contact lenses to microspheres and nanoparticles.

The following sections describe the current state of the art in fluorescent glucose sensing, from the molecular recognition elements (glucose receptors) to the energy transduction mechanisms and finally to the packaged devices—existing and prospective—containing encapsulated reagents. Furthermore, additional relevant subject matter on tissue optics is included for comparison of different approaches.

10.2.1 Molecular Recognition Agents for Glucose Assays

10.2.1.1 Protein-Based Glucose Receptors

Concanavalin A Concanavalin A is a globulin isolated from jack bean meal, possessing a pH-dependent tetrameric (pH $>$ 7) or dimeric (pH $<$ 7) structure and a deep pocket that binds saccharides. The sugar binding behavior requires the presence of metal ions Mn^{2+} or Ca^{2+};[27] physiological levels (1.183 mM)[28] are sufficient for complete activation of the lectin for saccharide binding (>1 mM). Con A was initially observed to form precipitates with glycogen and yeast mannan due to the multiple saccharide binding regions of the molecule.[29] This ability to bind sugars is not highly selective, and it has been noted that Con A will react with a large class of polysaccharides and sugars with varying association constants depending on the structural properties of the ligands. While the affinity for D-glucose is less than that of many other sugars (as measured by inhibition of α-D-mannopyranoside ($K_d = 1.7$ mM = 31 mg/dL, compared to $K_d = 0.71$ mM and $K_d = 0.34$ mM for D-fructose and maltose, respectively[30,31]), the potential for Con A as a receptor for β-D-glucose in biosensors has been explored due to the expected *in vivo* low levels of these competing molecules. Indeed, Con A was used in the first examples of nonenzymatic glucose biosensors.[15,16] More recent work on modification of Con A to increase solubility and stability for drug delivery applications showed functionalization of the lectin with up to five pendant poly(ethylene glycol) (PEG) groups per molecule increased the stability and also increased the affinity for glucose about five times; however, it is unclear whether this results in increased specificity for glucose, or if propensity for fructose binding is also altered as a result of PEGylation.[32]

As a result of its sugar binding capacity, Con A has an affinity for cell surface receptors, such as those on lymphocytes,[33,34] and, therefore, has biological activity. For lymphocytes, Con A attachment is mitogenic; binding to the cell surface also restricts mobility of other immunoglobulin receptors in the cell membrane[35] and inhibits phagocytosis by leukocytes.[36] As a result, Con A has severe toxicity issues;[37] however, recent research has suggested that the dose of protein that would potentially be released from implants would not pose a significant health risk.[38] Specifically, results from direct administration of Con A to rats indicated that even at 10 times the amount of Con A present in the implant, no significant toxicological or systemic response was observed.

Glucose Binding Engineered Proteins As the field of protein engineering advances rapidly, several researchers have adapted a glucose binding protein commonly found in the periplasmic space of *Escherichia coli*[39] for use in glucose sensor applications. These ellipsoidal proteins possess a single-substrate binding site in a cleft located between the two globular domains. Electron density maps have confirmed selective binding with β-D-glucose, with 13 strong hydrogen bonds linking polar side groups from both protein domains to the hydroxyls and ring oxygen of the sugar. Upon binding, the sugar becomes completely buried within the cleft, resulting from a hinge-bending motion that effectively surrounds the small ligand.[40] It has been shown that this process of trapping glucose and galactose induces a sufficient conformational change to allow optical transduction via incorporation of environment-sensitive dyes[41–43] or dual-labeling the proteins with energy transfer pairs.[44]

These receptors are technologically advantageous because of the ability to produce them in large quantities using bacterial bioreactors as well as the potential to directly modify the structure via protein engineering. However, the primary disadvantage of current forms of GBP is the high affinity for glucose: the dissociation constant is on the order of 1 μM,[41] which implies near-saturated receptors at physiological concentrations of glucose and, as a consequence, very low sensitivity to variations at these levels. Such receptors are attractive for monitoring approaches involving analysis of dilute fluid (e.g., extracted interstitial fluid, lacrimal fluid), but are not appropriate for true *in vivo* analysis. Although the majority of GBP-based sensors reported to date show very low sensitivity over physiological glucose levels, future developments in protein engineering techniques are likely to facilitate further advances in recognition molecules with tunable affinity constants. To make these receptors relevant for physiological monitoring, the K_d must be increased to at least the 1 mM range.

Enzymes In analogy to the enzyme-modified Clark electrode developed for glucose sensing,[45] enzymatic fluorescent sensors—optodes—have been demonstrated through a combination of selective enzymes and optical indicators.[46–48] In general, the active elements catalyze an oxidation–reduction reaction, and the consumption of a substrate or the resulting formation of a product is monitored using fluorescence.[48] Three well-known enzymes for reacting with glucose are glucose oxidases,[14] glucose dehydrogenases (GDH),[26,49] and hexokinases (HEX).[26,50,51] Each of these has been used, in different forms, to demonstrate potential for glucose sensing using intrinsic and extrinsic fluorescence changes.

One of the most widely researched enzymatic sensing schemes relies on GOx-catalyzed oxidation of glucose in the following reaction:[14]

$$\text{Glucose} + O_2 + H_2O \xrightarrow{\text{GOx}} \text{gluconic acid} + H_2O_2$$

Using this reaction, several opportunities are presented to indirectly monitor glucose levels, including fluorescence-based measurement of pH change due to gluconic acid formation, hydrogen peroxide produced, or oxygen consumed. Examples of indicators used in sensor prototypes based on each approach are described in a later section. Enzyme-based sensors are inherently very selective due to the stereospecific catalytic binding sites and are also fully reversible, making them ideal for use in biosensor applications. With the main difference being that optical indicators are nonconsuming, contrasted with surface reduction of oxygen typically used in Clark-type electrodes, the extensive literature on electrochemical biosensors can still be applied to optical devices in many cases, making available a broad spectrum of lessons and design ideas.[48] However, there are several drawbacks to using enzymatic schemes: (1) Enzymes tend to spontaneously deactivate (denature) over time, potentially resulting in drifts of sensitivity and/or range. (2) The consumptive nature of the enzyme could potentially result in local depletion of glucose levels, leading to ambiguous readings, as well as possible tissue damage due to excessive consumption of glucose and oxygen. (3) To obtain accurate indirect measurements of glucose, the reaction must remain glucose limited, requiring careful control of reaction–diffusion kinetics.

This latter point is a critical design aspect, and has been explored in detail from a theoretical point of view;[52–54] however, the key features can be understood intuitively by considering the relative concentrations of glucose and oxygen expected *in vivo*. Since both reactants must be present for the reaction to proceed and because the stoichiometry involves reaction of 1 mol of molecular oxygen per mole of glucose, it is obvious that the relative concentrations of the two substrates will determine which one is limiting. In blood and interstitial fluid, normal glucose levels are typically around 72–108 mg/dL (4–6 mM), whereas dissolved oxygen concentrations average in the 100–250 μM range. As a result, assuming near-equal diffusion rates, oxygen will be depleted rapidly and the range over which response is proportional to glucose will be limited to low concentrations. At higher concentrations, there will be no significant dependence of product formation rate or local oxygen levels on glucose. The result of this behavior is the need to limit glucose diffusion rate relative to oxygen, so that proportional changes can occur over a wider concentration range. This point has been established in regard to electrochemical sensors, and has been considered again more recently in the context of downscaling optical sensors to function at micro- and nanometer scales.[55]

Apoenzymes The intrinsic flavin fluorescence, which is environment sensitive, of GOx was shown to be an indicator of glucose binding events.[56] This work, which proved that sufficient conformational changes occur in the process of binding glucose, laid the foundation for the use of GOx as a receptor rather than a catalyst. The early findings showed changes in the intrinsic fluorescence over the low end of the physiologically relevant range (27–36 mg/dL). By labeling the enzyme with a fluorescein derivative, a glucose-responsive resonance energy transfer (RET) system, which exploits the distance-dependent energy transfer between the intrinsic flavine group and the attached fluorescein, responded with changes that correlated with glucose levels.[57] Yeast hexokinase was also shown to exhibit a 30% decrease in intrinsic tryptophan fluorescence during exposure to 216 mg/dL of glucose.[58] Using hexokinase labeled with an ANS derivative, a decrease in sensitivity was observed, contrary to results with GOx and GDH.[59] Furthermore, HEX-based transduction schemes are particularly vulnerable to excessive quenching when exposed to serum, indicating that a separation from the biological environment is necessary for *in vivo* applications.[26] In each of these cases, however, the fluorescence change varied with time after exposure to glucose, due to the enzymatic activity of the protein that continued to consume glucose.

In an effort to overcome the limitations of the catalytically active system and to improve poor understanding of the underlying mechanism of the previous systems, additional studies were performed on deactivated enzyme using apo-GOx.[60] In this work, apo-GOx was labeled with the environment-sensitive fluorophore 8-anilino-1-naphthalene sulfonic acid (ANS). It was found that the binding of glucose to apo-GOx results in a conformational change such that the steady-state intensity of ANS decreased 25% and the mean lifetime of ANS decreased about 40% over the range of 180–360 mg/dL glucose. This work showed that apo-GOx retained its high specificity and, similarly, the apo form of glucose dehydrogenase has been shown to retain glucose binding specificity. These results suggested the utility of deactivated enzymes as analyte-recognizing elements in optical sensors[61] and in many ways paved the way for advances in the development of a new reagentless biosensor genre.[21,62–64]

10.2.1.2 Aptamers Aptamers are nucleic acids that bind specific ligands, and they are being investigated as reagents for various applications of therapeutic and diagnostic value. Aptamers that bind medically relevant targets have been shown to modify cellular metabolism, and aptamer-based molecular receptors are being pursued for diagnostic assays to rival immunoassays.[65] Aptamers have been demonstrated as receptors in protein and nucleic acid biosensors.[66–68] As such, it is logical to expect that saccharide binding aptamers are on the horizon, at least in terms of exploratory research, but it does not appear that efforts to design glucose binding aptamers has yet met with any success.

10.2.1.3 Synthetic Receptors

Boronic Acid Derivatives Boronic acid compounds—weak Lewis acids consisting of one boron (electron deficient) atom and two hydroxyl group—form a reversible complex with *cis* 1,2- or 1,3-diols, such as the common biological carbohydrates glucose, galactose, and fructose (Figure 10.2).[69] Since the complex is reversible, the integration of boronic acid derivatives as recognition elements in glucose sensors is being aggressively explored. Many fluorescence-based readout systems are predicated upon the signaling of the diol binding event through modulation of fluorescence properties of a boronic acid compound, which are typically comprised of a binding moiety and a complementary fluorogenic group responding to photoinduced electron transfer, pH change, or another local interaction, discussed below in more detail.[70,71]

Boronic acid receptors have, in general, a principal disadvantage with respect to their potential for medical glucose monitoring: a lack of specificity. Typical values for comparing binding of fructose and glucose with phenylboronic acids indicate approximately 50-fold specificity for fructose at physiological pH.[72] Bisboronic acid configurations have been demonstrated to enhance glucose selectivity by appropriate orientation of the boronic acid compounds relative to the diol pairs on glucose as well as adjustment of length and type of chain used to link the boronic acid moieties greatly affects glucose selectivity, paving the way for the development of additional compounds with engineered affinity properties.[73,74] As a result, a BA derivative has been prepared with a glucose affinity 43- and 49-fold greater than that of fructose and galactose, respectively (D-glucose: $K_d = 0.68\ \text{mM} \approx 12\ \text{mg/dL} >$ D-fructose: $K_d = 29.4\ \text{mM} \approx 530\ \text{mg/dL} >$ D-galactose: $K_d = 33\ \text{mM} \approx 600\ \text{mg/dL}$); this should be sufficient for any *in vivo* monitoring purpose.[75]

Molecularly Imprinted Materials Hydrogels are insoluble, cross-linked networks of hydrophilic homo- or heterocopolymers, with the ability to take up

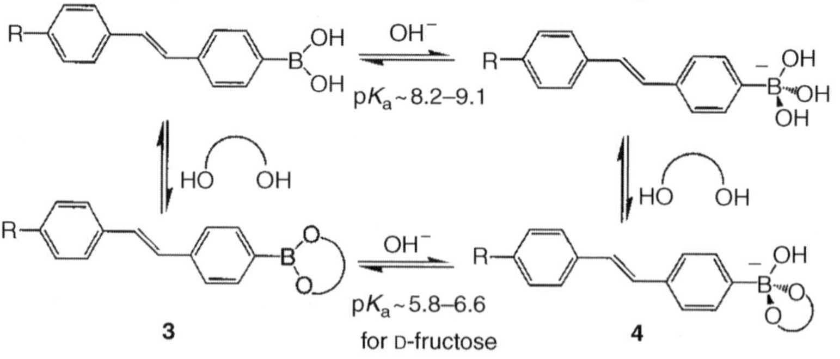

Figure 10.2 Equilibrium and conformation of different forms of the boronic acid group with and without sugar. Reprinted with permission from Ref. 69. Copyright 2001 American Chemical Society.

significant amounts of water. Recently, significant efforts in hydrogel engineering have been expended toward development of stimuli–responsive materials that reversibly swell or collapse as a result of local changes in temperature, pH, ionic strength, or other specific analyte concentration differences. These gels may be entirely artificial or may be "biohybrid" systems—those incorporating a protein within the polymer matrix to act as the molecular recognition element. Molecularly imprinted materials (MIMs; or MIPs, when the material is specifically a polymer) are completely artificial systems, which are constructed using a templating procedure. This is performed by first forming a complex between the template molecule (the target analyte, e.g., glucose) and the functional monomers or functional oligomers (or polymers).[76] These functionalities are selected based on their specific chemical structures that will interact with the template by covalent or noncovalent mechanisms, or both.[77–79] Once the complex is formed, the polymerization reaction is performed in the presence of a cross-linker and an appropriate solvent. Finally, the template molecules are removed, leaving "imprints" in the cross-linked material; the product is a heteropolymer matrix with specific recognition regions for the template molecule (Figure 10.3).

MIPs have been demonstrated to have several relevant features for biosensing, which could be useful in the development of fluorescent indicators for glucose. Several different forms of catalytically active ("artificial enzymes")[80–82] and affinity binding ("artificial antibodies")[9,83,84] MIPs have been proposed. In the latter case, the response to target binding is a swelling or shrinking due to the altered ionization of the gel (Figure 10.4). The optimization of processes and materials to achieve stable, reversibly responsive MIPs is an area of intense current research. Glucose-sensitive imprinted and environment-sensitive gels have been used as chemomechanical transducers, applied to microcantilevers that exhibited reversible response to changing glucose levels;[85,86] it does not appear, however, that these materials have been used along with fluorescence transduction for glucose sensing. A limitation of these entities as molecular receptors is their general sensitivity to pH, temperature, and ionic strength.[87–90] Although there have been reports of isomeric selectivity of MIPs,[76] follow-up studies showed that preferential uptake was instead a result of partitioning,[91] highlighting the importance of careful characterization of these materials to ensure true imprinting occurs.

10.3 TRANSDUCTION METHODS AND MATERIALS

Using the molecular receptors described above, a number of creative methods for transducing the binding of glucose using fluorescence have been demonstrated. These range from "direct" methods, in which glucose binding results in a change in fluorescence properties intrinsic to the receptor, to "indirect" methods involving monitoring of reaction products or cosubstrates using fluorescent indicators or displacement of fluorescent competing ligands binding from the receptor. As with the various receptors, each transduction approach has advantages and limitations, such that an ideal glucose assay for *in vivo* monitoring is yet to be developed.

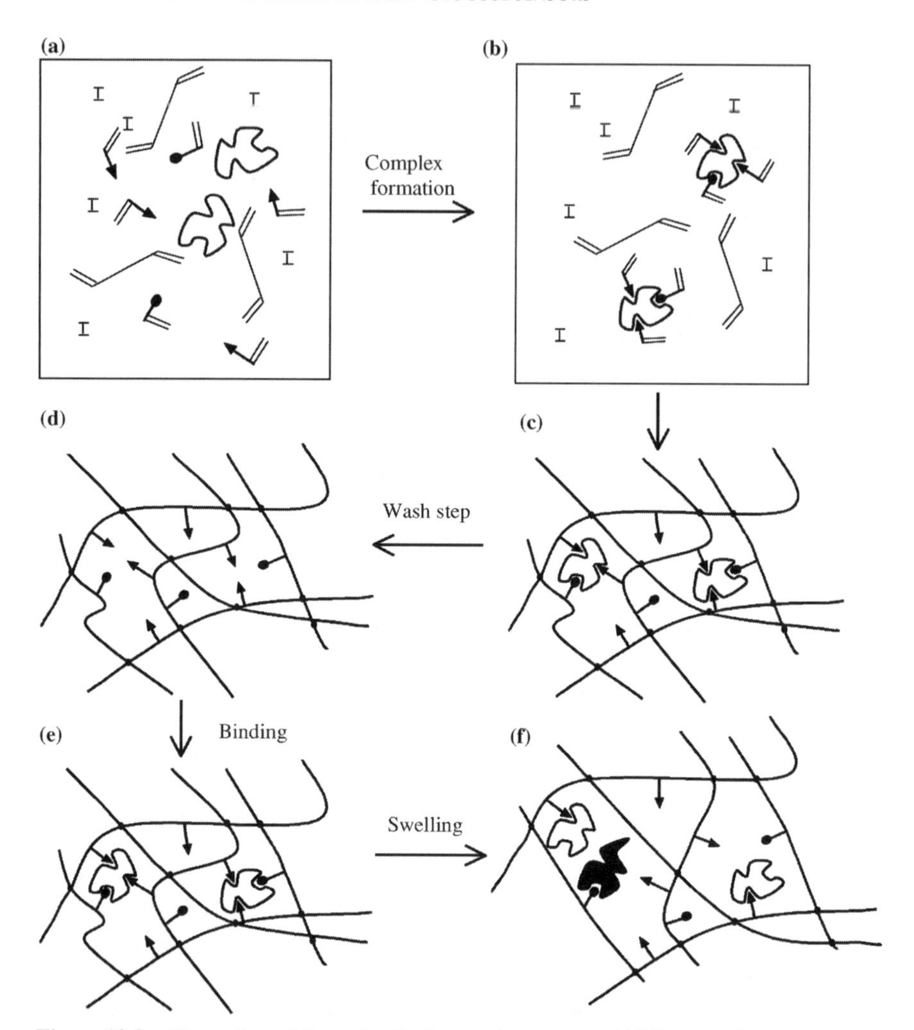

Figure 10.3 Illustration of the molecular imprinting process. (a) Mixture of template (target), functional monomer(s) (triangles and circles), cross-linker monomer, solvent, and initiator (I). (b) Formation of the prepolymerization complex via covalent or noncovalent interactions between template and functional monomers. (c) Network formation via polymerization and cross-linking. (d) Removal of original template by washing, leaving "imprints." (e) Rebinding of template to imprinted regions. (f) In systems with lower cross-link density, macromolecular chains may spatially rearrange resulting in regions of differing affinity and specificity (filled molecule is isomer of template). Reprinted with permission from Ref. 10. Copyright 2002 Elsevier.

10.3.1 Intrinsic Fluorescence

Proteins typically exhibit some level of ultraviolet-excited intrinsic fluorescence, due to the presence of aromatic amino acids tryptophan and tyrosine. These fluorogenic groups possess some inherent solvent sensitivity and, therefore, exhibit changes in net

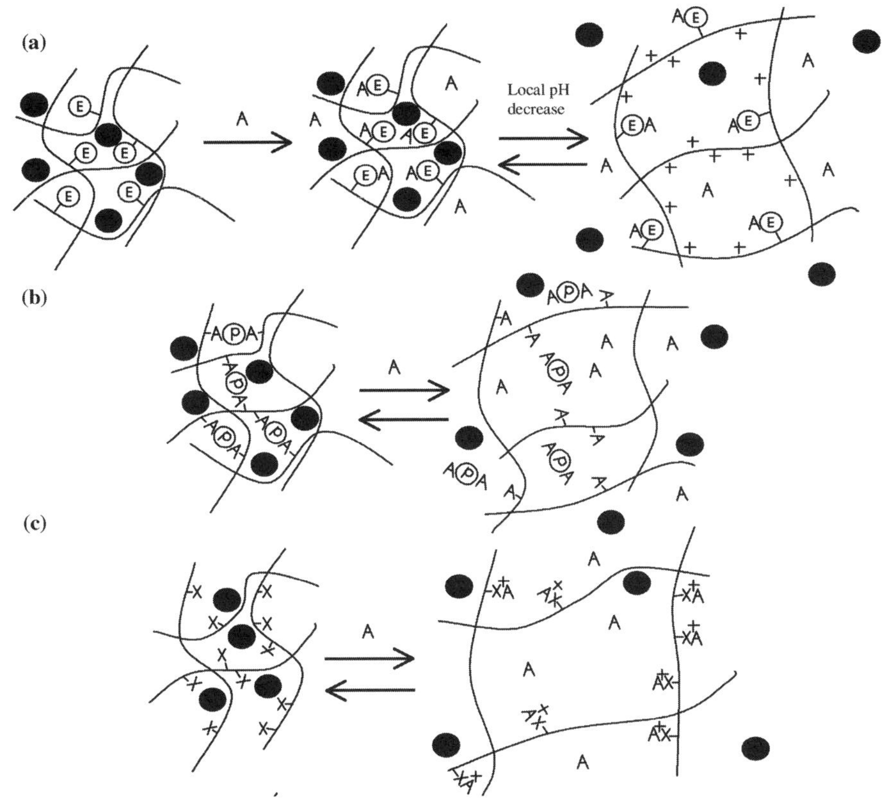

Figure 10.4 Illustration of potential mechanisms for swelling response of imprinted hydrogels to target analyte (images depict associated release of drug (•)). (a) pH-sensitive swelling induced by internal enzymatic reaction (E = immobilized enzyme). (b) Loss of effective crosslinks as analyte competes with immobilized analyte (or analogue) for binding positions on a recognitive protein (P). (c) Swelling resulting when analyte binds to a pendent functional group, forming an ionized complex. Reprinted with permission from Ref. 10. Copyright 2002 Elsevier.

quantum yield and spectral shifts upon conformational changes due to binding with targets. This phenomenon has been used to demonstrate binding of enzymes (and apoenzymes) with glucose, as noted above.[48,50,60,92,93] Unfortunately, however, the UV excitation required for this is poorly matched to optical fiber technology, is not suitable for transmission through tissue, and requires more expensive sources and detectors.

10.3.2 Exogenous Environment-Sensitive Fluorophores and Glucose Binding Proteins

While the changes in intrinsic fluorescence due to glucose binding present an optical challenge due to the low wavelengths required (250–300 nm),[60] the same approach of using environmental sensitivity has been extended through the use of exogenous solvent-sensitive probes. For example, the glucose-initiated conformational shift of the bacterial glucose binding protein was exploited for transduction of glucose concentration by inclusion of cysteine residues via site-directed mutagenesis and subsequent attachment of environment-sensitive fluorophores into the glucose binding site.[93,94] Two fluorophores were used, both of them showing an approximate change of 80% in fluorescence intensity at saturating levels of glucose. In follow-up work, a similar protein engineering approach was employed to strategically place cysteine for labeling with ANS, an environment-sensitive probe.[41] Results showed that ANS–GBP displayed a twofold decrease in intensity when exposed to saturating levels of glucose, with a dissociation constant of ~1 μM. The same report also showed that immobilization of ANS–GBP along with a long lifetime ruthenium complex onto the surface of a cuvette could be successfully used in low-frequency phase-modulation lifetime measurements to predict bulk glucose levels within the cuvette.

10.3.3 Energy Transfer

Fluorescence resonance energy transfer (FRET, or RET) refers to the nonradiative energy transfer of excited-state energy from an excited-state fluorophore (the donor, D) to another fluorophore (the acceptor, A). FRET occurs between donor and acceptor molecules when certain conditions are present, specifically (1) the fluorophores are in proximity, (2) the emission spectrum of donor overlaps with the excitation spectrum of the acceptor (Figure 10.5), and (3) the respective emission and excitation dipoles are sufficiently aligned.[95] FRET can be regarded as the interaction of transition dipoles of donor and acceptor groups, thus, the name fluorescence resonance energy transfer.[96,97]

A simplified theory of FRET is sufficient to describe affinity sensors used in fluorescence transduction of glucose concentrations. A key quantity that describes the potential FRET interaction between a donor–acceptor pair is the Förster distance, R_0, the distance at which half the donor molecules are quenched by the acceptor molecules. R_0 is proportional to several parameters of the fluorophores, in accordance with $R_0 = K^6\sqrt{\kappa^2 n^{-4}\,\phi_D J[\lambda]}$, where K is a constant. The variable κ^2 refers to the relative spatial orientation of the dipoles of D and A, taking on values from 0 to 4 for completely orthogonal dipoles and collinear and parallel transitional dipoles $\kappa^2 = 4$,

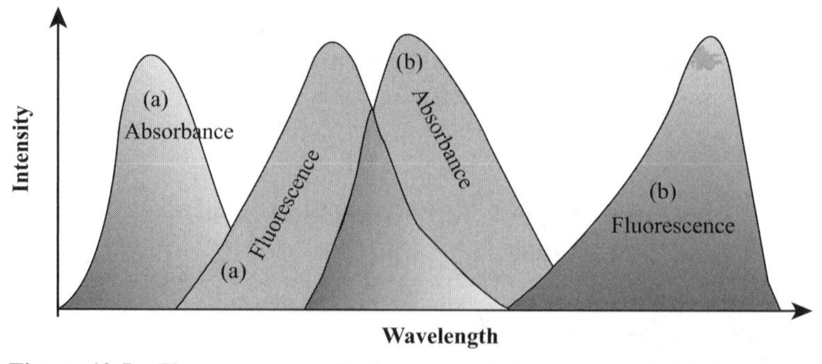

Figure 10.5 Fluorescence excitation and emission spectra of (a) donor and (b) acceptor molecules.

respectively. For random orientation (as is usually assumed, especially for solution-phase measurements), κ^2 value is set as 2/3. The remaining terms, n, ϕ_D, and $J[\lambda]$ correspond to the solvent refractive index, quantum yield of D in the absence of A, and the overlap integral, which measures the degree of overlap between the emission spectrum of D and the absorption spectrum of A, respectively. A key point from this discussion is that there must be significant spectral overlap for the dipoles to interact and for the proper excitation of A by D.

The energy transfer efficiency is directly proportional to the spectral overlap, and this also directly affects the Förster distance of a particular D–A pair. Figure 10.5 shows the D and A excitation and emission spectra in an ideal energy transfer system, wherein D and A have very distinct excitation spectra (so that A can only be excited by energy transfer and not by direct photon absorption at the wavelengths used to excite D)—the D emission and A excitation spectra overlap strongly—and the D and A emission maxima are well separated, so that the quenching of D fluorescence and the enhancement of A fluorescence can be individually measured.[98,99]

In practice, R_0 values vary significantly for different D–A FRET pairs, ranging from 40–80 Å. This distance must be comparable to the size of the proteins or other biomacromolecules being used for efficient energy transfer from D to A. The energy transfer (E) is given as the fraction of photons absorbed by D that are transferred to A and, therefore, is given as the ratio of transfer rate to the total decay rate of the donor, $E = ((k_T)/(\tau_D^{-1} + k_T))$ or $E = ((R_0^6)/(R_0^6 + r^6))$. It should be clear from these equations that when D and A are separated by Förster distance (R_0), the FRET is 50% efficient. However, another noticeable property is that the FRET efficiency highly depends on the distance between D and A molecules, as is shown graphically in Figure 10.6.

These mathematical and graphical representations serve to highlight the main advantages of using FRET as a transduction mechanism; it is highly sensitive to the distance between the two molecules, and the ratiometric nature allows variations in

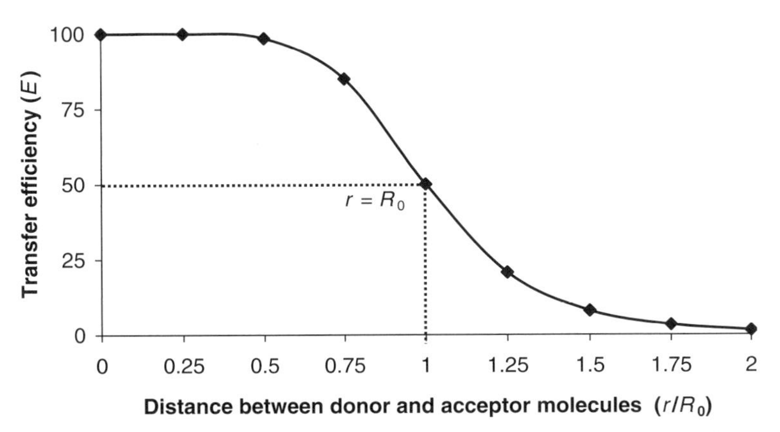

Figure 10.6 Distance-dependent efficiency of energy transfer between donor and acceptor molecules, where r is the intermolecular distance and R_0 is the Förster distance for the pair of molecules.

instrumental parameters, interrogated volume, and measurement configuration to be internally compensated.[100] In recent years, FRET has been applied in various fields of biochemistry, such as single-molecule FRET for detecting conformational changes and molecular interactions,[101] distance measurements in α-helical melittin,[102] protein folding measurements,[103] orientation of protein-bound peptide,[104] nanoscale biosensors using organic[105] and quantum dot[106] FRET donors, among others.

Most of the work toward realization of fluorescence affinity-based glucose sensors has focused on the development of competitive binding-based assays, which rely on changes in fluorescence emission properties modulated by fluorescence resonance energy transfer to optically transduce glucose levels.[55,64,107–109] Changes in FRET result from the competitive binding of a fluorescent ligand and analyte to receptor sites on lectins, glucose binding proteins, or deactivated glucose-specific enzymes, such that the ligand is displaced from the receptor when in the presence of the analyte, causing changes in energy transfer efficiency between the donor–acceptor pair, ultimately resulting in measurable shifts in mission spectra.[110,111]

There are essentially two types of sensors constructed using energy transfer for transduction. The first type (Figure 10.7a) involves donors and acceptors attached at appropriate sites on the same receptor molecule, and the distance between them is

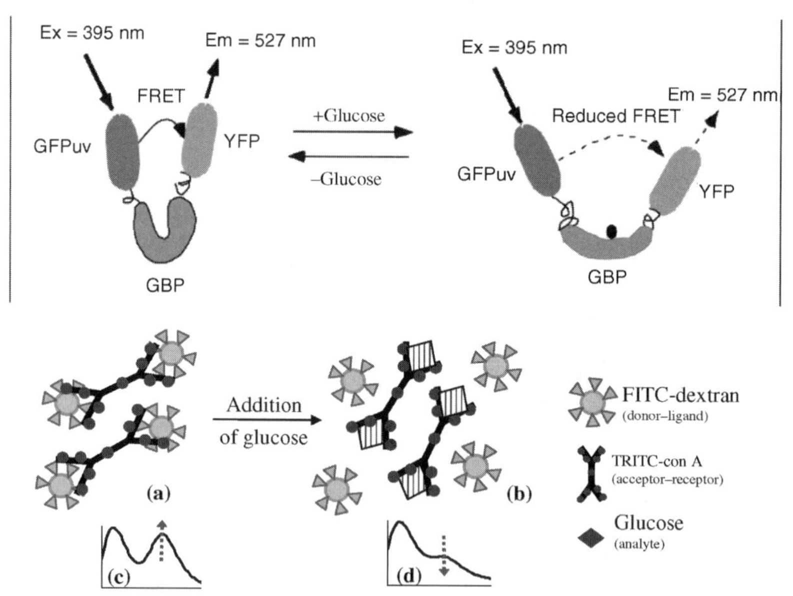

Figure 10.7 *Upper*: Design of a single-molecule glucose assay (glucose indicator protein (GIP)) based on FRET. The GBP adopts an "open" form in the presence of glucose (•), moving the two fluorescent protein domains away from the center of the GBP, resulting in a decrease in FRET.[112] *Lower*: Illustration of the combination of FRET and competitive binding. (a) FITC-dextran/TRITC-Con A complexes, and corresponding spectrum (c) exhibiting significant energy transfer; (b) displacement of dextran from Con A in the presence of glucose, and associated spectrum (d), possessing a strong donor peak relative to acceptor peak in the presence of glucose.

modulated by conformational changes that occur upon glucose binding. The second type (Figure 10.7b) requires the interaction between receptors and ligands that compete with the target analyte for the receptor binding site; in this case, the receptors and ligands are labeled with acceptor and donor groups, respectively, or vice versa, and energy transfer occurs when the ligand is bound to the receptor. Various embodiments of sensors have been developed using these approaches, as discussed in the following sections.

10.3.3.1 Single-Molecule RET Systems Using Dual-Labeled Engineered Proteins A single-molecule RET-based sensor was created using a GBP mutant by fusing two fluorescent reporter proteins (green fluorescent protein as the donor and yellow fluorescent protein as the acceptor) to the C and N terminus of the receptor, respectively.[112] This represents an efficient approach to assay reagent fabrication, since the fluorogenic groups are built directly into the probe at the stage of recombinant protein production, implying that large-scale production is possible and only immobilization of a single molecule type is necessary. During a glucose binding event, the spatial separation between the fluorescent proteins is altered, causing a change in RET efficiency. The reporters were encapsulated within microdialysis fibers, and experimental observations showed that the emission intensity at 527 nm was reduced upon exposure to solutions containing glucose. The response was shown to be reversible; however, a maximum intensity change was observed at 0.36 mg/dL, well below the physiological range. This again highlights the limitation of current bacterial glucose binding proteins as noted previously, extremely high glucose affinity.

10.3.3.2 Competitive Binding Systems The groundbreaking work of Schultz was the first example of a fluorescence-based competitive binding assay to monitor glucose, in which the glucose binding protein Con A was immobilized onto the lumen of a microdialysis fiber containing a high molecular weight fluorescein-labeled dextran (FITC-dextran) solution.[16] A single optical fiber (low numerical aperture) was interfaced with the dialysis fiber and was used to deliver and collect photons from the solution within the dialysis fiber. When in the presence of glucose, FITC-dextran molecules are displaced and diffused from the lumen into the core volume of the fiber, resulting in a rise in FITC emission intensity that could be correlated with glucose levels. Overall, a 60% change in intensity was observed over the range of 0–180 mg/dL of glucose; however, problems with drift occurred due to the lack of internal reference.

This concept was extended to a homogeneous fluorescence affinity assay with ratiometric functionality by employing RET measurements between the FITC-dextran and the rhodamine-labeled Con A (TRITC-Con A) within a microdialysis fiber.[113] In this scheme, a measurable ratiometric decrease in RET efficiency was observed as glucose-displaced FITC-dextran, resulting in a decreased acceptor emission intensity relative to donor emission intensity. Results indicated a linear correlation with glucose levels up to 200 mg/dL and 60% change in ratiometric intensity over the range of 0–900 mg/dL. A significant problem of irreversible aggregation of the Con A was noted within a few hours, complicating the development of this type of sensor without modification of the Con A to improve stability.

A novel scheme was then devised to eliminate the need for Con A immobilization on the microdialysis fiber wall while avoiding Con A aggregation.[111] In this work, Sephadex beads comprising pendent glucose moieties and two highly absorbing dyes were used. Safranin O and Pararosanilin were selected as dyes to block the excitation of Alexa Flour® 488, the fluorophore used to label Con A (AF488-Con A). When in the absence of glucose, excitation light passing through the beads is preferentially absorbed by the Safranin O and Pararosanilin, resulting in poor emission from AF488-Con A. Upon glucose diffusion into the fiber, glucose binding releases AF488-Con A from the beads, causing an increase in emission intensity that is well correlated with glucose levels. Additional efforts have extended the operating wavelength of this assay into the NIR (to potentially make a system more suitable for transdermal applications) by using alternative fluorophores.[110] However, studies performed to evaluate long-term use of both systems indicated leakage of the components from the membrane over time, resulting in poor long-term performance and ultimately proving that additional advances were necessary for future success.

Most recently, an *in vivo* investigation of a NIR Con A RET system was reported and the host response characterized.[114] In this work, Cy-7-labeled Con A and Alexa Flour 647-labeled dextran system were entrapped with a hollow microdialysis fiber immobilized on the tip of an optical fiber. The device was characterized *in vitro*, the result of which indicated a response range of 36–450 mg/dL. More important, the device was implanted subcutaneously and evaluated *in vivo* for 16 days. Results were promising, as the implant readout retained a high degree of correlation with blood glucose fluctuations (as measured by blood-draw methods). An increase in response time was observed at the end of the experimental period, with fibrous encapsulation cited as the cause.

A modification of previous assays using dextran was recently reported, in which a fourth-generation polyamidoamine (PAMAM) glycodendrimer was employed as the competing ligand. A solution-phase response of ~100% was observed for 0–360 mg/dL concentrations, with response time of 5 min.[109,115,116]

Recently, an alternative RET assay utilizing the inactive form of glucose oxidase (apo-GOx) as the target binding molecule has been proposed in an attempt to overcome the aforementioned difficulties associated with Con A-based assays.[21,63] This system retains the advantages of the competitive binding approach used with Con A—no consumption of the analyte or formation of by-products during the sensing process, and intrinsically ratiometric—and adds superior specificity and potentially more facile mass production by recombinant methods. Proof-of-concept work, paralleling earlier work, demonstrated that TRITC-apo-GOx and FITC-dextran provided a highly sensitive and specific response to β-D-glucose (approximately threefold greater than previous competitive binding systems) as well as a high degree of reversibility and stability. These findings make this assay an attractive alternative to those employing Con A.

10.3.4 Boronic Acid Indicators

Many fluorescent transduction systems are predicated upon the signaling of the diol binding event through modulation of fluorescence properties of the boronic acid compounds, which are typically comprised of a binding moiety and a fluorescent signaling moiety.[70] The first report on this concept presented anthrylboronic acid-based fluorescent saccharide sensors, successfully showing saccharide binding could signal a change in fluorescence properties.[71] The high apparent pK_a (~8.8) of the compound resulted in an emission intensity decrease with increasing saccharide concentration through a process termed chelation-enhanced quenching. While this initial work was indeed an important "proof of concept," anthrylboronic acid was found to be more sensitive to fructose binding events than glucose. The concept was then advanced through the integration of an amino group positioned between the boronic acid and the anthracene. The amino group effectively increased the compound's saccharide binding affinity by lowering the pK_a and enhanced photo-electron transfer, resulting in increased emission intensities with elevating saccharide levels (the opposite of which was seen with anthrylboronic acid).[72] However, preferential binding of fructose over glucose (approximately 50-fold specificity for fructose at physiological pH) was also observed in this case.

In an attempt to further increase binding affinity and enhance glucose selectivity, bisboronic acid compounds were developed and it was demonstrated that glucose selectivity may be enhanced by appropriately orienting the boronic acid compounds relative to the diol pairs on glucose.[74] Further work showed that linker chain length (and type) between the boronic acid moieties greatly affects glucose selectivity, paving the way for the development of additional compounds with engineered affinity properties.[73] Based on these findings, a derivative with a glucose selectivity 43-fold greater than that of fructose was then developed, increasing the likelihood that boronic acid-based sensors could indeed be designed for applications in glucose sensing.[116]

The effect of fluorophore choice on the selectivity, stability, and sensitivity of the BA-based sensors and the use of multiple pendant fluorophores to extend the emission wavelengths of the sensors through fluorescence resonance energy transfer was also explored.[117] This system utilized a phenanthrene donor and pyrene acceptor such that when excited with 299 nm light (the excitation maxima of phenanthrene), the entire system would emit at 417 nm (the emission maxima of pyrene). To compensate for source fluctuations and variations in probe concentration, ratiometric functionality was introduced by using 3-nitronaphthalic anhydride and 3-aminophenylboronic acid with various linker moieties to form a monoboronic acid sensor possessing bimodal emission spectra with peaks at 430 and 550 nm upon excitation with 377 nm light.[118] These results seemed to contradict earlier findings that demonstrated poor glucose selectivity with monoboronic acid derivatives. The authors suggest that competing factors may be simultaneously operating, such as conformational restriction between the phenylhydroxy-boronate–saccharide complex and its influence on the excited state, which could result in enhanced glucose selectivity.[8]

A more recent approach to designing direct fluorescent indicators for glucose involves two-component boronic acid sensors, where the fluorophore—usually anionic in nature—is quenched by a physically separate boronic acid-substituted viologen receptor. As the saccharide binds with the receptor, the quenching efficiency of the viologen is reduced, resulting in an increased emission intensity of the fluorophore (Figure 10.8).

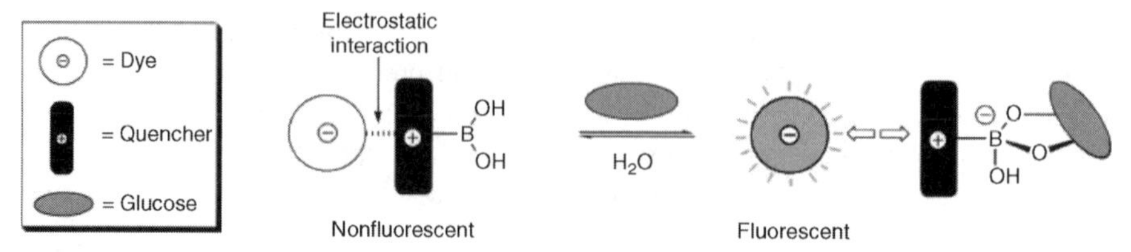

Figure 10.8 Two-component saccharide-sensing system using boronic acid receptor. Reprinted with permission from Ref. 170. Copyright 2006 American Chemical Society.

This facile approach readily allows the interchange of fluorophores without any modification of the receptor, a considerable advantage since transformations can cause unwanted changes in the photophysical dye properties. This work demonstrated that sensitivity and selectivity as well as optical properties of the boronic acid derivatives can be modulated by interchanging different viologens and fluorophores.[119,120] For these indicators immobilized in poly(HEMA) hydrogels attached to an optical fiber measurement system, the sensitivity was 0.1%/(mg/dL) over the range of 0–360 mg/dL, the response was fully reversible, and the response time was on the order of 5–10 min. An apparent limitation of this approach is that the dye/reagents may diffuse away from the gel; although the response was found to be very stable for up to 36 h under constant glucose challenges, it is unclear whether the reagent immobilization is sufficiently stable for long-term monitoring. Nevertheless, these findings are extremely encouraging and support optimism that BA-based sensors can be developed for *in vivo* monitoring applications.

10.3.5 Enzymatic Reactions with Indirect Indication

Recent advances in BA-based indicators have raised the hope that a single glucose indicator may be developed; however, reliable probes have not yet been developed to allow the direct monitoring of glucose over physiological levels with sufficient specificity and sensitivity. Due to the ready availability and high specificity, enzymes have been employed in many glucose-sensing schemes, where the coconsumption of nonglucose substrates and the generation of products are fluorescently tracked and used to indirectly determine glucose levels. In this section, we will focus on one specific case to highlight the general principles of fluorescence transduction of reactions between glucose and enzymes. The most widely researched enzymatic sensing scheme relies on GOx-catalyzed oxidation of glucose, described previously, in which oxygen and glucose are cosubstrates and the products formed are gluconic acid and hydrogen peroxide. In this reaction, several schemes could be exploited to indirectly monitor glucose levels using fluorescence readout: (1) measurement of proton production (changes in pH due to gluconic acid formation), (2) measurement of hydrogen peroxide produced, and (3) measurement of oxygen consumed.

10.3.5.1 Fluorescence-Based pH Transduction of Glucose Levels This approach to sensing using scheme (1) is conceptually simple: add a pH-sensitive fluorophore to an assay containing GOx and monitor the decrease in pH due to

GOx activity. However, this is difficult in practice because often the initial pH and the buffer capacity of the sample are unknown, making accurate calibrations difficult to obtain. Although little work has been done to develop this genre of glucose sensors, several significant contributions have been made.

The first example employed an immobilization of GOx and a pH-sensitive dye (1-hydroxypyrene-3,6,8-trisulfonate, HPTS) on the tip of an optical fiber.[44] As the local pH was reduced by gluconic acid production, measurable changes in fluorescence were observed. Using flow-through measurements with buffer solutions of various strengths, the optrode response time was determined to be 8–12 min, with a detection limit of 1.8–36 mg/dL glucose and response saturation at 36–54 mg/dL.

A highly innovative approach to fiber-optic probe development involved precise deposition of distinct regions of analyte-sensitive dye onto the distal end of an imaging bundle.[121] To accomplish this goal, the distal surface of the fiber was silanized to functionalize the fiber surface with polymerizable acrylate groups. A thin film of poly(hydroxyethyl methacrylate) doped with a pH-sensitive fluorescein derivative was deposited onto the functionalized fiber surface using a spin-coating method. Using spot-directed UV illumination through individual fibers, selected regions were photopolymerized, and upon rinsing, revealed a pH-sensitive optical array.

An alternative fiber-optic glucose sensor was then reported, where the pH change resulted in the swelling of an environment-sensitive poly(acrylamide) hydrogel.[122] In this device, the polyacrylamide gel was used to immobilize GOx and a rhodamine fluorochrome on the tip of an optical fiber. During exposure to glucose, a reduction in local pH due to gluconic acid production caused the polyacrylamide hydrogel to swell by changing the ionization state of the amine groups. Since the concentration of the fluorophore remains constant while the surrounding hydrogel begins to swell, a measurable decrease in rhodamine emission intensity was correlated with glucose concentrations up to 300 mg/dL. A recent report described the use of energy transfer to transduce pH changes in hydrogels, making the technique emission ratiometric while increasing the sensitivity.[123] However, it is not yet clear whether such swelling gels can exhibit sufficiently rapid and reversible response for real-time monitoring.

10.3.5.2 Fluorescence-Based H_2O_2 **Transduction of Glucose Levels** Using H_2O_2 as an indirect means to measure glucose levels has the advantage of minimal background interference because significant levels of hydrogen peroxide are not commonly found in body fluids.[124] However, most of these types of sensing assays require reaction coupling with peroxidase (POx). In such a system, hydrogen peroxide produced through glucose oxidation is then oxidized along with a noncolored substrate by POx, resulting in the formation of water and a colored/fluorescent product. By monitoring the resulting absorbance/fluorescence, correlations with glucose levels can be obtained. This approach has the significant advantage that peroxide is consumed, as it would otherwise attack the proteins and result in rapid loss of catalytic activity. However, because these assays are nonreversible and somewhat complex, they have not been implemented in long-term monitoring applications. For this technology to be considered further for long-term glucose monitoring, a reversible, reagentless probe for hydrogen peroxide would need to be developed.

A reversible hydrogen peroxide probe that functioned independent of POx was used to develop a fluorescent glucose-sensing assay.[124,125] This unique europium-based fluorochrome—europium(III)tetracycline (Eu_3TC)—is weakly fluorescent in an uncomplexed state; however, upon binding with hydrogen peroxide, a 15-fold increase in emission intensity is observed. In addition, a 100% increase in decay time (from 30 μs to 60 μs in the presence of hydrogen peroxide) indicates that this probe could also be successfully implemented into lifetime measurements. While excited at 405 nm, a trimodal emission spectra with peaks centered at 577, 592, and 616 nm can be used to extract glucose concentrations. A GOx-based glucose sensor using Eu_3TC was developed, using Eu_3TC and GOx immobilized within a polyacrylonitrile-co-polyacrylamide (Hypan) polymer matrix.[126] Results indicated a high degree of reversibility with sensitivity to glucose in the range of 1.8–90 mg/dL concentrations; however, approximately 2.5 h was required for the sensor to equilibrate following a 36 mg/dL glucose step, suggesting that the response time is far too long to be of practical value in monitoring. While this technology shows promise, additional advancements are necessary to significantly reduce response time and increase the analytical range before consideration for use in diabetic monitoring devices. In principle, solutions to these issues could be approached by changing the dimensions and materials in the immobilization matrix.

10.3.5.3 *Fluorescence-Based O_2* Transduction of Glucose Levels

Oxygen is one of the best-known collisional quenchers of fluorescence; therefore, many fluorophores exhibit, to some degree, oxygen sensitivity. In collisional quenching, the quencher contacts the fluorophores while in the excited state, returning the fluorophore to ground state without photon emission.[95] The process of collisional quenching is typically characterized by the Stern–Volmer equation:

$$\frac{F_0}{F} = \frac{\tau_0}{\tau} = 1 + k_q\tau_0[Q] = 1 + K_D[Q]$$

In this equation, F_0 and F are the fluorescence intensities in the absence and presence of the quencher, τ_0 and τ are the lifetimes of the fluorophore in the absence and presence of the quencher, k_q is the biomolecular quenching constant, and $[Q]$ is the concentration of the quencher (in this case, molecular oxygen). The Stern–Volmer quenching constant is K_D, and is calculated as the product of k_q and τ_0. The Smoluchowski equation describes the biomolecular quenching constant, k_q, and is given by

$$k_q = 4\pi Np(D_f + D_q) \cdot 10^3$$

where D_f and D_q are the diffusion coefficients of the quencher and fluorophore, respectively, N is Avogadro's number, and p is the probability of collision. Therefore, for oxygen indicators to be sensitive to low concentrations of oxygen, long lifetimes in the absence of the quencher (τ_0) must be exhibited (most oxygen indicators typically exhibit lifetimes greater than 100 ns). Although there are at present many oxygen indicators to choose from, it is important to consider that in physiological systems, oxygen concentration will vary. Thus, a glucose sensor based on O_2 transduction could become oxygen limited if not properly designed, losing complete glucose sensitivity. Nonetheless, early work showed the feasibility of using oxygen to indirectly determine concentrations of glucose, lactate, and alcohol;[127] however, it was the introduction of longer wavelength oxygen indicators that renewed interest in this type of sensor.[126]

One of the first successful reports of optically transduced glucose sensing via indirect oxygen monitoring employed GOx and decacyclene immobilized within a nylon membrane, and glucose-limited consumption of oxygen was measured via fiber-optic bundles.[128] As bulk glucose levels increased, local concentrations of oxygen deceased, allowing indirect monitoring of glucose concentration via decacyclene emission. Experiments were performed in a flow chamber where air-equilibrated buffer with varying glucose levels were exposed to the membrane. Results were promising, as a response time of 1–6 min and an analytical range of 1.8–360 mg/dL were demonstrated.

Following the development of ruthenium–ligand complexes, which demonstrated longer excitation/emission wavelengths than those of decacyclene, a surge of work was focused on developing oxygen-transduced glucose sensors. One of the first examples employed a ruthenium compound [tris(1,10-phenanthroline)-ruthenium (II)].[47] The ruthenium compound was incorporated into a silica gel, which was then placed onto the tip of an optical fiber. Oxygen sensitivity experiments performed on the optrode proved sensitivity over 0–1300 μM (750 torr). A layer of immobilized GOx was added onto the ruthenium-doped silica gel. To optically insulate the system, a surface coating of carbon black was adsorbed before testing in a flow chamber. The fiber was indeed sensitive to glucose concentrations up to 18 mg/dL, with response times around 6 min. Although these results were promising, the operational range would need to be greatly extended for this design to be useful in diabetic monitoring.

Additional work was done using luminescent porphyrin compounds (Pb and Pt) as probes in glucose-sensitive fibers.[129] In this work, porphyrin probes were selected as oxygen indicators due to higher sensitivity to oxygen than ruthenium compounds, leading to the hypothesis that a system with superior sensitivity over ruthenium-based systems could be developed. In this system, a porphyrin-doped polystyrene membrane was adsorbed to the surface of an optical fiber, onto which GOx was directly immobilized. Glucose sensitivity tests were performed by dipping the optrode into air-equilibrated solutions of glucose while monitoring the spectroscopic response in real time. The authors observed a 400% increase in porphyrin emission intensity at saturating levels of glucose, which was unfortunately below (18 mg/dL saturation) the physiologically relevant range. A probable reason for the poor analytical range is that the authors did not incorporate diffusion-limiting layers over the enzyme to modulate reaction–diffusion kinetics; therefore, local oxygen levels were rapidly depleted, resulting in a small range.

The ingenious approach of using imaging fiber bundles described above for pH-based monitoring was further extended for glucose sensing, where an optical fiber capable of monitoring glucose and oxygen continuously was developed.[130] In this work, oxygen-sensitive optical arrays were prepared by mixing Ru(4,7-diphenyl(Ph)-1,10-phen)$_3$Cl$_2$ (Ru(Ph$_2$phen)$_3^{2+}$), an oxygen-quenched ruthenium compound, in a siloxane copolymer. Glucose sensitivity experiments were performed under various bulk oxygen concentrations, and the results showed that glucose sensitivity and operational range could be affected—higher bulk oxygen levels extended the operational range since the sensor became oxygen limited at higher glucose levels. Response experiments showed sensitivity over the range of 10.8–360 mg/dL glucose, with a response time of approximately 20 s. In addition, the authors found a dramatic difference in step response properties of the sensing arrays with varying GOx-HEMA layer thicknesses, as expected for flux-based sensors. However, it is unclear whether this change in response is generated by the presence of a thicker diffusion barrier or the increase in enzyme concentration associated with the thicker GOx-HEMA layer. The implications of this study are profound for enzymatic fluorescence-based sensors, as they depict how enzyme concentration and/or membrane thickness, along with bulk oxygen levels, could drastically affect sensor response properties.

Additional significant work has been performed to examine the dependence of response properties on enzyme concentration and sensor size.[55,131,132] In this work, optical fibers with tip diameters ranging from 0.1–50 μm were prepared with a polyacrylamide coating containing $Ru(Ph_2phen)_3^{2+}$ and varying concentrations of GOx. It was found that an optimized response was observed when a 1:50 dilution of GOx to polyacrylamide was used during sensor fabrication, and an increase in response was observed until a 1:100 dilution was obtained. In this region of operation, glucose oxidation is kinetically limited, therefore the glucose and oxygen consumptions are directly proportional to the concentration of GOx within the polymer matrix. From 1:100 to 1:25, the reaction kinetics are governed by mass transport of analytes from the bulk into the sensor; therefore, an increasing concentration of enzyme has no effect on response properties since the reaction is now *diffusion limited.* The authors explain that at higher GOx concentrations (>1:25), the buffering capacity of the sensor is overcome, resulting in a local decrease of pH, which in turn decreases enzymatic activity. Another plausible explanation would be that the high levels of enzyme rapidly consume glucose on the outer boundary of the sensor, resulting in an insufficient decrease in oxygen levels within the sensors and, subsequently, a loss of sensitivity. A separate report showed a similar trend when the effect of enzyme concentration on device sensitivity within spherical sensors was examined.[53]

A similar experiment was performed using fiber diameter as the independent variable. In this experiment, fibers with diameters ranging from 0.1 to 50 μm were exposed to 90 mg/dL glucose and the response allowed to equilibrate. In this case, the results indicate that the oxygen response of the sensors is independent of diameter; however, clearly, the glucose response strongly depends on sensor diameter. Approximately, 20% increase in intensity is observed when the fiber diameter is decreased from 5 to 1 μm. The intensity increase was attributed to an increase in enzymatic activity per GOx molecule; for sensor diameters >10 μm, the authors state that enzymes trapped within the inner core of the sensors become "active" only after enzyme molecules toward the sensors' surface become deactivated.

Particle-based enzymatic glucose assays have also been developed, including both microspheres and nanoparticles, using similar principles to those described above, where an oxygen-quenched luminescent dye and GOx are entrapped within a micro/nanoscale container. As glucose diffuses into the sensor, local (internal) oxygen levels are proportionally reduced through GOx-initiated catalysis. The glucose-dependent oxygen levels are then relayed through the fluorescence emission of the oxygen reporter, providing an indirect means to monitor glucose levels.[17] Although the integration of GOx into the sensing scheme provides innate selectivity to the sensor as well as reversibility, controlling the reaction–diffusion kinetics of both glucose and oxygen is the key to the realization of optimized sensors.[133]

The first to report on preparing nanoscale spherical particles comprising GOx and O_2 indicators for use in intracellular monitoring of glucose was a variant of PEBBLE[134,135] (probes encapsulated by biologically localized embedding) technology, using nanoparticles prepared through a microemulsion polymerization technique.[136] The polyacrylamide-based PEBBLEs are polymerized in the presence of GOx, a sulfonated version of $Ru(Ph_2phen)_3^{2+}$ and an oxygen-insensitive fluorophore, resulting in sensors with a mean diameter of 45 nm. The sensors showed a dynamic range of 5.4–144 mg/dL, with a linear deviation occurring at glucose levels greater than 90 mg/dL, and exhibited approximately 100% change in ratiometric intensity from glucose depleted to saturated conditions and an approximate response time of 100 s; however, sensitivity to local oxygen was not discussed.

A significant advancement toward the development of an enzyme/oxygen indicator transduction scheme for glucose monitoring was the use of calcium alginate hydrogel microspheres (~50 μm particles) with specialized coatings.[52] These particles, doped with GOx and ruthenium-tris(4,7-diphenyl-1,10-phenanthroline) dichloride (Ru(dpp)), were coated with nanofilms deposited via the layer-by-layer self-assembly technique to control substrate mass transport properties. These particles exhibited a highly reversible and linear response to glucose over 0–600 mg/dL range.[53,137,138] While the maximum 10% change over the target range left something to be desired in terms of sensitivity, a key finding in this work was that the physical properties of the nanofilm coatings may be altered to "tune" the analyte transport properties, resulting in control of overall response characteristics such as sensitivity and linear range.[53]

The potential for this technology to succeed has been improved by the combination of the alginate with glycidyl silane, resulting in mesoporous microspheres.[139] Using this novel material combined with coimmobilized Pt(II) complexes and GOx, and transport-limiting nanofilms on the surface, tremendous increase in sensitivity to glucose was observed (in some cases, two orders of magnitude, relative to previous systems),[52,140] as well as excellent photostability and reversibility. More important, it was shown that by simply varying the surface-adsorbed nanofilm thickness by as little as 12 nm, the relative delivery of substrates into the sensor could be controlled, allowing the sensitivity (change in intensity ratio) to be tuned from 1–5%/mg/dL and upper range limits of 90–250 mg/dL. Although this work illustrated the ability to tune device sensitivity by altering substrate diffusivities, the measurement range fell far short of the desired 600 mg/dL, so additional experimental work is required to confirm the theoretical prediction that this range can be covered. Nonetheless, this work further support the idea that enzymatic sensors can work at the microscale.

As a final comment on fluorescent transduction methods, it is important to note that most of the work discussed above can be characterized as proof-of-concept research and almost exclusively employed steady-state fluorimetric analysis. Some potential general problems with intensity-based measurements include changes in optical path length, fluctuations in sampled fluorophore concentration, and photobleaching. Incorporation of reference fluorophores for ratiometric monitoring enables compensation for most of these artifacts, though differential photobleaching rates of indicator and reference emitter will still result in drifting calibration due to bleaching. Future development of these techniques for *in vivo* glucose transduction could likewise employ fluorescence lifetime analysis. Fluorescence lifetime—the average time at which the fluorophore remains in the excited state before returning to the ground state—is independent of fluorophore concentration and light scattering and absorption properties of the sample, making this measurement approach particularly well suited for *in vivo* monitoring.[95] This point was clearly demonstrated for a RET assay by experimental studies that monitored the decreased decay time of a donor fluorophore linked to Con A upon binding of acceptor-labeled α-D-mannoside.[141] Upon introduction of glucose into the system, the displacement of the labeled sugars results in a decrease in energy transfer and concomitant increase in the donor decay time. Results indicated that this assay scheme could be used with various donor–acceptor pairs, demonstrating the robustness and generality of this approach. Additional studies were performed on similar systems based on long lifetime fluorophores such as ruthenium complexes, reducing the need for high-frequency light modulation for phase lifetime measurements.[44,142] Instrumentation for lifetime analysis is more complicated and costly; therefore, ratiometric analysis with low-cost detectors and electronics is preferred for widespread adoption. However, advances in photonic technology have resulted in small-footprint commercially available phase fluorimeters (TauTheta, Inc.) and enabled even nanosecond lifetimes to be measured with relatively low-cost instrumentation;[41] thus, low-cost lifetime-based consumer products dedicated to glucose monitoring may be possible in the near future.

10.4 PACKAGING FOR *IN VIVO* USE

Putting sensing elements together in a format that is appropriate for *in vivo* use while retaining the ability to address the assay with light is a significant challenge to implementation of fluorescence-based sensing strategies. A number of candidate systems have been proposed and developed to varying degrees, from fluorescent contact lenses to tiny fiber-optic probes to injectable micro/nanoparticles and even fully implantable electro-optical systems. The following sections will discuss the progress and current state of the art in each of these areas and related key technical obstacles to success, concluding with a description of concepts for commercial products that are at various stages of development.

10.4.1 Contact Lenses

Contact lenses are attractive formats for fluorescence-based sensing, as they are optically transparent, easily accessible, and are sufficiently inexpensive to be disposable. A handheld device for patients to self-test has been developed to excite and measure fluorescence from lenses (Figure 10.9) and some limited clinical trials have been performed.[143,144] However, the efficacy of using tear fluid to manage diabetes

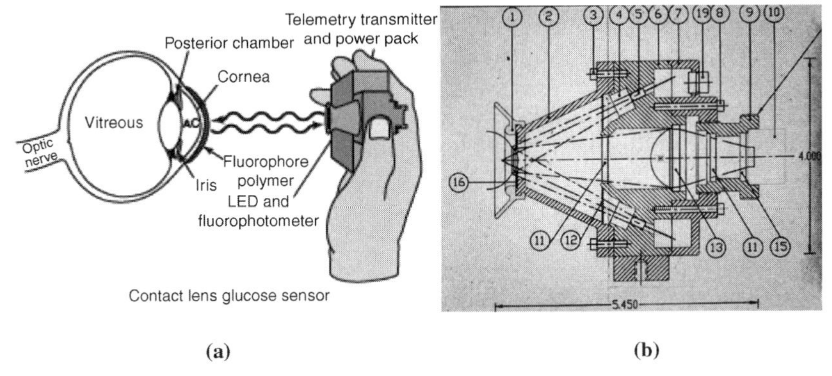

(a) (b)

Figure 10.9 (a) Diagram of the contact lens glucose sensor with a handheld photofluorometer. Excitation light is emitted by an LED, producing fluorescent light in the contact lens, which is sensed by the photofluorometer when it is brought to the eye. (b) Schematic of the handheld photofluorometer: (1) ocular to fit against the orbital rim, isolating the eye from ambient light; (2) injection-molded housing for the ocular; (3) set screw to adjust alignment of the ocular; (4) lens to focus Cree blue LED array on the eye; (5) blue (488 nm) LED circular array; (6) injection-molded main body housing; (7) injection-molded polyurethane back housing; (8) set screw to adjust alignment of the back housing; (9) rear housing for the sensor diode, battery, and integrated circuits; (10) sensor diode and rechargeable battery power supply; (11) 500 nm high pass barrier filters; (12) slit to direct blue LED on the eye; (13) integrated circuit for signal averaging and locking; (14) red central fixation LED to stabilize eye movement (not shown); (15) integrated circuit amplifiers; (16) slit at the eye to restrict blue light to the cornea and reduce autofluorescence; (17) telemetry transmitter (optional and not shown); (18) telemetry receiver at the insulin pump (optional and not shown); and (19) digital readout. Reprinted with permission from Ref. 143. Copyright 2006 Mary Ann Liebert, Inc.

has not been fully evaluated—the expected transport delays even under normal conditions will likely make implementation impractical, and uncontrollable variations such as eye hydration may significantly affect measurements. Still, a few examples of pursuit of this goal can be discussed.

Based on the ability to synthesize polymers incorporating boronic acid moieties, there is potential to directly produce glucose binding function into polymeric structures. As a result, the feasibility of creating contact lenses for continuous monitoring of glucose levels in tears has recently been investigated. Initial work has shown that boronic acid derivatives may be incorporated into an over-the-counter contact lenses made of poly(vinyl alcohol) hydrogels by diffusion-facilitated loading.[145] Lenses loaded with stilbene, polyene, and chalcone derivatives of BA were studied; while saccharide sensitivity was observed, the responses were drastically different from those obtained during solution-phase testing, attributed to pH differences between the testing environments. A boronic acid derivative containing various quinolinium and possessing a lower pK_a exhibited clear saccharide sensitivity.

The Con A–dextran competitive binding system was also implemented in contact lenses.[143,144] Using the same glucose assay described previously,[146,147] the sensing reagents were added to the contact lens precursor materials prior to polymerization. Fasting patients were given an oral glucose challenge and contact lens fluorescence was measured in parallel with venous blood glucose for a 3 h period. In some patients, contact lens fluorescence appeared to track blood glucose well while in others there was substantial divergence in the profiles. In the second report (one patient), fluorescence signals tracked blood values very well, except the ~20 min lag in both rising and falling values.[143] While these results are promising, no validation has been performed to show accuracy. A major concern with these particular lenses is the stability of the dextran in the hydrogel.

10.4.2 Disconnected Transducers

For *in vivo* minimally invasive monitoring, a system with a decoupled sensor and excitation/collection apparatus is the most ideal embodiment. Examples of this concept include fluorescent microspheres, hollow fiber membranes encapsulating sensing chemistry, and polymeric slabs doped with fluorescent reagents; in each case, the intention is to place the indicators in or beneath the skin with a simple surgical procedure, then communicate with the implants noninvasively with light using an external device (Figure 10.1).[138] In principle, the carriers are designed such that small molecular weight molecules like glucose are able to freely diffuse into the container, while permanently compartmentalizing the sensing assay, typically comprising larger macromolecules. The implants freely interact with interstitial fluid and, upon transdermal interrogation, emit spectra that are collected and analyzed to reveal interstitial glucose levels to the patient in a continuous or spot fashion.

These sensors analyze glucose levels in the interstitial fluid; therefore, before additional discussion, the relationship between glucose concentration in the interstitial fluid and blood should be addressed. Traditional glucose monitoring techniques and subsequent therapies are based on blood measurements, thus the relevance of interstitial measurements depends on the existence of a correlation between the two fluids. Under normal physiological conditions, changes in the glucose levels in the interstitial fluid and blood are highly correlated, with the exception of a lag time (~5–10 min).[148–152] In addition, all currently available FDA-approved minimally invasive devices extract glucose levels through sampling of the interstitial fluid; thus, for this chapter, glucose measurements by means of analyzing interstitial fluid will be considered an accurate method by which to manage diabetic therapies.

Numerous reports have demonstrated the ability to prepare functional glucose sensors by immobilizing GOx and an O_2 indicator within various species of gel slabs and use external excitation/collection to monitor sensor kinetics.[153] These slabs are relatively simple to prepare, although the compatibility of the reagents with the encapsulating material must be fully considered to ensure homogeneous distribution and stability, as well as appropriate diffusion rates for glucose and any cosubstrates or reaction products. In addition, as noted above, a number of different microsphere/nanoparticle formats have been proposed and investigated for use as glucose sensors. As the first example of this approach, a Con A–dextran competitive binding system with FRET transduction was shown to retain functionality while encapsulated within poly(ethylene glycol) particles.[147] In this work, TRITC-Con A and FITC-dextran were entrapped within poly(ethylene glycol) millispheres, where the addition of glucose resulted in FITC-dextran displacement and alterations in RET efficiency.

However, leaching of assay components, as well as long response times to changes in glucose, proved to be significant problems. This work represented the first reported attempt at encapsulating glucose-sensing reagents in a biocompatible hydrogel microsphere format.

Substantial additional work has followed, with various types of microsphere and microparticle encapsulations being demonstrated for both competitive binding and enzymatic approaches. Microsphere-based enzymatic sensors have been developed using various immobilizing matrices, ranging from alginate particles[54,140] to mesoporous hybrid polymer–silica microspheres.[139] The distinct advantages of using microscale particles for enzymatic sensing have been described,[53] and ongoing work in optimizing response sensitivity and range by controlling the size and material properties of the immobilization matrix is expected to yield information essential to getting these systems implant ready.

A novel approach to encapsulating competitive binding assays involved the use of nanoengineered microcapsule carriers.[64,154] A distinguishing feature of this work is the encapsulation of the sensing assay within *hollow* microscale capsules,[155] thereby allowing the free movement of the ligand, receptor, and analyte within the capsule during competitive displacement. To prepare the capsules, degradable inorganic microspheres were used as sacrificial templates onto which alternating layers of polyelectrolytes were adsorbed using the layer-by-layer self-assembly technique, followed by dissolution of the solid core.[137,156,157] The hollow capsules were loaded with the assay using simple diffusion, UV radiation was used to cross-link the polymers in the capsule walls, permanently entrapping the assay within the capsule. A fivefold increase in sensitivity over previously reported competitive binding systems, as well as high reversibility, was found.[64] In addition, a follow-up work using NIR-emitting fluorophores demonstrated the generality of the approach.[154]

10.4.2.1 Transdermal Monitoring Methods employing optical techniques in the direct transdermal probing of glucose levels are considered truly noninvasive because methods do not involve the harvesting of body fluids to determine glucose levels. These technologies involve the direct interrogation of skin tissue with light and detect the subsequent changes in absorption, transmission, reflection, and scattering properties to determine glucose levels. Minimally invasive methods would involve insertion of implants that are passive in terms of the host response (foreign body reaction) to their presence, and do not require a physical connection to the "outside world." Finally, invasive devices are those that involve a constant breach of the skin barrier or otherwise provide a pathway from the outside to the inside of the body, such as an optical fiber. In the latter cases, the fiber-optic cables serve as the light conduit and, therefore, represent the signal pathway between the chemo-optical transducer at the probe tip and the optical instrument. In contrast, communication between the optical instrumentation and the implanted transducers must occur by excitation and emission light traversing tissue; to better understand this interaction, a brief overview of skin optics along with how glucose affects optical characteristics is warranted.

Skin is a complex tissue having many nonhomogeneously distributed components that absorb and scatter light in a complex fashion, resulting in removal of photons by conversion to heat or redirection. Absorption occurs when the energy of the propagating photon is taken in by another molecule, resulting in an overall attenuation of intensity best described using the Beer–Lambert law. Scattering occurs when a propagating photon is forced to deviate from its trajectory by one or more optical nonuniformities (variations in refractive index) within the medium, and the anisotropy factor is used to determine the directional probability of the photon's new trajectory. The absorption (μ_a) and scattering (μ_s) coefficients, as well as the anisotropy factor, describe light propagation through a medium—skin, in this case. Early work in the field of tissue optics observed a pronounced decrease in the scattering and absorption coefficients of the various skin layers with wavelengths above 600 nm (Figure 10.10).[23] The term "optical window" was subsequently coined as the wavelength region between 600 and 1300 nm, which allows increased light propagation by avoiding significant scattering and absorbing from tissue components (<600 nm), while avoiding increased absorbance by water (>1300 nm).

It is noteworthy that changes in scattering and absorption properties correlated with glucose levels have been observed. In principle, glucose-induced changes in absorption properties are due to water displacement in the interstitium as glucose levels increase, while scattering properties are altered through refractive index changes from alterations in interstitial glucose levels.[158] However, the normal variations in physiological conditions such as skin temperature, hydration state, and even blood flow have been reported to affect the skin's optical properties, presenting a great challenge for the realization of reliable optical measurents.[159]

As light enters the skin, approximately 4% of the incident light is reflected back as a result of the refractive index mismatch between the air and the epidermis.[23] The remaining 96% enters the tissue, where it is attenuated due to successive

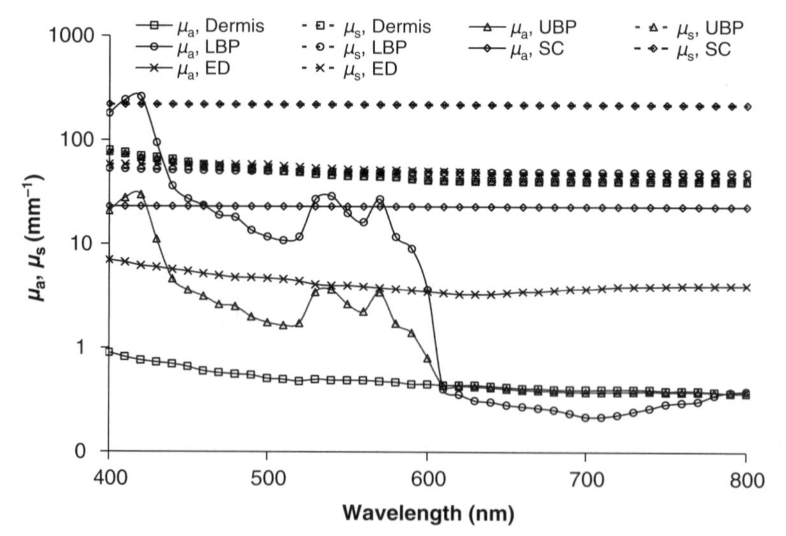

Figure 10.10 Scattering (μ_s) and absorption (μ_a) coefficients for the stratum corneum, epidermis, dermis, and blood-perfused dermis regions. SC, stratum corneum; ED, epidermis; UBP, upper blood plexus; LBP, lower blood plexus.[23]

absorption and scattering events, which contribute to diffusing the incident beam. The scattering and absorption properties of skin highly depend upon the wavelength of the incident light, such that light at UV (<300 nm) and IR (>1000 nm) wavelengths are strongly absorbed, while wavelengths in between tend to be strongly scattered (Figure 10.10).[23]

Tissue also contains some endogenous species that exhibit fluorescence, such as aromatic amino acids present in proteins (phenylalanine, tyrosine, and tryptophan), pyridine nucleotide enzyme cofactors (e.g., oxidized nicotinamide adenine dinucleotide, NADH; pyridoxal phosphate; flavin adenine dinucleotide, FAD), and cross-links between the collagen and the elastin in extracellular matrix.[100] These typically possess excitation maxima in the ultraviolet, short natural lifetimes, and low quantum yields (see Table 10.1 for examples), but their characteristics strongly depend on whether they are bound to proteins. Excitation of these molecules would elicit background emission that would contaminate the emission due to implanted sensors, resulting in baseline offsets or even major spectral shifts in extreme cases; therefore, it is necessary to carefully select fluorophores for implants. It is also noteworthy that the lifetimes are fairly short, such that use of longer lifetime emitters in sensors would allow lifetime-resolved measurements to extract sensor emission from overriding tissue fluorescence.

Based on these considerations, long wavelength fluorophores are generally preferred for implanted sensor applications. A recent report attempted to quantitatively compare the efficiency of optical interrogation of an implant containing luminescent materials.[160] In this report, an average attenuation constant (based on absorption of blood) was used in combination with the Beer–Lambert equation to calculate a relative figure of merit (FOM) for a series of fluorescent materials, from organic fluorophores to semiconductor quantum dots and carbon nanotubes. The FOM represents a relative intensity expected for the same input intensity and depth of implantation. The equation $\text{FOM} = \phi e^{-2\mu d}$ is intended to enable comparison of different fluorophores based on their differential quantum yield (ϕ) and the tissue attenuation properties at the excitation wavelength (μ). The values presented for

TABLE 10.1 Endogenous Fluorophores Found in Skin and Their Characteristics

Molecule	QY (%)	γ_{ex} (nm)	Extinction ($M^{-1}cm^{-1}$)	γ_{em} (nm)	Lifetime (ns)
Phenylalanine	2	260	~200	282	6.8
Tyrosine	13	275	~1400	304	3.6
Tryptophan	12	295	~5600	353	3.1
NADH	2	340	~6000	460	0.4–5[a]
Pyridoxal phosphate	10	330	~8000	400–535[b]	1–2
FAD	30	450	~1100	525	2.3
Collagen/elastin	10–40	333	0.3–1.5[c]	405	5.3/2.3

[a] The lifetime of NADH depends upon binding to proteins.
[b] The emission maximum of pyridoxal phosphate is strongly pH sensitive.
[c] Collagen/elastin extinction is presented as an effective absorption coefficient (cm^{-1}) at tissue concentrations.

TABLE 10.2 Comparison of Fluorescent Materials with Potential for Use in Fluorescent Biosensors

Luminescent material	QY (%)	γ_{ex} (nm)	Extinction ($M^{-1} cm^{-1}$)	$\mu_{ave(blood)}$ (cm^{-1})	FOM @ 1 cm/1 mm
Cy5[161]	27	620	250,000	31	$3.2 \times 10^{-28}/6.3 \times 10^{-4}$
Fluorescein[95]	95	496	75,000	135	$5.2 \times 10^{-118}/3.3 \times 10^{-12}$
Rhodamine 6G[162]	95	488	90,000	152.5	$3.3 \times 10^{-133}/1.1 \times 10^{-13}$
Rhodamine B[162]	31	514	90,000	150	$1.6 \times 10^{-131}/5.8 \times 10^{-14}$
Indocyanine green[163]	0.266	820	50,000	0.86	$4.7 \times 10^{-4}/2.2 \times 10^{-3}$
Indocyanine green[163]	1.14	830	50,000	0.8944	$1.9 \times 10^{-3}/9.6 \times 10^{-3}$
NIR QD[164]	13	840	580,000	0.9140	$2.1 \times 10^{-2}/1.1 \times 10^{-1}$
SWNT[165]	0.1	1042	500	0.5045	$3.6 \times 10^{-4}/9.1 \times 10^{-4}$
PtOEP[166]	50	530	75,000	300	$1.3 \times 10^{-261}/1.7 \times 10^{-26}$
PtOEP-K[166]	10	591	150,000	100	$1.4 \times 10^{-88}/3.3 \times 10^{-10}$

FOM in the paper, calculated at 1 cm depth, suggest the advantage of carbon nanotubes (CNT) and quantum dot (QDOT) systems over all organic materials, though indocyanine green (ICG) also exhibited reasonable properties.

The values for FOM have been recalculated here for purposes of comparison (Table 10.2), with the addition of two fluorophores currently gaining interest as ingredients for glucose sensors: the oxygen-quenched metalloporphyrins platinum octaethylporphyrin (PtOEP) and its ketone derivative (PtOEP-K). Note that extinction coefficients were obtained from several sources as indicated. The same FOM was also calculated for implants positioned at a depth of 1 mm, as shown in the last column of Table 10.2, and it should be immediately obvious that the depth of placement plays a critical role in determining the relative optical properties for the implants. At more superficial positions, for example, Cy5 holds an advantage over the single-walled carbon nanotubes (SWNT) because of the comparatively high quantum yield; the quantum dots are even more preferable at shallower depths for the same reason. This simple analysis serves to point out that the trade-off between the quantum yield and the excitation source penetration must be carefully considered.

Use of the simple Beer–Lambert equation as the relevant descriptor assumes that absorption is dominant, which is certainly not true, particularly at shorter wavelengths. However, even if this assumption is considered valid, it is immediately clear that other key parameters must be incorporated for fair comparison. First, blood is the dominant absorber only for certain regions of the spectrum, and blood cannot be said to occupy a 1 cm path length in skin, as blood content varies with depth and site. A second, and very important, consideration is the depth of implantation; indeed, skin is generally much thinner than 1 cm and the use of fluorescent implants has been proposed for the dermal layer, approximately 200–500 μm deep into the skin. This would place the reagents closer to the vascular bed and, in principle, also result in lesser transport-related differences between interstitial fluid and blood glucose levels. Finally, while the equation for FOM accounts for light attenuation along the two-way optical path between the tissue surface and the implant, it does not accurately depict the dependence of output intensity on the absorbing power (extinction) of the fluorophore, which along with the quantum yield determines the efficiency of converting excitation photons to luminescent photons.

To provide a clearer picture of the trade-offs between fluorophores, we can more accurately model the result using equations for (1) the light delivered to the sample (I_{d1}); (2) the amount of light absorbed by the sample (I_A); (3) the amount of absorbed light converted to fluorescence (I_f); and (4) the amount of fluorescence generated to reach the tissue surface (I_{fs}). First, the intensity delivered to the fluorophore placed at a depth d_1 is $I_{d_1} = I_0 e^{-\mu d_1}$. Second, the amount of that light absorbed by a bed of fluorophore with thickness d_2 is $I_A = I_{d_1}(1-e^{-cd_2\varepsilon})$, where c is the molar concentration and ε is the molar extinction coefficient (M/cm). Third, the fraction of this absorbed energy converted to fluorescence is equal to the quantum yield; therefore, the fluorescence intensity is found to be $I_f = \phi I_A$. Finally, assuming $d_2 \ll d_1$ and the optical properties remain constant despite the wavelength shift, the fluorescence intensity escaping from the tissue surface (I_{fs}) is equal to the emitted intensity (I_f) attenuated according to the Beer–Lambert relationship $I_{fs} = I_f\, e^{-\mu d_1}$. Thus, we can write the final equation relating the output fluorescence to the input as $I_{fs} = I_0\, e^{-\mu d_1}(1-e^{-cd_2\varepsilon})\phi e^{-\mu d_1}$, which simplifies to the relative intensity of $\text{FOM} = (I_{fs}/I_0) = \phi\, e^{-2\mu d_1}(1-e^{-cd_2\varepsilon})$. This should be a useful and more accurate FOM for comparing fluorescent materials intended for use in turbid media.

Using this expression for the FOM and the values presented in Table 10.1 for material properties, and assuming equal implant thickness of 200 μm and fluorophore concentration of 1 nM, the new FOM was calculated for depths up to 1 cm. Note that the concentration of 1 nM was chosen arbitrarily as a realistically low concentration; all calculations scale equally and, therefore, the comparisons are valid regardless of the actual concentration used. When the new FOM is calculated as a function of depth (Figure 10.11), an interesting and very important relationship is uncovered. At superficial depths, extinction power and quantum yield are dominant in determining the FOM, whereas the wavelength-dependent attenuation properties of the tissue become more important when implants are located at deeper positions. According to these estimates, SWNT possess no FOM advantage over any of the other materials when used at depths less than 200 μm; at depths greater than 1 mm, Cy5 and ICG still return more photons, and ICG continues to return more even above 1 cm. This advantage is more obvious when the extinction is properly considered, compared to the previous calculation where the FOM for Cy5 was lower than SWNT by 24 orders of magnitude. In summary, this simple model indicates that organic dyes are superior for superficial applications due to their high extinction and quantum yield. At this level of simulation, we can observe that as implant depths go beyond 1 mm, quantum dots and nanotubes begin to be more attractive.

It is worth noting that the above comparisons did not consider photodegradation. It is accurate to note that the inorganic emitters generally have superior stability over organic materials, and are attractive options in this regard; however, it is difficult to draw any fair comparisons of stability without specifying concentration- and intensity-normalized photodecomposition rate constants, and since these are not readily available for the luminophores discussed, this topic will not be discussed further herein.

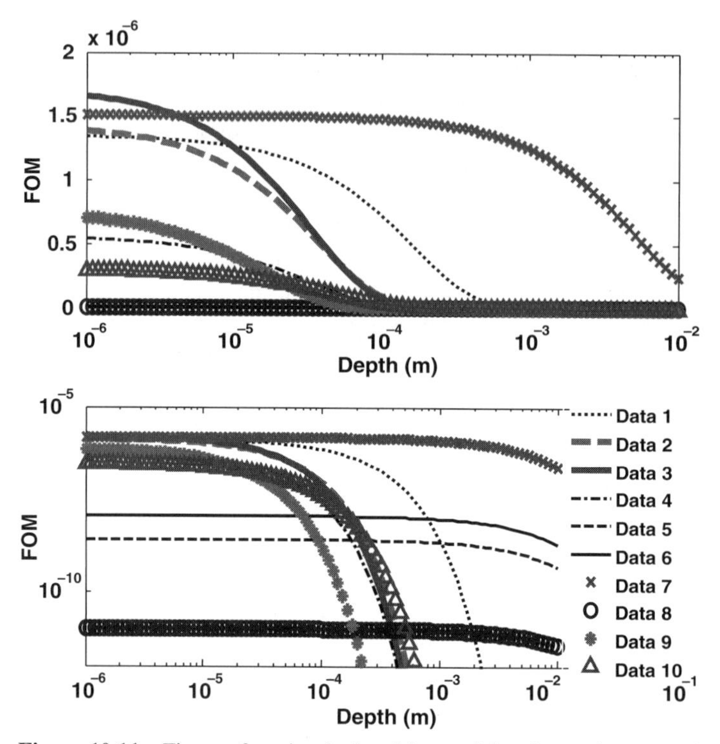

Figure 10.11 Figure of merit calculated for candidate fluorophores as a function of implantation depth. (a) Showing FOM on a linear *y*-scale. (b) Showing the same data on a log scale.

While such an analysis is instructive, it is also important to note that there are some significant shortcomings in using this approach to comparing materials for *in vivo* optical transduction. Specifically, scattering effects on light distribution in tissue, which are typically dominant for wavelengths below 600 nm, require modeling with more advanced techniques such as the diffusion approximation or Monte Carlo simulations. These methods have also been employed to assess the potential for glucose monitoring using fluorescent implants. For example, a Monte Carlo software package was developed to simulate photon propagation through skin tissue.[24] Simulations were used to track excitation photons as they propagate through the skin and, following the generation of a fluorescence event, track the propagation of the fluorescence photons throughout the tissue and escaping at the surface. Ultimately, a spatial distribution of sensor fluorescence at the skin surface was produced, which was able to aid in the design of an excitation/collection apparatus for implanted sensors as well as compare sensors with different optical properties. Simulation results pointed out that several factors ultimately dictated transdermal excitation/collection efficiency: (1) geometrical properties of the sensors and/or implanted sensor population, (2) depth at which the sensors were implanted, and (3) the separation distance between the excitation and collection optics.

The efficacy of exciting and collecting emission data from PEG hydrogel sensors implanted in an *in vivo* rat model was also assessed experimentally.[115]

Transdermal emission spectra were successfully collected from sensors implanted approximately 500 μm below the skin surface. In addition, a bolus injection of glucose into the tail vein resulted in measurable spectral changes from the implants. This important developmental step proved some feasibility of the smart tattoo concept as a means of monitoring glucose in a minimally invasive manner.

10.4.2.2 Tissue Oxygen The above comparisons of performance versus depth were based purely on optical considerations. A related concern for *in vivo* fluorescence-based glucose sensing, for enzymatic transduction that exhibits oxygen sensitivity, is the concentration of oxygen found in the local tissue. A brief description of the microcirculation within the skin is relevant to consider this point.[167] Blood flow is supplied to the skin through two plexuses: the upper and lower horizontal plexuses. The upper plexus is approximately 1–2 μm below the epidermis, while the lower horizontal plexus is at the dermal–subcutaneous junction. Ascending arterioles and descending venules are paired because they connect the two plexuses. Ascending from the upper plexus is a series of arterial capillaries that form the dermal papillary loops, which represent the main supply of nutrients to the skin. Sphincter-like smooth muscle cells are at the capillary–arteriole junction of the upper horizontal plexus, serving as a means to regulate the microcirculation. At the dermal–subcutaneous junction, the upper plexus directly connects to the lower plexus, which supplies nutrients to other structures in the skin such as hair follicles or sweat glands.

Oxygen is supplied to the dermal tissue through uptake from the atmosphere, as well as through oxygen transported through the capillaries.[168] Recent work has shown the contribution of oxygen from the atmosphere and the blood supply highly depends upon depth within the dermis. In this work, the authors produced a depth-dependent profile of intracutaneous oxygen levels assuming 90 μM and 0 μM (vasculature occlusion) oxygen within the hematogenic supply and 277 μM (air-equilibrated oxygen levels) oxygen at the air–skin interface (Figure 10.12).

By simulating the oxygen profile in the case of an occlusion within the vasculature, the authors determined that tissue as deep as 400 μm receives a portion of its oxygen supply from the atmosphere (Figure 10.12, trace B). However, at depths greater than 230 μm, a substantial portion of the oxygen is supplied by the vasculature. It is additionally important to note that oxygen levels within the dermis are highly depth dependent, such that oxygen levels at the epidermal–dermal junction are approximately 215 μM, in contrast to the oxygen levels at the papillary–reticular dermis junction, which are approximately 115 μM (overall, approximately 50% variation). For any sensors relying upon indirect measurements through oxygen transduction, bulk oxygen levels must be considered when selecting the appropriate oxygen probe as well as when analyzing sensor performance (local oxygen levels ultimately determine when the devices become oxygen limited).

10.4.3 Fiber-Optic Probes (Minimally Invasive)

Fiber probes represent a simpler challenge from the standpoint of pure sensor development than disconnected systems, in that reagent chemistry need only to be immobilized at the tip of an optical fiber. Probes can be inserted into samples,

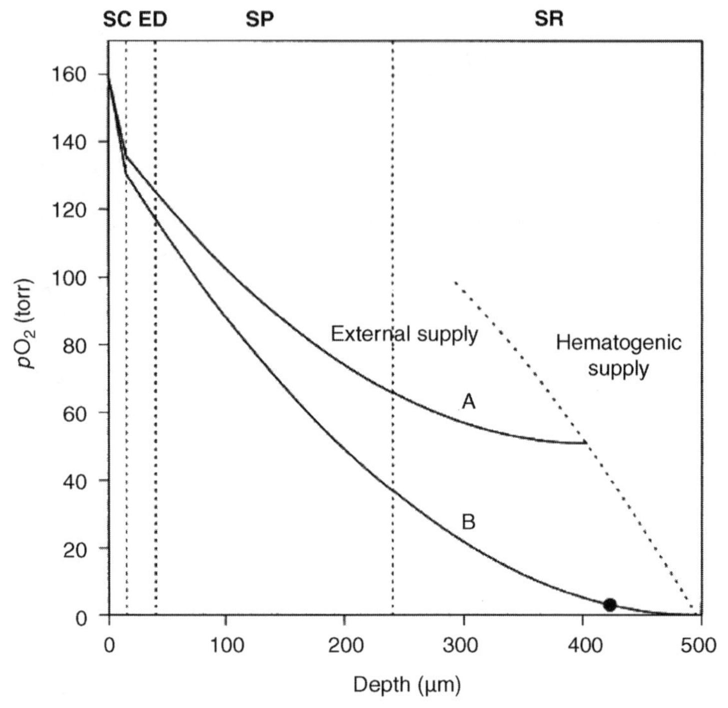

Figure 10.12 Theoretical estimation of the intracutaneous PO_2 profile. SC, stratum corneum; ED, epidermis; SP, papillary dermis; SR, reticular dermis. Reprinted with permission from Ref. 168. Copyright 2002 Blackwell Publishing.

including tissue, and removed as needed; all optical instrumentation may be designed to match the fiber. Fiber optics can also be easily deployed through catheters. However, the biomedical considerations of infection pathways presented by transcutaneous lines make this approach less attractive than fully implanted/disconnected systems for situations requiring constant measurements.

In many cases, new sensing chemistry concepts are validated with fiber-optic systems first before disconnected systems are attempted. In analogy to electrochemical sensors, initial work in this field was done using optical fibers with sensor components (enzymes, fluorescent pH or oxygen indicators) immobilized in transport limiting matrices.[129,131,169,180] Recent work has also demonstrated glucose sensors using boronic acid indicators trapped in hydrogels attached to fibers, with very positive results.[170]

10.5 ONGOING COMMERCIALIZATION ACTIVITIES

10.5.1 Sensors for Medicine and Science, Inc. (SMSI)

SMSI (Montgomery County, MD; www.s4ms.com/) is developing a fully implantable system that integrates fluorescence glucose-sensing chemistry, light source, detector, and electronics into a single package that will be placed under the skin using a large-bore needle called a trocar. While the exact chemistry is not disclosed, it is claimed to be nonconsuming and is, therefore, inferred to be based on affinity binding with either protein or synthetic receptors. The device will be powered by an external reader, which also acts as the receiver for measurement values transmitted via RF. This technology is apparently still in the testing phase. Using telemetry, the implant will communicate its glucose measurements to an external reader, without the need to draw blood.

10.5.2 GluMetrics

GluMetrics (Irvine, CA; www.glumetrics.com) is developing glucose sensors based on fluorescence that use a boronic acid glucose indicator. The first device, GlutCath, is intended to be used for continuous monitoring of hospitalized patients. As of August 2009, the product had been tested successfully in five patients but was still under development. It is a disposable catheter device that uses a fiber-optic cable incorporating a polymer containing a fluorescence assay using boronic acid as the receptor; the intended deployment is in peripheral veins, providing up to 48 h continuous analysis. They claim that the next development will comprise the sensing component of a closed-loop system.[171]

10.5.3 PreciSense, Inc.

PreciSense (PreciSense A/S, Hørsholm, Denmark; http://www.precisense.dk/) is developing a device that includes bioresorbable fluorescent microcapsules, a microcapsule placement unit, and a light detecting noninvasive reader unit. The glucose assay is based on FRET, using a "human glucose receptor" and dextran as the competing ligand, with long wavelength fluorophores and lifetime-based monitoring. Although results presented at recent diabetes meetings are encouraging, no details on planned public release are available at this time.

10.5.4 BioTex, Inc.

BioTex (Houston, TX; http://www.biotexmedical.com/) is working on an implantable polymer sensor containing glucose-sensitive near-infrared fluorescent chemistry. When implanted just below the surface of the skin, this "smart tattoo" responds to local glucose levels and a simple optical device can be used to record these levels. Major research efforts are being invested to develop a "smart tattoo" sensor with a functional lifetime of up to 6 months or even longer. Their recent work has been published, and preliminary experiments demonstrate reasonable long-term stability, no toxicity issues, and the results of *in vivo* monitoring are encouraging.[38,110,114,172,173]

10.6 SUMMARY/CONCLUDING REMARKS

In summary, a wide array of different techniques for fluorescence-based transduction of glucose concentration have been explored and have demonstrated potential for use in various formats for *in vivo* sensing. While no devices have yet been made available commercially, it appears that a number of techniques have been sufficiently developed to expect that clinically viable devices are just over the horizon and more activity in trials will be observed in the next few years. While the progress in fluorescence-based glucose sensing is significant, and many *in vitro* experimental validations have shown impressive results, whether these can be translated to survive *in vivo* for a sufficient time and made with a reliable process in a format that is amenable for self-testing remains to be seen. While fiber-optic probes are most likely to be implemented for clinical monitoring in the near future, disconnected and fully implantable systems still face significant hurdles to clinical use, particularly in the areas of biocompatibility and the effects of the host on sensor performance.

It appears that fluorescence techniques are poised to receive more serious consideration for accelerated development efforts. Key obstacles remaining include stability of receptors and fluorophores, challenges that will possibly be met partially by results of the intense efforts of molecular biology, polymer science, and nanotechnology. Advances in nanomaterials such as quantum dots will likely enable improvements in optical stability and choice of excitation/emission wavelengths for various transduction methods. Stabilization of natural and artificial enzymes and rendering immunogenic protein receptors "stealthy" may also aid the pursuit.

REFERENCES

1. Weiss S. Fluorescence spectroscopy of single biomolecules. *Science* 1999, 283, 1676–1683.
2. Anderson JM. In: Ratner BD, Hoffman AS, Schoen FJ, Lemons JE (Eds), *Biomaterials Science*, Vol. 1. Academic Press, 1996, pp. 165–173.
3. Kulcu E, Tamada JA, Reach G, Potts RO, Lesho MJ. Physiological differences between interstitial glucose and blood glucose measured in human subjects. *Diabetes Care* 2003, 26, 2405–2409.
4. Bremer TM, Edelman SV, Gough DA. Benchmark data from the literature for evaluation of new glucose sensing technologies. *Diabetes Technology & Therapeutics* 2001, 3, 409–418.
5. Koschwanez HE, Reichert WM. *In vitro, in vivo* and post explantation testing of glucose-detecting biosensors: current methods and recommendations. *Biomaterials* 2007, 28, 3687–3703.
6. ThomeDuret V, Reach G, Gangnerau MN, Lemonnier F, Klein JC, Zhang YN, Hu YB, Wilson GS. Use of a subcutaneous glucose sensor to detect decreases in glucose concentration prior to observation in blood. *Analytical Chemistry* 1996, 68, 3822–3826.
7. DiCesare N, Adhikari DP, Heynekamp JJ, Heagy MD, Lakowicz JR. Spectroscopic and photophysical characterization of fluorescent chemosensors for monosaccharides based on *N*-phenylboronic acid derivatives of 1,8-naphthalimide. *Journal of Fluorescence* 2002, 12, 147–154.
8. Cao H, Diaz DI, DiCesare N, Lakowicz JR, Heagy MD. Monoboronic acid sensor that displays anomalous fluorescence sensitivity to glucose. *Organic Letters* 2002, 4, 1503–1505.

9. Parmpi P, Kofinas P. Biomimetic glucose recognition using molecularly imprinted polymer hydrogels. *Biomaterials* 2004, 25, 1969–1973.
10. Byrne ME, Park K, Peppas NA. Molecular imprinting within hydrogels. *Advanced Drug Delivery Reviews* 2002, 54, 149–161.
11. Byrne ME, Park K, Peppas NA. Biomimetic Networks for Selective Recognition of Biomolecules, Materials Research Society, San Francisco, CA, 2002, pp. 193–199.
12. Alexeev VL, Sharma AC, Goponenko AV, Das S, Lednev IK, Wilcox CS, Finegold DN, Asher SA. High ionic strength glucose-sensing photonic crystal. *Analytical Chemistry* 2003, 75, 2316–2323.
13. Alexeev VL, Das S, Finegold DN, Asher SA. Photonic crystal glucose-sensing material for noninvasive monitoring of glucose in tear fluid. *Clinical Chemistry* 2004, 50, 2353–2360.
14. Wilson R, Turner APF. Glucose oxidase an ideal enzyme. *Biosensors & Bioelectronics* 1992, 7, 165–185.
15. Schultz JS, Sims G. Affinity sensors for individual metabolites. *Biotechnology & Bioengineering Symposium* 1979, 9, 65–71.
16. Schultz JS, Mansouri S, Goldstein IJ. Affinity sensor: a new technique for developing implantable sensors for glucose and other metabolites. *Diabetes Care* 1982, 5, 245–253.
17. Moschou EA, Sharma BV, Deo SK, Daunert S. Fluorescence glucose detection: advances toward the ideal *in vivo* biosensor. *Journal of Fluorescence* 2004, 14, 535–547.
18. Badugu R, Lakowicz JR, Geddes CD. Noninvasive continuous monitoring of physiological glucose using a monosaccharide-sensing contact lens. *Analytical Chemistry* 2004, 76, 610–618.
19. Badugu R, Lakowicz JR, Geddes CD. Fluorescence sensors for monosaccharides based on the 6-methylquinolinium nucleus and boronic acid moiety: potential application to ophthalmic diagnostics. *Talanta* 2005, 65, 762–768.
20. Badugu R, Lakowicz JR, Geddes CD. A wavelength-ratiometric fluoride-sensitive probe based on the quinolinium nucleus and boronic acid moiety. *Sensors and Actuators B* 2005, 104, 103–110.
21. Chinnayelka S, McShane MJ. Resonance energy transfer nanobiosensors based on affinity binding between apo-enzyme and its substrate. *Biomacromolecules* 2004, 5, 1657–1661.
22. Anderson RR, Parrish JA. The optics of human skin. *Journal of Investigative Dermatology* 1981, 77, 13–19.
23. Van Gemert MJC, Jacques SL, Sterenborg HJCM, Star WM. Skin optics. *IEEE Transactions on Biomedical Engineering* 1989, 36, 1146–1154.
24. McShane MJ, Russell RJ, Pishko MV, Cote GL. Glucose monitoring using implanted fluorescent microspheres. *IEEE Engineering in Medicine and Biology Magazine* 2000, 19, 36–45.
25. McShane MJ, Rastegar S, Pishko M, Cote GL. Monte Carlo modeling for implantable fluorescent analyte sensors. *IEEE Transactions on Biomedical Engineering* 2000, 47, 624–632.
26. Pickup JC, Hussain F, Evans ND, Rolinski OJ, Birch DJS. Fluorescence-based glucose sensors. *Biosensors & Bioelectronics* 2005, 20, 2555–2565.
27. Munske GR, Krakauer H, Magnuson JA. Calorimetric study of carbohydrate binding to concanavalin-A. *Archives of Biochemistry and Biophysics* 1984, 233, 582–587.

28. Fogh-Andersen N, Altura BM, Altura BT, Siggaard-Andersen O. Composition of interstitial fluid. *Clinical Chemistry* 1995, 41, 1522–1525.
29. Sumner JB, Howell SF. The identification of the hemagglutinin of the jack bean with concanavalin A. *Journal of Bacteriology* 1936, 32, 227–237.
30. Goldstein L, Hollerman C, Smith EE. Protein–carbohydrate interaction. II. Inhibition studies on interaction of concanavalin A with polysaccharides. *Biochemistry* 1965, 4, 876–883.
31. So LL, Goldstein IJ. Protein–carbohydrate interaction. XX. On number of combining sites on concanavalin A phytohemagglutinin of jack bean. *Biochimica et Biophysica Acta* 1968, 165, 398–404.
32. Kim JJ, Park K. Glucose-binding property of PEGylated concanavalin A. *Pharmaceutical Research* 2001, 18, 794–799.
33. Leon MA, Powell AE. Stimulation of human lymphocytes by concanavalin A. *Journal of the Reticuloendothelial Society* 1968, 5, 581.
34. Powell AE, Leon MA. Reversible interaction of human lymphocytes with mitogen concanavalin-A. *Experimental Cell Research* 1970, 62, 315–325.
35. Yahara I, Edelman GM. Restriction of mobility of lymphocyte immunoglobulin receptors by concanavalin-A. *Proceedings of the National Academy of Sciences of the United States of America* 1972, 69, 608–612.
36. Berlin RD. Effect of concanavalin-A on phagocytosis. *Nature* 1972, 235, 44–45.
37. Shoham J, Inbar M, Sachs L. Differential toxicity on normal transformed cells *in-vitro* and inhibition of tumour development *in-vivo* by concanavalin-A. *Nature* 1970, 227, 1244–1246.
38. Ballerstadt R, Evans C, McNichols R, Gowda A. Concanavalin A for *in vivo* glucose sensing: a biotoxicity review. *Biosensors & Bioelectronics* 2006, 22, 275–284.
39. Vyas NK, Vyas MN, Quiocho FA. Sugar and signal-transducer binding-sites of the *Escherichia coli* galactose chemoreceptor protein. *Science* 1988, 242, 1290–1295.
40. Sack JS, Saper MA, Quiocho FA. Periplasmic binding protein structure and function. Refined X-ray structures of the leucine/isoleucine/valine-binding protein and its complex with leucine. *Journal of Molecular Biology* 1989, 206, 171–191.
41. Tolosa L, Gryczynski I, Eichhorn LR, Dattelbaum JD, Castellano FN, Rao G, Lakowicz JR. Glucose sensor for low-cost lifetime-based sensing using a genetically engineered protein. *Analytical Biochemistry* 1999, 267, 114–120.
42. Ge XD, Tolosa L, Rao G. Dual-labeled glucose binding protein for ratiometric measurements of glucose. *Analytical Chemistry* 2004, 76, 1403–1410.
43. Ge XD, Tolosa L, Simpson J, Rao G. Genetically engineered binding proteins as biosensors for fermentation and cell culture. *Biotechnology and Bioengineering* 2003, 84, 723–731.
44. Tolosa L, Malak H, Raob G, Lakowicz JR. Optical assay for glucose based on the luminescence decay time of the long wavelength dye Cy5 (TM). *Sensors and Actuators B* 1997, 45, 93–99.
45. Clark LC, Lyons C. Electrode systems for continuous monitoring in cardiovascular surgery. *Annals of the New York Academy of Sciences* 1962, 102, 29–45.
46. Trettnak W, Leiner MJP, Wolfbeis OS. Fibre-optic glucose sensor with a pH optrode as the transducer. *Biosensors* 1989, 4, 15–26.
47. Moreno-Bondi MC, Wolfbeis OS, Leiner MJ, Schaffar BP. Oxygen optrode for use in a fiber-optic glucose biosensor. *Analytical Chemistry* 1990, 62, 2377–2380.

48. Wolfbeis OS. Fiber optic biosensing based on molecular recognition. *Sensors and Actuators B* 1991, B5, 1–6.
49. Curey TE, Goodey A, Tsao A, Lavigne J, Sohn Y, McDevitt JT, Anslyn EV, Neikirk D, Shear JB. Characterization of multicomponent monosaccharide solutions using an enzyme-based sensor array. *Analytical Biochemistry* 2001, 293, 178–184.
50. Hussain F, Birch DJ, Pickup JC. Glucose sensing based on the intrinsic fluorescence of sol–gel immobilized yeast hexokinase. *Analytical Biochemistry* 2005, 339, 137–143.
51. Pickup JC, Hussain F, Evans ND, Sachedina N. *In vivo* glucose monitoring: the clinical reality and the promise. *Biosensors & Bioelectronics* 2005, 20, 1897–1902.
52. Brown JQ. Modeling, design, and validation of fluorescent spherical enzymatic glucose microsensors using nanoengineered polyelectrolyte coatings. Louisiana Tech University, 2005.
53. Brown JQ, McShane MJ. Modeling of spherical fluorescent glucose microsensor systems: design of enzymatic smart tattoos. *Biosensors & Bioelectronics* 2006, 21, 1760–1769.
54. Brown JQ, Srivastava R, McShane MJ. Encapsulation of glucose oxidase and an oxygen-quenched fluorophore in polyelectrolyte-coated calcium alginate microspheres as optical glucose sensor systems. *Biosensors & Bioelectronics* 2005, 21, 212–216.
55. Rosenzweig Z, Kopelman R. Development of a submicrometer optical fiber oxygen sensor. *Analytical Chemistry* 1995, 67, 2650–2654.
56. Trettnak W, Wolfbeis OS. Fully reversible fibre-optic glucose biosensor based on the intrinsic fluorescence of glucose oxidase. *Analytica Chimica Acta* 1989, 221, 195–203.
57. Sierra JF, Galban J, De Marcos S, Castillo JR. Direct determination of glucose in serum by fluorimetry using a labeled enzyme. *Analytica Chimica Acta* 2000, 414, 33–41.
58. Maity H, Maiti NC, Jarori GK. Time-resolved fluorescence of tryptophans in yeast hexokinase-PI: effect of subunit dimerization and ligand binding. *Journal of Photochemistry and Photobiology B* 2000, 55, 20–26.
59. Maity H, Kasturi SR. Interaction of bis(1-anilino-8-naphthalenesulfonate) with yeast hexokinase: a steady-state fluorescence study. *Journal of Photochemistry and Photobiology B* 1998, 47, 190–196.
60. D'Auria S, Herman P, Rossi M, Lakowicz JR. The fluorescence emission of the apo-glucose oxidase from *Aspergillus niger* as probe to estimate glucose concentrations. *Biochemical and Biophysical Research Communications* 1999, 263, 550–553.
61. D'Auria S, Di Cesare N, Gryczynski Z, Gryczynski I, Rossi M, Lakowicz JR. A thermophilic apoglucose dehydrogenase as nonconsuming glucose sensor. *Biochemical and Biophysical Research Communications* 2000, 274, 727–731.
62. Chinnayelka S. Microcapsule biosensors based on competitive binding and fluorescence resonance energy transfer assays. Louisiana Tech University, 2005.
63. Chinnayelka S, McShane MJ. Glucose-sensitive nanoassemblies comprising affinity-binding complexes trapped in fuzzy microshells. *Journal of Fluorescence* 2004, 14, 585–595.
64. Chinnayelka S, McShane MJ. Microcapsule biosensors using competitive binding resonance energy transfer assays based on apoenzymes. *Analytical Chemistry* 2005, 77, 5501–5511.
65. Osborne SE, Matsumura I, Ellington AD. Aptamers as therapeutic and diagnostic reagents: problems and prospects. *Current Opinion in Chemical Biology* 1997, 1, 5–9.

66. Potyrailo RA, Conrad RC, Ellington AD, Hieftje GM. Adapting selected nucleic acid ligands (aptamers) to biosensors. *Analytical Chemistry* 1998, 70, 3419–3425.
67. Hamaguchi N, Ellington A, Stanton M. Aptamer beacons for the direct detection of proteins. *Analytical Biochemistry* 2001, 294, 126–131.
68. Manimala JC, Wiskur SL, Ellington AD, Anslyn EV. Tuning the specificity of a synthetic receptor using a selected nucleic acid receptor. *Journal of the American Chemical Society* 2004, 126, 16515–16519.
69. DiCesare N, Lakowicz JR. Spectral properties of fluorophores combining the boronic acid group with electron donor or withdrawing groups implication in the development of fluorescence probes for saccharides. *Journal of Physical Chemistry A* 2001, 105, 6834–6840.
70. Fang H, Kaur G, Wang B. Progress in boronic acid-based fluorescent glucose sensors. *Journal of Fluorescence* 2004, 14, 481–489.
71. Yoon J, Czarnik AW. Fluorescent chemosensors of carbohydrates. A means of chemically communicating the binding of polyols in water based on chelation-enhanced quenching. *Journal of the American Chemical Society* 1992, 114, 5874–5875.
72. James TD, Samankumara KRAS, Shinkai S. Novel photoinduced electron-transfer sensor for saccharides based on the interaction of boronic acid and amine. *Chemical Communications* 1994, 4, 477–478.
73. Appleton B, Gibson TD. Detection of total sugar concentration using photoinduced electron transfer materials: development of operationally stable, reusable optical sensors. *Sensors and Actuators B* 2000, 65, 302–304.
74. James TD, Samankumara KRAS, Iguchi R, Shinkai S. Novel saccharide-photoinduced electron transfer sensors based on the interaction of boronic acid and amine. *Journal of the American Chemical Society* 1995, 117, 8982–8987.
75. Karnati VV, Gao X, Gao S, Yang W, Ni W, Sankar S, Wang B. A glucose-selective fluorescence sensor based on boronic acid–diol recognition. *Bioorganic and Medicinal Chemistry Letters* 2002, 12, 3373–3377.
76. Wizeman WJ, Kofinas P. Molecularly imprinted polymer hydrogels displaying isomerically resolved glucose binding. *Biomaterials* 2001, 22, 1485–1491.
77. Wulff G. Molecular imprinting. *Annals of the New York Academy of Sciences* 1984, 434, 327–333.
78. Wulff G. Molecular recognition in polymers prepared by imprinting with templates. *ACS Symposium Series* 1986, 308, 186–230.
79. Ramstrom O, Ansell RJ. Molecular imprinting technology: challenges and prospects for the future. *Chirality* 1998, 10, 195–209.
80. Strikovsky A, Hradil J, Wulff G. Catalytically active, molecularly imprinted polymers in bead form. *Reactive and Functional Polymers* 2003, 54, 49–61.
81. Wulff G. Molecular imprinting: a way to prepare effective mimics of natural antibodies and enzymes. *Nanoporous Materials III* 2002, 141, 35–44.
82. Wulff G, Chong BO, Kolb U. Soluble single-molecule nanogels of controlled structure as a matrix for efficient artificial enzymes. *Angewandte Chemie, International Edition* 2006, 45, 2955–2958.
83. Ko DY, Lee HJ, Jeong B. Surface-imprinted, thermosensitive, core-shell nanosphere for molecular recognition. *Macromolecular Rapid Communications* 2006, 27, 1367–1372.

84. Ersoz A, Denizli A, Ozcan A, Say R. Molecularly imprinted ligand-exchange recognition assay of glucose by quartz crystal microbalance. *Biosensors & Bioelectronics* 2005, 20, 2197–2202.
85. Ji HF, Yan X, McShane MJ. Experimental and theoretical aspects of glucose measurement using a microcantilever modified by enzyme-containing polyacrylamide. *Diabetes Technology & Therapeutics* 2005, 7, 986–995.
86. Hilt JZ, Byrne ME, Peppas NA. Microfabrication of intelligent biomimetic networks for recognition of D-glucose. *Chemistry of Materials* 2006, 18, 5869–5875.
87. Brannonpeppas L, Peppas NA. Time-dependent response of ionic polymer networks to pH and ionic-strength changes. *International Journal of Pharmaceutics* 1991, 70, 53–57.
88. Amende MT, Hariharan D, Peppas NA. Factors influencing drug and protein-transport and release from ionic hydrogels. *Reactive Polymers* 1995, 25, 127–137.
89. Bell CL, Peppas NA. Water, solute and protein diffusion in physiologically responsive hydrogels of poly(methacrylic acid-g-ethylene glycol). *Biomaterials* 1996, 17, 1203–1218.
90. Peppas NA, Bures P, Leobandung W, Ichikawa H. Hydrogels in pharmaceutical formulations. *European Journal of Pharmaceutics and Biopharmaceutics* 2000, 50, 27–46.
91. Fazal FM, Hansen DE. Glucose-specific poly(allylamine) hydrogels: a reassessment. *Bioorganic & Medicinal Chemistry Letters* 2007, 17, 235–238.
92. Lakowicz JR, Gryczynski I, Tolosa L, Dattelbaum JD, Castellano FN, Li L, Rao G. Advances in fluorescence spectroscopy: multi-photon excitation, engineered proteins, modulation sensing and microsecond rhenium metal–ligand complexes. *Acta Physica Polonica A* 1999, 95, 179–196.
93. Scognamiglio V, Staiano M, Rossi M, D'Auria S. Protein-based biosensors for diabetic patients. *Journal of Fluorescence* 2004, 14, 491–498.
94. Marvin JS, Hellinga HW. Engineering biosensors by introducing fluorescent allosteric signal transducers: construction of a novel glucose sensor. *Journal of the American Chemical Society* 1998, 120, 7–11.
95. Lakowicz JR. *Principles of Fluorescence Spectroscopy*, 2nd edn. Kluwer Academic, New York, NY, 1999.
96. Dosremedios CG, Moens PDJ. Fluorescence resonance energy-transfer spectroscopy is a reliable "ruler" for measuring structural changes in proteins. Dispelling the problem of the unknown orientation factor. *Journal of Structural Biology* 1995, 115, 175–185.
97. Miller JN. Fluorescence energy transfer methods in bioanalysis. *Analyst* 2005, 130, 265–270.
98. Wu PG, Brand L. Resonance energy-transfer: methods and applications. *Analytical Biochemistry* 1994, 218, 1–13.
99. Birch DJS, Rolinski OJ. Fluorescence resonance energy transfer sensors. *Research on Chemical Intermediates* 2001, 27, 425–446.
100. Chirio-Lebrun MC, Prats M. Fluorescence resonance energy transfer (FRET): theory and experiments. *Biochemical Education* 1998, 26, 320–323.
101. Rasnik I, Mckinney SA, Ha T. Surfaces and orientations: much to FRET about? *Accounts of Chemical Research* 2005, 38, 542–548.
102. Goto Y, Hagihara Y. Mechanism of the conformational transition of melittin. *Biochemistry* 1992, 31, 732–738.

103. Cai K, Schirch V. Structural studies on folding intermediates of serine hydroxymethyltransferase using fluorescence resonance energy transfer. *Journal of Biological Chemistry* 1996, 271, 27311–27320.
104. Chapman ER, Alexander K, Vorherr T, Carafoli E, Storm DR. Fluorescence energy-transfer analysis of calmodulin–peptide complexes. *Biochemistry* 1992, 31, 12819–12825.
105. Mansouri S, Schultz JS. A miniature optical glucose sensor based on affinity binding. *Biotechnology* 1984, 2, 885–890.
106. Medintz IL, Clapp AR, Mattoussi H, Goldman ER, Fisher B, Mauro JM. Self-assembled nanoscale biosensors based on quantum dot FRET donors. *Nature Materials* 2003, 2, 630–638.
107. D'Auria S, DiCesare N, Staiano M, Gryczynski Z, Rossi M, Lakowicz JR. A novel fluorescence competitive assay for glucose determinations by using a thermostable glucokinase from the thermophilic microorganism *Bacillus stearothermophilus*. *Analytical Biochemistry* 2002, 303, 138–144.
108. Barone PW, Baik S, Heller DA, Strano MS. Near-infrared optical sensors based on single-walled carbon nanotubes. *Nature Materials* 2005, 4, 86–92.
109. Ibey BL, Beier HT, Rounds RM, Cote GL, Yadavalli VK, Pishko MV. Competitive binding assay for glucose based on glycodendrimer–fluorophore conjugates. *Analytical Chemistry* 2005, 77, 7039–7046.
110. Ballerstadt R, Polak A, Beuhler A, Frye J. *In vitro* long-term performance study of a near-infrared fluorescence affinity sensor for glucose monitoring. *Biosensors & Bioelectronics* 2004, 19, 905–914.
111. Ballerstadt R, Schultz JS. A fluorescence affinity hollow fiber sensor for continuous transdermal glucose monitoring. *Analytical Chemistry* 2000, 72, 4185–4192.
112. Ye KM, Schultz JS. Genetic engineering of an allosterically based glucose indicator protein for continuous glucose monitoring by fluorescence resonance energy transfer. *Analytical Chemistry* 2003, 75, 3451–3459.
113. Meadows DL, Schultz JS. Design, manufacture and characterization of an optical fiber glucose affinity sensor based on an homogeneous fluorescence energy transfer assay system. *Analytica Chimica Acta* 1993, 280, 21–30.
114. Ballerstadt R, Evans C, Gowda A, McNichols R. *In vivo* performance evaluation of a transdermal near-infrared fluorescence resonance energy transfer affinity sensor for continuous glucose monitoring. *Diabetes Technology & Therapeutics* 2006, 8, 296–311.
115. Ibey BL, Cote GL, Yadavalli V, Gant VA, Newmyer K, Pishko MV. Analysis of Longer Wavelength Alexa Fluor Dyes for Use in a Minimally Invasive Glucose Sensor, Institute of Electrical and Electronics Engineers Inc., Cancun, Mexico, 2003, p. 3446.
116. Gurpreet G, Lin N, Fang H, Wang B. Boronic acid-based fluorescence sensors for glucose monitoring. In: Geddes CD, Lakowicz JR (Eds), *Topics in Fluorescence Spectroscopy*, Vol. 11. Springer, New York, 2006, pp. 377–397.
117. Phillips MD, James TD. Boronic acid based modular fluorescent sensors for glucose. *Journal of Fluorescence* 2004, 14, 549–559.
118. Heagy MD. *N*-Phenylboronic acid derivatives of arenecarboximides as saccharide probes with virtual spacer design. In: Geddes CD, Lakowicz JR (Eds), *Topics in Fluorescence Spectroscopy*, Vol. 11. Springer, New York, 2006, pp. 1–20.
119. Cordes DB, Miller A, Gamsey S, Sharrett Z, Thoniyot P, Wessling R, Singaram B. Optical glucose detection across the visible spectrum using anionic fluorescent dyes and

a viologen quencher in a two-component saccharide sensing system. *Organic and Biomolecular Chemistry* 2005, 3, 1708–1713.

120. Cordes DB, Suri JT, Cappuccio FE, Camara JN, Gamsey S, Sharrett Z, Thoniyot P, Wessling RA, Singaram B. Two-component optical sugar sensing using boronic acid-substituted viologens with anionic fluorescent dyes. In: Geddes CD, Lakowicz JR (Eds), *Topics in Fluorescence Spectroscopy*, Vol. 11. Springer, New York, 2006, pp. 47–87.
121. Bronk KS, Walt DR. Fabrication of patterned sensor arrays with aryl azides on a polymer-coated imaging optical fiber bundle. *Analytical Chemistry* 1994, 66, 3519–3520.
122. McCurley MF. Optical biosensor using a fluorescent, swelling sensing element. *Biosensors & Bioelectronics* 1994, 9, 527–533.
123. Mack AC, Jinshu M, McShane MJ. Transduction of pH and glucose-sensitive hydrogel swelling through fluorescence resonance energy transfer. *IEEE Sensors* 2005, 2, 912–915.
124. Wolfbeis OS, Durkop A, Wu M, Lin Z. A Europium-ion-based luminescent sensing probe for hydrogen peroxide. *Angewandte Chemie, International Edition* 2002, 41, 4495–4498.
125. Wu M, Lin Z, Durkop A, Wolfbeis OS. Time-resolved enzymatic determination of glucose using a fluorescent europium probe for hydrogen peroxide. *Analytical and Bioanalytical Chemistry* 2004, 380, 619–626.
126. Duerkop A, Schaeferling M, Wolfbeis OS. Glucose sensing and glucose determination using fluorescent probes and molecular receptors. In: Geddes CD, Lakowicz JR (Eds), *Topics in Fluorescence Spectroscopy*, Vol. 11. Springer, New York, 2006, pp. 351–375.
127. Lubbers DW, Opitz N. Optical fluorescence sensors for continuous measurement of chemical concentrations in biological systems. *Sensors and Actuators* 1983, 4, 641–654.
128. Trettnak W, Leiner MJP, Wolfbeis OS. Optical sensors. Part 34. Fibre optic glucose biosensor with an oxygen optrode as the transducer. *Analyst* 1988, 113, 1519–1523.
129. Papkovsky DB. Luminescent porphyrins as probes for optical (bio)sensors. *Sensors and Actuators B* 1993, B11, 293–300.
130. Li L, Walt DR. Dual-analyte fiber-optic sensor for the simultaneous and continuous measurement of glucose and oxygen. *Analytical Chemistry* 1995, 67, 3746–3752.
131. Rosenzweig Z, Kopelman R. Analytical properties and sensor size effects of a micrometer-sized optical fiber glucose biosensor. *Analytical Chemistry* 1996, 68, 1408–1413.
132. Rosenzweig Z, Kopelman R. Analytical properties of miniaturized oxygen and glucose fiber optic sensors. *Sensors and Actuators B* 1996, B36, 475–483.
133. Zhu H, Srivastava R, Brown JQ, McShane MJ. Combined physical and chemical immobilization of glucose oxidase in alginate microspheres improves stability of encapsulation and activity. *Bioconjugate Chemistry* 2005, 16, 1451–1458.
134. Clark HA, Kopelman R, Tjalkens R, Philbert MA. Optical nanosensors for chemical analysis inside single living cells. 2. Sensors for pH and calcium and the intracellular application of PEBBLE sensors. *Analytical Chemistry* 1999, 71, 4837–4843.
135. Clark HA, Hoyer M, Philbert MA, Kopelman R. Optical nanosensors for chemical analysis inside single living cells. 1. Fabrication, characterization, and methods for intracellular delivery of PEBBLE sensors. *Analytical Chemistry* 1999, 71, 4831–4836.
136. Xu H, Aylott JW, Kopelman R. Fluorescent nano-PEBBLE sensors designed for intracellular glucose imaging. *Analyst* 2002, 127, 1471–1477.
137. Decher G. Fuzzy nanoassemblies: toward layered polymeric multicomposites. *Science* 1997, 277, 1232–1237.

138. McShane MJ. Microcapsules as "smart tattoo" glucose sensors: engineering systems with enzymes and glucose-binding sensing elements. In: Geddes CD, Lakowicz JR (Eds), *Topics in Fluorescence Spectroscopy*, Vol. 11. Springer, New York, 2006, pp. 131–163.
139. Stein EW, Grant PS, Zhu H, McShane MJ. Microscale enzymatic optical biosensors using mass transport limiting nanofilms. 1. Fabrication and characterization using glucose as a model analyte. *Analytical Chemistry* 2007, 79, 1339–1348.
140. Brown JQ, Srivastava R, Zhu H, McShane MJ. Enzymatic fluorescent microsphere glucose sensors: evaluation of response under dynamic conditions. *Diabetes Technology & Therapeutics* 2006, 8, 288–295.
141. Lakowicz JR, Maliwal B. Optical sensing of glucose using phase-modulation fluorimetry. *Analytica Chimica Acta* 1993, 271, 155–164.
142. Tolosa L, Szmacinski H, Rao G, Lakowicz JR. Lifetime-based sensing of glucose using energy transfer with a long lifetime donor. *Analytical Biochemistry* 1997, 250, 102–108.
143. March W, Lazzaro D, Rastogi S. Fluorescent measurement in the non-invasive contact lens glucose sensor. *Diabetes Technology & Therapeutics* 2006, 8, 312–317.
144. March WF, Mueller A, Herbrechtsmeier P. Clinical trial of a noninvasive contact lens glucose sensor. *Diabetes Technology & Therapeutics* 2004, 6, 782–789.
145. Badugu R, Lakowicz JR, Geddes CD. Development of smart contact lenses for ophthalmic glucose monitoring. In: Geddes CD, Lakowicz JR (Eds), *Topics in Fluorescence Spectroscopy*, Vol. 11. Springer, New York, 2006, pp. 399–429.
146. Ballerstadt R, Schultz JS. Competitive-binding assay method based on fluorescence quenching of ligands held in close proximity by a multivalent receptor. *Analytica Chimica Acta* 1997, 345, 203–212.
147. Russell RJ, Pishko MV, Gefrides CC, McShane MJ, Cote GL. A fluorescence-based glucose biosensor using concanavalin A and dextran encapsulated in a poly(ethylene glycol) hydrogel. *Analytical Chemistry* 1999, 71, 3126–3132.
148. Sieg A, Guy RH, Delgado-Charro MB. Noninvasive and minimally invasive methods for transdermal glucose monitoring. *Diabetes Technology & Therapeutics* 2005, 7, 174–197.
149. Steil GM, Rebrin K, Hariri F, Jinagonda S, Tadros S, Darwin C, Saad MF. Interstitial fluid glucose dynamics during insulin-induced hypoglycaemia. *Diabetologia* 2005, 48, 1833–1840.
150. Rebrin K, Steil GM. Can interstitial glucose assessment replace blood glucose measurements? *Diabetes Technology & Therapeutics* 2000, 2, 461–472.
151. Rebrin K, Steil GM, van Antwerp WP, Mastrototaro JJ. Subcutaneous glucose predicts plasma glucose independent of insulin: implications for continuous monitoring. *American Journal of Physiology* 1999, 277, E561–E571
152. Haluzik M, Kremen J, Blaha J, Anderlova K, Svacina S, Hendl J, Schaupp L, Ellmerer M, Pieber T. Relationship between glucose concentrations in interstitial fluid of subcutaneous adipose tissue and blood glucose in critically ill patients. *Diabetes* 2005, 54, A97.
153. Wu XJ, Choi MMF. Optical enzyme-based glucose biosensors. In: Geddes CD, Lakowicz JR (Eds), *Topics in Fluorescence Spectroscopy*, Vol. 11. Springer, New York, 2006, pp. 201–236.
154. Chinnayelka S, McShane MJ. Glucose sensors based on microcapsules containing an orange/red competitive binding resonance energy transfer assay. *Diabetes Technology & Therapeutics* 2006, 8, 269–278.

155. Zhu H, McShane MJ. Macromolecule encapsulation in diazoresin-based hollow polyelectrolyte microcapsules. *Langmuir* 2005, 21, 424–430.
156. Sukhorukov GB, Brumen M, Donath E, Mohwald H. Hollow polyelectrolyte shells: exclusion of polymers and donnan equilibrium. *Journal of Physical Chemistry B* 1999, 103, 6434–6440.
157. Radtchenko IL, Sukhorukov GB, Leporatti S, Khomutov GB, Donath E, Mohwald H. Assembly of alternated multivalent ion/polyelectrolyte layers on colloidal particles. Stability of the multilayers and encapsulation of macromolecules into polyelectrolyte capsules. *Journal of Colloid Interface Science* 2000, 230, 272–280.
158. Khalil OS. Spectroscopic and clinical aspects of noninvasive glucose measurements. *Clinical Chemistry* 1999, 45, 165–177.
159. Malin SF, Ruchti TL, Blank TB, Thennadil SN, Monfre SL. Noninvasive prediction of glucose by near-infrared diffuse reflectance spectroscopy. *Clinical Chemistry* 1999, 45, 1651–1658.
160. Barone PW, Parker RS, Strano MS. *In vivo* fluorescence detection of glucose using a single-walled carbon nanotube optical sensor: design, fluorophore properties, advantages, and disadvantages. *Analytical Chemistry* 2005, 77, 7556–7562.
161. Mujumdar RB, Ernst LA, Mujumdar SR, Lewis CJ, Waggoner AS. Cyanine dye labeling reagents: sulfoindocyanine succinimidyl esters. *Bioconjugate Chemistry* 1993, 4, 105–111.
162. Magde D, Rojas GE, Seybold P. Solvent dependence of the fluorescence lifetimes of xanthene dyes. *Photochemistry and Photobiology* 1999, 70, 737–744.
163. Malicka J, Gryczynski I, Geddes CD, Lakowicz JR. Metal-enhanced emission from indocyanine green: a new approach to *in vivo* imaging. *Journal of Biomedical Optics* 2003, 8, 472–478.
164. Kim S, Lim YT, Soltesz EG, De Grand AM, Lee J, Nakayama A, Parker JA, Mihaljevic T, Laurence RG, Dor DM, Cohn LH, Bawendi MG, Frangioni JV. Near-infrared fluorescent type II quantum dots for sentinel lymph node mapping. *Nature Biotechnology* 2004, 22, 93–97.
165. Hartschuh A, Pedrosa HN, Peterson J, Huang L, Anger P, Qian H, Meixner AJ, Steiner M, Novotny L, Krauss TD. Single carbon nanotube optical spectroscopy. *ChemPhysChem* 2005, 6, 577–582.
166. Papkovsky DB. New oxygen sensors and their application to biosensing. *Sensors and Actuators B* 1995, B29, 213–218.
167. Guyton AC, Hall JE. *Textbook of Medical Physiology*, 10th edn. Saunders, Philadelphia, 2000.
168. Stucker M, Struk A, Altmeyer P, Herde M, Baumgartl H, Lubbers DW. The cutaneous uptake of atmospheric oxygen contributes significantly to the oxygen supply of human dermis and epidermis. *Journal of Physiology (London)* 2002, 538, 985–994.
169. Wolfbeis OS, Oehme I, Papkovskaya N, Klimant I. Sol–gel based glucose biosensors employing optical oxygen transducers, and a method for compensating for variable oxygen background. *Biosensors & Bioelectronics* 2000, 15, 69–76.
170. Gamsey S, Suri JT, Wessling RA, Singaram B. Continuous glucose detection using boronic acid-substituted viologens in fluorescent hydrogels: linker effects and extension to fiber optics. *Langmuir* 2006, 22, 9067–9074.
171. Geertsma R, de Bruijn A, Hilbers-Modderman E, Hollestelle M, Bakker G, Roszek B. New and emerging medical technologies: a horizon scan of opportunities and risks.

Report No. 360020002, prepared for National Institute for Public Health and the Environment (RIVM), The Netherlands, 2007.

172. Ballerstadt R, Gowda A, McNichols R. Fluorescence resonance energy transfer-based near-infrared fluorescence sensor for glucose monitoring. *Diabetes Technology & Therapeutics* 2004, 6, 191–200.

173. Ballerstadt R, Kholodnykh A, Evans C, Boretsky A, Motamedi M, Gowda A, McNichols R. Affinity-based turbidity sensor for glucose monitoring by optical coherence tomography: toward the development of an implantable sensor. *Analytical Chemistry* 2007, 79, 6965–6974.

CHAPTER 11

THE USE OF SINGLE-WALLED CARBON NANOTUBES FOR OPTICAL GLUCOSE DETECTION

Paul W. Barone and Michael S. Strano

11.1 INTRODUCTION

Single-walled carbon nanotubes (SWNT) are cylindrical tubes of graphite with diameters on the order of 1 nm and lengths up to several millimeters.[1,2] Their nanometer scale dimensions give rise to a host of interesting phenomena. Depending on how the graphite sheet is rolled, SWNT can be either semiconducting or metallic.[1,3] While SWNT have been used in a wide number of applications, from composites to transistors, they have received a large amount of interest in regard to sensing. Due to their one-dimensional nature and the fact that all atoms are surface atoms, SWNT are especially sensitive to surface adsorption events. Metallic SWNT have been used to enhance performance of electrochemical sensors for the detection of a number of analytes ranging from DNA[4] to glucose.[5] Semiconducting SWNT have been used as field effect transistors for the detection of a wide variety of biomolecules; however, there is still some controversy about the nanotubes ability to transduce

In Vivo Glucose Sensing, Edited by David D. Cunningham and Julie A. Stenken

specific biomolecular events. The fairly recent discovery of SWNT near-infrared (NIR) fluorescence has opened a whole new avenue of sensing. This chapter will review the current state of SWNT-based fluorescence sensors as applied for glucose detection.

11.2 SINGLE-WALLED CARBON NANOTUBES

11.2.1 Optical Properties of SWNT

Single-walled carbon nanotubes have distinct optical absorption and fluorescence that are diameter dependent.[6,7] Changing how the graphite sheet making up the nanotubes is rolled affects the optical properties. Smaller diameter nanotubes will have larger band gaps and thus higher energy absorption and fluorescence than that of tubes with larger diameters, with nanotubes fluorescence ranging from 900 to 1600 nm.[6,7] Since all the atoms making up SWNT are surface atoms, SWNT are especially sensitive to their chemical environment, with proper functionalization rendering any response selective.[8] Changing the surfactant coating, required to solubilize SWNT in water, results in solvatochromic shifts in the fluorescence energy.[9] A tightly packed surfactant coating will give rise to higher energy emission than a coating that packs more loosely. Initial estimates of nanotubes quantum yield were around 0.1%;[6] however, recent studies have raised that estimate to 7%.[10] In addition, there is some evidence that the quantum yield of nanotubes is both length dependent and defect dependent, with nanotubes that are longer and with fewer defects having higher quantum yields.[11]

11.2.2 Advantages of SWNT for *In Vivo* Applications

The measured signal from an *in vivo* fluorescence-based sensor depends on both the quantum yield of the fluorophore, which is the ratio of emitted photons to absorbed photons, and the absorption of the excited and emitted light by the surrounding tissue.[12–15] While SWNT have a much lower quantum yield than many visible fluorophores, the most important factor for depth of implantation actually turns out to be the absorption coefficient of the surrounding medium.[16,17] A one-dimensional absorption–fluorescence model can be used to compare the suitability of fluorophores for *in vivo* applications;

$$I_s = I_0 \phi e^{-2\mu d - k\tau}$$

where I_s is the measured signal intensity, I_0 is the incident intensity, ϕ is the quantum yield, μ is the absorption coefficient of the medium and is wavelength dependent, k is the photobleaching rate constant for the fluorophore and is related to how quickly the fluorophore will photobleach, and τ is the total time of exposure for the device.[17] To maximize the signal measured from the device, the quantity $\phi e^{-2\mu d}$ must be maximized. Table 11.1 shows this value for a wide variety of common organic and nanoparticle fluorophores calculated using an averaged absorption coefficient for oxygenated and deoxygenated human whole blood. While visible fluorophores have

TABLE 11.1 Comparison Between Visible and NIR Fluorophores for Their Suitability Toward *In Vivo* Applications

Quantum yield (QY) standards	QY (%)	Conditions for QY measurement	Excitation (nm)	μ (oxy) (cm^{-1})	μ (deoxy) (cm^{-1})	$\phi e^{-2\mu d}$
Cy5[12]	27	PBS	620	2	60	3.20×10^{-28}
Fluorescein[14]	95	0.1 M NaOH, 22°C	496	150	120	5.23×10^{-118}
Rhodamine 6G[15]	95	Water	488	200	105	3.30×10^{-133}
Indocyanine green[13]	1.14	Blood (0.08 g/L)	830	1.01	0.7788	1.91×10^{-3}
Type II NIR QD[19]	13	PBS	840	1.05	0.778	2.09×10^{-2}
SWNT[6]	0.1	PBS	1042	0.889	0.12	3.65×10^{-4}

Absorption coefficients, μ, are shown for both oxygenated and deoxygenated blood. PBS, phosphate buffered saline.

more desirable quantum yields than nanoparticle systems, tissue absorption for visible wavelengths greatly attenuates the measured signal. For a fluorophore to have any real application in *in vivo* sensing, it must fluoresce in the NIR window where blood and tissue absorption and scattering is minimized.[16]

The measured signal from the device is only one important variable when discussing fluorophore viability. Equally important is the total fluorescence lifetime of the device, which is governed by the photobleaching rate of the fluorophore. A device will clinically fail when an error of 20% has occurred,[18] this makes the maximum working lifetime of the device:

$$\tau = -\frac{\ln(0.8)}{k}$$

Table 11.2 shows calculated estimates for sensor lifetimes based on some common visible and nanoparticle fluorophores, with photobleaching rate constants taken from the literature. Organic fluorophores are found to photobleach rapidly at a moderate fluence, and thus are not suitable for long-term use.[19,20] Quantum dots are vastly more photostable, but bleaching is still observed.[19] In contrast, SWNT are the only optical probes to date that do not exhibit photobleaching of any kind.[21] It is important to note that photobleaching rate constants depend on laser fluence, with higher fluence leading to faster photobleaching. A 600 mW/cm^2 fluence with a laser

TABLE 11.2 Calculated Sensor Lifetimes Based on the Fluorophore Photobleaching Rate Constants from the Literature

Fluorescent probe	Photobleaching rate constant (h^{-1})	Fluence (mW/cm^2)	Sensor lifetime
IR-Dye 78-CA[19]	250.92	600	3.2 s
Cy5[19]	20.52	600	39.1 s
Indocyanine green[20]	0.0412	28	5.4 h
Type II NIR QD[19]	0.0827	600	2.7 h
SWNT[21]	0	1.0×10^6	∞

spot size of 1 mm corresponds to a laser power of approximately 4.7 mW. For comparison, most commercial laser pointers have laser powers on the order of 1 mW and their use is not expected to lead to a marked increase in lifetime. Therefore, both their NIR fluorescence and photostability make SWNT ideal for long-term *in vivo* applications.

11.3 EFFECT OF BIOFOULING: SENSOR ARCHITECTURE MATTERS

A SWNT-based fluorescence sensor would ideally measure glucose directly. Such a method of operation is physically very different from the operation of the current state-of-the-art glucose sensors. As such, it will be useful to model both sensor responses in the presence of biofouling and note any differences. Biofouling of the sensor membrane is impossible to avoid using current technology,[22,23] and thus is the only effect we consider. As the membrane fouls, the reported glucose concentration decreases relative to that in the blood and eventually renders the sensor obsolete. We have modeled the biofouling as a simple first-order process where cells and proteins adsorb on the membrane surface, resulting in a decrease of the total membrane area,

$$A(t) = A_0 e^{-k_f t} \tag{11.1}$$

where $A(t)$ is the membrane area, A_0 is the initial membrane area, and k_f is the fouling rate constant.[17]

Current state-of-the-art electrochemical glucose sensors do not measure glucose directly. Instead, the device signal S is proportional to the flux J of glucose through a limiting membrane,

$$S \propto J = \frac{A(t)D}{L}(C_B - C_s) \tag{11.2}$$

where D is the diffusivity of glucose through the membrane, L is the membrane thickness, C_B is the glucose concentration in the bulk, and C_S is the glucose concentration at the sensor.

In contrast, a nonenzymatic optical sensor measures the analyte concentration directly and the measured signal is directly proportional to the glucose concentration.

$$\begin{gathered} S \propto C_s \\ \frac{dC_s}{dt} = \frac{A(t)D}{LV}(C_B - C_s) \end{gathered} \tag{11.3}$$

The effect of this subtle difference in device function can be seen when the measured signal in the presence of biofouling is modeled. As a model patient, we considered the transient response of an individual with basal insulin provided after each of the three daily meals. Blood glucose dynamics predicted by Sorensen was corrected for diffusion to subcutaneous tissue using the mass transport model of Schmidtke et al.[24,25] Figure 11.1 shows a model comparison between the sensor response of an electrochemical sensor and an optical sensor with an assumed

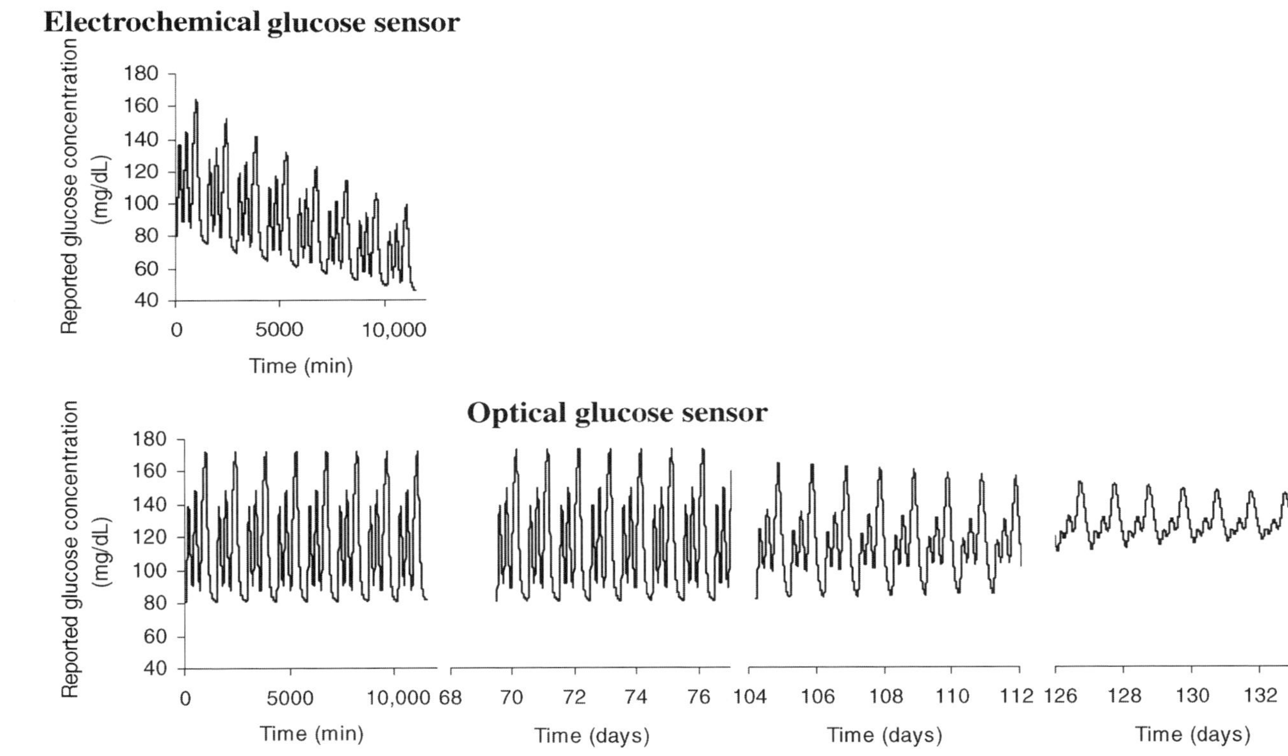

Figure 11.1 *Top*: Modeled response of a flux-based glucose sensor subject to a moderate biofouling rate. The sensor signal decreases immediately and is rendered useless after 3.7 days. *Bottom*: Modeled response of an optical glucose sensor subjected to the same fouling rate. The measured signal is not affected until a period of 76 days. Adapted with permission from Ref. 17.

biofouling rate of 5×10^{-5} min^{-1} and an initial membrane area of 2×10^{-6} cm^2. The electrochemical signal is directly proportional to flux and is immediately affected by fouling of the membrane, seen as a drift of the measured glucose. After 3.7 days, the sensor has lost 20% of its accuracy and must be recalibrated. In contrast, the optical sensor shows no immediate response to fouling of the membrane surface. Fouling of the membrane induces a mass transport resistance and will eventually cause a delay in sensor response; however, this will not occur until the transport of glucose across the membrane is commensurate with the rate of change in bulk glucose. In comparison to the electrochemical sensor, the optical sensor is theoretically stable for 76 days.

11.4 SWNT SENSOR DEVELOPMENT

11.4.1 Enzyme-Based SWNT Sensor

The main challenge facing development of solution phase, optical glucose sensors based on SWNT NIR fluorescence is that of providing the necessary colloidal stability while simultaneously affording a mechanism to selectively detect glucose. Nanotube colloidal stability is a concern, because SWNTs that are in van der Waals contact do not fluoresce.[6] Current methods to disperse nanotubes in solution involve high-intensity ultrasonication of nanotubes in a surfactant solution followed by ultracentrifugaiton.[6,26,27] Such processing guarantees the majority of nanotubes in solution will be individual and not aggregated. However, such processing is not compatible with many biological molecules.

Initial work in the design and fabrication of SWNT-based optical glucose sensors focused on the use of the enzyme glucose oxidase (GOx) in conjunction with a reaction mediator to control the nanotube fluorescence.[28] The sensing scheme is shown in Figure 11.2. The GOx is immobilized on the surface of the nanotube, providing the necessary functionality and stability to the nanotubes in solution. Potassium ferricyanide is irreversibly adsorbed to the nanotube surface, quenching the nanotube fluorescence. The addition of glucose will result in the production of hydrogen peroxide, as a reaction by-product. Ferricyanide will then undergo a redox reaction with hydrogen peroxide, decreasing the ferricyanide bound to the nanotube and resulting in a restoration of the nanotube fluorescence. While free ferricyanide has been shown to oxidize the reduced GOx, this is not a viable mechanism of SWNT fluorescence modulation due to the irreversible nature of ferricyanide adsorption to the SWNT surface.

Sensor fabrication occurs in two steps. The first step is the immobilization of GOx on the surface of the nanotube. This is accomplished by adding GOx to a solution of surfactant stabilized nanotubes and dialyzing away the surfactant. Dialysis is an ideal method for assembling enzymes on a nanotube surface, because the method allows retention of enzyme activity while simultaneously maintaining nanotube colloidal stability. The resulting GOx–SWNT solution exhibits a shift in the nanotube fluorescence indicative of the enzyme layer being less tightly packed around the nanotube than the surfactant layer. The second step is addition of ferricyanide to the GOx–SWNT solution. Adsorption of ferricyanide to the nanotube surface

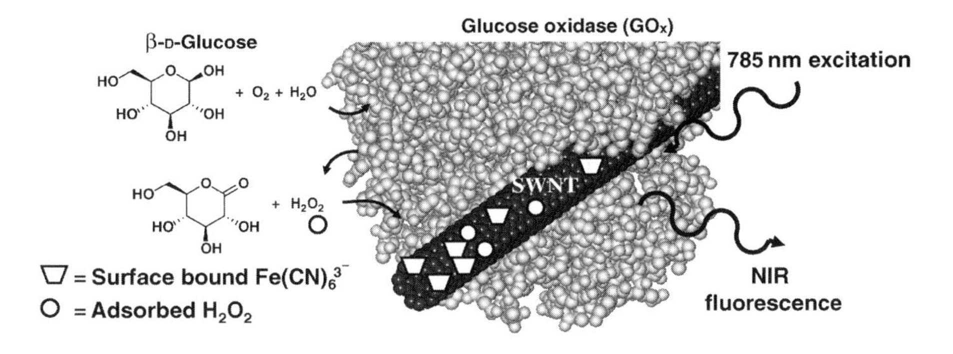

Figure 11.2 Schematic of GOx–SWNT-based glucose sensor. Glucose oxidase immobilized on the nanotube surface catalyzes the oxidation of glucose. The reaction by-product, hydrogen peroxide, then reacts with the reaction mediator, potassium ferricyanide, adsorbed to the nanotube surface resulting in an increase in SWNT fluorescence. Adapted with permission from Ref. 28.

results in a fluorescence attenuation, with 80% of the nanotube fluorescence being quenched at a ferricyanide concentration of 225 mM. It is important to emphasize that removal of free ferricyanide via dialysis does not result in fluorescence recovery, suggesting that the ferricyanide adsorbs irreversibly to the nanotube surface. Therefore, only a redox reaction of the ferricyanide adsorbed to the nanotube will result in fluorescence recovery.

Figure 11.3a shows the typical response of GOx–SWNT to the addition of 62.5 mM ferricyanide followed by subsequent glucose additions. The addition of

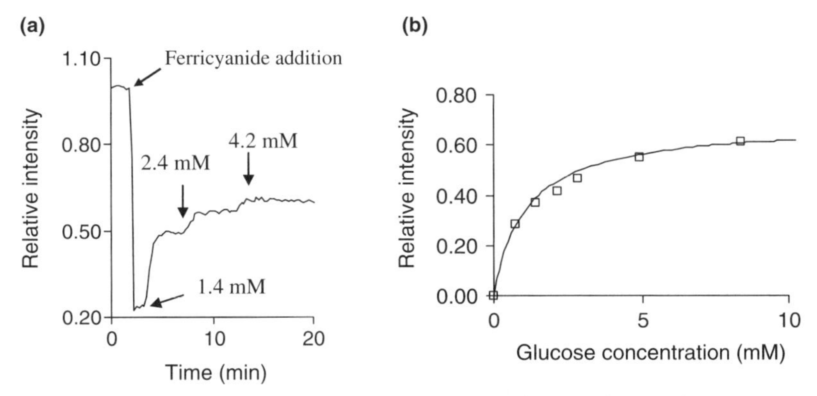

Figure 11.3 (a) Addition of 62.5 mM potassium ferricyanide to GOx–SWNT causes a decrease in fluorescence. Subsequent additions of glucose result in a fluorescence recovery. (b) Fluorescence recovery can be correlated to glucose concentration added and is found to follow a type I Langmuir adsorption isotherm. Adapted with permission from Ref. 28.

1.4 mM glucose causes a significant fluorescence recovery. Subsequent additions of glucose result in smaller and smaller degrees of fluorescence recovery. The fluorescence recovery can be plotted versus glucose concentration (Figure 11.3b), and can be seen to follow a type I adsorption isotherm. We calculate a detection limit of 34.7 μM with a working range up to 8 mM. Finally, there is no measurable response to glucose additions without the adsorbed ferricyanide, indicating that any local pH changes do not alter the GOx functionalized SWNT fluorescence.

While this sensor was the first instance of a SWNT-based optical sensor and a step forward in nanotube applications, the sensor does have some disadvantages. First and foremost, the sensor would operate as an optical analogue of a flux-based sensor in any *in vivo* setting by using a glucose-limiting membrane and as such suffers from the same drawbacks of those sensors mentioned earlier. The sensor also has a limited lifetime, as the ferricyanide is a limited resource, and will eventually be used to completion. Finally, the sensor is not regenerable and would need to be completely replaced after its lifetime has been reached.

11.4.2 SWNT Affinity Sensor

To avoid the problems of an enzyme-based sensor, it is necessary to completely redesign how the senor operates. Therefore, the next generation in SWNT sensor design is focused on utilizing protein competitive binding to make a SWNT-based glucose affinity sensor. Figure 11.4 shows the general sensor design.[29] The nanotube is first coated in dextran, which is a glucose analogue. The addition of concanavalin A (Con A), a lectin with four glucose binding sites, will induce protein-controlled

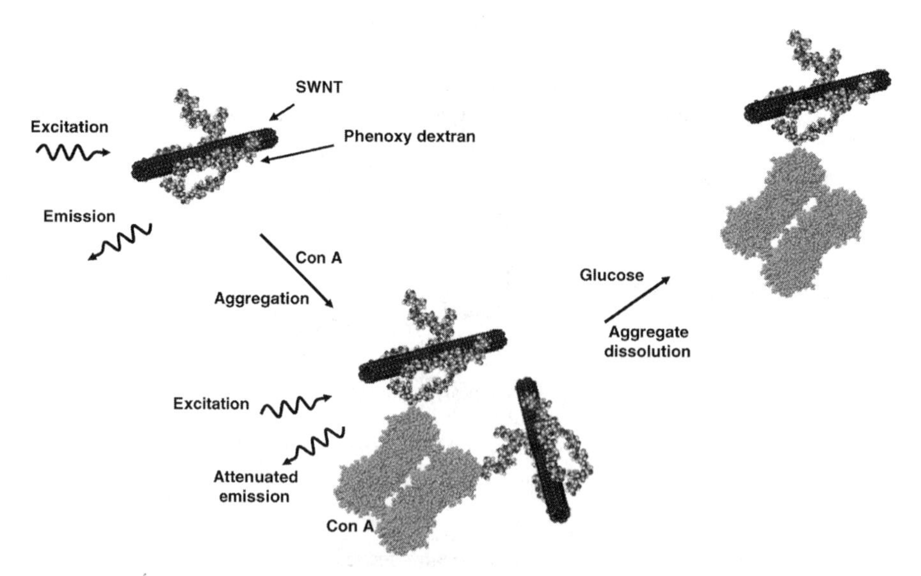

Figure 11.4 Schematic of SWNT affinity sensor. Dextran-coated SWNT aggregates upon addition of Con A, resulting in a fluorescence decrease. Addition of glucose induces aggregate dissolution and a fluorescence recovery. Adapted with permission from Ref. 29.

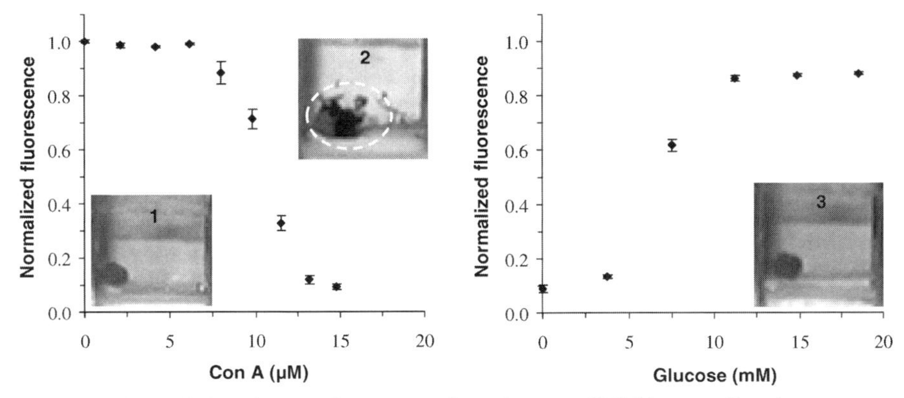

Figure 11.5 *Left*: Steady-state fluorescence from Dextran–SWNT versus Con A concentration. Picture 1 shows no visible nanotube aggregates before Con A is added, while a large nanotube aggregate is clearly visible in picture 2 (taken after 15 μM Con A). *Right*: Steady-state fluorescence of the sensor plotted versus glucose concentration. Picture 3 inset was taken after 20 mM glucose was added and clearly shows no visible aggregation. Adapted with permission from Ref. 29.

aggregation and a SWNT fluorescence decrease, as the number of nanotubes free in solution is decreased. The introduction of glucose will cause the equilibrium to shift and the dissolution of the aggregate in conjunction with a fluorescence recovery.

As nanotubes are inherently hydrophobic and dextran is a very hydrophilic polymer, it is first necessary to functionalize the dextran with hydrophobic phenoxy moieties.[29,30] Increasing the weight percent phenoxy content from 0 to 8 wt% results in the increase of the number of nanotube in solution, where dextran alone is not capable of suspending nanotubes in solution. Again, a gentle dialysis method is used to assemble dextran on the nanotube surface.

Addition of Con A to the phenoxy–dextran-stabilized nanotubes results in nanotube aggregation and a fluorescence decrease, as seen in Figure 11.5. A minimum amount of Con A is necessary before large enough nanotube aggregates are formed to cause a decrease in fluorescence. At a Con A concentration between 0 and 5.4 μM, the aggregate size is small enough to remain in solution and add to the total fluorescence signal, but at concentrations larger than 5.4 μM, the aggregates attain a size such that they begin to settle out of solution and are no longer excited by the laser. Additions of glucose reverse this effect. This is the first demonstration of a reversible affinity sensor using single-walled carbon nanotubes.[29]

It is important to think about how such a SWNT sensor would be employed *in vivo*. The sensing solution could be loaded into a dialysis capillary with a molecular weight cut off such that it is permeable to glucose, but not to nanotubes. Conceptually, this is shown in Figure 11.6. The capillary could then be coated in a hydrogel matrix filled with vascularization factors such as VEGF. Such a sensor could be implanted subdermally and then queried using a laser photodiode, with resulting nanotube fluorescence measured with a CCD camera.

Finally, there are growing concerns and uncertainty surrounding SWNT toxicity in the body. A recent report has highlighted this fact by demonstrating a

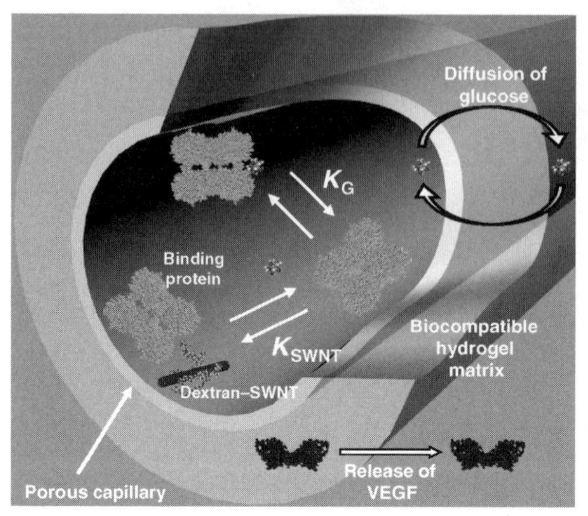

Figure 11.6 Dialysis capillary setup that could be used to employ the SWNT sensing system *in vivo*. The dextran–SWNT and Con A mixture is retained in the capillary while glucose is free to diffuse across the membrane. A biocompatible hydrogel, filled with VEGF, can be used to coat the capillary. Such a system could be implanted beneath the skin, with SWNT excitation from a laser photodiode and fluorescence detection from a CCD camera. Adapted with permission from Ref. 17.

similar immune response from long nanotube bundles ($>15\,\mu m$) compared to asbestos fibers.[31] However, there is a growing body of literature that points to the fact that nanotube cytotoxic properties appear situational, depending highly on the degree and type of functionalization, aggregation state, and the presence of metal catalyst particles remaining from synthesis. Pristine, untreated nanotubes are reported to cause oxidative stress and decrease cell viability,[32,33] with leftover catalyst particles thought to contribute to this effect.[34] The cytotoxicity can essentially be removed via functionalization, either covalent or noncovalent,[35,36] so that the nanotube displays hydrophilic groups. Nanotubes that have been noncovalently functionalized with polyethylene glycol have even been shown to cause no toxic effects in mice over a 4 month period.[37] These studies give strong evidence that nanotubes functionalized appropriately will find a place in biomedical applications such as *in vivo* sensors.

11.5 CONCLUSIONS AND FUTURE PROSPECTS

As this chapter has demonstrated, nanotubes are a unique and exciting material with potential applications toward *in vivo* glucose sensing. A sensor based on nanotube fluorescence would avoid a number of complications that current sensing technology faces. Single-walled carbon nanotube NIR fluorescence and photostability avoids *in vivo* transmission and sensor lifetime issues that most conventional fluorophores face. In addition, a SWNT sensor that detects glucose directly should not be as affected by biofouling as a flux-based sensor. We have demonstrated two prototype

SWNT-based glucose sensors: an enzyme-based sensor and an affinity sensor. These results are a promising beginning; however, there is much work left to do before the promise of SWNT *in vivo* sensors is realized. Investigation of new sensing modalities and optimization of current ones is necessary before the best nanotube-based sensing strategy is discovered.

REFERENCES

1. Dresselhaus MS, Dresselhaus G, Eklund PC. *Science of Fullerenes and Carbon Nanotubes*. Academic Press, San Diego, 1996.
2. Li QW, Zhang XF, DePaula RF, Zheng LX, Zhao YH, Stan L, Holesinger TG, Arendt PN, Peterson DE, Zhu YT. Sustained growth of ultralong carbon nanotube arrays for fiber spinning. *Advanced Materials* 2006, 18, 3160–3163.
3. Saito R, Dresselhaus MS, Dresselhaus G. *Physical Properties of Carbon Nanotubes*. Imperial College Press, London, 1998.
4. Wang SG, Wang RL, Sellin PJ, Zhang Q. DNA biosensors based on self-assembled carbon nanotubes. *Biochemical and Biophysical Research Communications* 2004, 325, 1433–1437.
5. Male KB, Hrapovic S, Liu YL, Wang DS, Luong JHT. Electrochemical detection of carbohydrates using copper nanoparticles and carbon nanotubes. *Analytica Chimica Acta* 2004, 516, 35–41.
6. O'Connell MJ, Bachilo SM, Huffman CB, Moore VC, Strano MS, Haroz EH, Rialon KL, Boul PJ, Noon WH, Kittrell C, Ma J, Hauge RH, Weisman RB, Smalley RE. Band gap fluorescence from individual single-walled carbon nanotubes. *Science* 2002, 297, 593–596.
7. Bachilo SM, Strano MS, Kittrell C, Hauge RH, Smalley RE, Weisman RB. Structure-assigned optical spectra of single-walled carbon nanotubes. *Science* 2002, 298, 2361–2366.
8. Durkop T, Getty SA, Cobas E, Fuhrer MS. Extraordinary mobility in semiconducting carbon nanotubes. *Nano Letters* 2004, 4, 35–39.
9. Moore VC, Strano MS, Haroz EH, Hauge RH, Smalley RE, Schmidt J, Talmon Y. Individually suspended single-walled carbon nanotubes in various surfactants. *Nano Letters* 2003, 3, 1379–1382.
10. Lefebvre J, Austing DG, Bond J, Finnie P. Photoluminescence imaging of suspended single-walled carbon nanotubes. *Nano Letters* 2006, 6, 1603–1608.
11. Heller DA, Mayrhofer RM, Baik S, Grinkova YV, Usrey ML, Strano MS. Concomitant length and diameter separation of single-walled carbon nanotubes. *Journal of the American Chemical Society* 2004, 126, 14567–14573.
12. Mujumdar R, Ernst L, Mujumdar S, Lewis C, Waggoner A. Cyanine dye labeling reagents: sulfoindocyanine succinimidyl esters. *Bioconjugate Chemistry* 1993, 4, 105–111.
13. Benson R, Kues H. Fluorescence properties of indocyanine green as related to angiography. *Physics in Medicine and Biology* 1978, 23, 159–163.
14. Lakowicz J. *Principles of Fluorescence Spectroscopy*, 2nd edn. Academic/Plenum Publishers, New York, Moscow, Dordecht, 1999.
15. Magde D, Rojas G, Seybold P. Solvent dependence of the fluorescence lifetimes of xanthene dyes. *Photochemistry and Photobiology* 1999, 70, 737.

16. Wray S, Cope M, Delpy DT, Wyatt JS, Reynolds EOR. Characterization of the near-infrared absorption-spectra of cytochrome-Aa3 and hemoglobin for the non-invasive monitoring of cerebral oxygenation. *Biochimica et Biophysica Acta* 1988, 933, 184–192.
17. Barone PW, Parker RS, Strano MS. *In vivo* fluorescence detection of glucose using a single-walled carbon nanotube optical sensor: design, fluorophore properties, advantages, and disadvantages. *Analytical Chemistry* 2005, 77, 7556–7562.
18. Heller A. Implanted electrochemical glucose sensors for the management of diabetes. *Annual Review of Biomedical Engineering* 1999, 1, 153–175.
19. Kim S, Lim Y, Soltesz EG, de Grand AM, Lee J, Nakayama A, Parker JA, Mihaljevic T, Laurence RG, Dor DM, Cohn LH, Bawendi MG, Frangioni JV. Near-infrared fluorescent type II quantum dots for sentinel lymph node mapping. *Nature Biotechnology* 2004, 22, 93–97.
20. Saxena V, Sadoqi M, Shao J. Degradation kinetics of indocyanine green in aqueous solution. *Journal of Pharmaceutical Science* 2003, 92, 2090–2097.
21. Graff RA, Swanson JP, Barone PW, Baik S, Heller DA, Strano MS. Achieving individual-nanotube dispersion at high loading in single-walled carbon nanotube composites. *Advanced Materials* 2005, 17, 980–984.
22. Clark H, Barbari TA, Stump K, Rao G. Histologic evaluation of the inflammatory response around implanted hollow fiber membranes. *Journal of Biomedical Materials Research* 2000, 52, 183–192.
23. Galeska I, Hickey T, Moussy F, Kreutzer D, Papadimitrakopoulos F. Characterization and biocompatibility studies of novel humic acids based films as membrane material for an implantable glucose sensor. *Biomacromolecules* 2001, 2, 1249–1255.
24. Sorensen JT, A physiologic model of glucose metabolism in man and its use to design and assess improved insulin therapies for diabetes, 1985.
25. Schmidtke D, Freeland A, Heller A, Bonnecaze R. Measurement and modeling of the transient difference between blood and subcutaneous glucose concentrations in the rat after injection of insulin. *Proceedings of the National Academy of Sciences of the United States of America* 1998, 95, 294–299.
26. Strano MS, Moore VC, Miller MK, Allen MJ, Haroz EH, Kittrell C, Hauge RH, Smalley RE. The role of surfactant adsorption during ultrasonication in the dispersion of single-walled carbon nanotubes. *Journal of Nanoscience and Nanotechnology* 2003, 3, 81–86.
27. Zheng M, Jagota A, Semke ED, Diner BA, McLean RS, Lustig SR, Richardson RE, Tassi NG. DNA-assisted dispersion and separation of carbon nanotubes. *Nature Materials* 2003, 2, 338–342.
28. Barone PW, Baik S, Heller DA, Strano MS. Near-infrared optical sensors based on single-walled carbon nanotubes. *Nature Materials* 2005, 4, 86–92.
29. Barone PW, Strano MS. Reversible control of carbon nanotube aggregation for a glucose affinity sensor. *Angewandte Chemie, International Edition* 2006, 45, 8138–8141.
30. Fournier C, Leonard M, Lecoqleonard I, Dellacherie E. Coating polystyrene particles by adsorption of hydrophobically-modified dextran. *Langmuir* 1995, 11, 2344–2347.
31. Poland CA, Duffin R, Kinloch I, Maynard A, Wallace WAH, Seaton A, Stone V, Brown S, Macnee W, Donaldson K. Carbon nanotubes introduced into the abdominal cavity of mice show asbestos-like pathogenicity in a pilot study. *Nature Nanotechnology* 2008, 3, 423–428.
32. Cui DX, Tian FR, Ozkan CS, Wang M, Gao HJ. Effect of single wall carbon nanotubes on human HEK293 cells. *Toxicology Letters* 2005, 155, 73–85.

33. Manna SK, Sarkar S, Barr J, Wise K, Barrera EV, Jejelowo O, Rice-Ficht AC, Ramesh GT. Single-walled carbon nanotube induces oxidative stress and activates nuclear transcription factor-kappa B in human keratinocytes. *Nano Letters* 2005, 5, 1676–1684.
34. Kagan VE, Tyurina YY, Tyurin VA, Konduru NV, Potapovich AI, Osipov AN, Kisin ER, Schwegler-Berry D, Mercer R, Castranova V, Shvedova AA. Direct and indirect effects of single walled carbon nanotubes on RAW 264.7 macrophages: role of iron. *Toxicology Letters* 2006, 165, 88–100.
35. Sayes CM, Liang F, Hudson JL, Mendez J, Guo WH, Beach JM, Moore VC, Doyle CD, West JL, Billups WE, Ausman KD, Colvin VL. Functionalization density dependence of single-walled carbon nanotubes cytotoxicity *in vitro. Toxicology Letters* 2006, 161, 135–142.
36. Kam NWS, O'Connell M, Wisdom JA, Dai H. Carbon nanotubes as multifunctional biological transporters and near-infrared agents for selective cancer cell destruction. *Proceedings of the National Academy of Sciences of the United States of America* 2006, 102, 11600–11605.
37. Schipper ML, Nakayama-Ratchford N, Davis CR, Kam NWS, Chu P, Liu Z, Sun X, Dai H, Gambhir SS. A pilot toxicology study of single-walled carbon nanotubes in a small sample of mice. *Nature Nanotechnology* 2008, 3, 216–221.

CHAPTER 12

INTRODUCTION TO SPECTROSCOPY FOR NONINVASIVE GLUCOSE SENSING

Wei-Chuan Shih, Kate L. Bechtel, Michael S. Feld, Mark A. Arnold and Gary W. Small

12.1 INTRODUCTION

Imagine the ability to obtain quantitative information from within the human body without collecting a drop of blood or any other representative sample. This is the goal of methods designed to measure glucose and other clinical analytes noninvasively

In Vivo Glucose Sensing, Edited by David D. Cunningham and Julie A. Stenken

with spectroscopy. The concept is to pass a band of selected light through a vascular region of the body and then extract the concentration of glucose, or some other targeted analyte, from an analysis of the resulting spectral information. The wavelengths and the type of data analysis used in such methods depend on the type of spectroscopy.

Pulse oximetry is a common example of this approach where light is used to determine noninvasively a clinical parameter.[1] In the case of pulse oximetry, two wavelengths of light are transmitted through tissue to measure the percent blood oxygen saturation. One wavelength is responsive to the deoxygenated form of hemoglobin (660 nm) and the other is responsive to oxyhemoglobin (940 nm).[2] A function is used to relate these intensities to the percentage of hemoglobin bound to oxygen. The ability to follow blood oxygen saturation continuously without collecting samples of blood has made this method extremely valuable in the clinical setting. Indeed, pulse oximetry is commonly used for a majority of hospital procedures, ranging from neonatal intensive care to outpatient adult care.[3]

The success and clinical utility of pulse oximetry can be misleading for the general applicability of this measurement strategy to other clinical analytes. It is critical to recognize that neither of the wavelengths used in the pulse oximetry measurement is specific for the targeted form of hemoglobin. The measured intensity at each wavelength is affected by the distribution of chemical components in the skin tissue and by the scattering properties of skin. Reliable measurements are only possible between people by using a relative measurement (intensity ratio) and by confining the measurement to signals associated with the arterial pulse.[4] Overall, pulse oximetry does not provide absolute concentration values, but is limited to relative concentrations of oxy- versus deoxyhemoglobin. The absolute noninvasive measurement of oxygen in blood or tissue is much more difficult to accomplish and is not routinely used in clinical medicine.[5]

Noninvasive glucose monitoring demands absolute glucose concentration measurements that match results obtained from conventional test strip technology. Absolute concentration measurements are complicated by the complexity of the sample matrix and variations of this matrix between individuals. Physical separations and selective chemical reactions are commonly used in analytical science to improve measurement accuracy. Such steps are not possible in a noninvasive analysis where all the selectivity information must be derived solely from the spectral information collected from the illuminated sample.

12.2 CLINICAL APPLICATIONS AND SIGNIFICANCE OF NONINVASIVE GLUCOSE MONITORING

Research to develop noninvasive glucose sensing technology is driven by the promise to provide *in vivo* glucose concentrations without the pain, cost, and inconvenience of test strip meters. The American Diabetes Association recommends a minimum of three self-monitoring blood glucose measurements daily for people with type 1 diabetes,[6] yet, on average, most people with type 1 diabetes perform fewer than one measurement a day.[7] Although the pain associated with test strip meters has decreased dramatically in recent years due to refinements in lancet devices and lower volume requirements for electrochemical meters, the pain and soreness associated with repeated fingersticks still impede frequent measurements for some individuals. Frequent monitoring is further discouraged by the cost of each test strip and the inconveniences associated with collecting and disposing of the blood sample. Testing in public is a significant inconvenience for many people with type 1 diabetes. A successful noninvasive glucose monitor that can provide clinically valuable information without pain, that can be reused without consuming a costly disposable test strip, and that can eliminate handling blood will result in more frequent measurements, tighter glycemic control, and fewer costly complications of the disease.

Frequent monitoring of glucose is becoming increasingly important for all critically ill patients, including patients with no prior history of diabetes. This point is highlighted by the pioneering findings published by Van den Berghe and coworkers,[8] where they report that tight glycemic control in critically ill patients reduced in-hospital mortality by 34%, bloodstream infections by 46%, acute renal failure requiring dialysis by 41%, the median number of red cell transfusions by 50%, and critical illness polyneuropathy by 44%. These powerful findings are corroborated by others.[8–11] Overall, the importance of controlling glycemia in critically ill patients has been increasingly recognized and many hospitals are establishing protocols for tight glycemic control through intensive insulin therapy for all hospital patients.[12]

Aggressive glycemic control requires frequent blood glucose measurements in order to deliver intravenous insulin for the purpose of maintaining glucose levels within a targeted therapeutic range. Presently, glucose levels are determined hourly from blood samples collected from an intravenous catheter. Clearly, continuous glucose monitoring would enable quicker response to abnormal glucose levels and afford the best glycemic control. A noninvasive method that eliminates sampling of blood and that give continuous, real-time analytical information would greatly enhance such treatments.

Besides glucose, other analytes of clinical value can be possibly quantified by noninvasive spectral analysis. *In vivo* concentrations of lactate and urea are examples. The concentration of lactate in blood is used clinically to follow intensive care treatments, to identify cardiac or liver failure, to determine hypoxia of tissues from atherosclerosis, and to detect bacterial infection. *In vivo* urea levels are valuable for optimizing hemodialysis treatments and tracking the accumulation of toxins for people with end-stage renal failure or recent kidney transplant recipients.

Spectroscopy-based noninvasive analysis for any of the clinical applications noted above provides a means to collect critical analytical information in a novel fashion compared to competing technologies. Alternatively, implants can be used, but operation of implanted sensors is confounded by biologic responses that degrade performance and demand *in situ* calibration techniques.

12.3 APPROACHES AND METHODS

Many approaches are documented in the peer-review and the patent literature for measuring glucose noninvasively in people.[13–15] These methods can be categorized as either indirect or direct. Indirect methods rely on an indirect property of glucose or the impact of glucose on a measured property. The most common example of an indirect method is the measurement of glucose based on changes in the scattering properties of skin.[16] The impact of blood glucose concentrations on the scattering properties of skin is well known and changes in the concentration of glucose in blood can be tracked by following the intensity of light scattered from skin. Indirect methods are generally confounded by a lack of selectivity because other endogenous components can likewise affect the monitored property.[17]

Direct methods are based on an intrinsic property of the glucose molecule and the measurement of this property. The best example of a direct measurement is vibrational spectroscopy as implemented by near-infrared absorption spectroscopy or Raman scattering spectroscopy. A summary of the major methods is provided in Section 12.5.

For all reported methods, multivariate analysis is used to enhance both the selectivity and the signal-to-noise ratio (SNR) of the measurement. Analytical signals for glucose typically overlap with signals that originate from other elements of the skin tissue matrix or that originate from environmental or instrumental sources of variance. Also, the component of the measured signal that originates from glucose is generally weak compared to the background noise. In both cases, multivariate methods of analysis can enhance measurement accuracy. Misuse of the various multivariate analysis methods, however, can create improper calibration models based on spurious correlations within the spectral data set.[18]

The underlying concepts are presented below for the multivariate calibration methods used in such measurements. A firm understanding of these methods is critical for a proper assessment of proposed methods for noninvasive glucose sensing.

12.4 MULTIVARIATE CALIBRATION METHODS

Extracting analyte concentrations from spectra of complex systems containing multiple analyte contributions with overlapping spectral features requires more information than is obtainable in a single wavelength measurement. Multivariate techniques take the full-range spectrum into account and exploit the multichannel (data at many wavelengths) nature of spectroscopic data to extract concentration information from analytes.[19,20]

Multivariate calibration is often treated as a black box because it can be mathematically complicated and conceptually challenging. Here, we present the fundamental concepts with minimum mathematics. The goal is to familiarize the reader with the basic principles of multivariate calibration and, more importantly, how to evaluate calibration results. More comprehensive treatment of this topic can be found in work dedicated to multivariate calibration methods and chemometrics.[19,21–23]

12.4.1 Introduction

The measured spectrum, $\mathbf{s}$, of a complex mixture can be written as a linear combination of analyte pure component spectra, $\mathbf{p}$, in proportion to the analyte concentrations, c:

$$\mathbf{s} = c_1 \cdot \mathbf{p}_1 + c_2 \cdot \mathbf{p}_2 + \cdots + c_n \mathbf{p}_n \tag{12.1}$$

(In this section, lowercase boldface type denotes a column vector and uppercase boldface type a matrix, and the superscript T denotes matrix transpose.) Multiple mixture spectra with varying analyte concentrations can be written together in matrix form as

$$\mathbf{S} = \mathbf{PC} \tag{12.2}$$

where $\mathbf{S}$ is a matrix of sample spectra with each spectrum occupying a column, $\mathbf{P}$ is a matrix of pure component spectra with each spectrum occupying a column, and $\mathbf{C}$ is a matrix of analyte concentrations with values for different spectra in columns.

For most spectroscopic applications, the goal of multivariate calibration is to predict the concentration of a given analyte(s) in a future (prospective) sample using only its measured spectrum and a previously determined model. To do this, the inverse calibration method is used in which equation (12.2) is rewritten as

$$\mathbf{C} = \mathbf{B}^{\mathrm{T}}\mathbf{S} \tag{12.3}$$

where $\mathbf{B}$ is a matrix of regression vectors, with each $\mathbf{b}$ for a given analyte occupying a column. In other words, the goal of (inverse) multivariate calibration is to obtain a "spectrum" of regression coefficients, $\mathbf{b}$, such that concentration of an analyte, c, can be accurately predicted by taking the scalar product of $\mathbf{b}$ with a prospective spectrum, $\mathbf{s}$:

$$c = \mathbf{b}^{\mathrm{T}}\mathbf{s} \tag{12.4}$$

The regression vector, $\mathbf{b}$, for each analyte is unique in an ideal noise-free linear system without component correlations (i.e., two or more analytes that vary together). Under realistic experimental conditions, however, only an approximation can be found for $\mathbf{b}$ under a set of experimental conditions.

A thorough multivariate calibration procedure encompasses three primary steps: (1) model building, (2) validation, and (3) prospective application, that is, prediction. Step 1 utilizes a set of data that include multiple spectra with known concentrations, called the calibration or training data, to calculate the $\mathbf{b}$ vector. Among the calibration data, a subset is reserved for validation and therefore is not used in determining the $\mathbf{b}$ vector. Step 2 uses these reserved data as examples on which to test the predictive capabilities of the $\mathbf{b}$ vector determined in step 1. Based on some prescribed criteria of optimality, for example, root mean square error of cross-validation (see Section 12.4.2), iterations can be performed between model building and validation until the best model is obtained. Step 3 is the prospective application of the resulting optimized model in which the optimal $\mathbf{b}$ vector is applied to future independent data to determine the analyte concentration. These primary steps in multivariate calibration are presented schematically in Figure 12.1.

The time required to complete the above described calibration procedure varies depending on the application. A critical factor in developing a robust calibration model is to incorporate all sources of spectral variance expected for future samples. The calibration procedure creates a model where the analyte-dependent signal is orthogonal to all nonanalyte-dependent sources of spectral variance. Significant time

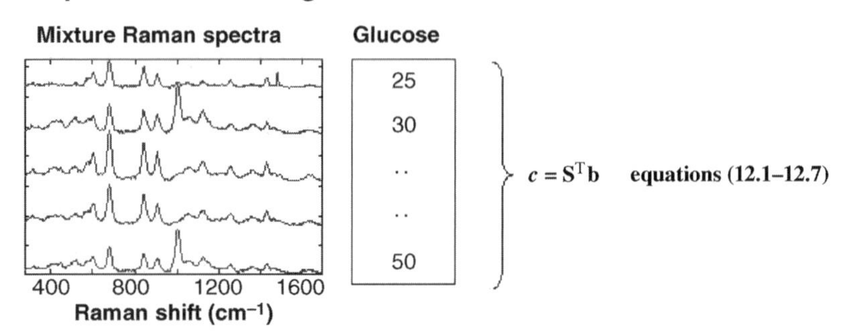

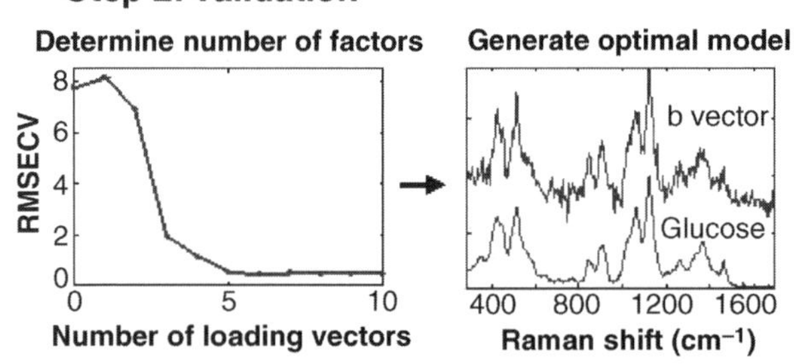

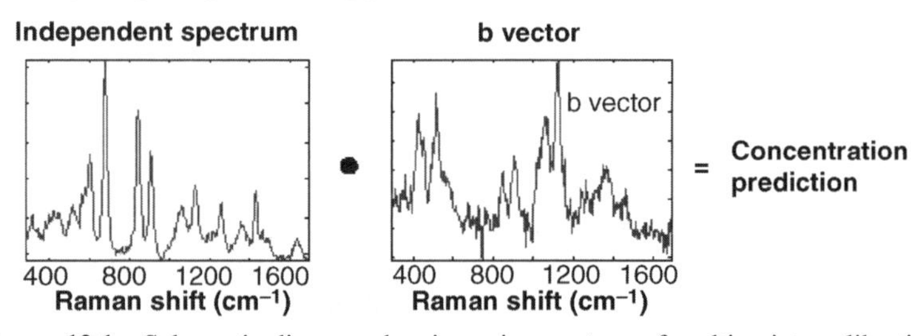

Figure 12.1 Schematic diagram showing primary steps of multivariate calibration.

can be required to incorporate all sources of spectral variance. In fact, it might be impractical to include all sources of spectral variance. For example, it might not be practical to establish the impact of changing the lamp in the spectrometer, even through this parameter could have a drastic impact on accuracy of predictions from the calibration model. Typically, several days are required to generate and collect spectra for a sufficient number of samples necessary to build a robust calibration model. Several more days are then required for testing of the model with independent sets of prospective samples.

12.4.2 Multivariate Calibration Models

There are two categories of calibration methods: explicit and implicit. Explicit methods utilize individual component spectra that can be measured or calculated.

Examples are ordinary least squares (OLS) and classical least squares (CLS). Explicit methods provide transparent models with easily interpretable results. However, highly controlled experimental conditions, high-quality spectra, and accurate concentration measurements of all components in the sample matrix may be difficult to obtain, particularly in biomedical applications.

When all of the individual component spectra are not known, implicit calibration methods are often adopted. Among these, factor analysis methods such as principal component regression (PCR)[24] and partial least squares (PLS)[25] are frequently used because they can function under conditions in which the number of spectra used for calibration is less than the number of wavelengths sampled. For example, a calibration set may include 30 spectra with each spectrum having 500 data points (wavelengths).

Unlike explicit methods, the performance of implicit methods cannot be simply judged by conventional statistical measures such as goodness of fit. As pointed out in the literature,[18] spurious effects such as system drift and covariations among constituents can be incorrectly interpreted as arising from the analyte of interest. This scenario has led to the development of hybrid methods in which elements of explicit and implicit techniques are combined to improve performance.

Commonly used multivariate calibration methods are described below in more detail.

12.4.2.1 Explicit Calibration Methods Ordinary least squares can be employed if spectra are known for all components in the sample matrix. For a pure component spectral matrix, $\mathbf{P}$, the regression matrix, $\mathbf{B}$, can be obtained by the pseudoinverse of $\mathbf{P}$ from equation (12.2):

$$\mathbf{B}^{\mathrm{T}} = (\mathbf{P}^{\mathrm{T}}\mathbf{P})^{-1}\mathbf{P}^{\mathrm{T}} \tag{12.5}$$

OLS is a simple, yet powerful explicit calibration technique from which the result can be easily interpreted with little ambiguity. However, the requirement that all spectral components be known reduces the application of OLS to quantitative biological spectroscopy.

In some cases, it may be difficult to chemically separate individual components in order to measure their spectra, but it may be possible to measure or estimate their concentrations. If so, CLS can be employed to obtain an estimate of the pure component spectral matrix, $\mathbf{P}$, through the pseudoinverse of $\mathbf{C}$ in equation (12.2):

$$\mathbf{P} = \mathbf{S}\mathbf{C}^{\mathrm{T}}(\mathbf{C}\mathbf{C}^{\mathrm{T}})^{-1} \tag{12.6}$$

OLS and CLS are complementary techniques. OLS calculates concentrations from a known set of component spectra, and CLS calculates component spectra from a known set of concentrations. The component spectra obtained by CLS can be used for OLS analysis of a new data set, as long as the two data sets have the same components.

12.4.2.2 Implicit Calibration Methods Explicit calibration methods demand complete knowledge of either spectral properties or concentration levels for all matrix components within a sample. Implicit calibration methods are best suited for cases where such complete information is not available. Implicit calibration schemes require only a set of calibration spectra, **S**, with each spectrum occupying a column of **S**, associated with several known concentrations of the analyte of interest that are expressed as a column vector, **c**. Developing an accurate regression vector, **b**, requires accurate values of **c** and **S**. The forward problem for the implicit calibration method is defined by the linear inverse mixture model for a single analyte:

$$\mathbf{c} = \mathbf{S}^{\mathrm{T}}\mathbf{b} \tag{12.7}$$

The goal of the calibration procedure is to use the set of data [**S**, **c**] to obtain an accurate **b** by inverting equation (12.7). The resulting **b** can then be used in equation (12.4) to predict the analyte concentration, c, of an independent prospective sample by measuring its spectrum, **s**. The "accuracy" of **b** is usually judged by its ability to correctly predict concentrations prospectively.

There are two primary difficulties in directly inverting equation (12.7). First, the system is usually underdetermined, which means there are more variables (e.g., wavelengths) than equations (e.g., number of calibration samples). Thus, direct inversion does not always yield a unique solution. Second, even if a pseudoinverse exists and results in a unique solution, the solution tends to be unstable because all measurements contain noise and error. That is, small variations in **c** or **S** can lead to large variations in **b**. Underdetermined and unstable models can be avoided by using data reduction methods, such as factor analysis, which reduce the dimensionality of the spectral data and much of the underlying noise within each spectrum.

Principal component regression and partial least squares are two widely used methods in the factor analysis category. PCR decomposes the matrix of calibration spectra into orthogonal principal components that best capture the variance in the data. These new variables eliminate redundant information and, by using a subset of these principal components, filter noise from the original data. With this compacted and simplified form of the data, equation (12.7) may be inverted to arrive at **b**.

PLS is similar to PCR with the exception that the matrix decomposition for PLS is performed on the covariance matrix of the spectra and the reference concentrations, while for PCR only the spectra are used. PLS and PCR have similar performance if noise in the spectral data and errors in the reference concentration measurements are negligible. Otherwise, PLS generally provides better analysis than PCR.[26]

An important advantage of implicit methods such as PLS and PCR over OLS or CLS is their ability to extract spectral components (called principal components in PCR, loadings in PLS, or, more generally, factors) without knowledge of the actual physical constituents comprising the spectrum. To a certain degree, this has encouraged users to treat implicit methods as a black box. However, the extracted spectral components are usually not identical to the physical constituent spectra. Thus, caution should be taken when attempting to identify features of suspected physical constituents in extracted spectral components.

Although PCR and PLS are powerful methods, they are not without their limitations. In particular, implicit calibration methods can be susceptible to chance correlations. Thus, when the calculated **b** is applied to a future spectrum in which those correlations are not present, increased error is likely. It may be possible to improve implicit calibration and limit spurious correlations by incorporating additional information about the system or analytes. This combination of features from implicit and explicit calibration methods is termed hybrid calibration.

12.4.2.3 Hybrid Calibration Methods Incorporating additional information into implicit models has been extensively pursued in many fields to enhance the functionality of calibration algorithms. In the chemical and applied spectroscopy literature, methods combining explicit and implicit schemes have been explored by Haaland et al.,[27] Wentzell et al.,[28] Berger et al.,[29] and Shih et al.[30] Haaland et al.[27] developed an augmented CLS/PLS hybrid algorithm that can incorporate nonlinearities such as temperature variations and known spectral components into the calibration process. This method was shown to outperform PLS when the independent prediction spectra included unmodeled spectral variation. Wentzell et al.[28] included information on measurement uncertainties in the decomposition of the calibration spectral data, thereby optimizing data extraction. This method was shown to outperform PCR and PLS when there is a nonuniform error structure. Berger et al.[29] and Shih et al.[30] utilized the pure component spectrum of the analyte of interest to build the calibration model with higher specificity. Berger et al. mathematically subtracted the pure component spectrum from the calibration data according to reference concentrations before performing PCR on the residuals. Shih et al. included the pure component spectrum as a nonlinear constraint in the regularized cost function. These methods were shown to outperform PLS, particularly when spurious correlations were present.

In principle, all of these hybrid methods outperform those without prior information. Depending on how the additional information is incorporated into the model algorithm, however, hybrid methods may reduce model performance by incorporating inaccuracies into the system.

12.4.3 Model Validation and Performance Evaluation

12.4.3.1 Model Validation Validation of the calibration model is crucial before prospective application. Two types of validation schemes can be adopted: internal and external. Internal validation, or cross-validation, is used when the number of calibration samples is limited. In cross-validation, a small subset of calibration data is withheld from the model building step. After the model is tested on these validation spectra, a different subset of calibration data is withheld and the **b** vector is recalculated. Various strategies can be employed for grouping spectra for calibration and validation. For example, a single sample is withheld in a "leave-one-out" scheme, and the calibration and validation process is repeated as many times as the number of samples in the calibration data set. In general, "leave-n-out" cross-validation can be implemented with n random samples chosen from a pool of calibration data.

The optimal model is determined by finding the minimum error between the extracted concentrations and the reference concentrations. Cross-validation is also used to determine the optimal number of model parameters, for example, the number of factors in PLS or principal components in PCR, and to prevent over- or underfitting. Technically, because the data sets used for calibration and validation are independent for each iteration, the validation is performed without bias. When a statistically sufficient number of spectra are used for calibration and validation, the chosen model and its outcome, the **b** vector, should be representative of the data.

When the calibration data are not limited, external validation, that is, prediction testing, can be employed. As opposed to internal validation, external validation tests the calibration model and optimizes the number of model parameters on data that never influences the model and therefore provides a more objective measure than internal validation.

12.4.3.2 Summary Statistics for Calibration Model and Prediction In determining the optimal model via cross-validation, the root mean square error of cross-validation (RMSECV) is calculated. RMSECV is defined as the square root of the sum of the squares of the differences between extracted and reference concentrations. The RMSECV is calculated for a particular choice of the number of model parameters. An iterative algorithm is often employed to vary the number of parameters and recalculate the RMSECV. The statistically significant minimum RMSECV and the corresponding number of model parameters are then chosen for determination of the final calibration model.

Another important statistic is the correlation coefficient (r) between the extracted and the reference concentrations. A higher correlation coefficient across a broad range of concentrations provides confidence that the calibration model is accurate.

The **b** vector chosen by the validation procedure can be employed prospectively to predict concentrations of the analyte of interest in independent data. Similar to the calculation of RMSECV, the root mean square error of prediction (RMSEP) for an independent data set is defined as the square root of the sum of the squares of the differences between predicted and reference concentrations.

For feasibility studies, RMSECV is a good indicator of performance as long as the number of calibration samples is statistically sufficient (see Section12.4.4.2). RMSEP, on the other hand, provides the ultimate objective metric by which any technology can be evaluated.

12.4.4 Is the Calibration Model Based on Analyte-Specific Information

Multivariate calibration models are often built on an underdetermined data set, that is, more wavelengths than samples. The use of powerful data reduction techniques, such as PCR and PLS, makes assessing the model validity an extremely important aspect of the analysis procedure. Here, we present four important criteria on which to judge the validity of results from multivariate calibration.

12.4.4.1 Theoretical and Practical Limits In spectroscopy, the analyte-specific signal is dependent on the number of analyte molecules sampled by the propagating light. Therefore, the effective path length (in transmission mode) and sampling volume (in reflection mode) of the light are important parameters in estimating the detection limit in turbid media. Modeling techniques such as diffusion theory[31] and Monte Carlo simulation[32] have been employed to calculate fluence distribution inside the sample and the angular and radial profiles of the transmitted or reflected flux. Simple simulations with synthetic data or experiments with tissue-simulating phantoms can be of great value in determining how close the theoretical limit can be realized in practice. In these studies, experimental conditions (e.g., signal-to-noise ratio, instrumental drifts, etc.) and tissue phantom composition (e.g., concentrations and distribution of component chemical species) can be precisely controlled and the model components well characterized in advance. Although *in vitro* experiments often present a "best-case" scenario, proving that the chosen technique and instrument can measure physiological levels of the selected analyte in tissue-simulating phantoms is necessary but not independently sufficient to justify the *in vivo* results.

12.4.4.2 Model Dimensionality In multivariate calibration, a large number of sample spectra can be reduced to fewer factors, otherwise termed principal components in PCR or loading vectors in PLS. In practice, only a subset of factors is significant in modeling the underlying analyte variations, while the others are more likely to be dominated by noise and measurement errors. Although an apparently lower RMSECV may be obtained by including more factors into the calibration model, the reduction in error may be fortuitous and the resulting model may have less predictive capability. Therefore, guidelines exist to help prevent overfitting the data. One such example is published by American Society of Testing and Materials on infrared multivariate calibration[33] and states that a minimum number of six independent samples are needed for each factor included in the model. Extra scrutiny is given to data analysis that does not properly address model dimensionality.

12.4.4.3 Chance or Spurious Correlations Multivariate calibration algorithms are powerful yet somewhat misleading if used without precaution. Owing to the nature of the underdetermined data set, any minute correlation present in the data may be picked up by the algorithm as legitimate analyte-specific variations. For example, Arnold et al.[18] measured the near-infrared absorption spectra of tissue phantoms devoid of glucose and used temporal glucose concentration profiles published by different research groups to demonstrate that the calibration model could produce an apparent correlation with glucose even though none was present. Calibration results such as these could actually satisfy multiple criteria for judging the validity of a calibration model. The lesson here is that chance or spurious correlations may be incorporated into the calibration model even when rigorous validation procedures have been followed. More seriously, if these chance or spurious correlations exist in future measurements, even positive prediction results could be based on nonanalyte-specific effects. Incorporation of prior or additional information by a hybrid method has been shown to provide more immunity to chance correlations.[27–30]

12.4.4.4 "Visualize" Analyte-Specific Signals The difficulty in visualizing analyte-specific information in biological spectra makes it challenging to verify the origin of the spectral information used by the calibration model and confirm that positive results are actually based on a specific analyte, such as glucose. However, some of this information can be obtained by examination of the **b** vector. The **b** vectors obtained from spectroscopic data contain spectral information of all model components and are not merely a collection of numbers. Under ideal, noise-free conditions the regression vector $\mathbf{b}_{ideal}$ can be explicitly derived from the model component spectra, or implicitly obtained from the calibration sample set. This $\mathbf{b}_{ideal}$ is termed the net analyte signal[34,35] or the OLS **b** vector, $\mathbf{b}_{OLS}$. Mathematically, it can be constructed by removing all parts of the analyte spectrum that are nonorthogonal to the spectral features associated with all nonanalytic components within the sample matrix.

Physical interpretation can be gained from considering two simple examples. First, $\mathbf{b}_{\text{ideal}}$ is identical (within a scaling factor) to the pure component spectrum of the analyte of interest, in the absence of chemical components with overlapping spectral features. In other words, if the analyte of interest is the only component that generates an optical signal, that is, absorption or emission, then $\mathbf{b}_{\text{ideal}}$ will be identical to the pure component spectrum of this analyte. So, the regression vector should "look" progressively more like the analyte spectrum as model complexity decreases or as spectral overlap is reduced for the matrix components. For example, when spectral overlap is low, such as in Raman spectroscopy, the spectral features of glucose have been identified in the experimentally derived **b** vector as supporting evidence.[36–38] Second, $\mathbf{b}_{\text{ideal}}$ is orthogonal to all other sources of spectral variance within the sample. This fact leads to a simple checkpoint called pure component selectivity analysis (PCSA).[39] In PCSA, the experimentally derived **b** vector multiplied by the analyte spectrum should give the concentration of the analyte, and **b** multiplied by the spectrum of a cosolute should equal zero.

In most cases, extensive overlap between spectral features associated with the analyte and individual matrix components precludes straightforward interpretation of the regression vector. Even with extensive spectral overlap, however, it is possible to determine the net analyte signal directly from the *in vivo* matrix and then compare this net analyte signal to the regression vector as direct evidence of analyte-specific information within the calibration model.[40]

Although a complete model is virtually never available for *in vivo* experiments, a good approximation is often obtainable. Therefore, a theoretically "correct" regression vector ($\sim\mathbf{b}_{\text{ideal}}$) should be calculated and examined for spectral abnormalities. An explanation must be provided to justify an experimentally derived regression vector that deviates far from $\mathbf{b}_{\text{ideal}}$.

Several papers in the literature have reported the successful measurement of glucose from *in vivo* human spectra collected noninvasively. Unfortunately, the validity of these reports often cannot be judged based on the supplied information. Clearly, the burden of proof is with the investigators to prove that glucose is indeed measured following these four criteria for judgment: (4) theoretical and practical limits, (5) dimensionality, (6) spurious correlations, and (7) analysis of the regression vector.

12.5 NONINVASIVE OPTICAL TECHNIQUES FOR GLUCOSE SENSING

A general summary is provided for direct and indirect optical techniques reported for noninvasive measurements of glucose that have appeared in the peer-reviewed literature within the past two decades. Direct approaches are based on glucose-specific intrinsic molecular properties such as optical absorption, Raman scattering, and optical rotation. Indirect approaches are based on glucose-induced changes in physiological or physical parameters such as the index of refraction and the scattering coefficient. Omar Khalil[14,41] has written two comprehensive reviews spanning the period from 1989 to 2003. It is not our intention here to conduct an exhaustive literature review, but to familiarize the reader with each technique. Fundamental principles and salient features of each technique will be followed by a summary of one or more representative studies, selected by considering the impact on the field and inclusion of sufficient supporting evidence. Complementary information, particularly related to near-infrared absorption spectroscopy, is available elsewhere.[13]

The cited literature includes a variety of study designs. In all, it is essential that glucose concentrations in the calibration samples vary over a wide range. For *in vitro* studies, this requirement can be satisfied by the original sample diversity or by spiking the samples with a glucose stock solution. For *in vivo* studies, one of the three different experimental protocols is typically followed: oral glucose or meal tolerance,[14] time randomized,[42] or glucose clamping.[43] To further increase the range of glucose variation, people or animals with diabetes have been involved in some of the studies.

A concern exists that calibration results based on glucose tolerance protocols are likely to be influenced by spurious or time-dependent correlations within the spectral data set.[18] This problem originates from the fact that many experimental parameters can change as a function of time, while glucose concentrations are also changing as a function of time. If these changing parameters have any measurable impact of the spectroscopic signal, then the statistical-based multivariate calibration model can be based on the statistical correlation between the glucose concentration and the varying parameter. In this case, the calibration model is based on a spurious correlation as opposed to spectroscopic information related specifically to the analyte of interest. Models based on spurious correlations cannot be robust and must be avoided. The best way to avoid such correlations is to randomize the concentration values as best as possible. Therefore, time-randomized and glucose clamping protocols offer a rigorous experimental design that should be used if possible. However, both time randomization and clamping techniques are relatively more costly and require higher compliance from the experimental subjects compared to more standard glucose tolerance tests.

For brevity, results from selected *in vitro* and *in vivo* studies employing either near-infrared absorption spectroscopy or Raman spectroscopy, the most commonly used techniques, are documented in Tables 12.1 and 12.2. In these tables, error estimates are reported with either "CV" or "P" in parentheses, indicating cross-validated or predicted results, respectively. For an explanation of these terms, please refer to Section 12.4.

12.5.1 Direct Approaches

12.5.1.1 Near-Infrared Absorption Spectroscopy Near-infrared absorption spectroscopy is one of the most widely pursued techniques for glucose sensing because of the relatively low cost of the necessary instrumentation, data with high signal-to-noise ratio, and the penetration depth of near-infrared light into biological tissue, which can reach depths of millimeter–centimeter in several windows within the 800–2500 nm (12,500–4000 cm^{-1}) near-infrared spectral region. Light at these wavelengths is absorbed by infrared-active overtone or combination vibrational

TABLE 12.1 Glucose Measurements Using NIR Absorption Spectroscopy

In vitro						
Authors	Spectral range (cm^{-1})	Mode	Sample	# Samples	Protocol	Approximate error (mM)
Haaland et al.[49]	6600–4250	Transmission	Whole blood	Various number from four individuals	Spiked	2 (CV)
Amerov et al.[51]	5000–4000	Transmission	Whole blood	80 Bovine blood	Spiked	0.96 (P)
Small et al.[50]	5000–4000	Transmission	Bovine plasma	69	Spiked	0.4–0.5 (P)
Hazen et al.[55]	5000–4000	Transmission	Human serum	242	Raw hospital samples	1.3 (P)
In vivo						
Author	Spectral range (cm^{-1})	Mode	Site	# Subjects	Protocol	Approximate error (mM)
Robinson et al.[48]	6600–4250	Transmission	Fingertip	1 Diabetic	Tolerance	1.1 (CV)
Marbach et al.[47]	9000–5500	Reflection	Inner lip	133	Time randomized	2.5–3.0 (CV)
Burmeister et al.[42]	7000–5000	Transmission	Tongue	5 Diabetics	Time randomized	>3 (P)
Samann et al.[53]	12,500–7407	Reflection	Fingertip	10 Diabetics		3.1–35.9 (P)
Maruo et al.[56]	6667–5556	Reflection	Forearm	2 Healthy	Tolerance	1–2 (CV)
Maruo et al.[54]	6579–5882	Reflection	Forearm	5 Healthy, 8 ICU	Tolerance	1.5 (P)
Olesberg et al.[40]	5000–4000	Transmission	Rat back	1	Clamp	2.2 (P)

TABLE 12.2 Glucose Measurements Using NIR Raman Spectroscopy

In vitro					
Authors	Excitation (nm)	Sample	# Samples	Protocol	Approximate error (mM)
Berger et al.[65]	830	Serum	66	No further sample preparation	1.5 (CV)
					1.7 (P)
Qu et al.[63]	785	Serum	24	Ultrafiltrated	0.38 (P)
Enejder et al.[62]	830	Whole blood	31	Preselected for hyperglycemia	1.2 (CV)
Rohleder et al.[64]	785	Serum	247	Ultrafiltrated	0.4 (P)
Pelletier et al.[68]	785	Aqueous humor	17	Measured within contact lens	1–1.5 (P)
In vivo					
Author	Excitation (nm)	Site	# Subjects	Protocol	Approximate error (mM)
Enejder et al.[37]	830	Forearm	17 Healthy	Tolerance	0.7–1.5 (CV)
Chaiken et al.[67]	785	Fingertip	25 Diabetics	Time randomized	1.2 (P)

transitions of molecules. Infrared-active transitions are associated with a change in the dipole moment of the molecule. Because overtone and combination transitions are much weaker than fundamental transitions, near-infrared light penetrates deep into living tissue. Tissue penetration depths at the longer wavelengths associated with fundamental vibrational absorption modes are limited to tens of microns due primarily to the strong absorption properties of water. Hence, these wavelengths of light cannot reach the dermis region and, therefore, cannot report clinically relevant levels of glucose in these tissues. The penetration depth in shorter wavelength regions is limited by hemoglobin absorption and light scattering. Thus, the near-infrared region is ideal for probing biological tissues.

The near-infrared spectral range can be roughly divided into three regions that have been explored for the application of noninvasive glucose sensing. The short-wavelength region (14,286–7300 cm^{-1}) includes numerous higher order transitions such as OH (10,654 cm^{-1}) and CH (8881 cm^{-1}) overtones. The spectral region 6500–5500 cm^{-1} corresponds to an OH and CH combination band (6510 cm^{-1}) and a CH overtone (5924 cm^{-1}). The longest wavelength region (5000–4000 cm^{-1}) includes the combination of CH stretching and bending transitions and CCH and OCH deformation bands at 4423 and 4300 cm^{-1}. These ring deformation bands may provide higher specificity for glucose as compared to the other regions. Additional consideration must be given to matrix components with strongly overlapping absorption spectra, such as water, fat, and hemoglobin, when various spectral ranges are employed.[41]

Glucose absorption at physiological concentrations is several orders of magnitude lower than that from water, which is the major background absorber in tissue. In addition, molecular overtone and combination bands are typically broad, leading to overlapping, yet distinctive, spectral features.[44] As a result of this spectral overlap, multivariate calibration techniques are required to extract quantitative analyte-specific information.[45]

Both dispersive and Fourier transform instrumentation are used for the collection of high signal-to-noise spectral data over the near-infrared spectrum. In both cases, a tungsten halogen lamp is used as the broadband light source, intermediate optic elements are used to deliver and collect the light, and a sensitive detector is required to measure the resulting light intensities. In a Fourier transform instrument, a Michelson interferometer is coupled with either a cryogenically cooled InSb detector or a thermoelectrically cooled InGaAs detector. Alternatively, a grating dispersion element can be used in conjunction with an InGaAs photodiode array. Both transmission and reflectance modes have been realized, frequently with fiber-optic probes. Recently, Olesberg et al.[46] demonstrated the use of a tunable diode laser that could significantly simplify instrumentation while increasing the signal-to-noise ratio as compared to lamp illumination.

Both reflection[47] and transmission modes[48] have been demonstrated. The reflection mode has the advantage of being a single-ended instrument; that is, the source and detector are on the same side of the measuring site. This facilitates optical probe design and allows for greater access to various tissue sites. The transmission mode requires the tissue site to be sandwiched between the source and detector optics and is therefore restricted by sample geometry and available space. However, the optical path length is better defined in the transmission mode than in the reflection mode, which may reduce error and simplify signal processing.

Early *in vitro* experiments in blood or plasma samples spiked with glucose demonstrated the detection capabilities of near-infrared absorption spectroscopy.[49,50] Significantly better results were obtained in plasma than in blood, due to higher optical throughput of plasma compared to whole blood, which is a highly scattering matrix. Results in scattering media illustrate that lower measurement errors are achieved in solutions with less scattering.[51,52] Basically, near-infrared measurement errors are inversely related to the SNR of the corresponding analytical spectra, and, in general, near-infrared spectra are collected under situations where spectrometer performance is limited by detector noise. Under detector-noise limited conditions, spectral SNR's increase with more radiant power at the detector and radiant power at the detector is degraded in scattering media compared to nonscattering media. Hence, larger measurement errors are observed in blood compared to plasma owing to increased light scattering and lower optical throughput of the whole-blood matrix.

Marbach et al.[47] analyzed *in vivo* diffuse reflectance spectra obtained from the inner lip of human subjects and discovered a lag time of $\sim$10 min for the glucose concentration in the optically probed volume relative to the blood glucose concentration. This time lag has a profound impact on the development of a noninvasive technique, as a significant portion of the spectroscopic signal originates from glucose molecules contained in tissue fluid other than blood. Samann et al.[53] evaluated the long-term accuracy of a near-infrared calibration algorithm, and the resulting wide range of errors demonstrated the need for very stable instrumentation and algorithms robust enough to accept changes in subject physiology. Maruo et al.[54] employed a novel numerically simulated calibration model to perform glucose predictions within several hours of the calibration phase. Using a glucose clamping protocol, Olesberg et al.[40] found spectral residuals similar to the glucose net analyte signal by removing principal components obtained during a fasting condition from spectra obtained during a hyperglycemic period. Olesberg's findings illustrate the presence of glucose-specific spectral information within noninvasive near-infrared spectra.

Table 12.1 summarizes results from the above and other selected near-infrared noninvasive experiments.

12.5.1.2 *Mid-Infrared Absorption Spectroscopy* To reduce the amount of spectral overlap, longer wavelength mid-infrared radiation in the 2.5–25 μm (4000–400 cm^{-1}) spectral range can be used to measure the fundamental vibrations of glucose. Mid-infrared tissue absorption spectra contain sharp peaks allowing for better molecular specificity. However, the absorption of water in this spectral range is orders of magnitude higher than in the near-infrared region, resulting in a much reduced penetration depth of light in biological tissue on the order of 10–100 μm. Hence, glucose-containing fluid cannot be easily sampled in *in vivo* applications.

A mid-infrared absorption instrument generally consists of a Fourier transform design with the same basic components as noted above for the Fourier transform near-infrared spectrometers (broadband light source, Michelson interferometer, and detector optimized for the mid-infrared spectral region.)

Most of the reported mid-infrared research was carried out *in vitro*, including those using dried samples,[57] to reduce water absorption. Although mid-infrared absorption spectroscopy can extract clinical information from blood or serum,[58] its penetration depth is severely limited and therefore restricted to near-surface measurements in tissue. Heise and Marbach[59] attempted to measure glucose in oral mucosa and concluded that no clear evidence exists that glucose can be detected. The inability to penetrate deep into human tissue has lead the Heise group to explore an alternative use of mid-infrared spectroscopy for measuring glucose in subcutaneous samples of interstitial fluids collected by microdialysis sampling techniques.[60]

12.5.1.3 Near-Infrared Raman Spectroscopy Fundamental vibrational states of molecules can be probed with any wavelength of light by Raman spectroscopy. In spontaneous Stokes Raman scattering, incident light scattered off a molecule is shifted to longer wavelengths, with the difference in energy corresponding to vibrational transitions of the molecule. The selection rules for Raman scattering and infrared absorption are different, but the molecular information is the same. Infrared-active vibrational transitions alter the dipole of the molecule, whereas Raman-active vibrational transitions alter the polarizability of the molecule. Further comparisons between near-infrared absorption spectroscopy and Raman spectroscopy can be made concerning the data characteristics.[61] Near-infrared spectroscopy offers high SNR data but with broad, overlapping spectral features. Raman spectroscopy offers sharp spectral features, but weak signals result in lower SNR data. In this chapter, we focus on spontaneous Raman scattering. Resonant techniques such as surface-enhanced Raman scattering are discussed in Chapter 15.

Because Raman shifts are independent of excitation wavelength, near-infrared radiation (typically 785 or 830 nm) is chosen for deeper penetration into biological tissue and to reduce the laser-induced fluorescence background that renders UV–visible spontaneous Raman measurements impractical. However, near-infrared tissue fluorescence is still several orders of magnitude larger than physiological glucose Raman signals. Although Raman spectra possess sharp, distinct peaks, the background fluorescence signal can often be a major limitation to this technique because of its spectral variation and the associated detector noise. Multivariate calibration is required to extract glucose-specific concentration information.

A near-infrared Raman instrument consists of a laser source, optic elements for light delivery and collection, and either a Fourier transform spectrometer with an InGaAs detector, a grating with a photodiode array, or a grating with a cooled CCD detector. Owing to the intrinsically weak Raman signal, special considerations are often given to the design of the collection optics. A microscope objective is the most common collection optic used; however, a paraboloidal mirror has also been utilized to increase light collection.[62]

In vitro measurements have been performed in filtered blood serum,[63,64] blood serum,[65] and whole blood[63]. Rohleder et al.[64] discovered that measurements from serum are greatly improved by ultrafiltration to remove macromolecules that cause intense Raman background and subsequently impair measurement accuracy. Results from whole blood have greater error than results from filtered or unfiltered serum but are still within the clinically acceptable range.[63] Lambert et al.[66] performed measurement in human aqueous humor, simulating measurements in the eyes.

Laser dosimetry concerns may preclude some *in vivo* applications. To date, only two groups[37,67] have reported successful *in vivo* studies on human subjects. Enejder et al.[37] reported measurements of glucose concentrations in 17 nondiabetic volunteers following an oral glucose tolerance protocol. Results based on individual and multiple volunteers demonstrated that a glucose-specific calibration model was likely obtained. Chaiken et al.[67] reported the acquisition of Raman spectra of whole blood *in vivo* using tissue modulation. Glucose concentrations were subsequently extracted via analysis of a particular spectral range of the whole-blood spectra. A calibration model derived from one individual was able to generate meaningful predictions on independent data.

Table 12.2 summarizes these and other selected Raman spectroscopy studies. Most published accounts utilize the same spectral range ($\sim$300–1800 cm^{-1}), and thus specific ranges are not specified here.

12.5.1.4 Optical Activity and Polarimetry Glucose is a chiral molecule that rotates the polarization of incident light.[69] The rotation angle can be measured by polarimetry and is related to the glucose concentration. Polarimetry is often performed at a single wavelength, therefore avoiding the need for multivariate calibration.[70] However, the analysis can be complicated when dealing with birefringent turbid biological tissue.[71]

12.5.2 Indirect Approaches

Light scattering in biological tissue is largely a result of refractive index mismatches across physical boundaries. In the diffusion approximation,[72] the reduced scattering coefficient can be expressed as a function of the number density and diameter of spherical scatterers and the refractive index mismatch between the scatterers and the surrounding medium.[73] It is known that concentration variations of tissue osmolytes change the index mismatch between the extracellular fluid and structural scatterers such as cell membranes and protein matrix, therefore creating measurable differences in the tissue scattering coefficient. Glucose has a much greater effect in altering refractive index compared to other tissue osmolytes, such as potassium chloride, sodium chloride, and urea.[74,75]

12.5.2.1 Diffuse Reflectance Spectroscopy Instead of analyzing the frequency response of the light to extract chemical information as is done in reflectance-mode absorption spectroscopy, diffuse reflectance spectroscopy extracts the bulk absorption and scattering coefficients by fitting the spectrum to a particular model.[76]

A steady-state diffuse reflectance spectroscopy instrument typically includes a broadband light source, intermediate optics, spatially separated delivery–collection optical fiber probes,[77] and a CCD-based grating spectrometer. Frequency-based approaches have also been pursued.[78] Correlations between the glucose concentration and the tissue transport scattering coefficient have been observed.[77,78]

12.5.2.2 Optical Coherence Tomography Optical coherence tomography (OCT) has been used to detect glucose and other analyte concentrations in biological tissue.[73,79–81] Based on interferometry, OCT provides reasonable range (~mm) and depth resolution (~10 μm) for localized tissue reflectance measurements. An OCT system consists of a broadband light source, intermediate optics, a Michelson interferometer, fiber-optic probes, and detector. In particular, Larin et al.[80] demonstrated a correlation between the slope of the OCT signal versus depth and glucose concentration in 15 healthy individuals. It is suggested that glucose-induced local changes in the index of refraction are related to the slope of the OCT signal.

12.5.3 Other Approaches

Thermal emission is based on measuring the fundamental absorption bands of glucose at ~10 μm, using the body's naturally emitted infrared radiation as the source. The detection equipment is similar to that used for infrared absorption spectroscopy. Malchoff et al.[82] reported the evaluation of a prototype that measures the infrared emission from the tympanic membrane.

Photoacoustic spectroscopy is an alternative method to detect absorption in liquids and gases or refractive index changes. The sample is excited by short nanosecond to picosecond laser pulses. Light absorption and subsequent localized heating generates detectable ultrasonic waves, which can be picked up by piezoelectric transducers. Photoacoustic spectroscopy has been used to measure glucose concentrations *in vivo*,[83] but no advantage was shown over near-infrared absorption spectroscopy. Photoacoustic spectroscopy can also be used as an indirect approach that detects refractive index changes.

12.6 CRITERIA FOR SUCCESSFUL NONINVASIVE GLUCOSE MEASUREMENTS

A set of critical criteria can be identified for a successful noninvasive glucose monitor regardless of the instrumental or spectroscopic approach. These criteria center on selectivity of the analytical measurement, SNR of the instrumentation, physical and chemical properties of the measurement site, and robustness of the calibration model. As a group, these criteria impact measurement accuracy and their demonstration largely establishes the feasibility of a given technological approach.

Selectivity is of paramount importance for a successful noninvasive glucose measurement. The collected spectroscopic information must contain a selective signature for glucose relative to all components within the matrix that can impact the spectroscopic signal. This criterion is true regardless of whether the measurement is based on a direct or indirect approach. The basis of chemical selectivity must be established and clearly distinguished. Analyte distinction is particularly important given the heavy use of multivariate statistical methods of analysis and the propensity to either overmodel the data or base calibrations on spurious correlations. An emphasis on the net analyte signal and methods to verify analyte specificity from multivariate calibration models is promising.

Instrument performance must be sufficient to enable the collection of the selective glucose signature in a reliable manner relative to background noise. Ultimately, the SNR of the instrumentation defines the limit of detection for glucose and detailed experimental results are needed to establish the level of SNR that is necessary to measure glucose at clinically relevant concentrations. Tissue phantoms provide an excellent means to establish the relationship between the instrumental SNR and the limit of detection. Instrumentation must then be designed to provide this level of performance for spectral data collected noninvasively from living tissue.

The physical and chemical properties of the measurement site greatly influence accuracy of noninvasive clinical measurements. Noteworthy physical parameters include thickness, scattering properties, and temperature of the tissue at the measurement site. Chemical issues center on the molecular makeup of the tissue (water, protein, fats, amino acids, glycolytic structures, etc.) and the heterogeneous distribution of these chemical components throughout the measurement site.

Thickness of the tissue is critical because it determines the number of analyte molecules detected, thereby defining the limit of detection for a given instrumental SNR. Basically, the sampled tissue layer must be sufficiently thick to provide enough analyte molecules within the optical path to generate a signal that can be distinguished from the background noise. If tissue layers are too thick, however, excessive scattering can result in a loss of raw signal and the SNR is reduced.

Distribution of the tissue structure within a measurement site must also be understood and its impact on the analytical measurement established. For spectroscopic methods, the incident light enters the tissue and interacts with a multitude of cellular and noncellular structures as it propagates through the tissue. Depending on the measurement site and type of spectroscopy, the measured light might interact with a wide variety of tissue structures. For skin measurements, the light can potentially interact with the epidermis, dermis, and subcutaneous tissue, each of which presents a different chemical composition. If one assumes that the glucose is primarily located in the aqueous fraction of the tissue, then the tissue matrix can be roughly divided into three compartments, the interstitial fluid (outside the cells), intracellular fluid (inside the cells), and blood. Although the concentration of glucose in the interstitial fluid is known to match that in blood under steady-state conditions, the concentration of glucose can vary greatly in these three compartments depending on the mass transport and reaction kinetic properties of different regions within the tissue matrix. Clearly, many factors can influence the concentration of glucose in the localized region of the measurement.

Finally, the robustness of the calibration is a critical parameter that must be established before any noninvasive measurement technology will be useful. Important issues include (1) the ability to collect reliable spectra for each measurement; (2) a protocol required to establish a working calibration model; (3) time stability of a calibration model; and (4) sensitivity of the calibration model to external factors, such as ambient temperature and vibrations. These issues go far beyond demonstrating the feasibility of a noninvasive spectroscopic sensor but pertain to its eventual practical implementation.

REFERENCES

1. Flewelling R. Noninvasive optical monitoring. *The biomedical engineering handbook*, CRC Press/IEEE Press, Boca Raton, FL, 1995, pp. 1346–1356.
2. McMorrow RC, Mythen MG. Pulse oximetry. *Current Opinion in Critical Care* 2006, 12, 269–271.
3. Jubran A. Pulse oximetry. *Intensive Care Medicine* 2004, 30, 2017–2020.
4. Mendelson Y. Pulse oximetry—theory and applications for noninvasive monitoring. *Clinical Chemistry* 1992, 38, 1601–1607.
5. Ferrari M, Mottola L, Quaresima V. Principles, techniques, and limitations of near infrared spectroscopy. *Canadian Journal of Applied Physiology* 2004, 29, 463–487.
6. American Diabetes Association. Self-monitoring of blood glucose. *Diabetes Care* 1994, 17, 81–86.
7. Ruggiero L, Glasgow R, Dryfoos JM, Rossi JS, Prochaska JO, Orleans CT, Prokhorov AV, Rossi SR, Greene GW, Reed GR, Kelly K, Chobanian L, Johnson S. Diabetes self-management. Self-reported recommendations and patterns in a large population. *Diabetes Care* 1977, 20, 568–576.
8. Van den Berghe G, Wouters P, Weekers F, Verwaest C, Bruyninckx F, Schetz M, Vlasselaers D, Ferdinande P, Lauwers P, Bouillon R. Intensive insulin therapy in critically ill patients. *New England Journal of Medicine* 2001, 345, 1359–1367.
9. Conner TM, Flesner-Gurley KR, Barner JC. Hyperglycemia in the hospital setting: the case for improved control among non-diabetics. *Annals of Pharmacotherapy* 2005, 39, 492–501.
10. Lien LF, Angelyn Bethel M, Feinglos MN. In-hospital management of type 2 diabetes mellitus. *Medical Clinics of North America* 2004, 88, 1085–1105.
11. Krinsley JS. Effect of an intensive glucose management protocol on the mortality of critically ill adult patients. *Mayo Clinic Proceedings* 2004, 79, 992–1000.
12. Lien LF, Angelyn Bethel M, Feinglos MN. In-hospital management of type 2 diabetes mellitus. *Medical Clinics of North America* 2004, 88, 1085–1105.
13. Arnold MA, Small GW. Perspectives in analytical chemistry: noninvasive glucose sensing. *Analytical Chemistry* 2005, 77, 5429–5439.
14. Khalil OS. Non-invasive glucose measurements at the dawn of the new millennium: an update. *Diabetes Technology & Therapeutics* 2004, 6, 660–697.
15. Heise HM. Non-invasive monitoring of metabolites using near infrared spectroscopy: state of the art. *Hormone and Metabolic Research* 1996, 28, 527–534.
16. Kohl M, Cope M, Essenpreis M, Boecker D. Influence of glucose concentration on light scattering in tissue simulating phantoms. *Optics Letters* 1994, 19, 2170–2172.
17. Kohl M, Essenpreis M, Cope M. The influence of glucose concentration upon the transport of light in tissue simulating phantoms. *Physics in Medicine and Biology* 1995, 40, 1267–1287.
18. Arnold MA, Burmeister JJ, Small GW. Phantom glucose calibration models from simulated noninvasive human near-infrared spectra. *Analytical Chemistry* 1998, 70, 1773–1781.
19. Martens H, Naes T. *Multivariate Calibration*. John Wiley & Sons, New York, 1989.
20. Geladi P, Kowalski BR. Partial least-squares regression: a tutorial. *Analytica Chimica Acta* 1986, 185, 1–17.

21. Brereton RG. *Chemometrics: Data Analysis for the Laboratory and Chemical Plan*. Wiley, Hoboken, NJ, 2003.
22. Otto M. *Chemometrics: Statistics and Computer Application in Analytical Chemistry*. Wiley-VCH, New York, 1999.
23. Massart DL. *Handbook of Chemometrics And Qualimetrics*. Elsevier, New York, 1997.
24. Gunst RF, Mason RL. *Regression Analysis and Its Application: A Data-Oriented Approach*. Marcel Dekker, New York, 1980.
25. Wold S, Martin H, Wold H. *Lecture Notes in Mathematics*. Springer-Verlag, Heidelberg, 1983.
26. Thomas EV, Haaland DM. Comparison of multivariate calibration methods for quantitative spectral analysis. *Analytical Chemistry* 1990, 62, 1091–1099.
27. Haaland DM, Melgaard DK. New classical least-squares/partial least-squares hybrid algorithm for spectral analyses. *Applied Spectroscopy* 2001, 55, 1–8.
28. Wentzell PD, Andrews DT, Kowalski BR. Maximum likelihood multivariate calibration. *Analytical Chemistry* 1997, 69, 2299–2311.
29. Berger AJ, Koo TW, Itzkan I, Feld MS. An enhanced algorithm for linear multivariate calibration. *Analytical Chemistry* 1998, 70, 623–627.
30. Shih WC, Bechtel KL, Feld MS. Constrained regularization: hybrid method for multivariate calibration. *Analytical Chemistry* 2007, 79, 234–239.
31. Ishimaru A. *Wave Propagation and Scattering in Random Media*. Academic Press, New York, 1978.
32. Wang LH, Jacques SL, Zheng LQ. MCML—Monte Carlo modeling of light transport in multilayered tissues. *Computer Methods and Programs in Biomedicine* 1995, 47, 131–146.
33. ASTM. Standard practice for infrared, multivariate, quantitative analysis, Global Engineering Documents, Philadelphia, 1995.
34. Lorber A. Error propagation and figures of merit for quantification by solving matrix equations. *Analytical Chemistry* 1986, 58, 1167–1172.
35. Lorber A, Faber K, Kowalski BR. Net analyte signal calculation in multivariate calibration. *Analytical Chemistry* 1997, 69, 1620–1626.
36. Enejder AMK, Koo TW, Oh J, Hunter M, Sasic S, Feld MS, Horowitz GL. Blood analysis by Raman spectroscopy. *Optics Letters* 2002, 27, 2004–2006.
37. Enejder AMK, Scecina TG, Oh J, Hunter M, Shih WC, Sasic S, Horowitz GL, Feld MS. Raman spectroscopy for noninvasive glucose measurements. *Journal of Biomedical Optics* 2005, 10, 031114.
38. Berger AJ, Koo TW, Itzkan I, Horowitz G, Feld MS. Multicomponent blood analysis by near-infrared Raman spectroscopy. *Applied Optics* 1999, 38, 2916–2926.
39. Arnold MA, Small GW, Xiang D, Qui J, Murhammer DW. Pure component selectivity analysis of multivariate calibration models from near-infrared spectra. *Analytical Chemistry* 2004, 76, 2583–2590.
40. Olesberg JT, Liu LZ, Van Zee V, Arnold MA. *In vivo* near-infrared spectroscopy of rat skin tissue with varying blood glucose levels. *Analytical Chemistry* 2006, 78, 215–223.
41. Khalil OS. Spectroscopic and clinical aspects of noninvasive glucose measurements. *Clinical Chemistry* 1999, 45, 165–177.
42. Burmeister JJ, Arnold MA, Small GW. Noninvasive blood glucose measurements by near infrared transmission spectroscopy across human tongues. *Diabetes Technology & Therapeutics* 2000, 2, 5–16.

43. Defronzo RA, Tobin JD, Andres R. Glucose clamp technique: method for quantifying insulin-secretion and resistance. *American Journal of Physiology* 1979, 237, E214–E223.

44. Chen J, Arnold MA, Small GW. Comparison of combination and first overtone spectral regions for near infrared calibration models for glucose and other biomolecules in aqueous solutions. *Analytical Chemistry* 2004, 76, 5405–5413.

45. Small GW. Chemometrics and near infrared spectroscopy: avoiding the pitfalls. *Trends in Analytical Chemistry* 2006, 25, 1057–1066.

46. Olesberg JT, Arnold MA, Mermelstein C, Schmitz J, Wagner J. Tunable laser diode system for noninvasive blood glucose measurements. *Applied Spectroscopy* 2005, 59, 1480–1484.

47. Marbach R, Koschinsky T, Gries FA, Heise HM. Noninvasive blood glucose assay by near-infrared diffuse reflectance spectroscopy of the human inner lip. *Applied Spectroscopy* 1993, 47, 875–881.

48. Robinson MR, Eaton RP, Haaland DM, Koepp GW, Thomas EV, Stallard BR, Robinson PL. Noninvasive glucose monitoring in diabetic patients: a preliminary evaluation. *Clinical Chemistry* 1992, 38, 1618–1622.

49. Haaland DM, Robinson MR, Koepp GW, Thomas EV, Eaton RP. Reagentless near-infrared determination of glucose in whole blood using multivariate calibration. *Applied Spectroscopy* 1992, 46, 1575–1578.

50. Small GW, Arnold MA, Marquardt LA. Strategies for coupling digital filtering with partial least-squares regression: application to the determination of glucose in plasma by Fourier-transform near-infrared spectroscopy. *Analytical Chemistry* 1993, 65, 3279–3289.

51. Amerov AK, Chen J, Small GW, Arnold MA. Scattering and absorption effects in the determination of glucose in whole blood by near-infrared spectroscopy. *Analytical Chemistry* 2005, 77, 4587–4594.

52. Green CE, Wiencek JM, Arnold MA. Multivariate calibration models for lysozyme from near-infrared transmission spectra in scattering solutions. *Analytical Chemistry* 2002, 74, 3392–3399.

53. Samann A, Fischbacher C, Jagemann KU, Danzer K, Schuler J, Papenkordt L, Muller UA. Non-invasive blood glucose monitoring by means of near infrared spectroscopy: investigation of long-term accuracy and stability. *Experimental and Clinical Endocrinology & Diabetes* 2000, 108, 406–413.

54. Maruo K, Oota T, Tsurugi M, Nakagawa T, Arimoto H, Hayakawa M, Tamura M, Ozaki Y, Yamada Y. Noninvasive near-infrared blood glucose monitoring using a calibration model built by a numerical simulation method: trial application to patients in an intensive care unit. *Applied Spectroscopy* 2006, 60, 1423–1431.

55. Hazen KH, Arnold MA, Small GW. Measurement of glucose and other analytes in undiluted human serum with near-infrared transmission spectroscopy. *Analytica Chimica Acta* 1998, 371, 255–267.

56. Maruo K, Tsurugi M, Tamura M, Ozaki Y. *In vivo* noninvasive measurement of blood glucose by near-infrared diffuse-reflectance spectroscopy. *Applied Spectroscopy* 2003, 57, 1236–1244.

57. Diessel E, Willmann S, Kamphaus P, Kurte R, Damm U, Heise HM. Glucose quantification in dried-down nanoliter samples using mid-infrared attenuated total reflection spectroscopy. *Applied Spectroscopy* 2004, 58, 442–450.

58. Kim YJ, Hahn S, Yoon G. Determination of glucose in whole blood samples by mid-infrared spectroscopy. *Applied Optics* 2003, 42, 745–749.

59. Heise HM, Marbach R. Human oral mucosa studies with varying blood glucose concentration by non-invasive ATR-FT-IR-spectroscopy. *Cellular and Molecular Biology* 1998, 44, 899–912.
60. Vogt O, Damm U, Heise HM. Towards reagent-free continuous blood glucose monitoring for the critical care environment using micro-dialysis and infrared spectrometry. Proceedings of the IASTED Internal Conference, Biomedical Engineering, Innsbruck, Austria, 2005, pp. 305–308.
61. Ren M, Arnold MA. Comparison of multivariate calibration models for glucose, lactate and urea from near infrared and Raman spectra. *Analytical and Bioanalytical Chemistry* 2007, 387, 879–888.
62. Enejder AMK, Koo TW, Oh J, Hunter M, Sasic S, Feld MS, Horowitz GL. Blood analysis by Raman spectroscopy. *Optics Letters* 2002, 27, 2004–2006.
63. Qu JNY, Wilson BC, Suria D. Concentration measurements of multiple analytes in human sera by near-infrared laser Raman spectroscopy. *Applied Optics* 1999, 38, 5491–5498.
64. Rohleder D, Kiefer W, Petrich W. Quantitative analysis of serum and serum ultrafiltrate by means of Raman spectroscopy. *Analyst* 2004, 129, 906–911.
65. Berger AJ, Koo TW, Itzkan I, Horowitz G, Feld MS. Multicomponent blood analysis by near-infrared Raman spectroscopy. *Applied Optics* 1999, 38, 2916–2926.
66. Lambert JL, Morookian JM, Sirk SJ, Borchert MS. Measurement of aqueous glucose in a model anterior chamber using Raman spectroscopy. *Journal of Raman Spectroscopy* 2002, 33, 524–529.
67. Chaiken J, Finney W, Knudson PE, Weinstock RS, Khan M, Bussjager RJ, Hagrman D, Hagrman P, Zhao YW, Peterson CM, Peterson K. Effect of hemoglobin concentration variation on the accuracy and precision of glucose analysis using tissue modulated, noninvasive, *in vivo* Raman spectroscopy of human blood: a small clinical study. *Journal of Biomedical Optics* 2005, 10, 031111.
68. Pelletier CC, Lambert JL, Borchert M. Determination of glucose in human aqueous humor using Raman spectroscopy and designed-solution calibration. *Applied Spectroscopy* 2005, 59, 1024–1031.
69. Barron LD. *Molecular Light Scattering and Optical Activity*, 2nd edn. Cambridge University Press, Cambridge, UK; New York, 2004, xxii, 443 pp.
70. Cameron BD, Cote GL. Noninvasive glucose sensing utilizing a digital closed-loop polarimetric approach. *IEEE Transactions on Biomedical Engineering* 1997, 44, 1221–1227.
71. Wan QJ, Coté GL, Dixon IB. Dual-wavelength polarimetry for monitoring glucose in the presence of varying birefringence. *Journal of Biomedical Optics* 2005, 10, 024029.
72. Ishimaru A. *Wave Propagation and Scattering in Random Media*. Academic Press, New York, 1978.
73. Graaff R, Aarnoudse JG, Zijp JR, Sloot PMA, Demul FFM, Greve J, Koelink MH. Reduced light-scattering properties for mixtures of spherical-particles: a simple approximation derived from Mie calculations. *Applied Optics* 1992, 31, 1370–1376.
74. Weast RC. *CRC Handbook of Chemistry and Physics*. CRC Press, Cleveland, OH, 1977, v.
75. Larin KV, Akkin T, Esenaliev RO, Motamedi M, Milner TE. Phase-sensitive optical low-coherence reflectometry for the detection of analyte concentrations. *Applied Optics* 2004, 43, 3408–3414.
76. Farrell TJ, Patterson MS, Wilson B. A diffusion-theory model of spatially resolved, steady-state diffuse reflectance for the noninvasive determination of tissue optical properties *in vivo*. *Medical Physics* 1992, 19, 879–888.

77. Bruulsema JT, Hayward JE, Farrell TJ, Patterson MS, Heinemann L, Berger M, Koschinsky T, Sandahl-Christiansen J, Orskov H. Correlation between blood glucose concentration in diabetics and noninvasively measured tissue optical scattering coefficient. *Optics Letters* 1997, 22, 190–192.
78. Maier JS, Walker SA, Fantini S, Franceschini MA, Gratton E. Possible correlation between blood glucose concentration and the reduced scattering coefficient of tissues in the near-infrared. *Optics Letters* 1994, 19, 2062–2064.
79. Ghosn MG, Tuchin VV, Larin KV. Depth-resolved monitoring of glucose diffusion in tissues by using optical coherence tomography. *Optics Letters* 2006, 31, 2314–2316.
80. Larin KV, Eledrisi MS, Motamedi M, Esenaliev RO. Noninvasive blood glucose monitoring with optical coherence tomography: a pilot study in human subjects. *Diabetes Care* 2002, 25, 2263–2267.
81. Esenaliev RO, Larin KV, Larina IV, Motamedi M. Noninvasive monitoring of glucose concentration with optical coherence tomography. *Optics Letters* 2001, 26, 992–994.
82. Malchoff CD, Shoukri K, Landau JI, Buchert JM. A novel noninvasive blood glucose monitor. *Diabetes Care* 2002, 25, 2268–2275.
83. MacKenzie HA, Ashton HS, Spiers S, Shen YC, Freeborn SS, Hannigan J, Lindberg J, Rae P. Advances in photoacoustic noninvasive glucose testing. *Clinical Chemistry* 1999, 45, 1587–1595.

CHAPTER 13

NEAR-INFRARED SPECTROSCOPY FOR NONINVASIVE GLUCOSE SENSING

Mark A. Arnold, Jonathon T. Olesberg, and Gary W. Small

In Vivo Glucose Sensing, Edited by David D. Cunningham and Julie A. Stenken

13.1 INTRODUCTION TO NEAR-INFRARED SPECTROSCOPY

The near-infrared region of the electromagnetic spectrum begins just beyond the red wavelengths of the visible spectrum and ends at the beginning of the mid-infrared spectrum. In wavelength and wavenumber units, this corresponds to 700–2500 nm and 14,285–4000 cm^{-1}, respectively. This wide spectral range incorporates absorption features that originate from low-energy electronic transitions as well as combination and overtone transitions associated primarily with vibrational modes of N–H, O–H, and C–H bonds.[1–3]

The absorption properties of water are critical for near-infrared spectroscopy in biological matrices. Water possesses several prominent absorption bands throughout the near-infrared spectrum.[4,5] Given the high concentration of water in many biological matrices, these absorption bands define a set of three spectral windows through which near-infrared spectra can be collected in aqueous matrices. These windows correspond to the (1) short-wavelength near-infrared (14,285–8500 cm^{-1}), (2) first-overtone (7500–5500 cm^{-1}), and (3) combination (5000–4000 cm^{-1}) spectral ranges. The combination spectral range corresponds to molecular transitions associated with combinations of stretching and bending vibrations of C–H, N–H, and O–H bonds. The first-overtone region corresponds to the first overtone of stretching vibrations and the short-wavelength spectral range corresponds to higher order combination and overtone transitions. The nature of these combination and overtone transitions results in weak absorptions and broad spectral features compared to the fundamental mid-infrared absorption bands.

As one progresses from combination spectra to first-overtone spectra to short-wavelength near-infrared spectra, the absorption bands diminish in magnitude and broaden. Weaker and broader absorption bands impact spectroscopic measurements in two critical ways. First, broader bands result in greater spectral overlap between sample components, which negatively impacts both the selectivity and sensitivity of analytical measurements. Second, lower absorptivities result in lower signals for the sample components, which negatively impact the measurement sensitivity and limit of detection. On the other hand, lower absorptivities for solute and solvent molecules alike provide for longer optical paths through aqueous samples. Generally, sample thicknesses are limited to several millimeters for the combination spectra, 10 mm for first-overtone spectra, and several centimeters for short-wavelength near-infrared spectra. To some extent, the longer optical path lengths that are possible for the first-overtone spectra can counterbalance lower solute absorptivities in terms of measurement sensitivity compared to combination spectra.[6]

Although composed of weak and overlapping spectral features, near-infrared spectra can be used to extract analytical information from complex sample matrices. Chemical sensing with in-line near-infrared spectroscopy is a general technique that can be used to quantify multiple analytes in complex matrices, often without reagents or sample pretreatment.[7–9] Applications are widespread in the food sciences, agricultural industry, petroleum refining, and process analytical chemistry.[10–13] These activities demonstrate that near-infrared spectroscopy can provide selective and accurate quantitative measurements both rapidly and nondestructively.

Biomedical applications of near-infrared spectroscopy are limited in comparison. Near-infrared wavelengths of light are used for pulse oximetry[14–18] and are proposed for imaging tumors in soft tissue.[19–22] Pulse oximetry is an enormously successful technology that monitors the percentage of hemoglobin saturated with oxygen. Tumor detection by scattered near-infrared photons is an experimental technology with promise as a potential alternative to mammograms for the early detection of breast cancer. Both of these technologies rely on light in the short-wavelength region of the near-infrared spectrum (0.7–1.0 μm). Light at these wavelengths is not strongly absorbed by the components of living tissue, so this light penetrates several centimeters into tissue and permits either the probing of blood or the detection of tumors.

13.2 CONCEPT OF NONINVASIVE SPECTROSCOPIC MEASUREMENTS

Noninvasive spectroscopic measurements involve passing a selected band of near-infrared radiation through a vascular region of tissue. As this light propagates through the tissue, photons are both scattered and absorbed by the tissue matrix.[23] Light exiting the tissue is then collected and the concentration of the component of interest is determined from an analysis of the resulting tissue spectrum. Typically, a multivariate calibration method is used to relate features within the noninvasive tissue spectrum and the desired analyte concentration, as described in Chapter 12.

13.2.1 Measurement Configurations

Several optical configurations are reported for collecting noninvasive near-infrared spectra of living tissue.[24,25] Transmission, diffuse reflectance, transflectance, and photoacoustic spectroscopic configurations have been used and the basic features of each are illustrated in Figure 13.1. For transmission measurements, the probing radiation propagates completely through a layer of tissue with the collected photons exiting the tissue from the opposite side relative to the incident light. This configuration demands access to both sides of the tissue matrix, which can limit the measurement site.

In a diffuse reflectance configuration, a fraction of the incident photons are backscattered by the tissue matrix and directed to the detection optics.[26] The penetration depth of the detected photons depends on several factors, including the wavelength of the incident radiation, the angle of incidence, dimensions of the scattering bodies within the tissue matrix, and the refractive index mismatch between these scattering bodies and the surrounding matrix (i.e., interstitial fluid). Spectroscopic information embedded within a typical diffuse reflectance measurement is heavily weighted toward the outer surface of the sample. For diffuse reflectance spectra of skin, the collected light is obtained primarily from the epidermis layer of the skin tissue, which has little interstitial fluid and is not ideal for measuring *in vivo* levels of glucose.

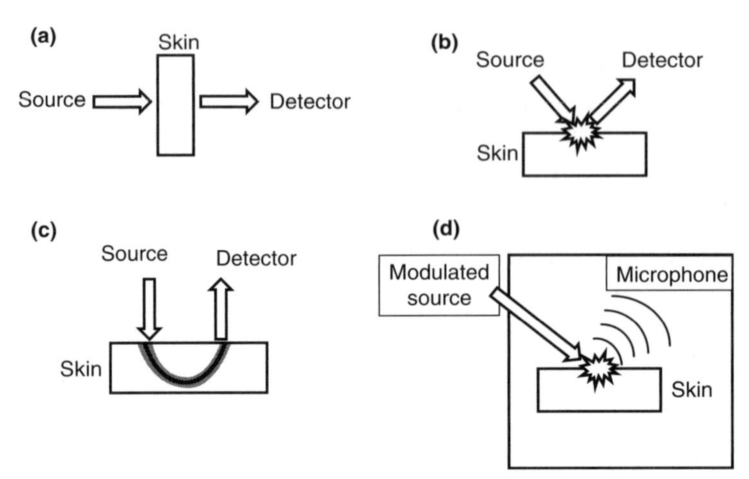

Figure 13.1 Schematic diagram of different measurement configurations: (a) transmission, (b) diffuse reflectance, (c) transflectance, and (d) photoacoustic.

A second form of diffuse reflectance spectroscopy (the transflectance measurement geometry) is commonly used to probe more deeply into the tissue matrix and collect spectra from the dermis layer of the skin. For a transflectance measurement, a set of optical fibers is used to launch the incident light into the skin matrix and also to collect a fraction of the diffuse reflected light from the skin.[27] As illustrated in Figure 13.1c, the source fiber is physically separated from the detector fiber by an optimized distance. In this case, the collected light corresponds to multiscattered photons that propagate through the dermis layer of the skin. The separation distance between the source and detector fibers controls the penetration depth of the collected photons. The depth of penetration increases as the distance between the fibers increases. It also follows that the intensity of the collected light decreases significantly with an increase in the distance between the source and detector fibers owing to an increase in the number of scattering events before collection by the detector fiber. In practice, a ring of detection fibers is commonly used to improve the signal-to-noise ratio (SNR) of transreflectance measurements.

Finally, photoacoustic spectroscopy is reported for the collection of tissue spectra.[28,29] In these measurements, the sample is located in a sealed chamber and the incident beam is modulated at an acoustic frequency. Absorption of light generates localized heating that creates an acoustic wave that is detected by a sensitive microphone. Photoacoustic measurements are generally ideal for highly scattering or opaque samples. As noted for diffuse reflectance spectroscopy, photoacoustic measurements are heavily weighted to the surface, which may be limiting for noninvasive *in vivo* measurements.

13.2.2 Measurement Selectivity

Any practical clinical application of noninvasive near-infrared spectroscopy demands selective quantification of the targeted matrix component. For spectroscopic measurements, selectivity originates from the unique spectral features of the targeted analyte relative to the other chemical components of the matrix that absorb the propagating near-infrared radiation. As noted above, molecular absorptions associated with near-infrared spectra are weak and overlapping. For a given analyte and sample matrix, the relative strengths and degree of overlap of the absorption bands dictate the viability of a near-infrared measurement.

Individual near-infrared absorption bands have absorptivities on the order of 10^{-4} AU/mm/mM for peak absorption bands in aqueous matrices.[30] Such low absorptivities limit detection to the major components within skin tissue. As a general rule of thumb, substances must be present at concentrations above 1 mM to be quantified by near-infrared spectroscopy. Although such low absorptivities greatly restrict the number of possible analytes one can measure in clinical samples, the inability to measure chemicals present below millimolar concentrations enhances selectivity by rendering measurements insensitive to many different types of endogenous molecules. Only the major chemical components of these biological samples must be considered for selectivity purposes.

The overlapping nature of near-infrared spectra complicates the determination of a particular component in the presence of others. Indeed, it is rare to find a single wavelength for which absorption of light can be attributed to a single matrix component. As a result, multivariate calibration methods are commonly used to quantify a targeted species in complex samples. As described in Chapter 12, multivariate methods generally provide selectivity while filtering out high-frequency noise, thereby enhancing both accuracy and precision. Factor-based regression methods are most commonly used to achieve a calibration vector that selectively quantifies the targeted analyte.

13.2.3 Optimizing Measurement Parameters

Several key parameters must be considered to achieve selectivity and optimal analytical performance. At a minimum, spectral range, signal-to-noise ratio, and sample thickness or optical path length must be considered. In practice, these parameters are interdependent and their optimization can be difficult to realize. A detailed analysis of these parameters has been published.[31]

Spectral range is a major parameter, particularly for quantitative measurements in aqueous samples. As noted above, the high concentration of water in biological samples creates three regions for analysis. Although many literature reports indicate that spectra were collected from over all these regions simultaneously, it is not possible to establish one set of conditions that is ideal for collecting spectra from aqueous samples over the full near-infrared spectrum. Different spectral regions demand different sample thicknesses for optimal performance.[6] The best sample thickness is determined by a compromise between optical throughput and measurement sensitivity.[6,32] According to the Beer–Lambert law, an increase in the sample thickness increases the sensitivity between the degree of light absorbance and analyte concentration. With an absorbing solvent (e.g., 55 M water), the radiant power of the transmitted light will decrease exponentially as a function of the optical path length. In many cases, near-infrared spectroscopic measurements are limited by detector noise and as such the signal-to-noise ratio decreases with a decrease in the radiant power at the detector. As for any analytical measurement, the sensitivity and limit of detection are directly proportional to the SNR measurement. Overall, the ideal sample thickness is determined as a compromise between the linear relationship between sensitivity and sample thickness versus the inverse exponential relationship between SNR and sample thickness.[31]

At the limit of detection, the number of analyte molecules within the optical path must be sufficiently high to generate an absorbance that can be distinguished from the underlying spectral noise. The number of detected molecules is controlled by both the analyte concentration and sample thickness, and the magnitude of absorbance is controlled by the inherent absorptivity of the relevant functional groups on the analyte molecule. For noninvasive glucose measurements, the targeted concentration range is controlled by the clinical chemistry and the absorptivity is set at each wavelength by the molecular structure of glucose, which leaves sample thickness as the sole adjustable parameter to enhance measurement sensitivity. Ultimately, the ideal sample thickness depends upon the instrumental SNR for a targeted limit of detection.

13.3 ANALYTICAL MEASUREMENTS IN BIOLOGICAL MIXTURES

The analytical utility of near-infrared spectroscopy can be demonstrated by an analysis of mixtures composed of glucose, lactate, urea, alanine, ascorbate, and triacetin in a pH 6.8 aqueous phosphate buffer.[6] The chemical structures of these test compounds are presented in Figure 13.2. These components were selected to represent different classes of molecules expected in typical biological matrices. Glucose represents carbohydrates; lactate represents small organic acids; urea is a

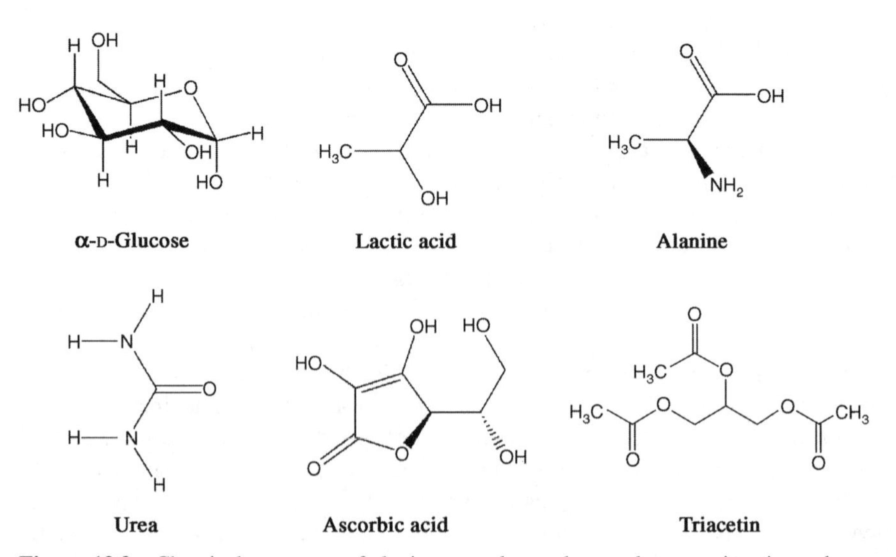

Figure 13.2 Chemical structures of alanine, ascorbate, glucose, lactate, triacetin, and urea.

natural component of blood and interstitial fluid; alanine represents amino acids; ascorbate is a common biological interference for electrochemical biosensor measurements of glucose; and triacetin represents triglycerides.

13.3.1 Near-Infrared Spectra

Individual near-infrared absorption spectra are presented in Figure 13.3 for each of the mixture components. Figure 13.3 presents absorption spectra over the first-overtone and combination spectral regions, respectively. Each spectrum was collected from 100 mM solution of the selected solute dissolved in a pH 6.8 phosphate buffer solution and absorbance was calculated relative to a reference spectrum of the blank phosphate buffer.

Several noteworthy features of near-infrared spectroscopy are illustrated in the spectra presented in Figure 13.3. First, these spectra illustrate the large degree of spectral overlap that is commonly associated with near-infrared spectra. There is no single wavelength or wavenumber that can be used to monitor glucose selectively in solutions composed of these solutes. This strong spectral overlap calls into question the selectivity of analytical methods based on near-infrared spectra. Successful quantitative near-infrared measurements demand the use of multivariate analysis methods to distinguish the target analyte from the matrix components.

The spectra in Figure 13.3 also highlight the small absorption coefficients associated with these absorption bands. As noted above, molar absorptivities are generally on the order of $10^{-4}\,\mathrm{mM}^{-1}\,\mathrm{mm}^{-1}$ for combination wavelengths. In the first-overtone region, absorptivities decrease to the range of $10^{-5}\,\mathrm{mM}^{-1}\,\mathrm{mm}^{-1}$.[30] Such low absorptivities, coupled with millimeter thick samples, demand high signal-to-noise ratios to achieve submillimolar limits of detection. An instrumental SNR on the order of 50,000 is required to monitor the tens of microabsorbance units expected under such conditions. For example, an absorbance value of 10^{-4} is expected for a 1 mm thick sample with a concentration of 1 mM when the molar absorptivity is $10^{-4}\,\mathrm{mM}^{-1}\,\mathrm{mm}^{-1}$. The corresponding ratio of the measured versus incident intensity of light (I/I_0) for this magnitude of absorbance is 1.0002303, where $-\log(I/I_0) = 10^{-4}$. To distinguish I from I_0 under these conditions, the intensity measurement requires a precision of 0.000023 or 10% of the measured intensity value. The associated SNR is 43,000 or 1.0002303/0.000023 for this measurement.

The tiny absorbance values associated with quantitative near-infrared spectroscopy render these measurements extremely sensitive to slight variations in spectrometer alignment. Indeed, slight differences in incident radiant power between the sample and reference spectra create small positive or negative offsets along the absorbance axis. Such offsets are commonly observed for near-infrared spectra of aqueous solutions, as is apparent in the spectra presented in Figure 13.3.

Spectral absorptivities for water over both the first-overtone and combination wavelength regions are similar in magnitude compared to the absorptivities of most organic solutes (e.g., glucose, lactate, among others).[4,30] This fact combined with the high concentration of water in aqueous solutions explains the strong background absorption that characterizes spectra from aqueous solutions. Part of the analytical

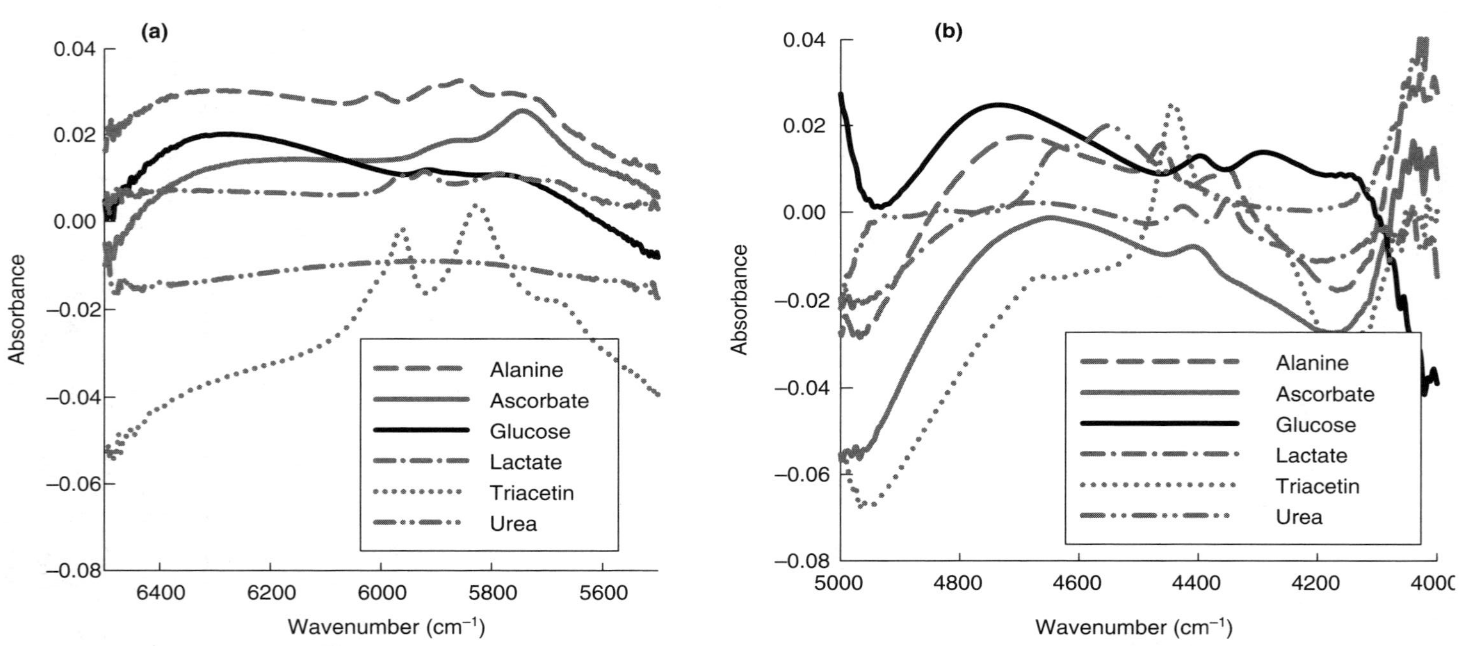

Figure 13.3 Absorption spectra of alanine, ascorbate, glucose, lactate, triacetin, and urea over the first-overtone (a) and combination (b) spectral ranges.

challenge is to measure the analyte absorption accurately against this strong background. Indeed, near-infrared spectra of aqueous samples are highly sensitive to parameters that influence the underlying water absorption spectra, such as solution temperature.[33]

Another consequence of the strong absorption properties of water is the spectral impact of the displacement of water by dissolved solutes. Generally, in absorption spectroscopy, the solvent is selected not to absorb over the wavelength range of interest. When the absorption properties of the solvent are negligible, any displacement of solvent molecules from the optical path by the dissolution of solute molecules has a negligible effect on the measured spectrum. For near-infrared spectra of aqueous solutions, however, the absorption spectrum depends heavily on the degree of water displacement by solutes in the sample.

As glucose is dissolved in an aqueous buffer solution, for example, a fraction of water molecules within the optical path are displaced by glucose molecules. The number of water molecules displaced depends on the specific molar volume of the solute. For glucose, it has been determined that 6.24 molecules of water are displaced for each molecule of glucose dissolved in the solution.[30] The change in the magnitude of light transmittance before and after the dissolution of glucose depends on both the number of displaced water molecules and the relative magnitude of the absorptivities between glucose and water. The observed absorbance, A, can be estimated by the Beer–Lambert law:

$$A = \varepsilon_g b c_g - \varepsilon_w f_{w/g} b c_g \tag{13.1}$$

where ε_g and ε_w are the absorptivities of glucose and water, respectively, at the monitored wavelength, b is the optical path length, c_g is the concentration of glucose, and $f_{w/g}$ is the water displacement coefficient for glucose.[30] This equation indicates that the difference between the molar absorptivity of glucose and the product of the absorptivity of water and the water displacement coefficient $(\varepsilon_g - \varepsilon_w f_{w/g})$ controls whether the solution becomes more or less transparent at the monitored wavelength. If the molar absorptivity of glucose at the monitored wavelength is greater than the product of the water absorptivity and the water displacement coefficient $(\varepsilon_g > \varepsilon_w f_{w/g})$, then the absorbance increases with higher glucose concentrations. If, on the other hand, the molar absorptivity of glucose is less than the product of the water absorptivity and the water displacement coefficient $(\varepsilon_g < \varepsilon_w f_{w/g})$, then the absorbance decreases and the intensity of the transmitted incident radiation increases. Finally, no change in the transmitted light intensity is observed when the molar absorptivity of glucose equals the product of the water absorptivity and the water displacement coefficient $(\varepsilon_g = \varepsilon_w f_{w/g})$.[30]

The third noteworthy feature of near-infrared spectra presented in Figure 13.3 is the uniqueness of the spectral patterns for each analyte. Although the spectral features are highly overlapping, the spectrum for glucose is notably unique relative to the others. The uniqueness of each spectrum provides the selectivity that is required for sound analytical measurements. However, the extensive overlap dictates that an analysis of the full spectrum is needed to extract the unique spectral signature for the targeted analyte relative to the sample matrix. Powerful multivariate methods are available for this purpose, as described in Chapter 12.

13.3.2 Calibration Models for Glucose

The analytical utility of near-infrared spectroscopy can be illustrated by establishing and characterizing calibration models for the measurement of glucose in a set of aqueous solutions composed of glucose, lactate, urea, ascorbate, alanine, and triacetin. In this experiment, 80 different samples were prepared with randomized concentrations of each component ranging from 1 to 35 mM.[6] Solute concentrations were randomized to minimize covariance between the different solute concentrations. All solutions were prepared in a pH 6.8 phosphate buffer solution.

Near-infrared spectra were collected with a Nicolet 670 Nexus Fourier transform (FT) spectrometer.[6] Solutions were thermally equilibrated at $37.0 \pm 0.1°C$ prior to collecting spectra. The following two unique sets of near-infrared spectra were collected: (1) first-overtone spectra (6500–5500 cm^{-1}) with a 7.5 mm optical path length and (2) combination spectra (5000–4000 cm^{-1}) with a 1.5 mm optical path length. Single-beam spectra were collected in triplicate as 256 coadded interferograms that were Fourier transformed to produce spectra with 1.94 cm^{-1} point spacing.

Glucose calibration models were generated individually from first-overtone and combination spectra by using the partial least squares (PLS) algorithm presented in Chapter 12. Both spectral range and number of model factors were optimized to give the lowest measurement errors. This model optimization process involves a random splitting of the spectral data into two subsets of data, where one is used to train or build the calibration model and the other is used exclusively to validate model performance.[34] It is important that all spectra for a given sample go together either into the training or validation data sets.

The ability to quantify glucose is illustrated in the concentration correlation plots presented in Figure 13.4, where Figure 13.4a and b presents results for first-overtone and combination spectra, respectively. In this figure, the concentration of glucose predicted from the optimized PLS calibration model is plotted as a function of the known concentration of glucose in each prepared sample. Ideally, these two

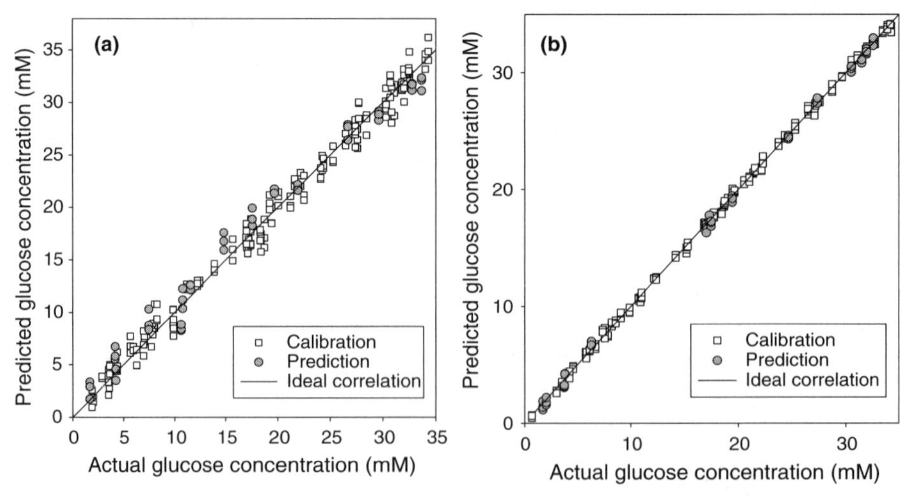

Figure 13.4 PLS calibration models for glucose from near-infrared spectra collected over the first-overtone (a) and combination (b) spectral ranges.

concentrations are equal and the points fall on the unity line with a slope of one and a zero y-intercept. The ideal response is provided in Figure 13.4 as the solid lines. The points in Figure 13.4 are coded to distinguish points used to generate the calibration model (open squares) and those used solely for model validation (gray circles).

Both calibration models represented in Figure 13.4 can accurately predict the concentration of glucose in this complex six-component mixture. In both cases, the predicted concentrations fall along the unity line. Both models demonstrate scatter about the unity line and the degree of scatter is noticeably less for the model constructed from combination spectra. The standard error of prediction (SEP) is a key metric for these calibration models, where the SEP is given by the root mean square of the difference in the predicted and known concentrations, as indicated in equation(13.2):

$$\mathrm{SEP} = \sqrt{\frac{\Sigma(C_\mathrm{p} - C_\mathrm{r})^2}{N_\mathrm{p}}} \tag{13.2}$$

where C_p is the predicted analyte concentration from the calibration model, C_r is the reference or known concentration, and N_p is the number of predicted concentrations. The optimized calibration model from first-overtone spectra uses a spectral range of 6030–5780 cm^{-1} and nine PLS factors to provide an SEP of 1.12 mM. The optimized calibration model from combination spectra uses a spectral range of 4560–4100 cm^{-1} and 11 factors to provide an SEP of 0.45 mM. The smaller SEP for combination spectra is consistent with less scattering about the unity line in Figure 13.4b.

13.3.3 Calibration Models for Other Solutes

An attractive feature of near-infrared spectroscopy is the ability to extract information from multiple analytes from a single spectrum. The same spectra used above to measure glucose in this mixture of six components can be used to generate independent calibration models for the other matrix constituents.

Table 13.1 summarizes results from calibration models generated for each analyte. Several interesting observations are evident from the values in this table.

TABLE 13.1 PLS Calibration Models for the Six-Component Mixture

	First-overtone spectra			Combination spectra		
Analyte	Spectral range (cm^{-1})	Rank[a]	SEP (mM)	Spectral range (cm^{-1})	Rank	SEP (mM)
Glucose	6030–5650	9	1.12	4560–4100	11	0.45
Lactate	6220–5780	9	0.33	4440–4280	8	0.27
Urea	6150–5950	7	7.33	4630–4440	6	0.19
Ascorbate	5980–5580	8	0.45	4560–4000	12	0.53
Alanine	6080–5800	8	0.25	4540–4350	7	0.20
Triacetin	5920–5620	6	0.26	4510–4400	7	0.18

[a] Number of latent variables or factors in the PLS calibration model.

First, the ideal spectral range is different for each analyte. Second, a significant number of factors is needed for each calibration model, ranging from 6 factors for urea and triacetin to 12 factors for ascorbate. Third, the SEP is very high for the urea calibration model generated from first-overtone spectra. In this case, the SEP is 7.33 mM, which is clearly the largest for all these models. In fact, urea has no absorption bands within the monitored first-overtone spectral region, so accurate urea predictions should not be possible from these spectra. This lack of ability to predict urea concentrations from first-overtone spectra demonstrates the effectiveness of the experimental design to minimized concentration correlations between solutes within these mixture solutions.

The SEP values in Table 13.1 further demonstrate that models generated from combination spectra are generally superior to those generated from first-overtone spectra. The superior performance by PLS models based on combination spectra is most notable for glucose and urea. As noted above, urea has no absorption features in the first-overtone region, so a vastly superior performance from combination spectra is expected and observed. For glucose, the superior performance of the combination spectra is related to the distinction of the near-infrared spectra in these two spectral regions. A visual comparison of the spectra in Figure 13.3 reveals greater distinction for the combination spectra compared to the first-overtone spectra.

13.3.4 Net Analyte Signal Vectors and Spectral Range

The spectral comparison between first-overtone and combination spectra noted above can be quantified by comparing the net analyte signal (NAS)[35,36] for glucose over these two spectral ranges.[6] The NAS corresponds to that portion of the analyte spectrum that is orthogonal, or unique, relative to all nonnoise sources of spectral variance within the data set. The sources of spectral variance within the mixture samples include each of the solutes (except urea in the first-overtone range) and variations in the spectral background. Determining the NAS involves treating the individual solution spectrum of each component as a vector. In addition, spectral variance contributed by a series of background spectra can be characterized as a series of orthogonal vectors through a simple principal component analysis, as described in Chapter 12. The spectral vector for glucose will have a component that is colinear and a component that is orthogonal to the subspace defined by the spectral vectors for each cosolute in the mixture matrix and the principal component vectors for the background spectra. The component of the glucose spectral vector that is orthogonal to this subspace is the NAS for glucose.

The length of the NAS vector can be normalized for both concentration and optical path length to give a term that directly relates to the sensitivity of the multivariate calibration model.[31] When little of the analyte spectrum overlaps with the matrix subspace variance, the degree of orthogonality is larger and the magnitude of the NAS vector is greater. Conversely, the more the analyte spectrum directly overlaps with spectral features of the matrix components, the more the analyte spectral vector will be colinear with the matrix subspace vectors and the smaller will be the magnitude of the NAS vector.

Lengths of the NAS for glucose relative to lactate, urea, alanine, ascorbate, and triacetin have been reported for both the first-overtone and combination spectral regions.[6] Values are 9.2×10^{-5} and 3.5×10^{-4} AU/mM^{-1} mm^{-1} for the first-overtone and combination spectral regions, respectively. These values confirm the enhanced sensitivity and superior performance for the PLS calibration models generated from combination spectra.

Overall, the results presented in Table 13.1 illustrate that near-infrared spectroscopy is capable of excellent selectivity despite strong spectral overlap of the chemical components within a sample. Glucose can be measured selectively in the presence of lactate, urea, alanine, ascorbate, and triacetin. Furthermore, superior analytical performance is obtained from combination spectra as opposed to first-overtone near-infrared spectra. The SEP for glucose improves by nearly 2.5-fold in comparison to first-overtone spectra. This improved performance is realized because of greater distinction of the glucose spectrum over the combination wavelengths compared to the first-overtone wavelengths, as illustrated by comparing Figure 13.3a and b. Prediction errors improved more drastically for urea because of a lack of absorption bands over the first-overtone wavelengths.

13.4 MEASUREMENT SELECTIVITY

It is interesting to consider the limits of selectivity for near-infrared spectroscopy, given the broad and overlapping nature of the principal absorption bands. Certainly, a critical test of selectivity is the ability to distinguish glucose, sucrose, and maltose. Sucrose is a disaccharide composed of glucose and fructose subunits and maltose is a disaccharide composed of two glucose subunits. The absorption spectra for these carbohydrates are presented in Figure 13.5. All three spectra consist of three absorption bands centered at approximately 4300, 4400, and 4750 cm^{-1}. Only slight differences in the exact position, width, and magnitude of these absorption bands are

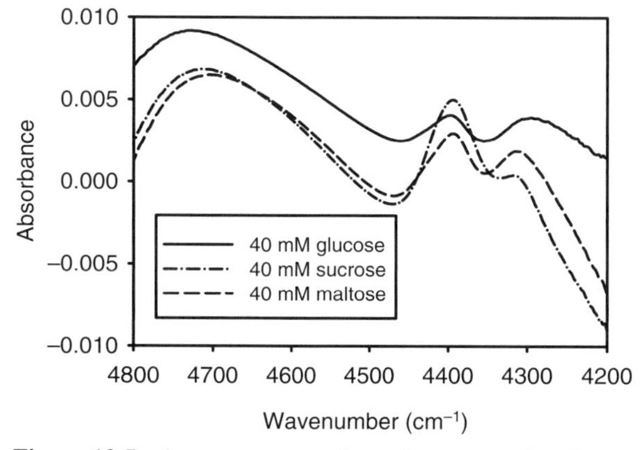

Figure 13.5 Pure component absorption spectra for glucose, sucrose, and maltose over the combination spectral range.

TABLE 13.2 PLS Models for Glucose, Sucrose, and Maltose from Combination Near-Infrared Spectra[37]

Analyte	Spectral range (cm^{-1})	Rank[a]	SEP (mM)
Glucose	4420–4220	5	0.73
Sucrose	4500–4300	5	0.17
Maltose	4550–4400	5	0.35

[a] Number of latent variables or factors in the PLS calibration model.

evident. Are selective measurements possible given the extent of spectral overlap between glucose, sucrose, and maltose?

13.4.1 Selectivity from PLS Calibration Models

Calibration models were established for glucose, sucrose, and maltose from combination near-infrared spectra collected from 50 unique samples composed of glucose, sucrose, and maltose with concentrations ranging 0.5–50 mM.[37] A random concentration design was used to minimize concentration correlations between these solutes and solutions were prepared in a pH 6.8 phosphate buffer solution. Near-infrared spectra were collected as 128 coadded interferograms over the 5000–4000 cm^{-1} spectral range and the processed single-beam spectra had a point spacing of 0.964 cm^{-1}. More details of the experimental design can be found elsewhere.[37]

Results from optimized PLS calibration models for each component are presented in Table 13.2 and Figure 13.6. Both spectral range and number of factors were optimized individually for each component[37] and the results are summarized in Table 13.2. The corresponding concentration correlation plots are presented in Figure 13.6.

Results from the PLS analysis reveal excellent selectivity for each of the mixture components. In each case, the training and validation data fall on the unity line with no indication of systematic errors. The lowest SEP is obtained for sucrose (SEP = 0.17 mM), followed by maltose (SEP = 0.35 mM) and glucose (SEP = 0.73 mM). Such a ranking is expected based on differences in the chemical structure, where sucrose is most distinctive given the presence of a fructose subunit. Prediction errors are twice as high for glucose relative to maltose, which is consistent with twice the number of glucose moieties per mole of maltose compared to glucose.

13.4.2 Pure Component Selectivity Analysis

Although the findings presented in Table 13.2 and Figure 13.6 indicate excellent selectivity, a more rigorous analysis of selectivity is warranted given the chemical similarity and extreme spectral overlap between glucose, sucrose, and maltose. A pure component selectivity analysis (PCSA) can be used to further characterize the selectivity of the PLS calibration model.[31,37]

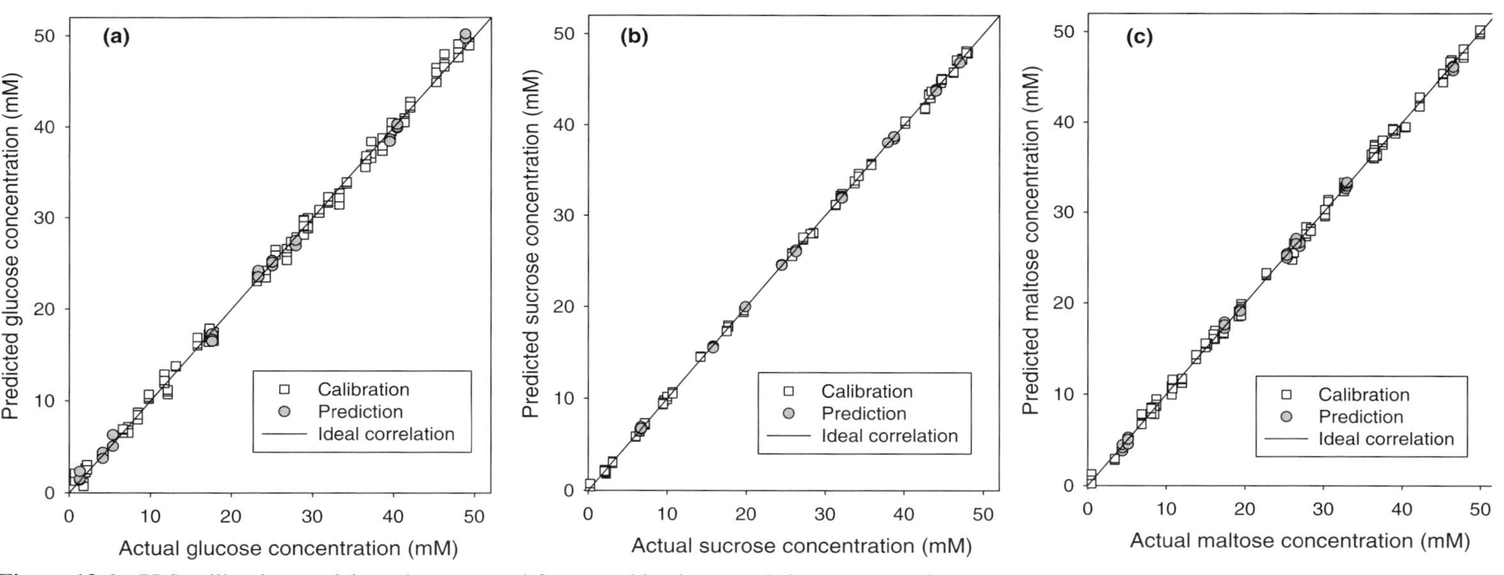

Figure 13.6 PLS calibration model results generated from combination near-infrared spectra for glucose (a), sucrose (b), and maltose (c).

In the PCSA method, the calibration model is generated in the normal way by using a set of calibration spectra to establish the multivariate relationship between spectral variance and analyte concentration. These calibration spectra originate from samples that encompass all the chemical variance expected for subsequent prediction or unknown samples. For the PCSA method, a set of pure component aqueous spectra is collected for the analyte and all other major components of the chemical matrix. Pure component spectra are generated from solutions with only the component of interest and no other cosolutes. Solutions are prepared over a range of concentrations of this pure component. For example, 1, 5, 10, 20, and 40 mM standard solutions might be prepared individually for glucose, sucrose, and maltose. These pure component spectra are sequentially subjected to the multivariate calibration model and the output is evaluated. Ideally, the output from all nonanalyte solutes should be zero, which would indicate that none of the spectral features associated with this nonanalyte component is used by the PLS model to predict the analyte concentration. Likewise, output from the analyte pure component spectra should produce concentrations that match analyte concentrations in these pure component solutions. Correct concentrations from the analyte pure component spectra indicate that all the required calibration information resides within the analyte spectrum. In this case, the analyte spectrum provides the chemical basis of selectivity for the measurement.

An example of the PCSA method is provided in Figure 13.7 for the measurement of glucose in the mixtures of glucose, sucrose, and maltose.[37] Figure 13.7 shows

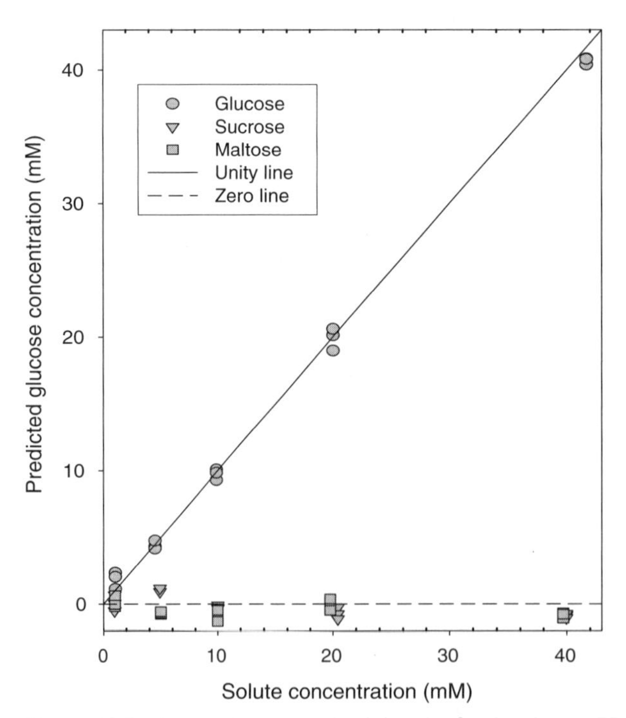

Figure 13.7 Pure component selectivity plot for the glucose PLS calibration models relative to glucose (circle), sucrose (triangle), and maltose (square).

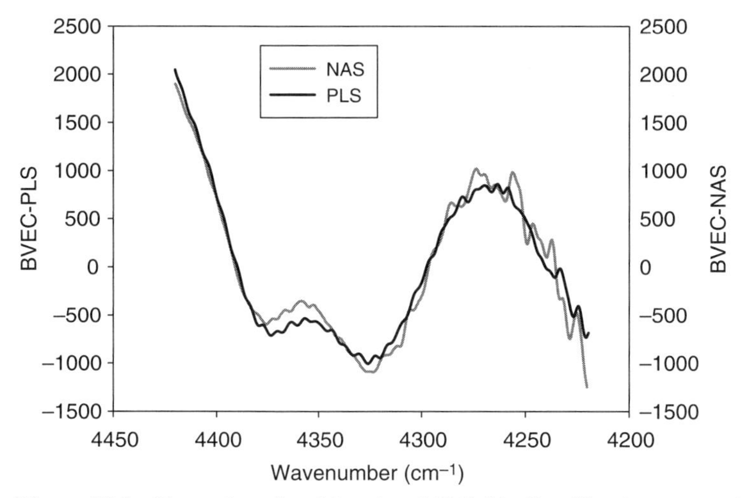

Figure 13.8 Net analyte signal (gray) and PLS (black) calibration vectors for the measurement of glucose in the presence of sucrose and maltose.

the result of the PCSA where input spectra from only sucrose or only maltose elicit no response from the PLS calibration model for glucose, while accurate glucose predictions are generated from pure component solutions composed solely of glucose. The solid lines and dashed lines presented in this plot correspond to the ideal unity and zero slope lines, respectively.

The selective response to glucose in this PCSA plot indicates that the PLS calibration model is selective for glucose and unaffected by the spectral features of sucrose and maltose. In terms of the spectra, this selectivity implies that the quantitative measurements are based on the NAS for glucose relative to sucrose and maltose. In this way, the calibration model is orthogonal to the spectral features of both sucrose and maltose. Indeed, Figure 13.8 shows the PLS calibration vector superimposed on the calculated NAS for glucose relative to sucrose and maltose. These vectors overlap, which indicates that the PLS calibration vector corresponds to the NAS, thereby defining the chemical basis of selectivity for this case.

The concept of the PCSA method is general and this method should be applicable to many types of multivariate calibration techniques. As near-infrared and other spectroscopic methods are developed further for noninvasive *in vivo* clinical measurements, it is critical to understand the chemical basis of measurement selectivity. Unfortunately, calibration models generated from multivariate statistics are typically accepted without further investigation. Application of the PCSA method can help to establish the chemical or spectroscopic basis of predicted concentrations.

13.5 NEAR-INFRARED SPECTROSCOPY OF LIVING SKIN

Living tissue is a nonideal matrix for spectroscopy. The transmission of light through living tissue is complicated by numerous phenomena related to complex interactions between the propagating light with the heterogeneous nature of the tissue matrix.

These interactions greatly complicate the basic spectroscopy of skin and amplify the difficulty of measuring glucose noninvasively.

13.5.1 Spectral Properties of Rat and Human Skins

Living skin can be both highly scattering and highly absorbing depending on the wavelengths of light used for the noninvasive measurement.[38,39] Across the near-infrared spectrum, shorter wavelengths tend to be more highly scattering and less absorbing. Over the combination spectral range (5000–4000 cm^{-1}), skin is less scattering and more absorbing compared to either the first-overtone or shorter wavelength regions of the near-infrared spectrum. Scattering is created by the interaction of the propagating light wave with a multitude of interfaces with differences in refractive index. Absorption is caused primarily by interactions with water within the tissue matrix. In general, scattering and absorption processes attenuate the incident light by two to three orders of magnitude over the combination wavelengths.

Numerous skin sites are possible for noninvasive near-infrared measurements.[40] Examples proposed in the literature include the inner lip mucosa,[26] finger,[41] forearm,[27,42] and tongue.[43] In all cases, the principal spectral features of the noninvasive spectra correspond to the extent of scattering superimposed on absorption bands that originate from water, protein, and fatty tissue within the skin matrix.[44,45] To a first approximation, noninvasive near-infrared spectra of skin can be fitted to a Beer–Lambert function that incorporates the additive features of absorption due to water, fat, β-sheet protein, and type III collagen, with additional terms for tissue scattering (offset) and small temperature variations (a slope term). This function takes the following form:

$$A_{\text{skin}} = \beta_{\text{w}} A_{\text{water}} + \beta_{\text{P1}} A_{\text{P1}} + \beta_{\text{P2}} A_{\text{P2}} + \beta_{\text{f}} A_{\text{fat}} + \beta_{\text{S}} A_{\text{slope}} + \beta_0 A_{\text{offset}} \tag{13.3}$$

where the terms A_{skin}, A_{water}, A_{protein1}, A_{protein2}, A_{fat}, A_{slope}, and A_{offset} represent the absorbance spectra corresponding to skin, water, β-sheet protein (P1), type III collagen protein (P2), fat, temperature induced changes in the underlying water spectrum, and scattering and the corresponding β_x terms represent regression coefficients for the fit. The magnitudes of these regression coefficients are determined by applying multiple linear regression to a set of pure component spectra that represent the different skin components. The magnitude of the regression coefficient indicates the physical amount of each component in the optical path relative to the amount of each component that was used to generate the pure component spectrum. For example, if the pure component spectrum of water used in the regression analysis was generated from a 1 mm thick sample of water and the best fit regression coefficient is 0.6, then the effective aqueous path length in the tissue spectrum is 0.6 mm.[31]

The ability to model noninvasive near-infrared spectra of skin with the model presented in equation (13.3) is illustrated in Figure 13.9. In this figure, the measured and fitted absorption spectra are plotted for noninvasive spectra collected from across the human and rat skins. For the most part, the fitted spectrum matches the measured spectrum, which indicates the major components of the skin spectra are incorporated into the model. The largest degree of mismatch is evident around 4300 cm^{-1}, which

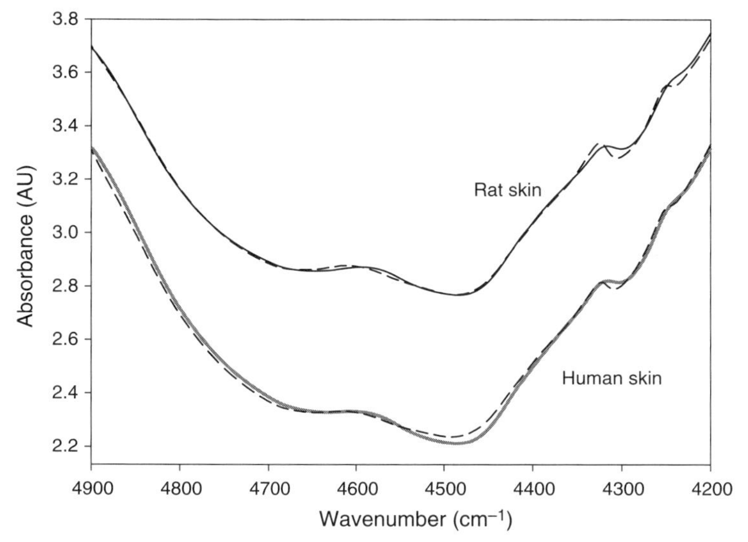

Figure 13.9 Noninvasive absorption spectra relative to air collected from human (gray) and rat (black) skins with each fitted spectrum shown as the broken black line.

represents the C–H bond region and is often associated with long alkyl chains, such as fat-type molecules.

13.5.2 Measurement Site Considerations

The measurement site for collecting noninvasive spectra is critical for the ultimate success of the measurement. Results from many different *in vitro* experiments to characterize the ability to measure glucose from near-infrared spectra collected over the combination wavelengths indicate that the most reliable, glucose-specific information originates from the spectral region encompassing the 4400 and 4300 cm^{-1} absorption bands of glucose (see Figure 13.3b). Fatty tissue possesses strong absorption bands near these glucose bands, which adversely affects the ability to quantify glucose in living tissue.[45] For this reason, a measurement procedure that eliminates fatty tissue is ideal for noninvasive measurements.

Skin is composed of the following layers: (1) epidermis, (2) dermis, and (3) subcutaneous space. The epidermis is a thin (100 μm) protective outer layer of skin with no adipose cells and essentially no vascular flow. The lack of vascular flow results in little glucose that tracks blood glucose.[46] The dermis layer is thicker (approximately 1 mm) and is void of adipose cells (fatty tissue). In addition, the concentrations of glucose in dermis tissue track the concentration of glucose in blood, based on measurements of harvested interstitial fluid.[47] As a result, dermis tissue is well suited for noninvasive measurements. Interstitial fluid in the subcutaneous space is also known to track blood glucose levels and this layer is easy to reach mechanically, which makes this region well suited for implanted glucose biosensors.

Two distinctive strategies are reported for eliminating fatty tissue in noninvasive spectra. The first is to restrict the measured light from reaching the fatty regions of the subcutaneous tissue. This approach used a transflectance optical geometry (see Figure 13.1) to collect multiscattered photons that traverse only the dermis component of the skin matrix.[48] The optical measurement can be restricted to the dermis layer of skin by controlling the distance between the incident and detector fibers placed on the skin surface.

The second strategy for eliminating fatty tissue from the noninvasive measurement is to restrict the measurement to regions of the body with minimal subcutaneous fat. The tongue is one potential measurement site that possesses very little fatty tissue. Results from noninvasive measurements across the human tongue have been reported where the near-infrared spectra were collected over the first-overtone region of the near-infrared spectrum.[43] A statistical analysis of this noninvasive tongue data suggests that glucose can be measured from such spectra, but the prediction errors are too high for clinical purposes.

The skin on the back of the hand is another potential measurement site that is void of fat-containing adipose tissue.[49] This skin can be pinched and light can be transmitted across this fold of skin to give a noninvasive spectrum composed primarily of dermis tissue. The human spectrum in Figure 13.9 is derived from the skin on the back of the hand from a human volunteer. As discussed above, this spectrum consists of spectral features related to water, scattering, protein, and fat. The magnitude of the fat absorption is very low, compared to the other matrix components.

13.5.3 Animal Model for Noninvasive Glucose Sensing

An accurate animal model is important for the advancement of technology for noninvasive glucose sensing. Such a model enables control over important aspects of the experiment, such as control over *in vivo* glucose concentrations and the ability to set the rate and direction of glycemic changes.

The thin skin on the upper shoulder area of Sprague Dawley rats is a suitable model for noninvasive human measurements. Although this rat skin is not completely void of fatty tissue, transmission spectra collected through this skin show only small absorption features attributed to fat. Typical absorbance spectra for human skin (back of the hand) and rat skin are provided in Figure 13.9 for comparison. Most important, the shapes of the human and rat spectra are very similar, with only slight differences noted in the fat absorption regions between 4400 and 4200 cm^{-1}. This similarity in shape is critical because it indicates that the major chemical components are represented in both cases at reasonably consistent relative amounts. A striking difference between these two spectra is the offset where the overall attenuation is much higher for the rat skin relative to the human skin. The reason for this offset can be found by comparing the fitted spectra.

As noted above, raw noninvasive spectra can be fitted to the model in equation (13.3) and the regression terms can be compared to give an indication of the differences or similarities in spectra from human and rat skins. Such an analysis of the spectra in Figure 13.9 reveals that the coefficients are similar for the rat and human spectra except for the coefficients that correspond to scattering (β_o) and water (β_w). The rat skin is more highly scattering compared to the human skin. The scattering coefficient is nearly twice that for the human skin. The human skin spectra, however, has a higher content of water compared to the rat skin. A typical effective aqueous layer thickness is 1.0 mm for human skin and 0.5 mm for rat skin. These differences are significant because the ability to measure glucose depends on the aqueous layer thickness and the SNR measurement. The longer aqueous path length for the human skin will enhance the sensitivity of noninvasive measurements. Likewise, the lower scattering coefficient for human skin will provide higher intensities of light at the detector, which will directly correspond to higher SNRs for these measurements. These results reveal that rat skin is a conservative model for human skin. In other words, it should be easier to measure glucose, or other clinical analytes, in human skin relative to rat skin.

13.6 NONINVASIVE GLUCOSE MEASUREMENTS

Attempts to measure glucose concentrations noninvasively in rat skin are motivated by the success of the human tongue measurements and the realization that noninvasive spectra collected over the combination wavelengths should provide better analytical glucose measurements on the basis of *in vitro* results. Noninvasive spectral data from this animal model can be used to verify the existence of glucose-specific information within spectra collected noninvasively from the living skin. Such an approach is a critical step on the path toward successful noninvasive sensing of glucose in human subjects.

13.6.1 Noninvasive Near-Infrared Spectra Collected Across the Rat Skin

In a typical experiment, noninvasive spectra are collected continuously as the *in vivo* glucose concentration is controlled by intravenous infusion of fluids.[50] Glucose concentrations are held low and constant for the first 3 h. Then, glucose concentrations are increased to more than 30 mM. Glucose is maintained at this level for approximately 1 h, after which the glucose concentration is allowed to decay naturally to basal levels over the next hour. During the course of the experiment, arterial blood samples are collected periodically for analysis by a standard reference method.

The photograph presented in Figure 13.10 shows a typical interface used to collect these noninvasive spectra. Light is incident on one side of the skinfold and a fraction of the transmitted light is collected directly from across the input fiber. Bundles of low-hydroxy silica fibers are used to deliver and collect the near-infrared radiation for the measurement. For the experiments described here, the noninvasive spectra were collected with a Fourier transform spectrometer set for a resolution of 16 cm^{-1} and 128 coadded interferograms. Each recorded spectrum required approximately 60 s to acquire and save. A total of 370 spectra were collected over a period of nearly 7 h while *in vivo* glucose concentrations varied from 6 to 33 mM (108–594 mg/dL).

By knowing the concentration of glucose in the arterial blood as a function of the experiment time course, it is possible to assign a glucose concentration to each noninvasive spectrum. Accurate glucose assignment, however, requires consideration

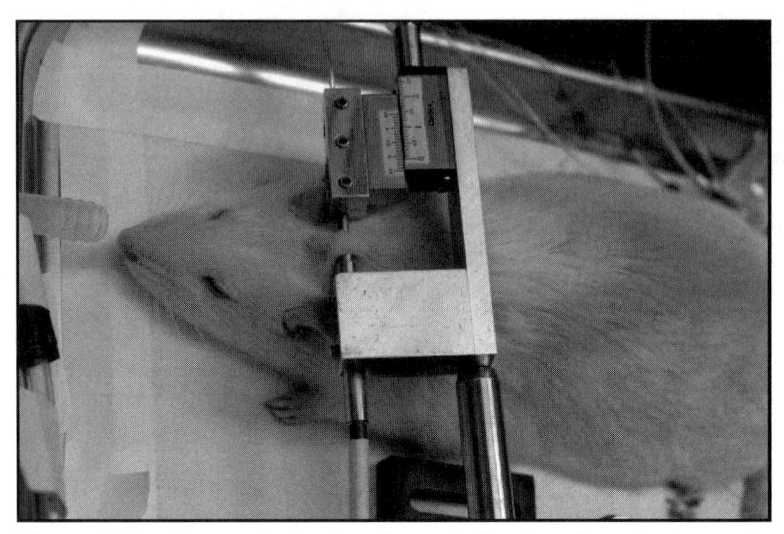

Figure 13.10 Photograph showing fiber-optic interface for collecting noninvasive near-infrared spectra from a rat skin model.

of a lag time between the concentrations of glucose in blood and interstitial fluid at the measurement site.[51] Reports by many research groups indicate that a step change in the concentration of glucose in blood requires some time before the corresponding concentration of glucose is observed either in the subcutaneous tissue,[52,53] or in the dermis layer of skin.[54] Although the exact magnitude of this lag time varies by individual, 10 min is a reasonable estimate.

13.6.2 PLS Calibration Model for Noninvasive Glucose Measurements

The noninvasive spectra collected from the rat skin can be assigned glucose concentrations on the basis of the reference measurements. They can then be subjected to a routine PLS multivariate analysis, as described above. The results of such an analysis are presented in Figure 13.11 where the concentrations of glucose predicted by the calibration model are superimposed on the measured arterial blood values as a function of time. The squares and circles are used to distinguish the spectra used to train the PLS model and those used for validation purposes, respectively.

In general, both prediction and calibration points fall along the arterial time profile (solid line). In this particular experiment, the spectrometer-to-tissue interface remained on the animal continuously, except for a period from 13:00 to 13:45 when the interface was repeatedly removed and reset between each measurement. Clearly, resetting the interface generates a significant source of variation. Nevertheless, the PLS model appears to accurately model glucose from these noninvasive near-infrared spectra.

Although the data in Figure 13.11 suggest that glucose is being measured accurately, a conventional PLS analysis gives *no proof that the predictions are based*

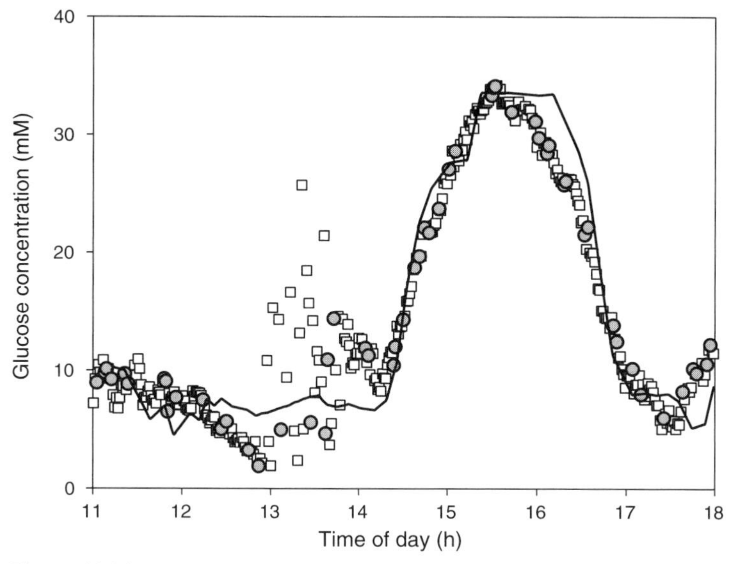

Figure 13.11 PLS calibration model results for the noninvasive measurement of glucose in rat skin indicating spectra used for training (squares) and validation (circles) of the model.

on the spectral properties of the glucose molecule. As noted in Chapter 12, the PLS algorithm has a propensity to utilize spurious correlations between nonglucose-dependent spectral variance and glucose concentrations within a data matrix.[55] For this reason, it is critical to examine the noninvasive data further to find direct evidence of glucose specificity.

13.6.3 Impact of Incorrect Glucose Assignments

One way to test the validity of a calibration model is to evaluate the impact of rearranging the reference concentrations.[56] The glucose assignments used to create the PLS calibration model described in Figure 13.11 are based on a 10 min lag time between the arterial blood glucose values and the assigned tissue glucose levels. This set of assignments might be labeled "correct glucose concentrations." The glucose assignments can be easily altered to assign incorrect glucose concentrations to each spectrum. For example, the glucose transient might be arbitrarily shifted forward by 2 h, so each spectrum will be assigned an incorrect glucose concentration.

Repeating the routine PLS analysis with the "incorrect glucose concentrations" produces the results presented in Figure 13.12. Again, the individual points are coded to distinguish spectra used for training and validating the calibration model and the model results are superimposed on the glucose profile used to make the glucose assignments. The prediction pattern of this "false" calibration model is striking as it appears to accurately predict the assigned, yet incorrect, glucose concentrations. Because the glucose assignments are incorrect, the resulting calibration model cannot possibly be predicting glucose concentrations on the basis of glucose-specific information.

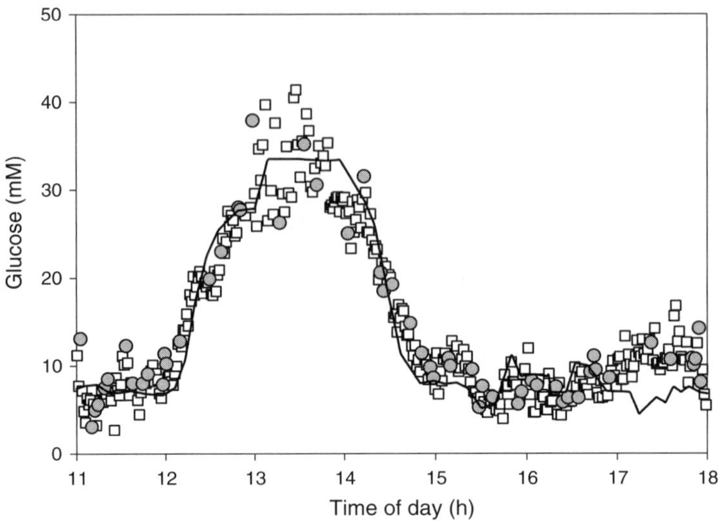

Figure 13.12 PLS calibration model for noninvasive glucose measurements based on incorrect glucose concentration assignments indicating spectra used for training (squares) and validation (circles) purposes.

More importantly, the false calibration model represented in Figure 13.12 calls into question the validity of the calibration model with the correct glucose assignments (Figure 13.11). Without further analysis, it is impossible to conclude that glucose is being measured in Figure 13.11 owing to the known incorrectness of the model represented in Figure 13.12.

13.6.4 Analysis of Residual Spectra

A residual spectrum analysis can be used to provide a better understanding of the origin of the analytical information used to generate Figure 13.11.[50] A residual spectrum highlights the components of a given spectrum that differ from a set of reference spectra. For the set of noninvasive spectra collected across the rat skin, all the noninvasive spectra collected during the first 3 h of the experiment can be used to establish the nonglucose-dependent spectral variance. Indeed, the concentration of glucose was purposely held constant during this initial 3 h period, which means that glucose did not contribute to the spectral variance observed over this period. If one assumes that the nonglucose-dependent spectral variance measured during this initial period accurately represents the natural *in vivo* variance at all glucose concentrations, then a residual spectrum can be calculated using a vector projection analysis to remove the nonglucose-dependent spectral variance from the high glucose concentration spectra.[50] The nonglucose-dependent spectral variance is characterized by performing a principal component analysis on the initial baseline spectra. These principal components are then removed from all subsequent spectra with elevated glucose concentrations. The resulting tissue residual spectrum provides a measure of the spectral features in the tissue that are distinct relative to the normal *in vivo* variations observed during the euglycemic period.

A series of residual spectra is presented in Figure 13.13 for the hyperglycemic transient. Spectra collected within the shaded regions at the beginning and end of the experiment correspond to background spectra and were used to establish the non-glucose-dependent spectral variance. The residual spectra are shown as insets on top of the arterial glucose concentration profile with indications as to the time at which they were taken. Around 14:00, the residuals are very small (on the level of spectral noise), which is expected because the glucose concentration has not changed significantly compared to the glucose levels in the background spectra. However, as the glucose concentration increases, a small residual emerges that grows until the glucose concentration reaches its maximum value around 15:30. The residual absorbance spectrum then diminishes in amplitude as the glucose concentration returns to the original level.

The residual plots in Figure 13.13 reveal a spectral structure that is present during the hyperglycemic period but is not described by the normal nonglucose tissue variations observed in the first 3 h of the experiment. The shape of the residual spectrum remains fixed as the concentration of glucose varies along the time course of the glucose transient. The magnitude of the spectral structure varies, however, according to the glucose concentration. Clearly, these tissue residual spectra are a direct consequence of the presence of glucose in the animal's tissue.

It is noteworthy that the shape of this tissue residual spectrum does not resemble the pure component spectrum of glucose. This fact is highlighted in Figure 13.14a in which the average residual spectrum from a group of high glucose-containing tissue spectra is superimposed on a pure component solution spectrum of glucose. In fact, this spectral residual should not be the same as the pure component glucose spectrum, even if the absorption spectrum of glucose is the origin of this spectral residual. Again, the residual spectrum corresponds to the component of the collected tissue spectrum that is orthogonal to the nonglucose-dependent spectral variance. A more meaningful exercise is to compare this tissue residual spectrum to the residual spectrum determined between the pure component glucose absorption spectrum and the nonglucose-dependent spectral variance. Figure 13.14b shows these two residual spectra and the degree of overlap strongly suggests that the measured tissue residual spectrum originates directly, and exclusively, from the glucose spectrum. In this sense, the measured tissue residual spectrum is the spectral signature for glucose relative to all *in vivo* sources of spectral variance captured in this data set.

By definition, the residual spectrum determined from the pure component glucose spectrum is the NAS for glucose in this particular tissue matrix.[35] As described above, the NAS corresponds to the orthogonal component of the glucose absorption spectrum relative to the nonglucose-dependent spectral variance.

13.6.5 Net Analyte Signal Calibration Model

If the residual spectrum generated from the pure component glucose absorption spectrum (Figure 13.14b) accurately represents the NAS for glucose, then it should be possible to calculate a calibration model for glucose based only on the glucose absorption spectrum and the nonglucose variance observed in the first 3 h. The NAS spectrum can be converted into a calibration spectrum by scaling it to give the proper

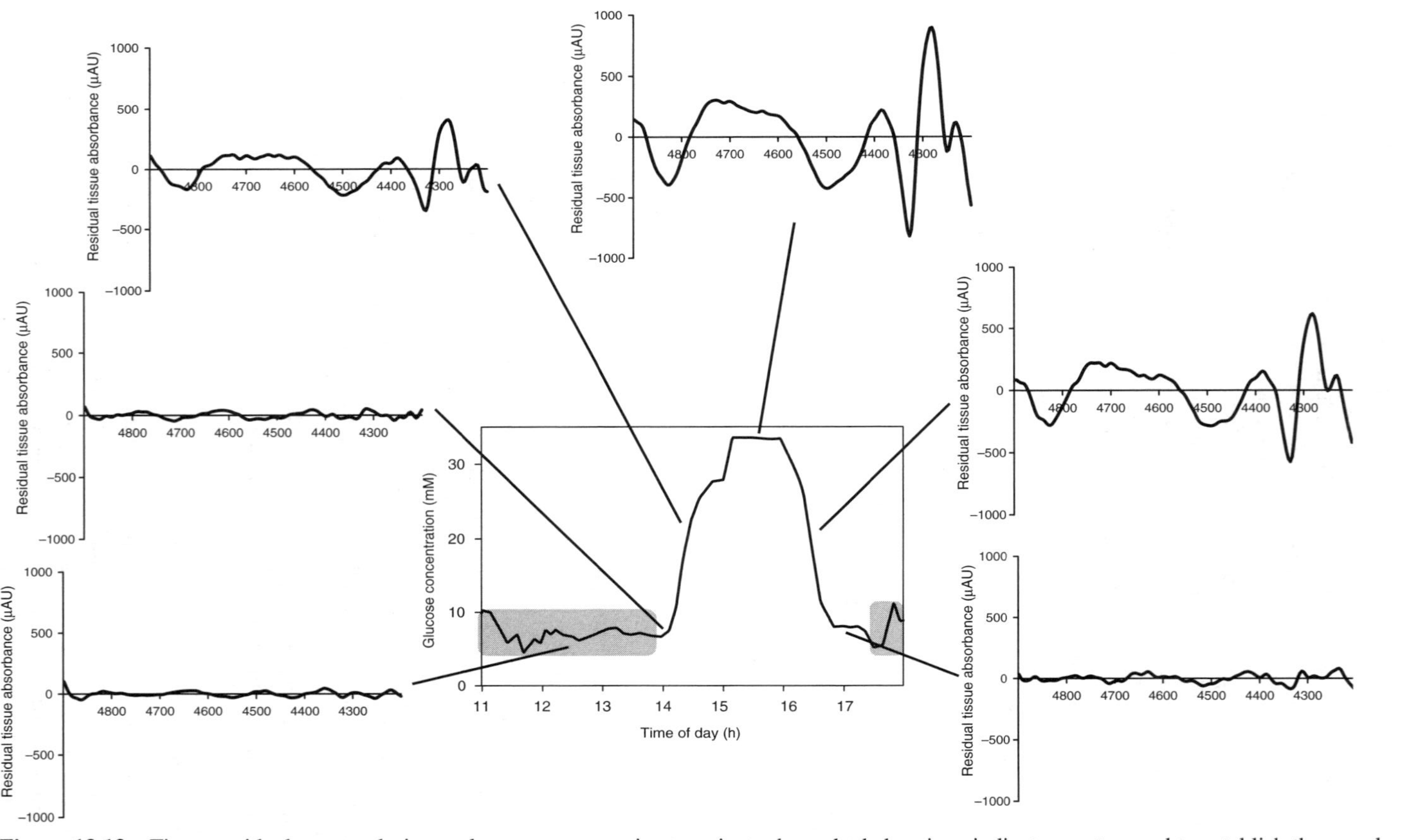

Figure 13.13 Tissue residual spectra during a glucose concentration transient where shaded regions indicate spectra used to establish the nonglucose-dependent spectral variance.

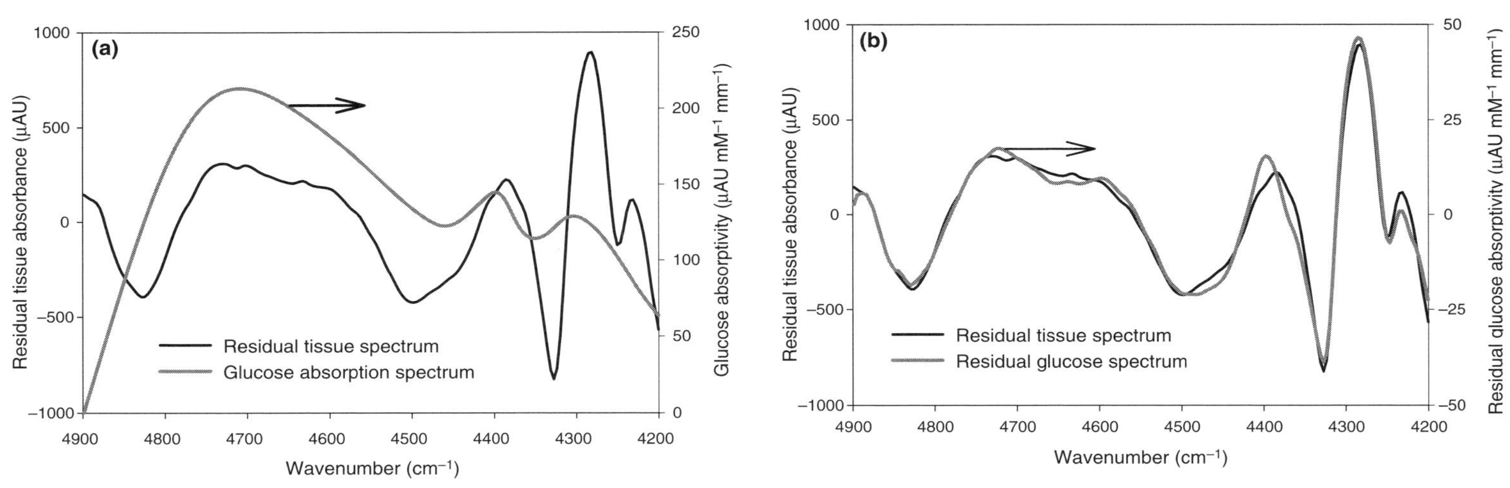

Figure 13.14 Comparison of tissue residual spectrum determined at high tissue glucose concentration (black) to (a) pure component solution absorption spectrum of glucose (gray) and (b) residual pure component glucose absorption spectrum (gray).

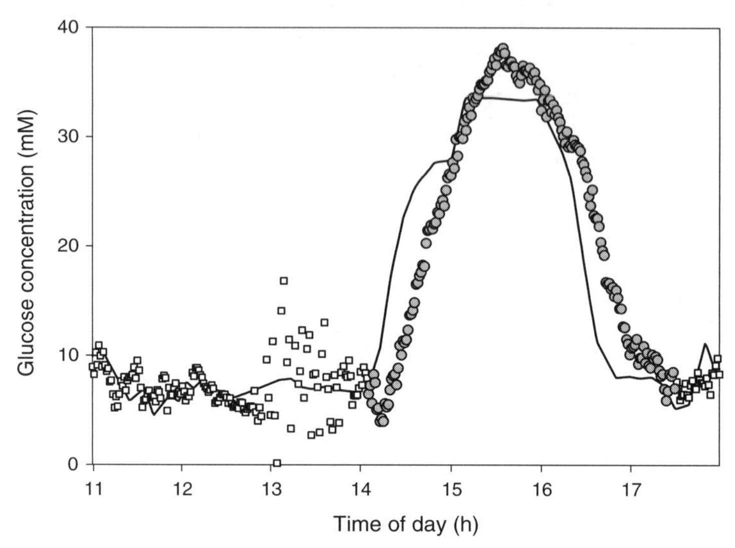

Figure 13.15 Noninvasive glucose concentration predictions from the net analyte signal calibration model compared to arterial blood glucose concentrations (black line). Squares indicate spectra used to establish nonglucose-dependent spectral variations for the net analyte signal calculation.

units of concentration for a known aqueous path length. The results plotted in Figure 13.15 show the raw predictions from a calibration vector based solely on the glucose NAS. The square and circles represent absolute glucose concentration predictions from the NAS calibration model and the black line corresponds to the arterial reference. The squares coincide with the spectra used to establish the nonglucose-dependent spectral variance under constant glucose concentrations and the circles represent spectra collected during the hyperglycemic transient. No time delay has been incorporated into these predictions, which accounts for the evident lag before increases and after decreases in arterial blood glucose levels.

The plot in Figure 13.15 clearly indicates that the predicted glucose concentrations track the actual glucose changes with a high degree of accuracy. Again, these predictions are based on the residual glucose spectrum, or NAS, relative to the nonglucose spectral variance. It is critical to underscore the point that these glucose predictions were obtained with absolutely no statistical regression analysis and, therefore, should not be attributed to spurious correlations, but to genuine glucose-specific information. In other words, the observed residual tissue spectra have both the *shape* and *amplitude* that would be expected based on the addition of glucose to the animal.[50]

13.6.6 Comparison of PLS and NAS Calibration Vectors

The analysis of the noninvasive PLS calibration models presented in Sections 13.6.2 and 13.6.3 called into question the validity of the model demonstrated in Figure 13.11. Because an apparently functional calibration model is possible from incorrect glucose assignments (Figure 13.12), it is not possible to believe the model generated when the correct glucose assignments are used. Without further analysis, it is impossible to

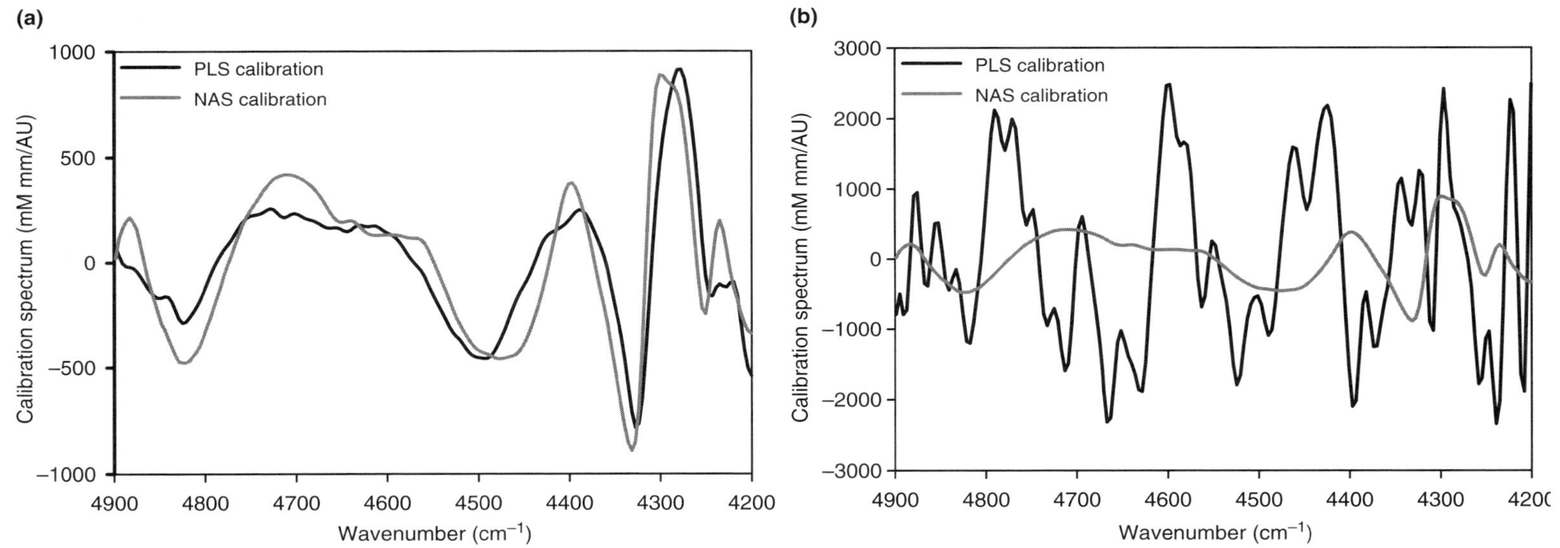

Figure 13.16 Comparison of PLS calibration vectors (black) and net analyte signal calibration vector (gray) when the PLS model is based on correct (a) and incorrect (b) glucose concentration assignments.

conclude that glucose is being measured in Figure 13.11 owing to the known incorrectness of the model represented in Figure 13.12.

It is useful to view the calibration vectors associated with these PLS calibration models and to compare these vectors to the NAS calibration vectors presented in Section 13.6.5. These calibration vectors are presented in Figure 13.16. In Figure 13.16a, the NAS calibration model is compared to the PLS calibration model based on the correct glucose assignments. Figure 13.16b, on the other hand, compares the NAS calibration vector to the false PLS calibration model, which is based on the incorrect glucose assignments.

The similarity in the calibration models presented in Figure 13.16a strongly suggests that the PLS results obtained with the correct glucose assignments are based on glucose-specific spectral information. Indeed, the PLS calibration vector corresponding to the correct glucose assignments is very similar in both shape and magnitude compared to the NAS calibration vector. The PLS calibration model based on incorrect glucose assignments has no similarity to the NAS calibration vector, as one would expect given the lack of glucose-specific information in this false PLS calibration model. This detailed comparison of these calibration models provides a clear method to distinguish between a PLS model based on glucose information versus that based on correlated nonglucose-specific spectral variations. These findings represent direct evidence that glucose-specific information is embedded within noninvasive near-infrared spectra.

13.7 CONCLUSIONS AND FUTURE DIRECTIONS

The ability to measure glucose noninvasively in living tissue is a complex and difficult analytical problem. Many groups, both in academia and industry, have attempted to measure glucose by noninvasive optical methods. These groups have discovered that it is relatively easy to monitor a signal that changes in magnitude during a transitory variation in blood glucose concentrations. These changes in signal can be correlated to glucose concentrations in either a univariate or multivariate manner. These correlations are based principally on the secondary effects of glucose, such as variations in the refractive index of interstitial fluid within the tissue matrix. Many of the early papers published in this area were not measuring glucose as was originally believed at the time the papers were written.

The field of noninvasive glucose sensing has progressed to the point where simple correlations are no longer acceptable, but more rigorous evidence is needed to claim successful noninvasive measurements. The information provided in this chapter is designed to illustrate several critical points in this regard. First, glucose-specific information is available within near-infrared spectra collected noninvasively from the living tissue. This analytical information is embedded within a complex signal that incorporates spectral variations from a wide variety of sources. Second, near-infrared spectroscopy has the fundamental potential to provide excellent selectivity for sound analytical measurements in complex biological matrices. Although high selectivity is possible with near-infrared spectroscopy, selectivity cannot be guaranteed by simply applying multivariate statistical methods to near-infrared spectral data. Spurious correlations within the raw or processed data can easily generate false calibration models. Third, methods are available to more deeply probe the origin of the chemical information that is used for noninvasive spectral measurements. These methods, and others to be developed in the future, are critical to our understanding of the chemical basis of noninvasive glucose measurements.

Clearly, the fact that glucose-specific information is available within noninvasive near-infrared spectra is a major advance. The main question for the next generation of researchers is to establish how this analytical information can be obtained in a reliable and practical manner.

Much of the work described in this chapter was supported by grants from the National Institutes of Diabetes and Digestive and Kidney Diseases (DK-60657 and DK-02925). The corresponding content is solely the responsibility of the authors and does not necessarily represent the official views of the National Institute of Diabetes and Digestive and Kidney Diseases or the National Institutes of Health.

REFERENCES

1. Burns DA, Ciurczak EW. *Handbook of Near-Infrared Analysis*. Marcel Dekker, New York, 1992.
2. Siesler H, Ozaki Y, Kawata S, Heise M. *Near Infrared Spectroscopy: Principles, Instruments, and Applications*. Wiley-VCH, Weinheim, 2002.
3. Ozaki Y, Morita S, Du Y. Spectral analysis. In: Ozaki Y, McClure WF, Christy AA (Eds), *Near-Infrared Spectroscopy in Food Science and Technology*. Wiley-Interscience, Hoboken, NJ, 2007, pp. 47–72.
4. Bayly JG, Kartha VB, Stevens WH. The absorption spectra of liquid phase H_2O, HDO and D_2O from 0.7 μm to 10 μm. *Infrared Physics* 1963, 3, 211–222.
5. Jensen PS, Bak J, Andersson-Engels S. Influence of temperature on water and aqueous glucose absorption spectra in the near- and mid-infrared regions at physiologically relevant temperatures. *Applied Spectroscopy* 2003, 57, 28–36.
6. Chen J, Arnold MA, Small GW. Comparison of combination and first overtone spectral regions for near infrared calibration models for glucose and other biomolecules in aqueous solutions. *Analytical Chemistry* 2004, 76, 5405–5413.
7. Chang SY, Wang NS. Monitoring polymerization reactions by near-IR spectroscopy. *ACS Symposium Series* 1995, 598, 147–165.
8. Dumitrescu OR, Baker DC, Foster GM, Evans KE. Near infrared spectroscopy for in-line monitoring during injection molding. *Polymer Testing* 2005, 24, 367–375.
9. McClure WF. 204 years of near infrared technology: 1800–2003. *Journal of Near Infrared Spectroscopy* 2003, 11, 487–518.
10. Santos AF, Silva FM, Lenzi MK, Pinto JC. Monitoring and control of polymerization reactors using NIR spectroscopy. *Polymer-Plastics Technology and Engineering* 2005, 44, 1–61.
11. Reich G. Near-infrared spectroscopy and imaging: basic principles and pharmaceutical applications. *Advanced Drug Delivery Reviews* 2005, 57, 1109–1143.
12. Workman J, Koch M, Veltkamp D. Process analytical chemistry. *Analytical Chemistry* 2005, 77, 3789–3806.

13. Workman JJ. Review of process and non-invasive near-infrared and infrared spectroscopy: 1993–1999. *Applied Spectroscopy Reviews* 1999, 34, 1–89.
14. Wahr JA, Tremper KK. Noninvasive oxygen monitoring techniques. *Critical Care Clinics* 1995, 11, 199–217.
15. Severinghaus JW, Kelleher JF. Recent developments in pulse oximetry. *Anesthesiology* 1992, 76, 1018–1038.
16. Bowes WA, Corke BC, Hulka J. Pulse oximetry: a review of the theory, accuracy and clinical applications. *Obstetrics and Gynecology* 1989, 74, 541–546.
17. Kelleher JF. Pulse oximetry. *Journal of Clinical Monitoring* 1989, 5, 37–62.
18. Mendelson Y. Pulse oximetry: theory and applications for noninvasive monitoring. *Clinical Chemistry* 1992, 38, 1601–1607.
19. Nioka S, Chance B. NIR spectroscopic detection of breast cancer. *Technology in Cancer Research and Treatment* 2005, 4, 497–512.
20. Heffer E, Pera V, Schutz O, Siebold H, Fantini S. Near-infrared imaging of the human breast: complementing hemoglobin concentration maps with oxygenation images. *Journal of Biomedical Optics* 2004, 9, 1152–1160.
21. Gu YQ, Chen WR, Xia M, Jeong SW, Liu H. Effect of photothermal therapy on breast tumor vascular contents: noninvasive monitoring by near-infrared spectroscopy. *Photochemistry and Photobiology* 2005, 81, 1002–1009.
22. Tromberg BJ, Shah N, Lanning R, Crussi A, Espinoza J, Pham T, Svacsand L, Butler J. Non-invasive *in vivo* characterization of breast tumors using photon migration spectroscopy. *Neoplasia* 2000, 2, 26–40.
23. Yamada Y. Fundamental studies of photon migration in biological tissues and their application to optical tomography. *Optical Review* 2000, 7, 366–374.
24. Heise HM. Non-invasive monitoring of metabolites using near infrared spectroscopy: state of the art. *Hormone and Metabolic Research* 1996, 28, 527–534.
25. Rolfe P. *In vivo* near-infrared spectroscopy. *Annual Review of Biomedical Engineering* 2000, 2, 715–754.
26. Heise HM, Bittner A, Marback R. Near-infrared reflectance spectroscopy for noninvasive monitoring of metabolites. *Clinical Chemistry and Laboratory Medicine* 2000, 38, 137–145.
27. Maruo K, Tsurugi M, Tamura M, Ozaki Y. *In vivo* noninvasive measurement of blood glucose by near-infrared diffuse–reflectance spectroscopy. *Applied Spectroscopy* 2003, 57, 1236–1244.
28. Weiss R, Yegorchikov T, Shusterman A, Raz I. Noninvasive continuous glucose monitoring using photoacoustic technology: results from the first 62 subjects. *Diabetes Technology & Therapeutics* 2007, 9, 68–74.
29. MacKenzie HA, Ashton HS, Spiers S, Shen Y, Freeborn SE, Hannigan J, Lindberg J, Rae P. Advances in photoacoustic noninvasive glucose testing. *Clinical Chemistry* 1999, 45, 1587–1595.
30. Amerov AK, Chen J, Arnold MA. Molar absorptivities of glucose, water and other biological molecules over the first overtone and combination regions of the near infrared spectrum. *Applied Spectroscopy* 2004, 58, 1195–1204.
31. Arnold MA, Small GW. Perspectives in analytical chemistry: noninvasive glucose sensing. *Analytical Chemistry* 2005, 77, 5429–5439.
32. Hazen KH, Arnold MA, Small GW. Measurement of glucose in water with first overtone near infrared spectra. *Applied Spectroscopy* 1998, 52, 1597–1605.

33. Hazen KH, Arnold MA, Small GW. Temperature insensitive near infrared measurements of glucose in aqueous matrices. *Applied Spectroscopy* 1994, 48, 477–483.
34. Small GW. Chemometrics and near-infrared spectroscopy: avoiding the pitfalls. *Trends in Analytical Chemistry* 2006, 25, 1057–1066.
35. Lorber A. Error propagation and figures of merit for quantification by solving matrix equations. *Analytical Chemistry* 1986, 58, 1167–1172.
36. Lober A, Faber K, Kowalski BR. Net analyte signal calculation in multivariate calibration. *Analytical Chemistry* 1997, 69, 1620–1626.
37. Arnold MA, Small GW, Xiang D, Qiu J, Murhammer DW. Pure component selectivity analysis of multivariate calibration models from near infrared spectra. *Analytical Chemistry* 2004, 76, 2583–2590.
38. Troy TL, Thennadil SN. Optical properties of human skin in the near infrared wavelength range of 1000 to 2200 nm. *Journal of Biomedical Optics* 2001, 6, 167–176.
39. Tseng SH, Grant A, Durkin AJ. *In vivo* determination of skin near-infrared optical properties using diffuse optical spectroscopy. *Journal of Biomedical Optics* 2008, 13, 014016.
40. Khalil OS. Non-invasive glucose measurements at the dawn of the new millennium: an update. *Diabetes Technology & Therapeutics* 2004, 6, 660–697.
41. Yamakoshi K, Yamakoshi Y. Pulse glucometry: a new approach for noninvasive blood glucose measurements using instantaneous differential near-infrared spectrophotometry. *Journal of Biomedical Optics* 2006, 11, 054028.
42. Khalil OS, Yeh SJ, Lowery MG, Wu X, Hanna CF, Kantor S, Jeng TW, Kanger JS, Bolt RA, de Mul FF. Temperature modulation of the visible and near infrared absorption and scattering coefficients of human skin. *Journal of Biomedical Optics* 2003, 8, 191–205.
43. Burmeister JJ, Arnold MA, Small GW. Noninvasive blood glucose measurements by near infrared transmission spectroscopy across human tongues. *Diabetes Technology & Therapeutics* 2000, 2, 5–16.
44. Burmeister JJ, Chung H, Arnold MA. Phantoms for noninvasive blood glucose sensing with near-infrared transmission spectroscopy. *Photochemistry and Photobiology* 1998, 67, 50–55.
45. Burmeister JJ, Arnold MA. Evaluation of measurement sites for noninvasive blood glucose sensing with near-infrared transmission spectroscopy. *Clinical Chemistry* 1999, 45, 1621–1627.
46. Rao G, Guy RH, Glikfeld P, LaCourse WR, Leung L, Tamada J, Potts RO, Azimi N. Reverse iontophoresis: noninvasive glucose monitoring *in vivo* in humans. *Pharmaceutical Research* 1995, 12, 1869–1873.
47. Collison ME, Stout PJ, Glushko TS, Pokela KN, Mullin-Hirte DJ, Racchini JR, Walter MA, Mecca SP, Rundquist J, Allen JJ, Hilgers ME, Hoegh TB. Analytical characterization of electrochemical biosensor test strips for measurement of glucose in low-volume interstitial fluid samples. *Clinical Chemistry* 1999, 45, 1665–1673.
48. Malin SF, Ruchti TL, Blank TB, Thennadil SN, Monfre SL. Noninvasive prediction of glucose by near-infrared diffuse reflectance spectroscopy. *Clinical Chemistry* 1999, 45, 1651–1658.
49. Marks R. Mechanical properties of the skin. In: Goldsmith LA (Ed.), *Biochemistry and Physiology of the Skin*. Oxford University Press, New York, 1983.
50. Olesberg JT, Liu L, Van Zee V, Arnold MA. *In vivo* near-infrared spectroscopy of rat skin tissue with varying blood glucose levels. *Analytical Chemistry* 2006, 78, 215–223.

51. Fraser DM. *Biosensors in the Body: Continuous In-Vivo* Monitoring. John Wiley & Sons, New York, 1997.
52. Schoonen AJM, Wientjes KJ. A model for transport of glucose in adipose tissue to a microdialysis probe. *Diabetes Technology & Therapeutics* 2003, 5, 589–598.
53. Wilson GS, Gifford R. Biosensors for real-time *in vivo* measurements. *Biosensors & Bioelectronics* 2005, 20, 2388–2403.
54. Stout P, Pokela K, Mullins-Hirte D, Hoegh T, Hilgers M, Thorp A, Collison M, Glushko T. Site-to-site variation of glucose in interstitial fluid samples and correlation to venous plasma glucose. *Clinical Chemistry* 1999, 45, 1674–1675.
55. Arnold MA, Burmeister JJ, Small GW. Phantom glucose calibration models from simulated noninvasive human near infrared spectra. *Analytical Chemistry* 1998, 70, 1773–1781.
56. Arnold MA, Liu L, Olesberg JT. Selectivity assessment of noninvasive glucose measurements based on analysis of multivariate calibration vectors. *Journal of Diabetes Science and Technology* 2007, 1, 454–462.

CHAPTER 14

NONINVASIVE GLUCOSE SENSING WITH RAMAN SPECTROSCOPY

Wei-Chuan Shih, Kate L. Bechtel, and Michael S. Feld

In Vivo Glucose Sensing, Edited by David D. Cunningham and Julie A. Stenken

14.1 INTRODUCTION TO RAMAN SPECTROSCOPY

Light that is scattered from a molecule is primarily elastically scattered; that is, the incident and the scattered photons have the same energy. A small probability exists, however, that a photon is scattered inelastically, resulting in either a net gain or loss of energy of the scattered photon. This inelastic scattering, discovered by Raman and Krishna,[1] allows fundamental molecular vibrational transitions to be measured at any excitation wavelength.

Raman scattering is a coherent one-step process in which one photon is exchanged for another through interaction with a molecule. Schematically, the Raman process is depicted as a molecule in an initial vibrational state proceeding to a higher or lower vibrational state through excitation to a "virtual state," with simultaneous scattering of a new photon from this state. The difference in energy between the incident and the scattered photon is equal to the energy difference between the initial and final vibrational states of the molecule. A loss in photon energy is termed Stokes Raman scattering and a gain in photon energy is termed anti-Stokes Raman scattering.[2] These processes are depicted in Figure 14.1.

Not all vibrational transitions can be accessed by Raman scattering. Raman-active transitions are those associated with a change in polarizability of the molecule. In classical terms, this can be viewed as a perturbation of the electron cloud of the molecule.

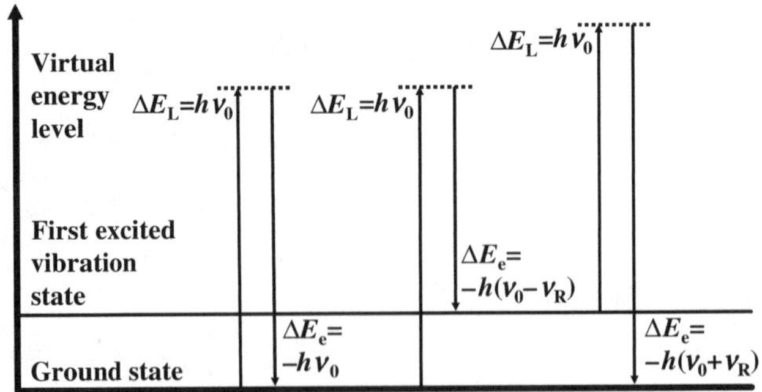

Figure 14.1 Energy diagram for Rayleigh, Stokes Raman, and anti-Stokes Raman scattering.

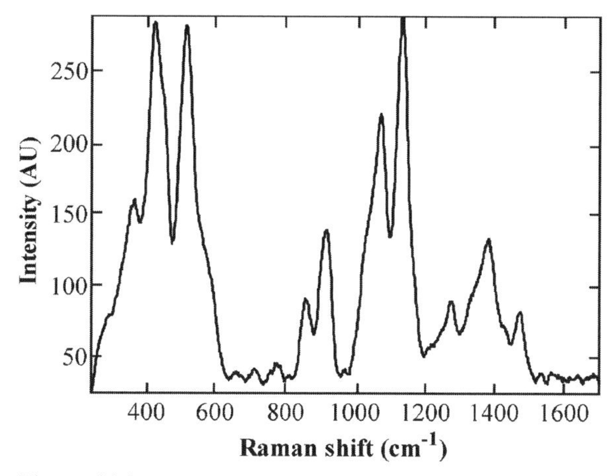

Figure 14.2 A Raman spectrum consists of scattered intensity plotted versus energy. This figure uses glucose water solution measured in a quartz cuvette as an example.

A Raman spectrum is a plot of scattered light intensity versus energy shift (also called Raman shift) reported in wavenumbers (cm^{-1}). An example spectrum of aqueous glucose is shown in Figure 14.2. To convert from a wavenumber shift to wavelength, the incident wavelength must be known. For example, a 600 cm^{-1} Raman shift occurs at 873.5 nm if the excitation wavelength is 830 nm or at 823.8 nm if the excitation wavelength is 785 nm.

In this chapter, we focus on nonresonant spontaneous Raman scattering. A special case of Raman scattering, surface-enhanced Raman spectroscopy (SERS), is discussed in Chapter 15.

14.2 BIOLOGICAL CONSIDERATIONS FOR RAMAN SPECTROSCOPY

14.2.1 Using Near-Infrared Radiation

Raman shifts are independent of excitation wavelength, and thus Raman spectroscopy offers the flexibility to select a suitable excitation wavelength for a specific application. The choice of NIR excitation for probing biological tissue is justified by three advantageous features: low-energy optical radiation, deep penetration, and reduced background fluorescence. Light in the NIR region is nonionizing and therefore does not pose the same exposure risk as X-ray radiation. Additionally, NIR light penetrates relatively deep into the tissue, on the order of millimeters to centimeters in some spectral windows. These depths are possible due to reduced elastic scattering, which decreases at longer wavelengths, and the lack of significant absorption bands in this region. Fluorescence is also much lower in the NIR region as compared to shorter wavelengths, thus allowing the less intense Raman bands to be resolved.

Figure 14.3 illustrates the absorption spectra of major endogenous tissue absorbers, namely, water, skin melanin, hemoglobin, and fat. Also shown is the

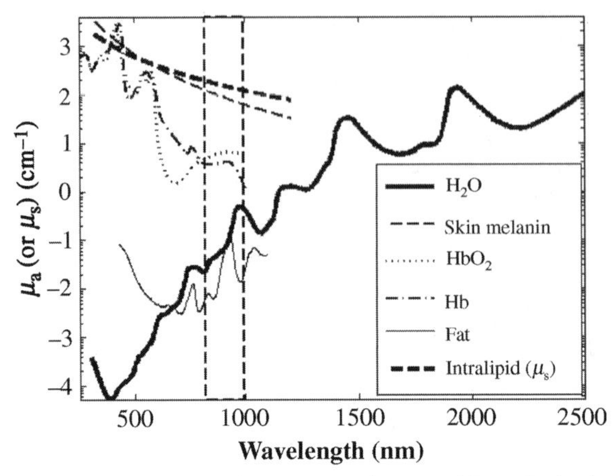

Figure 14.3 Absorption spectra of water, skin melanin, hemoglobin, and fat. Also shown is the scattering spectrum of 10% intralipid, a lipid emulsion often used to simulate tissue scattering. Data are obtained from http://omlc.ogi.edu/spectra/index.html.

scattering spectrum of 10% intralipid, a lipid emulsion often used to simulate tissue scattering. The "diagnostic window," in which a group of minima exists, is outlined.

A final consideration for the selection of excitation wavelength in Raman spectroscopy is the efficiency of the silicon-cased charge-coupled device (CCD) detector. Due to silicon absorption, CCD detectors are prohibitively inefficient above 1000 nm. As a result, 785 nm or, more recently, 830 nm are often chosen as the excitation wavelength to fully exploit the "diagnostic window" while retaining an acceptable quantum efficiency detector.

14.2.2 Background Signal in Biological Raman Spectra

Although greatly reduced in intensity as compared to UV–visible excitation, NIR Raman spectra of biological samples are often accompanied by a strong background, generally attributed to fluorescence. Macromolecules such as proteins and lipids are thought to contribute to the fluorescence background.[3] Although Raman bands are clearly distinguished above the background, the presence of the background results in higher shot noise and therefore decreases the signal-to-noise ratio. Furthermore, the background decreases as a function of time with accompanying spectral variation, which may or may not be attributed to photobleaching. This variation interferes with the multivariate analysis necessary for these types of measurements. Thus, it is desirable to either reduce the background during data collection or remove it in preprocessing without introducing artifacts. Most background removal methods in the literature are based on low-order polynomial fitting and subsequent subtraction. Many researchers have found that a fifth-order polynomial best approximates the shape of the background.[4–7] However, because of the inevitable introduction of spectral artifacts, some researchers have found that removing the background does not improve calibration results obtained from multivariate analysis.[8]

14.2.3 Heterogeneities in Human Skin

Uniform analyte distribution is often a good assumption for liquid samples such as blood serum or even whole blood if stirring is continuous. For biological tissue, human skin in particular, heterogeneity is a major factor. Detailed morphological structures and molecular constituents of skin have been studied using confocal Raman spectroscopy.[9]

The skin is a layered system with two principle layers: epidermis and dermis. The epidermis is the outmost layer of skin and itself consists of multiple layers such as the stratum corneum, stratum lucidum, and stratum granulosum. The major constituent of human epidermis is keratin, comprising approximately 65% of the stratum corneum. The dermis is also a layered tissue composed of mainly collagen and elastin. Blood capillaries are present in the dermis, and thus this region is targeted for optical analysis. However, it has been suggested that the majority of the glucose molecules sampled by a noninvasive optical technique are present in the interstitial fluid (ISF), which is found throughout tissue but tends to collect at the epidermis– dermis interface.[10]

14.3 QUANTITATIVE CONSIDERATIONS FOR RAMAN SPECTROSCOPY

14.3.1 Minimum Detection Error Analysis

If all component spectra in a mixture sample are known, the minimum detection error, Δc, can be calculated via a simple formula derived by Koo et al.:[11]

$$\Delta c = \frac{\sigma}{s_k}\mathrm{olf}_k \tag{14.1}$$

The first factor on the right-hand side, σ, describes the noise in the measured spectrum and the second factor, s_k, describes the signal strength, calculated as the norm of the kth model component. The last factor, olf_k, is termed the "overlap factor" and can take on values between 1 and ∞.

The overlap factor indicates the amount of nonorthogonality (overlap) between the kth model component and the other model components. Mathematically, the overlap factor is equal to the inverse of the correlation coefficient between the kth component spectrum, $\mathbf{s}_k$, and the OLS regression vector, $\mathbf{b}_{\mathrm{OLS}}$:

$$\mathrm{olf}_k = \frac{1}{\mathrm{corr}(\mathbf{b}_{\mathrm{OLS}}, \mathbf{s}_k)} \tag{14.2}$$

The OLS regression vector, also called the net analyte signal, is the part of the kth component spectrum that is orthogonal to all interferents. It is equivalent to the kth component spectrum when no interferents exist.

Correlation between two vectors is calculated by

$$\mathrm{corr}(u, v) = \frac{\sum_{i=1}^{n}(u_i - \bar{u})(v_i - \bar{v})}{\sum_{i=1}^{n}(u_i - \bar{u})^2 \sum_{i}^{n}(v_i - \bar{v})^2} \tag{14.3}$$

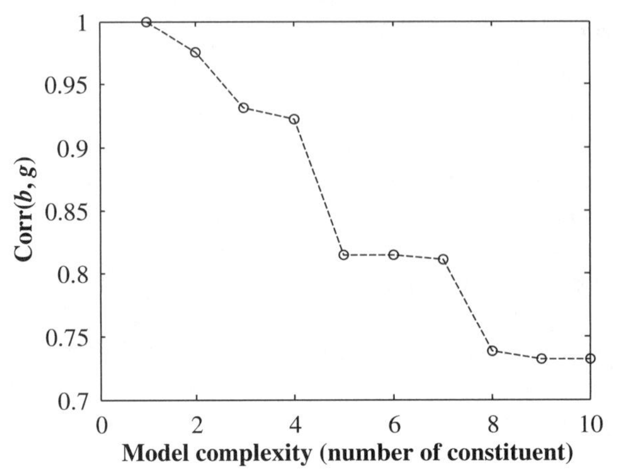

Figure 14.4 Correlation between the OLS regression vector and the glucose spectrum versus model complexity.

In the absence of interferents, the correlation coefficient is equal to 1 and therefore $\text{olf}_k = 1$. In this case, the minimum detection error is defined solely on the basis of signal-to-noise considerations. When interferents exist, the correlation coefficient is always smaller than 1 and therefore olf_k is always larger than 1. The minimum detection error approaches infinity when there is complete overlap and the analyte signal is indistinguishable from interfering components.

To estimate the overlap factor for Raman measurements of glucose in skin, we have used a 10-component skin-mimicking model. Beginning with the spectrum of glucose, spectra of other constituents with strong Raman signals including collagen type I, keratin, triolein, actin, collagen type III, cholesterol, phosphatidylcholine, hemoglobin, and water were added one at a time to increase the model complexity. At each addition, the correlation between $\mathbf{b}_{\text{OLS}}$ and the glucose spectrum was calculated that ranged from 1 to 0.73 as shown in Figure 14.4. Therefore, the overlap factor for glucose and skin is estimated at 1.4.

The high molecular specificity of Raman spectra results in less spectral overlap than other modalities such as NIR absorption spectroscopy, thus enabling the detection of low signal strength components such as glucose.

14.3.2 Multivariate Calibration

As discussed previously, although Raman spectroscopy provides good molecular specificity, spectral overlap is inevitable with the presence of multiple constituents. Furthermore, the glucose Raman signal is only 0.3% of the total skin Raman signal.[12,13] Taken into consideration with the varying fluorescence background and random noise, it is not feasible to quantify the glucose signal by recording the skin Raman spectrum at only a few wavelengths. For quantitative analysis, multivariate techniques, which utilize the full-range spectra, are employed. In multivariate calibration, a set of calibration spectra and the associated glucose concentrations are used to calculate a regression vector. This regression vector, or **b** vector, can be applied to a future independent spectrum with unknown glucose content to extract the concentration.[14–16] An introduction to multivariate techniques is described in Chapter 12.

14.4 INSTRUMENTATION

As discussed previously, background fluorescence impedes observation of Raman signal from biological tissue using UV–visible excitation wavelengths. To overcome this limitation, NIR excitation was employed with Fourier transform spectrometers in the late 1980s.[17] With the advent of high quantum efficiency CCD detectors and holographic diffractive optical elements, researchers have increasingly employed CCD-based dispersive spectrometers.[3,18–24] The advantages of dispersive NIR Raman spectroscopy are that compact solid-state diode lasers can be used for excitation, the imaging spectrograph can be *f*-number matched with optical fibers for better throughput, and cooled CCD detectors offer shot-noise limited detection.

As a tutorial for the selection of building blocks for a Raman instrument with high collection efficiency, we present a summary of the key design considerations.

14.4.1 Excitation Light Source

Laser excitation at one of two wavelengths, 785 and 830 nm, is most common. The trade-off lies in that excitation at lower wavelengths has a higher efficiency of generating Raman scattering but also generates more intense background fluorescence. The current trend is toward the use of external cavity laser diodes because they are compact and of relatively low cost. In other embodiments, argon-ion laser pumped titanium–sapphire lasers are used extensively. The titanium–sapphire laser can provide higher power output with broader wavelength tunability, but is bulkier (several feet in dimension) and more expensive to maintain than diode lasers.

Because Raman scattering occurs at the same energy shift regardless of the excitation wavelength, narrowband excitation must be used to prevent broadening of the Raman bands. Furthermore, the wings of the laser emission (amplified spontaneous emission) can extend beyond the cutoff wavelength of the notch filter used to suppress the elastically scattered light and obscure low-wavenumber Raman bands. This problem is most severe in high-power diode lasers and a holographic band-pass or interference laser line filter, with attenuation greater than 6 OD (optical density) is usually required. Finally, for quantitative measurements a photodiode is often needed to monitor the laser intensity to correct for fluctuations.

14.4.2 Light Delivery

The filtered laser light can be delivered to the sample either through free space or through an optical fiber. In the free-space embodiments, beam shaping is usually performed to correct for astigmatism and other laser light artifacts. The incident light at the sample can be either focused or collimated, depending on collection considerations. For biological tissue, the total power per unit area is an important consideration, and thus spot size on the tissue is an oft-reported parameter.

Raman probes constructed from fused silica optical fibers have gained much attention recently. Typically, low-OH content fibers are utilized to reduce the fiber fluorescence. The probe design also includes filters at the distal end to suppress the fused silica Raman signal from the excitation fiber and suppress the elastically scattered light entering the collection fibers.[25] Commercial probes are now available and they offer ruggedness and easy access to samples with various special or geometrical constraints.

14.4.3 Light Collection

As Raman scattering is a weak process, photons are precious and high collection efficiency is desired for a higher signal-to-noise ratio. Specialized optics such as Cassegrain microscope objectives and nonimaging paraboloidal mirrors have been employed to increase both the collection spot size and the effective numerical aperture of the optical system.[10]

The majority of photons that exit the air–sample interface are elastically scattered and remain at the original laser wavelength. This light must be properly attenuated, or it will saturate the entire CCD detector. Holographic notch filters are extensively employed for this purpose and can attenuate the elastically scattered light to greater than 6 OD, while passing the Raman photons with greater than 90% efficiency. However, notch filters are very sensitive to the incident angle of light and thus provides less attenuation to off-axis light. In some instances, the size of the notch filter is one of the determining factors of the throughput of an instrument.

Specular reflection, light that is elastically scattered without penetrating the tissue, is also undesirable. Strategies such as oblique incidence,[26] 90° collection geometry,[3] and a hole in the collection mirror have been realized to reduce its effect.[27]

14.4.4 Light Transport

After filtering out most of the elastically scattered light, the Raman scattered light must be transported to the spectrograph with minimum loss. To match the rectangular shape of the entrance slit of a spectrograph, the originally round shape of the collected light can be relayed by an optical fiber bundle with the receiving end arranged into a round shape and the exiting end arranged linearly.[22]

14.4.5 Spectrograph and Detector

In dispersive spectrographs for Raman spectroscopy, transmission holographic gratings are often used for compactness and high dispersion. Holographic gratings can be custom blazed for specific excitation wavelengths and provide acceptable efficiency. In addition, liquid nitrogen cooled and more recently thermoelectric-cooled CCD detectors offer high sensitivity and shot-noise limited detection in the

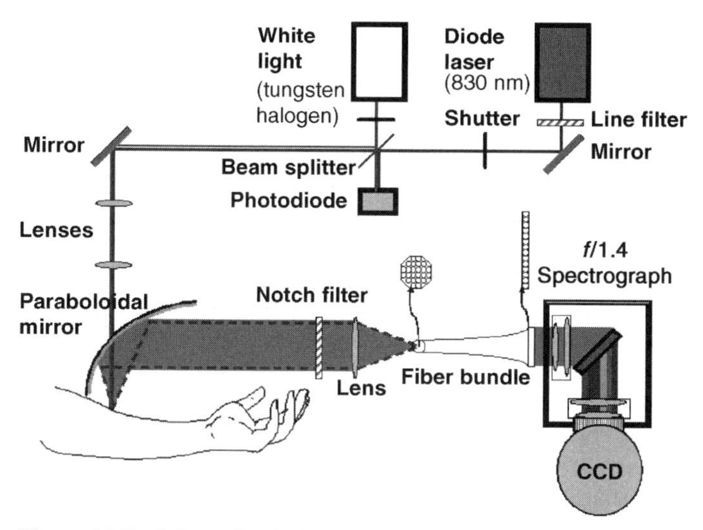

Figure 14.5 Schematic of a free-space Raman instrument for noninvasive glucose measurements used at the MIT Spectroscopy Laboratory.

near-infrared wavelength range up to ~1 μm. These detectors can be controlled using programs such as Labview to facilitate experimental studies.

To increase light throughput in Raman systems, the CCD chip size can be increased vertically to match the spectrograph slit height. However, large-format CCD detectors show pronounced slit image curvature that must be corrected in preprocessing (described below).

As an example of these design considerations, Figure 14.5 shows a schematic of the high-throughput Raman instrument currently used in our laboratory. We opted for free space delivery of the excitation light through a small hole in an off-axis half-paraboloidal mirror. Backscattered Raman light is collimated by the mirror and passed through a 2.5 in. holographic notch filter to reduce elastically scattered light. The Raman photons are focused onto a shape-transforming fiber bundle with the exit end serving as the entrance slit of an *f*/1.4 spectrometer. The prefiltering stage of the spectrometer was removed to reduce fluorescence and losses from multiple optic elements. The back-thinned, deep depletion, liquid nitrogen-cooled CCD is 1300 × 1340 pixels, height matched to the fiber bundle slit. This instrument was specifically designed for high sensitivity measurements in turbid media.

14.5 DATA PREPROCESSING

After data collection, various preprocessing steps are undertaken to improve data quality. The preprocessing steps chosen can lead to different calibration results; therefore, it is important for researchers to thoroughly document the exact steps taken.

Frequently employed procedures are described in the subsequent sections.

14.5.1 Image Curvature Correction

Increase of usable detector area is an effective way to improve light throughput in Raman spectroscopy employing multichannel dispersive spectrographs. However, owing to out-of-plane diffraction the entrance slit image appears curved.[28] Direct vertical binning of detector pixels without correcting the curvature results in degraded spectral resolution.

Various hardware approaches, such as employing curved slits[26,28] or convex spherical gratings, have been implemented.[29] In the curved slit approaches, fiber bundles have been employed as shape transformers to increase Raman light collection efficiency. At the entrance end, the fibers are arranged in a round shape to accommodate the focal spot, and at the exit end in a curved line, with curvature opposite that introduced by the remaining optical system. This exit arrangement serves as the entrance slit of the spectrograph and provides immediate first-order correction of the curved image, as described below. However, for quantitative Raman spectroscopy, with substantial change of the image curvature across the wavelength range of interest ($\sim$150 nm) and narrow spectral features, this first-order correction is not always satisfactory.

As an alternative to the hardware approach, software can be employed to correct the curved image, with potentially better accuracy and flexibility for system modification. In our past research, we have developed a software method using a highly Raman-active reference material to provide a sharp image on the CCD.[8] Using the curvature of the slit image at the center wavelength as a guide, we determine by how many pixels in the horizontal direction each off-center CCD row needs to be shifted in order to generate a linear vertical image. This pixel shift method, as well as the curved-fiber-bundle hardware approach, ignores the fact that the slit image curvature is wavelength dependent. The resulting spectral quality of the pixel shift method is thus equivalent to the curved-fiber-bundle hardware approach.[28] This issue becomes more important when large CCD chips and high-NA spectrographs are employed for increasing the throughput of the Raman scattered light.

Recently, a software approach using multiple polystyrene absorption bands was developed for infrared spectroscopy.[30] In this section, we present a similar method that was developed concurrently, which calibrates on multiple Raman peaks to generate a curvature map. This curvature mapping method shows significant improvement over first-order correction schemes.

The curvature mapping method requires an initial calibration step. In calibration, a full-frame image is taken with a reference material that has prominent peaks across the spectral range of interest, for example, acetaminophen (Tylenol) powder. We chose nine prominent peaks across the wavelength range of interest, as depicted by the arrows in Figure 14.6.

The calibration algorithm generates a map of the amount of shift for each CCD pixel and a scale factor to maintain signal conservation in each CCD row. Once the map and the scale factor are generated, usually when the system is first set up, the correction algorithm can be applied to future measurements.

Significant improvement is observed from the pixel shift method to the curvature mapping method, especially toward either side of the CCD, as can be

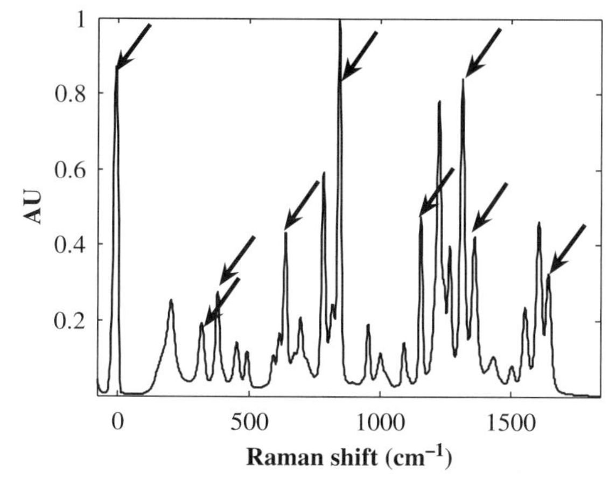

Figure 14.6 Raman spectrum of acetaminophen powder, used as the reference material in the calibration step. Nine prominent peaks used as separation boundaries are indicated by arrows.

seen by comparing Figure 14.7c and e. The overall linewidth reduction in 14 prominent peaks is 7% (FWHM). Such improvement is significant considering that the equivalent slit width is ~360 μm. If a narrower slit is employed for better spectral resolution, the overall linewidth reduction will be more significant. Note that the images were taken with 5-pixel CCD hardware vertical binning to reduce the amount of data, since the curvature is barely noticeable within such a short range. The error introduced by the hardware binning is much less than 1 pixel, and thus negligible.

14.5.2 Spectral Range Selection

Multivariate calibration methods attempt to find spectral components based on variance. The presence of a spectral region with large nonanalyte-specific variations may bias the algorithm and cause smaller analyte-specific variance to be overlooked. Therefore, the "fingerprint" region from approximately 300–1700 cm^{-1} is often chosen for analysis.

14.5.3 Cosmic Ray Removal

Cosmic rays hit the CCD array at random times with arbitrary intensity, resulting in spikes at individual pixels. When the array is summed and processed, sharp spectral features of arbitrary intensities may appear in the Raman spectra. These artifacts are typically removed before multivariate calibration.

One approach is based on the assumption that the spectrum does not change its intensity from frame to frame other than due to noise and cosmic rays. Therefore, by comparing multiple neighboring frames, a statistical algorithm can be used to identify cosmic rays. Another solution compares adjacent pixels in the same spectrum and detects abrupt jumps in intensity from pixel to pixel. Once a cosmic ray contaminated pixel is identified, its value can be replaced by the average of neighboring pixels.

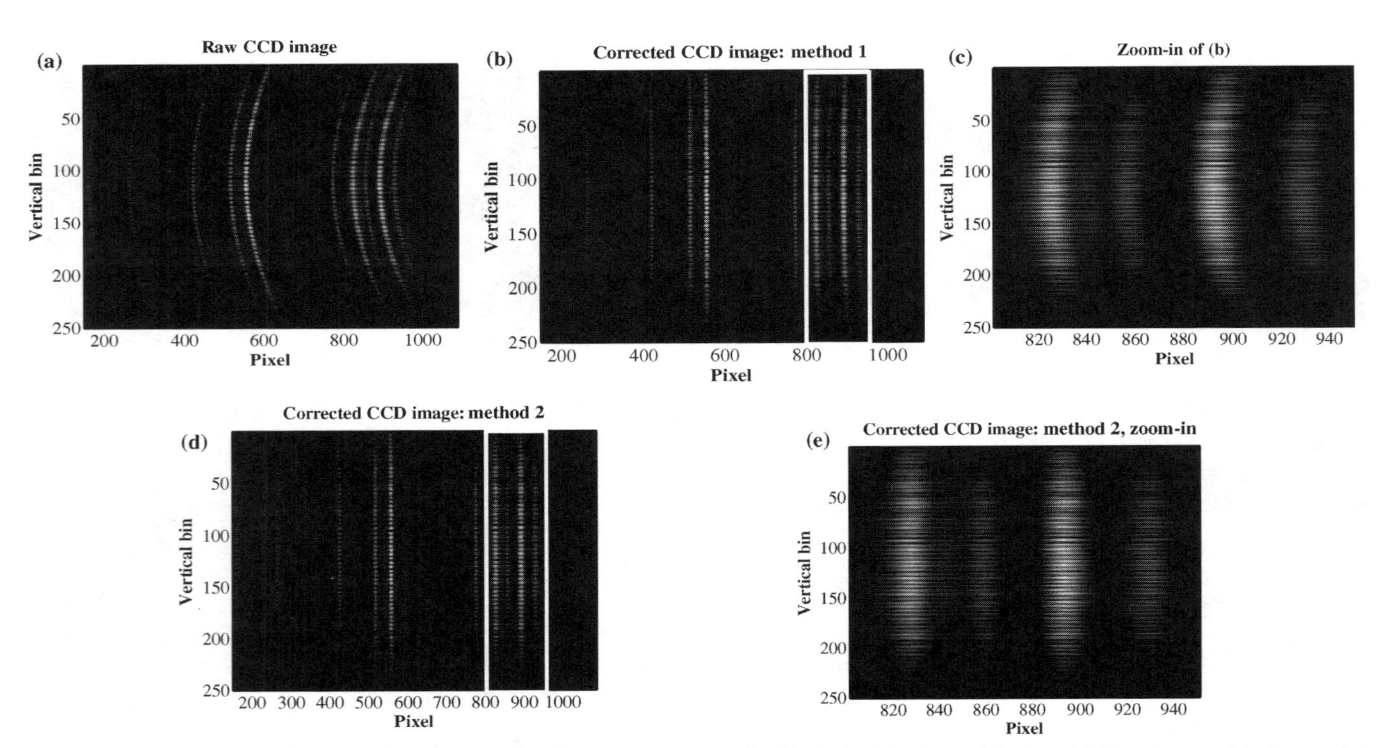

Figure 14.7 CCD image of acetaminophen powder. Images were created with 5-pixel hardware binning. (a) Raw image, (b) after applying pixel shift method, (c) zoom-in of the box in (b), (d) after applying curvature mapping method, and (e) zoom-in of the box in (d).

14.5.4 Background Subtraction

As mentioned in the biological considerations section, the background signal in Raman spectra is one of the limiting factors in determining the detection limit. Background removal techniques only approximate the shape of the background, and therefore improvement in further quantitative analysis is often limited. However, for qualitative analysis, background-removed spectra provide better interpretation of the underlying constituents.

14.5.5 Random Noise Rejection and Suppression

Photon shot-noise limited performance can be achieved using a liquid nitrogen-cooled CCD camera. When a detector is shot-noise limited, the random noise can be estimated by the square root of the measured intensity. The most effective way to increase the signal-to-noise ratio (SNR) under shot-noise limited conditions is to increase the integration time of the CCD or the throughput of the instrument. However, extending the integration beyond a certain timescale offers no extra benefit as other errors such as system drifts begin to dominate performance.[10] Once the data are collected, signal processing is the only way to further enhance the SNR. Pixel binning along the wavelength axis is one method of increasing the SNR and results indicate an optimal number exists for binned pixels.[10] However, the drawback to this method is degradation in spectral resolution. More commonly employed are Savitzky–Golay smoothing algorithms, which retains the data length.

14.5.6 White Light Correction and Wavelength Calibration

When spectra collected from different instruments or on different days are to be compared, white light correction and wavelength calibration are required. White light correction is performed by dividing the Raman spectra to a spectrum from a calibrated light source, for example, a calibrated tungsten halogen lamp measured under identical conditions. Combinatorial spectral responses of the optical components, the diffraction grating, and the CCD camera can be effectively removed and thus reveal more of the underlying Raman spectral features. Wavelength calibration is performed to transform the pixel-based axis into a wavelength-based (or wavenumber-based) axis, allowing for comparison of Raman features across instruments and time.

14.6 *IN VITRO* STUDIES

In the following sections, we review the application of Raman spectroscopy to glucose sensing *in vitro*. *In vitro* studies have been performed using human aqueous humor (HAH), filtered and unfiltered human blood serum, and human whole blood, with promising results. Results in measurement accuracy are reported in root mean squared error values, with RMSECV for cross-validated and RMSEP for predicted values. The reader is referred to Chapter 12 for discussion on these statistics.

14.6.1 Aqueous Humor of the Eye

Lambert et al. have explored the use of Raman spectroscopic measurements of glucose present in the aqueous humor of the eye.[20,21,31] This is undoubtedly an excellent portal for optical measurements with potential advantages such as easy access and less complex fluid composition. In spite of these advantages, a spectroscopic measurement in the eye carries the risk of injury if the probing light is too intense. Therefore, dosimetry and a foolproof light delivery method are important concerns for *in vivo* human studies.

Recently they demonstrated *in vitro* predictive capability of Raman spectroscopic measurements using a PLS calibration model derived from an artificial model.[21,31] Human aqueous humor was used as the *in vitro* sample for prediction and artificial aqueous humor (AAH) was used to construct the calibration model. In the AAH model, five analytes including glucose at physiological concentrations were designed to vary independently with little correlation between any two analytes. The main advantage of using an AAH model is to break the glucose–lactate correlation present in HAH (correlation coefficient $r \sim 0.4$). The sample was placed in a contact lens for measurement by a Raman instrument using 785 nm excitation and a microscope objective with 180° geometry for Raman signal collection. Each spectrum was obtained with excitation power ~100 mW and integration time ~150 s.

Lambert et al. obtained an RMSEP of approximately 1–1.5 mM with $R^2 \sim 0.99$. Glucose spectral features were clearly observed in the second PLS factor, further supporting the calibration accuracy.[31] They pointed out future directions such as focusing on demonstrating safety and efficacy in humans. In addition, although the time delay between the blood and aqueous humor glucose concentrations was previously measured invasively using a rabbit model,[32] the implication to human subjects is yet well understood.

14.6.2 Blood Serum

14.6.2.1 Unprocessed Samples Our laboratory began investigating noninvasive blood analysis using Raman spectroscopy in the mid-1990s.[33–35] The first biological sample study was conducted on serum and whole-blood samples from 69 patients over a 7-week period.[23] No sample processing or selection criteria were employed, with the exception of locating a few samples with extreme glucose concentrations to represent the glucose levels of a range of diabetes patients. An 830 nm, diode laser was employed for excitation and a microscope objective for light collection. The laser power at the sample was ~250 mW and the integration time for each spectrum was equivalent to 300 s. The glucose measurement results in serum were quite encouraging, with PLS calibration providing an RMSECV of 1.5 mM. However, the glucose measurement results in whole-blood result were not satisfactory because of reduced signal from the high turbidity. Glucose spectral features were identified in both the PLS weighting vector and the **b** vector, supporting that the calibration model was based on glucose. For detailed discussion on calibration, please refer to Section 12.4.

14.6.2.2 With Ultrafiltration Qu et al.[3] described the use of Raman spectroscopy for glucose measurements in human serum samples after ultrafiltration, a process to remove macromolecules. Ultrafiltration can effectively eliminate Raman signals from large protein molecules that dominate unfiltered serum samples, thus significantly improving the signal-to-noise ratio and therefore the detection limit. Nevertheless, it is time consuming and requires extra sample preparation.

The experimental setup employed 785 nm excitation with a 90° collection geometry. Each spectrum was obtained with excitation power ~300 mW and integration time equivalent to 2.5 min. Because filtered serum is nearly transparent at 785 nm, excitation of Raman scattering is effectively along the entire laser path, creating a line source in the cuvette. Thus, the authors surmise that better collection efficiency could be obtained with optics designed specifically for this type of source, as opposed to the standard spherical lens they employed.

Regardless of potential collection efficiency improvements, they obtained an RMSEP of 0.38 mM. However, the PLS calibration model was obtained using 30 samples with 12 factors retained for the development of the regression vector. Without reported evidence of glucose spectral features, it is difficult to determine whether the data were overfit. The model was applied to 24 samples that were not in the calibration set, giving some justification to the analysis.

Rohleder et al.[36] described measurement of glucose in both serum and filtered serum from 247 blood donors. A commercial spectrometer was used to acquire the spectra with 785 nm excitation and a double holographic grating covering a wavelength range of 785–1082 nm. Spectra were obtained with excitation power ~200 mW at the sample and integration time ~300 s. PLS calibration models were generated based on 148 samples and employed to predict the concentrations of the remaining 99 samples. They obtained an RMSEP of 0.95 mM ($R^2 \sim 0.97$) and 0.34 mM ($R^2 \sim 0.996$) in unfiltered and filtered serum samples, respectively.

14.6.3 Whole Blood

The main difficulty for measurement in whole blood as compared to serum is the much higher absorption and scattering of whole blood attributed to the presence of hemoglobin and red blood cells, respectively. The combined effect is a ~4× decrease in collected analyte Raman signal.

To confirm the reason for poor whole-blood results compared to serum was the 4× decrease in Raman signal, a subsequent whole-blood study was performed by Enejder et al.[22] in our laboratory. They demonstrated the feasibility of measuring multiple analytes in 31 whole-blood samples with laser intensity and integration time similar to the previously mentioned serum study.[23] A 4× increase in Raman signal collection was achieved by employing a paraboloidal mirror and a shape-transforming fiber bundle for better collection efficiency, as depicted in Figure 14.5. PLS leave-one-out cross-validation was performed and an RMSECV of 1.2 mM was obtained. The number of PLS factors compared to the number of samples raises the concern of overfitting. However, glucose spectral features were identified in the regression vector, providing more confidence that the model was based on glucose.

14.7 *IN VIVO* STUDIES

14.7.1 Tissue Modulation Approach

Chaiken et al.[24,37,38] proposed using Raman spectroscopy to measure glucose *in vivo* with a technique called "tissue modulation," that is, continuously press/unpress the measurement site with a mechanical apparatus. The basic principle is that during the "press" phase, blood is expelled from the measuring site and thus the spectrum is considered as nearly devoid of blood. During the "unpress" phase, the spectrum is considered to be a sum of both blood and other tissue constituents.

In a recent report,[24] the difference spectrum between pressed and unpressed phases was interpreted as the whole-blood spectrum. Each spectrum was obtained with excitation power ~31 mW and integration time ~100 s. Apparent glucose signal and blood volume factor were extracted by summing over 686–375 cm^{-1} and 1800–1000 cm^{-1} in the difference spectrum, respectively. Integrated normalized unit (INU) was then defined as the ratio of the apparent glucose signal to the blood volume factor. They claimed that 375–686 cm^{-1} contains the most glucose information and 1800–1000 cm^{-1} contains mostly fluorescence plus Raman signal from other tissue constituents, indicative of blood volume.

Eighteen spectroscopic samples paired with fingerstick reference measurements were collected from an individual over 2 days with a time-randomized protocol. A calibration model was built by fitting a line through the plot of INU versus reference glucose concentration. The model was then applied to 31 samples collected over the next 14 weeks (7 additional samples from the same individual and 24 from different individuals). After rejecting 11 outliers (~35%), they obtained a correlation coefficient (r) of 0.8 and a standard deviation of ~1.2 mM.

Their work presents an interesting idea to isolate the glucose-containing blood spectrum by tissue modulation. The result suggests that the INU correlates to not only the reference glucose concentration, but also the hemoglobin, which raises the concern of specificity. This technique can potentially be useful if several issues can be addressed. First, summation over 686–375 cm^{-1} obviously includes contributions from interferents, and it is not clear how much error that introduces. In addition, hemoglobin is not the only substance that fluoresces, and thus it is unclear why the calculated blood volume factor could be representative of actual blood volume. Furthermore, it is likely that most glucose Raman signal measured from skin originates from glucose molecules in the interstitial fluid. It is therefore unclear if the results were indeed based on blood glucose as claimed by the authors.

14.7.2 Direct Approach

In our laboratory, Enejder et al. conducted a transcutaneous study on 17 nondiabetic volunteers using a version of the instrument depicted in Figure 14.5.[8] Spectra were collected from the forearm of human volunteers in conjunction with an oral glucose tolerance test protocol, involving the intake of a high-glucose containing fluid, after which the glucose levels were elevated to more than twice that found under fasting conditions. Periodic reference glucose concentrations were obtained from fingerstick blood samples and subsequently analyzed by a Hemocue device. The glucose concentrations for all volunteers ranged from 3.8 to 12.4 mM (~68–223 mg/dL).

Raman spectra in the range 1545–355 cm^{-1} were selected for data analysis. An average of 27 (461/17) spectra were obtained for each individual with a 3 min integration time per spectrum. Each spectrum was obtained with excitation power ~300 mW and integration time equivalent to 3 min. Spectra from each volunteer were analyzed using PLS with leave-one-out cross-validation, with eight factors retained for development of the regression vector. For one subject, a mean absolute error (MAE) of 7.8% (RMSECV ~ 0.7 mM) and an R^2 of 0.83 were obtained.

An additional method for determining the influence of glucose in the calibration is to examine the results of calibration formed by combining together data sets from a number of volunteers. The error is expected to rise as data from more volunteers are added. However, a limited rise would indicate that the signal from the common analyte, glucose, is strong enough to be seen in the presence of other variations. A calibration was generated on data comprising 244 samples from a group of 9 volunteers whose individual calibration quality appeared to be relatively high. Such a calibration resulted in an MAE of 12.8% with $R^2 \sim 0.7$, while combining all 17 volunteers gave MAE ~16.9% (RMSECV ~1.5 mM). The number of PLS factors compared to the number of spectra is in danger of overfitting in an individual calibration. However, the grouping schemes involving 9 (244 spectra) and 17 (461 spectra) volunteers utilized 17 and 21 factors, respectively, which is more acceptable. Another encouraging piece of evidence was that multiple glucose spectral features were identified in the regression vectors, as exemplified by strong correlation between the **b** vector and glucose spectrum, indicating that glucose was a major contributor in the calibration.

This study was an initial evaluation of the ability of Raman spectroscopy to measure glucose noninvasively with the focus on determining its capability in a range of subjects. The protocol did not include measurements on the volunteers over a number of days, and thus independent data was not obtained. Furthermore, oral glucose tolerance test protocols are susceptible to correlation with the fluorescence background decay, which likely decays as a result of photobleaching, although this has not been confirmed. This correlation may enhance the apparent prediction results. Therefore, more studies, preferably involving glucose clamping performed on different days, are required.

14.8 TOWARD PROSPECTIVE APPLICATION

The results from the *in vitro* and *in vivo* studies reviewed in the previous sections are very encouraging. They demonstrate the feasibility of building glucose-specific *in vivo* multivariate calibration models based on Raman spectroscopy. To bring this technique to the next level, prospective application of a calibration algorithm on independent data with clinically acceptable detection results needs to be demonstrated. From our perspective, this objective requires advances to be made in extracting glucose information without spurious correlations to other system components and correcting for variations in subject tissue morphology and color. We have developed new tools to address these issues. Specifically, a novel multivariate calibration technique with higher analyte specificity that is more robust against interferent covariation or chance correlation was developed. This technique, constrained regularization (CR), is described in Section 14.8.1.2. Also, a new correction method to compensate for turbidity-induced sampling volume variations across sites and individuals was developed. This method, intrinsic Raman spectroscopy (IRS), is introduced in Section 14.8.2.4. Additionally, other considerations for successful *in vivo* studies, such as reference concentration accuracy, optimal collection site determination, and so on, will be discussed in the context of future directions.

14.8.1 Analyte-Specific Information Extraction Using Hybrid Calibration Methods

Multivariate calibration methods are in general not analyte specific. Calibration models are built based on correlations in the data, which may be owing to the analyte or to systematic or spurious effects. One way to effectively boost the model specificity is through incorporation of additional analyte-specific information such as its pure spectrum. Hybrid methods merge additional spectral information with calibration data in an implicit calibration scheme. In the following section, we present two of these methods developed in our laboratory.

14.8.1.1 Hybrid Linear Analysis Hybrid linear analysis (HLA) was developed by Berger et al.[39] First, analyte spectral contributions are removed from the sample spectra by subtracting the pure spectrum according to reference concentration measurements. The resulting spectra are then analyzed by principal component analysis with significant principal components extracted. These principal components are subsequently used as basis spectra to perform an orthogonalization process on the pure analyte spectrum. The orthogonalization results in a **b** vector that is essentially the portion of the pure analyte spectrum that is orthogonal to all interferent spectra, akin to the net analyte signal.

HLA was implemented experimentally *in vitro* with a three-analyte model composed of glucose, creatinine, and lactate. Significant improvement over PLS was obtained owing to the incorporation of the pure glucose spectrum in the algorithm development. However, because HLA relies on the subtraction of the analyte spectrum from the calibration data, it is highly sensitive to the accuracy of the spectral shape and its intensity. For complex turbid samples in which absorption and scattering are likely to alter the analyte spectral features in unknown ways, we find that the performance of HLA is impaired. Motivated by advancing transcutaneous measurement of blood analytes *in vivo*, constrained regularization was developed as a more robust method against inaccuracies in the pure analyte spectra.

14.8.1.2 Constrained Regularization (CR) To understand constrained regularization, multivariate calibration can be viewed as an inverse problem. Given the inverse mixture model for a single analyte

$$\mathbf{c} = \mathbf{S}^{\mathrm{T}}\mathbf{b} \tag{14.4}$$

the goal is to invert equation (14.4) and obtain a solution for **b**. Factor-based methods such as principal component regression (PCR) and partial least squares (PLS) summarize the calibration data, [**S**, **c**], using a few principal components or loading vectors. Whereas CR seeks a balance between model approximation error and noise propagation error by minimizing the cost function, Φ,[40]

$$\Phi(\Lambda, \mathbf{b}_0) = \left\|\mathbf{S}^{\mathrm{T}}\mathbf{b} - \mathbf{c}\right\|^2 + \Lambda\|\mathbf{b} - \mathbf{b}_0\|^2 \tag{14.5}$$

with $\|\mathbf{a}\|$ the Euclidean norm (i.e., magnitude) of **a**, and $\mathbf{b}_0$ a spectral constraint that introduces prior information about **b**. The first term of Φ is the model approximation error, and the second term is the norm of the difference between the solution and the constraint, which controls the smoothness of the solution and its deviation from the constraint. If $\mathbf{b}_0$ is zero, the solution is the common regularized solution. For $\Lambda = 0$ the least squares solution is then obtained. In the other limit, in which Λ goes to infinity, the solution is simply $\mathbf{b} = \mathbf{b}_0$.

A reasonable choice for $\mathbf{b}_0$ is the spectrum of the analyte of interest because that is the solution for **b** in the absence of noise and interferents. Another choice is the net analyte signal[41] calculated using all of the known pure analyte spectra. Such flexibility in the selection of $\mathbf{b}_0$ is owing to the manner in which the constraint is incorporated into the calibration algorithm. For CR, the spectral constraint is included in a nonlinear fashion through minimization of Φ, and is thus termed a "soft" constraint. On the other hand, there is little flexibility for methods such as HLA, in which the spectral constraint is algebraically subtracted from each sample spectrum before performing PCA. We term this type of constraint a "hard" constraint.

In numerical simulations and experiments with tissue phantoms, we found that with CR the RMSEP is lower than methods without prior information, such as PLS, and is less affected by analyte covariations. We further demonstrated that CR is more robust than HLA when there are inaccuracies in the applied constraint, as often occurs in complex or turbid samples such as biological tissue.[27]

An important lesson learned from the study is that there is a trade-off between maximizing prior information utilization and robustness concerning the accuracy of such information. Multivariate calibration methods range from explicit methods with maximum use of prior information (e.g., OLS, least robust when accurate model is not obtainable), hybrid methods with an inflexible constraint (e.g., HLA), hybrid methods with a flexible constraint (e.g., CR), and implicit methods with no prior information (e.g., PLS, most robust, but is prone to be misled by spurious correlations). We believe CR achieves the optimal balance between these ideals in practical situations.

14.8.2 Sampling Volume Correction Using Intrinsic Raman Spectroscopy

Sample variability is a critical issue in prospective application. For optical technologies, variations in tissue optical properties such as absorption and scattering coefficients can create distortions in measured spectra. This section provides a brief overview of techniques to correct turbidity-induced spectral and intensity distortions in fluorescence and Raman spectroscopy, respectively. In particular, photon migration theory is presented as an analytical tool to model diffuse reflectance, fluorescence, and Raman scattering arising from turbid biological samples. Monte Carlo simulation is introduced as an effective and statistically accurate tool to numerically model light propagation in turbid media. Using the photon migration model and Monte Carlo simulations, preliminary results of intrinsic Raman spectroscopy are presented.

14.8.2.1 Optical Properties Biological Tissue Light propagation in turbid media such as biological tissue is governed by elastic scattering and absorption of the media. Elastic scattering is a phenomenon in which the direction of the photon is changed but not its energy, usually owing to discontinuities in material properties (e.g. refractive index) in the media, and absorption is the conversion of light energy into another form of energy (usually thermal energy). Most analytical and numerical models employ macroscopic optical properties, including the absorption coefficient, μ_a (cm^{-1}), the scattering coefficient, μ_s (cm^{-1}), the single scattering angle θ, and the elastic scattering anisotropy, $g = \langle \cos\theta \rangle$, average cosine of the single scattering angle. The absorption and scattering coefficients are the probability of a photon being absorbed or scattered per unit path length. The sum of μ_a and μ_s is called the total attenuation coefficient, μ_t, with its inverse defined as the mean free depth. The phase function is a probability density function of the scattering deflection angle, describing the probability of a scattering angle at which single scattering event occurs. For example, the Heyney–Greenstein phase function[42] is often used to approximate tissue scattering. In general, these optical properties are wavelength dependent.

Optical properties of biological tissue are known to be affected by physiological conditions, tissue morphology, and laser irradiation. Different levels of hematocrit (red blood cells) in whole blood cause different absorption (hemoglobin) and scattering (red cells) properties. Similarly, different skin layer thickness, morphology, and melanin content cause optical turbidity to vary. Such turbidity variations exist across different tissue sites or individuals, and are generally slowly varying in time. On the other hand, laser irradiation can cause shorter timescale temporal variations in turbidity, typically as a result of heating.[43]

A limiting factor in noninvasive optical technology is variations in the optical properties of samples under investigation that result in spectral distortions[44–48] and sampling volume (effective optical path length) variability.[49–54] These variations will impact a noninvasive optical technique not only in interpretation of spectral features, but also in the construction and application of a multivariate calibration model if such variations are not accounted for. As a result, correction methods need to be developed and applied before further quantitative analysis. For Raman spectroscopy, relatively few correction methods appear in the literature, and most of them are not readily applicable to biological tissue.[55–59]

In fluorescence spectroscopy, however, diffuse reflectance correction of spectral distortions in biological media has been studied extensively. Analytical models based on photon migration theory,[44] diffusion theory,[46,60,61] as well as empirical models,[62] have been reported to obtain "intrinsic fluorescence." In the following, we will review a particular correction method based on photon migration theory for fluorescence spectroscopy and introduce its Raman counterpart.

14.8.2.2 Corrections Based on Photon Migration Theory Light propagation in turbid media can be described by the radiative transfer equation.[63] However, the analytical solution to this integro-differential equation can be found only for very special conditions and approximations. The most extensively studied approximation is diffusion theory, which is used to model photons that experience multiple scattering events.[63] Another very useful approximation is photon migration theory, developed by Wu et al.[44,45] This method employs probabilistic concepts to describe the scattering of light and to set up a framework that allows the calculation of the diffuse reflectance from semi-infinite turbid media. The total diffuse reflectance from a semi-infinite medium can be written as

$$R_d \approx \sum_{n=1}^{\infty} f_n(g) \times a^n \tag{14.6}$$

with $f_n(g)$ the photon escape probability distribution, n the number of scattering events before escaping, g the scattering anisotropy, and a the albedo $(\mu_s/(\mu_s + \mu_a))$. Two fundamental assumptions are made: the photon escape probability distribution of a semi-infinite medium only depends on the number of scattering events and anisotropy; the lineshape of the escape probability distribution can be approximated by exponential function, that is, $f_n(g) = k(g)\mathrm{e}^{-k(g)n}$. These assumptions are validated by Monte Carlo modeling.

In the same paper, Wu et al. derived an analytical equation relating measured fluorescence (F) to the intrinsic fluorescence (IF), defined as the fluorescence as measured from a optically thin slice of tissue, through diffuse reflectance (R):[44,45]

$$\mathrm{IF} \approx \frac{1}{(1-a_\mathrm{x})} F \frac{a_\mathrm{x}-a_\mathrm{m}}{R_\mathrm{x}-R_\mathrm{m}} \tag{14.7}$$

with IF the intrinsic fluorescence, a the albedo, and R the diffuse reflectance. Subscripts x and m denote quantities at the excitation and emission wavelengths, respectively.

This equation and its variants have been employed to recover turbidity-free fluorescence spectra from various types of tissues. The correction facilitates interpretation of underlying fluorophores and consequently improves the accuracy of disease diagnosis.[44,45,48]

The same general principle that applies for intrinsic fluorescence should hold true for Raman spectroscopy as well. Unlike in fluorescence spectroscopy, spectral distortion owing to prominent absorbers is less of an issue in the NIR wavelength range. However, for quantitative analysis the turbidity-induced sampling volume variations become very significant and usually dominate over spectral distortions.

An equation analogue to equation (14.14) can be derived for the intrinsic Raman signal (IR) under semi-infinite conditions (sample extends into a half plane and all unabsorbed photons eventually exit the air/sample interface):

$$\mathrm{IR} \approx \mu_{\mathrm{t,x}} \mathrm{Ram} \frac{a_\mathrm{x}-a_\mathrm{R}}{R_\mathrm{x}-R_\mathrm{R}} \tag{14.8}$$

Because most Raman instruments rely on a notch filter to prevent CCD saturation from elastically scattered light, diffuse reflectance at the excitation wavelength is not directly available. Monte Carlo simulations and experimental results show that the intrinsic Raman signal for arbitrary samples, as well as collection geometries, can be more conveniently described by

$$\mathrm{IR} \approx m_{\mathrm{t,x}} \frac{\mathrm{Ram}}{(a + bR_{\mathrm{m}}^{\mathrm{c}})} \tag{14.9}$$

Parameters in equation (14.9) can be experimentally calibrated and employed to obtain the intrinsic Raman signal.

14.8.2.3 Monte Carlo Method Monte Carlo simulation is a statistical tool based on macroscopic optical properties that are assumed to extend uniformly over small units of tissue volume (i.e., a voxel). A predefined grid is employed to simulate photon–tissue interaction sites. The mean free path of the photon–tissue interaction sites typically ranges 10–1000 μm. This method does not consider the details of energy distribution within voxels. Photons are treated as classical particles, and the wave features are neglected.[63,64] Since its early introduction as a tool to simulate photon elastic scattering, capabilities such as polarization,[65,66] temporal resolution,[67] fluorescence,[68] and Raman scattering[10] have been developed. Details of the Monte Carlo simulation for diffuse reflectance (the core program) are well documented in the literature.[64]

14.8.2.4 Intrinsic Raman Spectroscopy To test equation (14.8), the product (Ram × μ_t) is plotted versus the ratio $(R_x - R_m)/(a_x - a_m)$ in Figure 14.8 using results from Monte Carlo simulations. The intrinsic Raman signal can be obtained from the slope of the linear fit. Note that equation (14.8) is only legitimate when the semi-infinite condition holds, but expression equation (14.16) should be valid for any

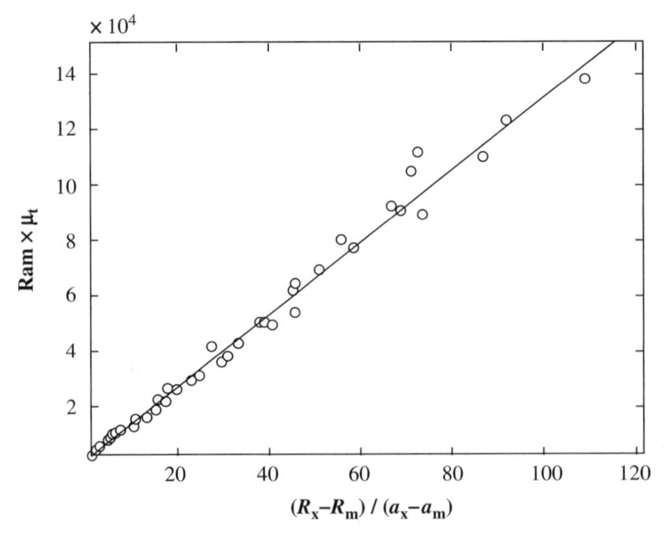

Figure 14.8 (Ram × μ_t) versus $(R_x - R_m)/(a_x - a_m)$. The slope is the intrinsic Raman signal.

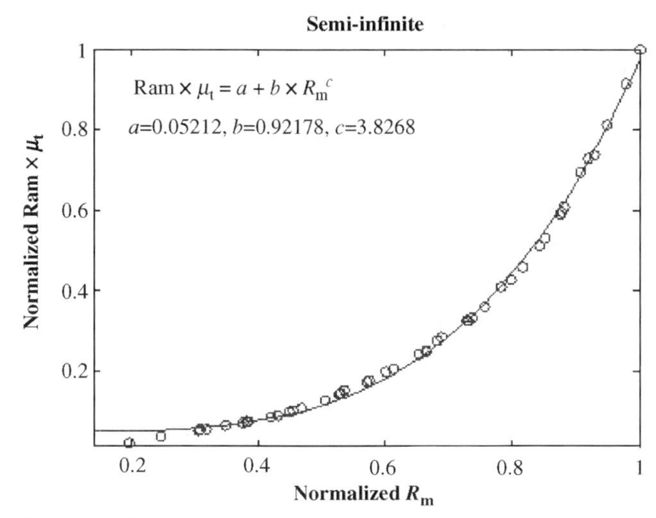

Figure 14.9 (Ram × μ_t) versus R_m. The fit to the curve can be used to correct for sampling volume variation.

sample geometry. To test equation (14.16), the product (Ram × μ_t) is plotted versus R_m in Figure 14.9 using results from Monte Carlo simulations. The fit to this curve is the intrinsic Raman signal and can be used to correct for sampling volume variations. It can be seen that this expression fits less well in the presence of high absorption (lower Raman and reflectance). However, such high absorption cases in general are rare in biological tissue in the NIR spectral region.

To apply IRS, one needs to know μ_t of the samples. Extraction of optical properties has been studied by many researchers.[69–73] The majority of methods are based on diffusion theory or variants of it. Our laboratory extracts optical properties from biological tissue routinely in other wavelength ranges, and a similar method could be employed for this purpose.[73]

14.8.3 Other Considerations and Future Directions

The results presented above are promising. However, for noninvasive Raman spectroscopy to be applied prospectively with clinically acceptable accuracy, several additional modifications/improvements/advances need to be implemented. We address these below.

14.8.3.1 Accurate Reference Concentration Measurements An additional factor that greatly affects the performance of the calibration algorithm is the accuracy of the reference measurements. In spectroscopic techniques such as Raman, a large portion of the collected glucose signal likely originates from the glucose molecules in the ISF. In addition, it is well known that the interstitial glucose lags the plasma glucose concentration from 5 to 30 min in humans.[74] As a result, using plasma glucose as the reference concentration may introduce errors. Methods of extracting interstitial fluid for glucose reference measurements should be explored.

14.8.3.2 Background Signal and its Variations Over Time As mentioned earlier, the intense background, typically described as fluorescence, can limit the detection accuracy in three aspects: the noise associated with the background decreases the SNR; the changes in its spectral shape over time confounds the calibration algorithm; and its intensity variations over time introduces nonanalyte-specific correlation into the calibration model. Unfortunately, the background-associated noise cannot be removed by background removal techniques. Furthermore, it has been found that removing the background using polynomial fitting does not improve calibration results, potentially owing to nonanalyte-specific artifacts.[8] Thus, methods to reduce the background signal at its origin should be explored. One approach may be using prephotobleaching combined with intentional motion by, for example, scanning the illumination spot around an area slightly larger than the spot itself. With such a scheme, the apparent background can be lower to start with, and the photobleaching can be reduced.

14.8.3.3 Optimal Probing Depth Through Accurate Sample Positioning The probing depth and sample positioning are critical for optimal collection of glucose-specific Raman scattered photons and calibration transfer. In experiments, the optimal probing depth can be estimated from extracted optical properties, and therefore the correct distance between the sample and the collection optic can be determined for each measurement site. To address this, a fundamental study of morphological and layer structures at the probing site should be carried out with a computer-controlled three-axis precision stage, as has been done on particular parts of skin.[9] Because most Raman scatterers have specific spatial distribution in skin, such as keratin in the epidermis, collagen in the dermis, and so on, a two-layer model can be developed and utilized. Given such distinctive spatial distributions between keratin and collagen, we can obtain information about the probing depth and even layer thickness by comparing the relative magnitude of keratin and collagen Raman signals. By knowing the exact sampling volume and its coverage of various skin morphological structures, we can estimate how much of the glucose-containing region (*dermis* in the two-layer model) is sampled. This information can effectively lead to better reference concentrations, improving the calibration accuracy.

14.8.3.4 Motion Artifacts and Skin Heterogeneity A key component to obtaining accurate and robust calibrations is the sample interface. The sample interface should ideally limit motion while maintaining a constant pressure and temperature. One approach to combat inadvertent motion artifacts is to intentionally build motion into the calibration model. This can be achieved by scanning the laser spot within a larger area.

14.8.3.5 Optimal Data Collection Site Individual calibration models based on cross-validation can be established for several candidate sites such as forearm, fingernail, and so on, and the results can be compared. The minimum detection error analysis can also be employed to evaluate different sites.

14.9 CONCLUSION

Quantitative Raman spectroscopy is a promising technique for noninvasive glucose sensing. From its early development with *in vitro* studies by several groups, *in vivo* studies have been realized with the aid of more advanced instrumentation and calibration algorithms. The *in vivo* studies performed to date have demonstrated the feasibility of obtaining glucose-specific multivariate calibration models. For Raman spectroscopy to be a viable clinical technique, successful prospective studies must be carried out. From our perspective, breakthroughs have to be made in the following directions: enhancing glucose specificity, correcting for diversity across individuals, accurate reference concentration measurements, reducing the fluorescence background, sample positioning and interface, and optimal site determination.

In this chapter, we presented our research efforts addressing the first two issues with constrained regularization and intrinsic Raman spectroscopy, respectively. These techniques will play a critical role in prospective studies involving multiple sites/subjects/days. We are currently planning for a multiple-subject and multiple-day *in vivo* study, first on dogs and then on humans. We believe these new developments together with a robust sample interface will enable us to demonstrate prospective applicability.

REFERENCES

1. Raman CV, Krishnan KS. A new type of secondary radiation. *Nature* 1928, 121, 501–502.
2. McCreery RL. *Raman Spectroscopy for Chemical Analysis*. John Wiley & Sons, New York, 2000.
3. Qu JNY, Wilson BC, Suria D. Concentration measurements of multiple analytes in human sera by near-infrared laser Raman spectroscopy. *Applied Optics* 1999, 38, 5491–5498.
4. Baraga JJ, Feld MS, Rava RP. Rapid near-infrared Raman-spectroscopy of human tissue with a spectrograph and CCD detector. *Applied Spectroscopy* 1992, 46, 187–190.
5. Gornushkin IB, Eagan PE, Novikov AB, Smith BW, Winefordner JD. Automatic correction of continuum background in laser-induced breakdown and Raman spectrometry. *Applied Spectroscopy* 2003, 57, 197–207.
6. Lieber CA, Mahadevan-Jansen A. Automated method for subtraction of fluorescence from biological Raman spectra. *Applied Spectroscopy* 2003, 57, 1363–1367.
7. Vickers TJ, Wambles RE, Mann CK. Curve fitting and linearity: data processing in Raman spectroscopy. *Applied Spectroscopy* 2001, 55, 389–393.
8. Enejder AMK, Scecina TG, Oh J, Hunter M, Shih WC, Sasic S, Horowitz GL, Feld MS. Raman spectroscopy for noninvasive glucose measurements. *Journal of Biomedical Optics* 2005, 10, 031114.
9. Caspers PJ, Lucassen GW, Puppels GJ. Combined *in vivo* confocal Raman spectroscopy and confocal microscopy of human skin. *Biophysical Journal* 2003, 85, 572–580.
10. Koo T-W. Measurement of blood analytes in turbid biological tissue using near-infrared Raman spectroscopy. PhD Thesis, Massachusetts Institute of Technology, Cambridge, 2001.

11. Scepanovic O, Bechtel KL, Haka AS, Shih WC, Koo TW, Feld MS. Determination of uncertainty in parameters extracted from single spectroscopic measurements. *Journal of Biomedical Optics* 2007, 12, 064012.
12. Roe JN, Smoller BR. Bloodless glucose measurements. *Critical Reviews in Therapeutic Drug Carrier Systems* 1998, 15, 199–241.
13. Guyton AC, Hall JE. *Human Physiology and Mechanisms of Disease*, 6th edn. Saunders, Philadelphia, 1997.
14. Martens H, Naes T. *Multivariate Calibration*. John Wiley & Sons, New York, 1989.
15. Geladi P, Kowalski BR. Partial least-squares regression: a tutorial. *Analytica Chimica Acta* 1986, 185, 1–17.
16. Kowalski BR, Lorber A. Recent advances in multivariate calibration. *Abstracts of Papers of the American Chemical Society* 1988, 196, 100-Anyl.
17. Hirschfeld T, Chase B. FT-Raman spectroscopy: development and justification. *Applied Spectroscopy* 1986, 40, 133–137.
18. Wang Y, Mccreery RL. Evaluation of a diode–laser charge coupled device spectrometer for near-infrared Raman spectroscopy. *Analytical Chemistry* 1989, 61, 2647–2651.
19. Qu JY, Shao L. Near-infrared Raman instrument for rapid and quantitative measurements of clinically important analytes. *Review of Scientific Instruments* 2001, 72, 2717–2723.
20. Lambert JL, Morookian JM, Sirk SJ, Borchert MS. Measurement of aqueous glucose in a model anterior chamber using Raman spectroscopy. *Journal of Raman Spectroscopy* 2002, 33, 524–529.
21. Lambert JL, Pelletier CC, Borchert M. Glucose determination in human aqueous humor with Raman spectroscopy. *Journal of Biomedical Optics* 2005, 10, 031110.
22. Enejder AMK, Koo TW, Oh J, Hunter M, Sasic S, Feld MS. Horowitz GL Blood analysis by Raman spectroscopy. *Optics Letters* 2002, 27, 2004–2006.
23. Berger AJ, Koo TW, Itzkan I, Horowitz G, Feld MS. Multicomponent blood analysis by near-infrared Raman spectroscopy. *Applied Optics* 1999, 38, 2916–2926.
24. Chaiken J, Finney W, Knudson PE, Weinstock RS, Khan M, Bussjager RJ, Hagrman D, Hagrman P, Zhao YW, Peterson CM, Peterson K. Effect of hemoglobin concentration variation on the accuracy and precision of glucose analysis using tissue modulated, noninvasive, *in vivo* Raman spectroscopy of human blood: a small clinical study. *Journal of Biomedical Optics* 2005, 10, 031111.
25. Motz JT, Hunter M, Galindo LH, Gardecki JA, Kramer JR, Dasari RR, Feld MS. Optical fiber probe for biomedical Raman spectroscopy. *Applied Optics* 2004, 43, 542–554.
26. Huang ZW, Zeng HS, Hamzavi I, McLean DI, Lui H. Rapid near-infrared Raman spectroscopy system for real-time *in vivo* skin measurements. *Optics Letters* 2001, 26, 1782–1784.
27. Shih W-C, Bechtel KL, Feld MS. Constrained regularization: hybrid method for multivariate calibration. *Analytical Chemistry* 2007, 79, 234–239.
28. Zhao J. Image curvature correction and cosmic removal for high-throughput dispersive Raman spectroscopy. *Applied Spectroscopy* 2003, 57, 1368–1375.
29. Chrisp MP. Convex diffraction grating imaging spectrometer. U.S. Patent 5,880, 834, 1999.
30. Pelletier I, Pellerin C, Chase DB, Rabolt JF. New developments in planar array infrared spectroscopy. *Applied Spectroscopy* 2005, 59, 156–163.
31. Pelletier CC, Lambert JL, Borchert M. Determination of glucose in human aqueous humor using Raman spectroscopy and designed-solution calibration. *Applied Spectroscopy* 2005, 59, 1024–1031.

32. Cameron BD, Baba JS, Cote GL. Measurement of the glucose transport time delay between the blood and aqueous humor of the eye for the eventual development of a noninvasive glucose sensor. *Diabetes Technology & Therapeutics* 2001, 3, 201–207.
33. Berger AJ, Wang Y, Sammeth DM, Itzkan I, Kneipp K, Feld MS. Aqueous dissolved gas measurements using near-infrared Raman spectroscopy. *Applied Spectroscopy* 1995, 49, 1164–1169.
34. Berger AJ, Wang Y, Feld MS. Rapid, noninvasive concentration measurements of aqueous biological analytes by near-infrared Raman spectroscopy. *Applied Optics* 1996, 35, 209–212.
35. Berger AJ, Itzkan I, Feld MS. Feasibility of measuring blood glucose concentration by near-infrared Raman spectroscopy. *Spectrochimica Acta Part A—Molecular and Biomolecular Spectroscopy* 1997, 53, 287–292.
36. Rohleder D, Kiefer W, Petrich W. Quantitative analysis of serum and serum ultrafiltrate by means of Raman spectroscopy. *Analyst* 2004, 129, 906–911.
37. Chaiken J, Finney WF, Yang X, Knudson PE, Peterson KP, Peterson CM, Weinstock RS, Hagrman D. Progress in the noninvasive *in-vivo* tissue-modulated Raman spectroscopy of human blood. *Proceedings of SPIE* 2001, 4254, 216–227.
38. Chaiken J, Peterson CM. Tissue modulation process for quantitative noninvasive *in vivo* spectroscopic analysis of tissues. U.S. Patent 5,880,834, 2001.
39. Berger AJ, Koo TW, Itzkan I, Feld MS. An enhanced algorithm for linear multivariate calibration. *Analytical Chemistry* 1998, 70, 623–627.
40. Bertero M, Boccacci P. *Introduction to Inverse Problems in Imaging*. Institute of Physics Publishing, Bristol, UK, 1998.
41. Lorber A. Error propagation and figures of merit for quantification by solving matrix equations. *Analytical Chemistry* 1986, 58, 1167–1172.
42. Henyey L, Greenstein J. Diffuse radiation in the galaxy. *Astrophysical Journal* 1941, 93, 70–83.
43. Welch AJ, van Gemert MJC. *Optical-Thermal Response of Laser-Irradiated Tissue*. Plenum Press, 1995.
44. Wu J, Feld MS, Rava RP. Analytical model for extracting intrinsic fluorescence in turbid media. *Applied Optics* 1993, 32, 3585–3595.
45. Zhang QG, Muller MG, Wu J, Feld MS. Turbidity-free fluorescence spectroscopy of biological tissue. *Optics Letters* 2000, 25, 1451–1453.
46. Zhadin NN, Alfano RR. Correction of the internal absorption effect in fluorescence emission and excitation spectra from absorbing and highly scattering media: theory and experiment. *Journal of Biomedical Optics* 1998, 3, 171–186.
47. Gardner CM, Jacques SL, Welch AJ. Fluorescence spectroscopy of tissue: recovery of intrinsic fluorescence from measured fluorescence. *Applied Optics* 1996, 35, 1780–1792.
48. Muller MG, Georgakoudi I, Zhang QG, Wu J, Feld MS. Intrinsic fluorescence spectroscopy in turbid media: disentangling effects of scattering and absorption. *Applied Optics* 2001, 40, 4633–4646.
49. Weersink R, Patterson MS, Diamond K, Silver S, Padgett N. Noninvasive measurement of fluorophore concentration in turbid media with a simple fluorescence/reflectance ratio technique. *Applied Optics* 2001, 40, 6389–6395.
50. Arnold MA, Small GW. Noninvasive glucose sensing. *Analytical Chemistry* 2005, 77, 5429–5439.

51. Cote GL, Lec RM, Pishko MV. Emerging biomedical sensing technologies and their applications. *IEEE Sensors Journal* 2003, 3, 251–266.
52. Khalil OS. Spectroscopic and clinical aspects of noninvasive glucose measurements. *Clinical Chemistry* 1999, 45, 165–177.
53. Diamond KR, Farrell TJ, Patterson MS. Measurement of fluorophore concentrations and fluorescence quantum yield in tissue-simulating phantoms using three diffusion models of steady-state spatially resolved fluorescence. *Physics in Medicine and Biology* 2003, 48, 4135–4149.
54. Pogue BW, Burke G. Fiber-optic bundle design for quantitative fluorescence measurement from tissue. *Applied Optics* 1998, 37, 7429–7436.
55. Nijhuis TA, Tinnemans SJ, Visser T, Weckhuysen BM. Operando spectroscopic investigation of supported metal oxide catalysts by combined time-resolved UV–VIS/Raman/on-line mass spectrometry. *Physical Chemistry Chemical Physics* 2003, 5, 4361–4365.
56. Waters DN. Raman spectroscopy of powders: effects of light absorption and scattering. *Spectrochimica Acta Part A—Molecular and Biomolecular Spectroscopy* 1994, 50, 1833–1840.
57. Aarnoutse PJ, Westerhuis JA. Quantitative Raman reaction monitoring using the solvent as internal standard. *Analytical Chemistry* 2005, 77, 1228–1236.
58. Kuba S, Knozinger H. Time-resolved *in situ* Raman spectroscopy of working catalysts: sulfated and tungstated zirconia. *Journal of Raman Spectroscopy* 2002, 33, 325–332.
59. Tinnemans SJ, Kox MHF, Nijhuis TA, Visser T, Weckhuysen BM. Real time quantitative Raman spectroscopy of supported metal oxide catalysts without the need of an internal standard. *Physical Chemistry Chemical Physics* 2005, 7, 211–216.
60. Patterson MS, Pogue BW. Mathematical model for time-resolved and frequency-domain fluorescence spectroscopy in biological tissue. *Applied Optics* 1994, 33, 1963–1974.
61. Finlay JC, Foster TH. Recovery of hemoglobin oxygen saturation and intrinsic fluorescence with a forward-adjoint model. *Applied Optics* 2005, 44, 1917–1933.
62. Finlay JC, Conover DL, Hull EL, Foster TH. Porphyrin bleaching and PDT-induced spectral changes are irradiance dependent in ALA-sensitized normal rat skin *in vivo*. *Photochemistry and Photobiology* 2001, 73, 54–63.
63. Ishimaru A. *Wave Propagation and Scattering in Random Media*. Academic Press, New York, 1978.
64. Wang LH, Jacques SL, Zheng LQ. MCML—Monte Carlo Modeling of Light Transport in Multilayered Tissues. *Computer Methods and Programs in Biomedicine* 1995, 47, 131–146.
65. Ramella-Roman JC, Prahl SA, Jacques SL. Three Monte Carlo programs of polarized light transport into scattering media: part I. *Optics Express* 2005, 13, 4420–4438.
66. Ramella-Roman JC, Prahl SA, Jacques SL. Three Monte Carlo programs of polarized light transport into scattering media: part II. *Optics Express* 2005, 13, 10392–10405.
67. Hielscher AH, Liu HL, Chance B, Tittel FK, Jacques SL. Time-resolved photon emission from layered turbid media. *Applied Optics* 1996, 35, 719–728.
68. Swartling J, Pifferi A, Enejder AMK, Andersson-Engels S. Accelerated Monte Carlo models to simulate fluorescence spectra from layered tissues. *Journal of the Optical Society of America A—Optics Image Science and Vision* 2003, 20, 714–727.
69. Farrell TJ, Patterson MS, Wilson B. A diffusion theory model of spatially resolved, steady-state diffuse reflectance for the noninvasive determination of tissue optical properties *in vivo*. *Medical Physics* 1992, 19, 879–888.

70. Dam JS, Pedersen CB, Dalgaard T, Fabricius PE, Aruna P, Andersson-Engels S. Fiber-optic probe for noninvasive real-time determination of tissue optical properties at multiple wavelengths. *Applied Optics* 2001, 40, 1155–1164.
71. Doornbos RMP, Lang R, Aalders MC, Cross FW, Sterenborg HJCM. The determination of *in vivo* human tissue optical properties and absolute chromophore concentrations using spatially resolved steady-state diffuse reflectance spectroscopy. *Physics in Medicine and Biology* 1999, 44, 967–981.
72. Nichols MG, Hull EL, Foster TH. Design and testing of a white-light, steady-state diffuse reflectance spectrometer for determination of optical properties of highly scattering systems. *Applied Optics* 1997, 36, 93–104.
73. Zonios G, Perelman LT, Backman VM, Manoharan R, Fitzmaurice M, Van Dam J, Feld MS. Diffuse reflectance spectroscopy of human adenomatous colon polyps *in vivo*. *Applied Optics* 1999, 38, 6628–6637.
74. Boyne MS, Silver DM, Kaplan J, Saudek CD. Timing of changes in interstitial and venous blood glucose measured with a continuous subcutaneous glucose sensor. *Diabetes* 2003, 52, 2790–2794.

CHAPTER 15

SURFACE-ENHANCED RAMAN SPECTROSCOPY FOR GLUCOSE SENSING

Nilam C. Shah, Jonathan M. Yuen, Olga Lyandres, Matthew R. Glucksberg, Joseph T. Walsh and Richard P. Van Duyne

In Vivo Glucose Sensing, Edited by David D. Cunningham and Julie A. Stenken

15.1 BACKGROUND

15.1.1 Introduction

Frequent monitoring of glucose levels is critically important to the management of diabetes and reduction of complications. The detection of glucose is a highly researched area in the quest for a sensor that can give direct, real-time measurements, accurately, in real time, with minimum invasiveness. Several optical techniques are being explored for glucose detection, including infrared spectroscopy, polarimetry, polyacrylamide hydrogels, and photonic crystals.[1–5] However, there are limitations to these techniques including interfering water absorption, overlapping signals from competing analytes, and complications from indirect measurement. Surface-enhanced Raman spectroscopy (SERS) offers an alternative approach to detecting glucose that can overcome these limitations. Previously SERS has not been widely used for bioanalysis due to several factors including the requirement that molecules should be within a few nanometers of a roughened metal surface and the spectral complexity of biological media. While the use of SERS-active labels resulted in successful detection of proteins and antigens *in vitro*,[6–8] *in vivo* direct detection of bioanalytes remains a challenge. This chapter outlines the development of a sensitive, portable, robust SERS-based glucose sensor that overcomes many of the obstacles related to using SERS to detect bioanalytes. We start with a simplified *in vitro* scheme to examine fundamental properties of the glucose sensor such as reversibility, stability, and temporal response. Then, we demonstrate quantitative detection of glucose in an aqueous environment and in a simulated biological environment (i.e., bovine plasma). Furthermore, we show successful quantitative detection of glucose *in vivo* by implanting a SERS substrate into the interstitial space of a rat. Finally, we compare the SERS-based glucose sensor to a commercially available glucose sensor.

15.1.2 Raman Spectroscopy

When light interacts with molecules, it is either absorbed, transmitted, or scattered. The scattered component can either be elastic (i.e., Rayleigh) or inelastic (i.e., Raman). Raman scattered photons are shifted in frequency from the incident photons by the amount of energy transferred to the molecule to excite its vibrational modes. This process is referred to as normal Raman scattering (NRS). The normal Raman spectrum is a plot of the intensity of Raman scattered light versus the energy difference between incident and scattered photons and is extremely sensitive to the molecular structure of the target molecule.[9] The individual bands in a normal Raman spectrum are characteristic of specific vibrational motions. As a result, every molecular species has its own unique Raman spectrum. This molecular specificity of Raman spectroscopy is utilized to create a glucose sensor. The molecular specificity of NRS for six common metabolic analytes found in the interstitial space is clearly illustrated in Figure15.1. For example, analytes whose molecular structures differ only in the position of a single atom like glucose (Figure15.1, upper left) and galactose (Figure15.1, middle left) are easily distinguished. Even more

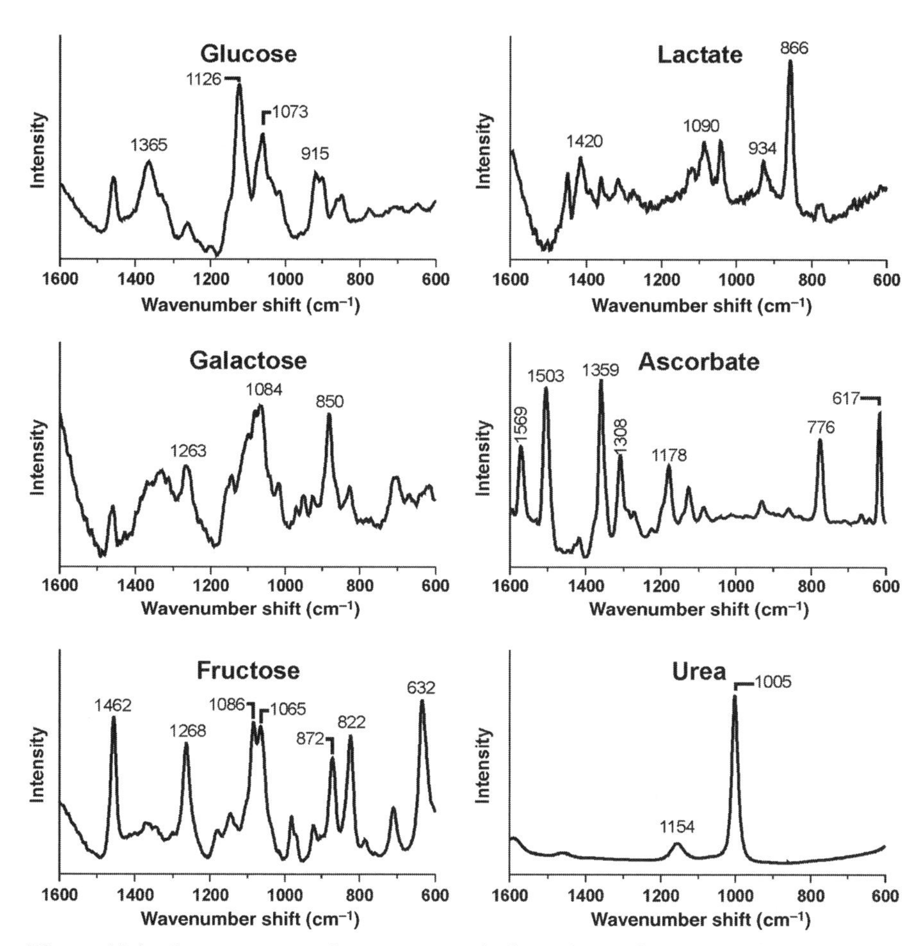

Figure 15.1 Raman spectra of common metabolic analytes of the interstitium. Saturated aqueous solutions, $\lambda_{ex} = 532$ nm, $P = 30$ mW, and $t = 2$ min.

dramatic is the ability to distinguish between isomers such as glucose (Figure15.1, upper left) and fructose (Figure15.1, lower left), providing significant selectivity with the analysis.

The application of Raman spectroscopy to the measurement of glucose is not new. It has been shown that normal Raman spectroscopy can readily detect physiological concentrations of glucose *in vitro* from a simulated aqueous humor solution.[10] Using partial least squares (PLS) analysis, Lambert et al. were able to predict glucose levels ranging from 50 mg/dL (2.8 mM, hypoglycemic) to 1300 mg/dL (72.2 mM, severe diabetic) with a standard error of 24.7 mg/dL (1.4 mM). Berger et al. were able to detect glucose concentrations with an accuracy of 26 mg/dL (1.4 mM) in serum and 79 mg/dL (4.4 mM) in whole blood using PLS.[11] These are promising results and show the potential of Raman spectroscopy for glucose measurements; unfortunately, the light exposure in both experiments was significantly higher than is permissible in a biological specimen.[12] High-power lasers (100–300 mW) and long acquisition times (as long as 5 min) are required because of the small normal Raman scattering cross section of glucose, 5.6×10^{-30} cm^2/molecule/stearadians.[9]

The Raman scattering cross section for one of the most intense vibrational modes known, the $n1$ totally symmetric ring breathing mode of liquid benzene, is only 1.06×10^{-29} cm^2/molecule/stearadian. Since glucose has an even smaller normal Raman scattering cross section, due to the extremely weak interaction between an input (viz., excitation) photon and the analyte molecule, the signal will be weak. Thus, to get adequate signal-to-noise, NRS requires relatively high concentrations, high laser power (100–300 mW), and long integration times (as long as 5 min). Typically, the limit of detection (LOD) for NRS is of order 10–100 mM. Ongoing efforts to develop Raman technology for noninvasive glucose sensing are presented in the Chapter 14.

Raman optical activity (ROA) spectroscopy is highly sensitive to structural changes, while Raman difference spectroscopy is capable of detecting small differences in Raman signal at low concentrations of analytes. In both these techniques, however, the resultant difference signals are very small and long data acquisition times are required if signals from physiologic concentrations of analytes are to be obtained with precision or accuracy.[13,14] Therefore, these approaches are not ideal for a rapid, robust, and clinical analysis.

One solution to overcome the low signal intensity of NRS is to use resonance Raman spectroscopy (RRS), which can increase the Raman cross section. In RRS the excitation wavelength of the incident light is chosen to be coincident with an intense molecular electronic absorption transition (viz., chromophore) in the target molecule. RRS works extremely well for heme proteins such as cytochrome *c* using visible light and for direct excitation of amino acids in proteins using ultraviolet light. RRS has been proven to be a powerful tool in biology for examining protein structure. Furthermore, in its time-resolved variant, it is one of the few biophysical techniques capable of structural insight into the femtosecond timescale. The sensitivity enhancement factors associated with RRS can be as large as 10^4–10^5 so that the typical LOD for RRS can be as low as 10^{-6}–10^{-5} M. Unfortunately, glucose has no visible or accessible ultraviolet chromophores and would require excitation in the deep ultraviolet region ($\lambda_{ex} \sim 200$ nm) of the spectrum. Ultraviolet excitation is unlikely to be appropriate for *in vivo* sensing due to photodamage of DNA and absorption by water.

Another solution to the insensitivity of NRS was discovered by Van Duyne in 1977 and is known as SERS.[15] When an analyte molecule is adsorbed on or is located within 0–2 nm of a roughened silver or gold metal surface, the intensity of normal Raman scattering by the molecule can be amplified by enhancement factors of 10^6–10^8. In the case of resonant analyte molecules, the combined SERS and RRS enhancement factors can be as high as 10^{14}–10^{15} for single molecules and is known as surface-enhanced resonance Raman scattering (SERRS). Although the details of the enhancement mechanism(s) responsible for such large enhancements are not yet understood,[16–18] recent experiments have shown that by careful control of nanoparticle structure, the SERS enhancement factors due to the electromagnetic enhancement mechanism may be increased to at least 10^8 for nonresonant adsorbates and to 10^{10}–10^{15} for resonant adsorbates.[19] SERS is most intense on silver, but other metals, such as copper and gold, also function well as SERS-active surfaces. SERS signals are optimized when the energy of the localized surface plasmon resonance (LSPR) lies within the energy range of the Raman scattered photons. Excitation of the LSPR excites the conduction band electrons of the metal. The oscillating electrons set up an amplified electromagnetic field surrounding the nanostructured surface. Molecules affected by the local electromagnetic field exhibit stronger Raman excitation and scattering probabilites.[20]

15.1.3 Localized Surface Plasmon Resonance

The signature optical property of a noble metal nanostructure is the LSPR. This resonance occurs when the correct wavelength of light strikes a noble metal nanostructure, causing the plasma of conduction band electrons to oscillate collectively. The term LSPR is used to emphasize that this collective oscillation is localized near the surface of the nanoparticle and to differentiate it from propagating surface plasmons that are referred to simply as surface plasmons. The two consequences of LSPR excitation are (1) selective photon absorption and (2) generation of locally enhanced or amplified electromagnetic fields at the nanoparticle surface. The wavelength of the LSPR depends on many factors including the material, size, shape, spacing, and dielectric environment of the nanoparticle. Many research groups are currently exploring the size-dependent optical properties of noble metal nanoparticles to exploit LSPR in applications with optical filters,[21,22] substrates for surface-enhanced spectroscopies,[23–34] biosensors,[35–37] bioprobes,[38,39] chemical sensors,[40,41] and optical devices.[40–45]The LSPR for noble metal nanoparticles in the twenty to few hundred nanometer size range occurs in the visible and IR regions of the spectrum and can be measured by UV–visible–IR extinction spectroscopy.[46] The resulting SER spectrum is related intricately to the spectral location of the LSPR. We have previously shown that we can fabricate nanostructured surfaces with the LSPR matched to the excitation wavelength using nanosphere lithography.[47] An atomic force micrograph of a nanostructured surface with the LSPR matched to 785 nm excitation is shown in Figure15.2. This structure consists of silver film-over-nanospheres (AgFON) on a copper substrate. The fabrication procedure of these substrates is described below.

15.1.4 Nanosphere Lithography

Nanosphere lithography (NSL) is an inexpensive fabrication technique developed in the Van Duyne lab to produce nanoparticle arrays with precisely controlled shape, size, and interparticle spacing, and—accordingly—precisely controlled LSPRs.[48] The production of nanoparticles by NSL begins with the self-assembly of size-monodispersed nanospheres to form a two-dimensional colloidal crystal deposition mask: nanospheres are deposited on the substrate surface, allowed to diffuse freely, seek their lowest energy conformation, and as the solvent evaporates the nanospheres are drawn together in a hexagonally close-packed pattern (Figure15.3a). Following self-assembly of the nanosphere mask, a noble metal or other material is deposited by thermal evaporation, electron beam deposition (EBD), or pulsed laser

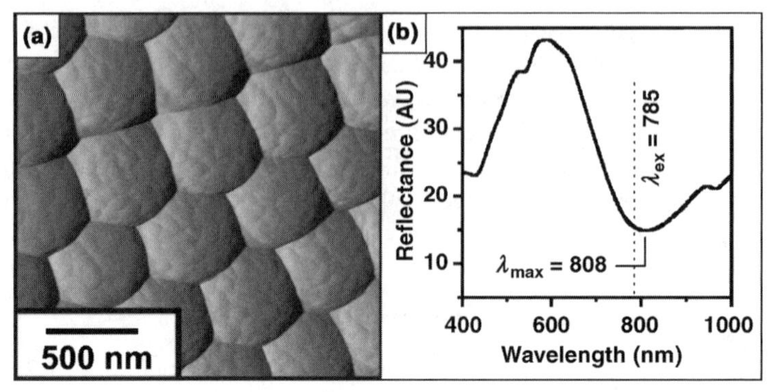

Figure 15.2 (a) Atomic micrograph of silver film-over-nanospheres (AgFON). (b) Reflectance spectrum of the AgFON nanostructure optimized with the LSPR matched to 785 nm excitation.

deposition (PLD) from a source normal to the substrate through the nanosphere mask to a controlled mass thickness (typically in the range of 20–80 nm). Subsequently, the nanosphere mask is removed by sonicating the entire sample in a solvent, thus leaving behind nanoparticles. The tetrahedral nanoparticles are shown in Figure15.3b, deposited through the nanosphere mask on to the substrate. Previous work has demonstrated that the LSPR of NSL-derived nanoparticles depends on nanoparticle material, size, shape, and interparticle spacing as well as substrate, solvent, dielectric thin-film overlayers, and molecular adsorbates.[46] The size of these nanoparticles is easily tuned to have a peak LSPR that matches the excitation source. NSL can be used to produce monodispersed, reproducible, and material-general nanoparticles. Tetrahedral Ag nanoparticle arrays produce enhancement factors on the order of 10^8.[19,49] NSL can also be utilized to fabricate many new nanostructure derivatives such as silver film-over-nanospheres (AgFONs), which are used for the glucose sensor (Figure15.2). AgFONs exhibit enhancement factors on the order of 10^7.[47] However, AgFONs have a significant increase in surface area accessible to analyte molecules increasing the total signal intensity and making them more suitable for glucose sensing. AgFONs are fabricated by depositing a layer of metal

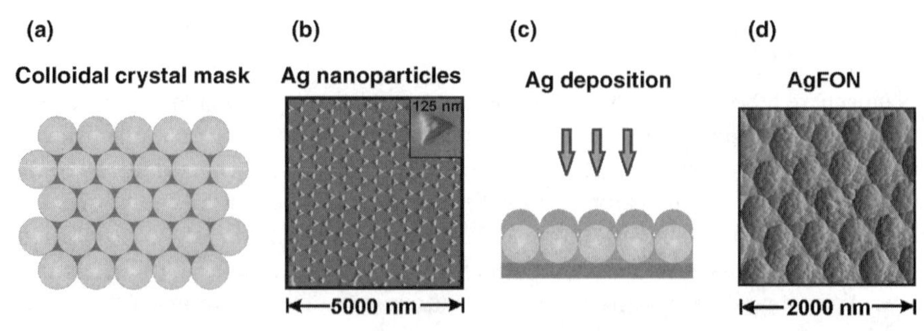

Figure 15.3 (a) Depiction of nanosphere monolayer, (b) atomic force micrograph of resulting tetrahedral nanoparticle array formed after deposition of metal and removal of nanospheres, (c) schematic of AgFON fabrication, and (d) an atomic force micrograph of the AgFON.

over the nanosphere mask with mass thickness of approximately a half of the sphere diameter as shown in Figure15.3c. The metal is deposited through and over the spheres such that the Ag film is in contact with the supporting substrate. These contacts serve as heat sinks for the surface and prevent photothermal damage to the surface. The resulting surface (Figure15.3d) can then be functionalized and used for SERS sensing.

15.1.5 Partitioning of Glucose

When a Raman-active molecule is located close to nanostructured noble metal surfaces, the ensemble-averaged Raman signal increases by up to eight orders of magnitude[49] for nonresonant molecules. Under some conditions, the Raman signal from single molecules can be enhanced by 14 or 15 orders of magnitude in the case of resonant adsorbates.[17,18] Theoretical analysis suggests that molecules confined within the decay length of the electromagnetic fields, namely, 0–2 nm, will exhibit SER spectra even if they are not chemisorbed.[20] For the SERS-active surfaces used in glucose detection, the decay length at which the intensity decreases by a factor of 10 was calculated to be 2.8 nm.[47] SERS possesses many desirable characteristics as a tool for the chemical analysis of *in vivo* molecular species including high specificity, attomole to high zeptomole mass sensitivity, micromolar to picomolar concentration sensitivity, and interfacial generality;[50] however, glucose affinity to bare metal surfaces is low. Therefore, it is essential to partition glucose close to the surface to detect it with SERS. To increase the number of probed glucose molecules, we need to facilitate their interaction with the metal surface by using a self-assembled monolayer (SAM), in a manner analogous to high-performance liquid chromatography (HPLC).[51–55] In addition, the partition layer also needs to bind glucose reversibly such that the fluctuations in glucose levels can be reflected accurately and the sensor does not saturate within the physiological concentration range. We have explored several different partition layers to increase the interaction of glucose with the metal surface and to ensure reversible binding of glucose.

Our early work indicated that decanethiol (DT)[56] and tri(ethylene glycol)-terminated alkanethiol (EG3)[57] would be promising partition layers. However, upon further investigation it was found that DT was hydrophobic and not compatible with an aqueous environment, while EG3 is a challenge to synthesize, making its availability limited. Recently, a new mixed monolayer, consisting of DT and mercaptohexanol (MH), has been explored.[58] This mixed DT/MH SAM has dual hydrophobic and hydrophilic properties that make it ideal for *in vivo* use. Figure15.4a depicts the steps involved in making the DT/MH-functionalized AgFON. We hypothesize, using a space-filling model, that the DT/MH SAM creates a pocket that brings glucose closer to the noble metal surface, thus enhancing the SERS signal (Figure15.4b). A SERS spectrum of the DT/MH AgFON is shown in Figure15.4c. We have demonstrated reversibility, stability, rapid time response, and quantitative detection of glucose with the DT/MH SAM both *in vitro* and *in vivo* as detailed below.

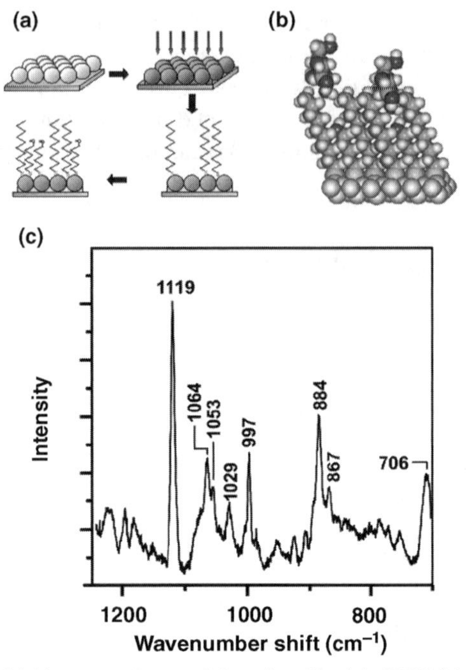

Figure 15.4 Decanethiol/mercaptohexanol-functionalized AgFON fabrication, space-filling model, and spectrum. (a) AgFONs were prepared by depositing a film of metal onto a layer of self-assembled nanospheres. The AgFON was then functionalized by incubating in ethanolic solutions of mercaptohexanol for 45 min and then in decanethiol overnight. (b) Space-filling model of glucose partitioning into the DT/MH-functionalized AgFON. (c) SERS spectrum of a DT/MH-functionalized AgFON. $\lambda_{ex} = 785$ nm, $P = 55$ mW, and $t = 2$ min.

15.2 EXPERIMENTAL PROCEDURES

15.2.1 AgFON Substrate Preparation

Copper 18 mm diameter discs were utilized as substrates for glucose detection. After cleaning, approximately 10 μL of the nanosphere suspension (4% solids, 390 nm diameter) was drop coated onto each copper substrate and allowed to dry in ambient conditions.[58] The substrates were then mounted into an electron beam deposition system for metal deposition (Kurt J. Lesker, Clairton, PA). Silver metal films ($d_m = 200$ nm) were deposited over and through the sphere masks on the substrates.[58,59]

The AgFON surfaces were then functionalized with a SAM composed of decanethiol and mercaptohexanol. The AgFONs were first incubated in 1 mM DT in ethanol for 45 min and then transferred to 1 mM MH in ethanol for at least 12 h. The SAM-functionalized surfaces were then mounted into a small-volume flow cell for SERS measurements.

15.2.2 SERS Apparatus

A titanium–sapphire laser (CW Ti:Sa, model 3900, Spectra Physics, Mountain View, CA) pumped by a solid-state diode laser, λ_{ex} 532 nm (model Millenia Vs, Spectra Physics) was used to generate λ_{ex} of 785 nm as described previously.[58] The setup consisted of laser line and high-pass filters for the corresponding wavelengths used. Furthermore, a lens was used to focus the beam onto the sample and a collection lens was used to focus scattered light onto the entrance slit of the single-grating monochromator (Acton Research Scientific, Trenton, NJ). For initial studies of reversibility, stability, and time response of the sensor, 532 nm excitation was used. For all *in vivo* studies, and *in vitro* studies conducted in a biological environment (such as plasma), 785 nm excitation was used because NIR excitation minimizes autofluorescence of proteins.[58] A small-volume flow cell was used to control the external environment of the AgFON surfaces.

15.2.3 Quantitative Multivariate Analysis

All data processing was performed using MATLAB (MathWorks, Inc., Natick, MA) and PLS_Toolbox (Eigenvector Research, Inc., Manson, WA). Prior to analysis, the spectra were smoothed with a second-order polynomial and window size of 9. The slowly varying background, commonly seen in SERS experiments, was removed by subtracting a fourth-order polynomial fit. This method greatly reduced varying background levels with minimum effect on the SERS peaks. Chemometric analysis was performed using the PLS method and leave-one-out (LOO) cross-validation algorithm.[57,60] PLS is an inverse calibration method that does not require *apriori* knowledge of all the components in the system.[61] To construct the calibration, known concentrations of the analyte are used. The regression vector, **b**, is then determined based on the spectra matrix, **R**, and concentration of the training set, **c**:

$$\mathbf{c} = \mathbf{R}\mathbf{b}$$

The regression vector shows the vibrational features of the analyte (in our case glucose) and is used to predict the concentration of an unknown sample.

To calibrate the SERS sensor, a training set was acquired that consisted of spectra of solutions with glucose concentration in the physiological range (10–450 mg/dL). The root mean square error of calibration (RMSEC) was calculated to gauge how well the data fit the calibration model. To further ensure a valid model, it is important to test the calibration model with an independent validation set.[61] Once an independent validation set of spectra was acquired, the resulting regression vector was used to predict concentrations. Finally, the root mean squared error of prediction (RMSEP) was calculated to gauge the accuracy of the prediction.

15.2.4 Temporal Response Analysis

At the conclusion of an experiment, a DT/MH-functionalized AgFON was removed after being incubated in bovine plasma or implanted in the rat for 5 h and placed in a flow cell containing bovine plasma to simulate the *in vivo* environment.

High and low concentrations of glucose dissolved in plasma were injected into the cell to cause step changes in glucose concentration. PeakFit 4.12 software (Systat Software Inc, Richmond, CA) was used to process the data. Using Matlab software a fourth-order polynomial was subtracted from the baseline SERS spectra to remove the background signal. A linear best-fit baseline correction and Savitzky–Golay smoothing were also applied in PeakFit. The data were fit to the superposition of the Lorentzian amplitude line shapes to determine the amplitude of the Raman bands. The data were then iteratively fit to an exponential curve to determine the $1/e$ time constant.

15.2.5 Surgical Implantation and Spectroscopic Measurements

All surgical procedures followed animal study protocols filed with Northwestern University's Institutional Animal Care and Use Committee (IACUC). Adult, male Sprague Dawley rats (300–500 g, $N=4$) were anesthetized using Nembutal (sodium pentobarbital, Abbott Laboratories, Abbott Park, IL) with an initial dose of 50 mg/kg. Rats are an accepted animal model for metabolic disorders including diabetes.[62] They are also commonly used for experimental glucose measurements. Animal anesthesia was maintained by additional hourly administrations of Nembutal (25 mg/kg). After initial anesthetization, the animals were checked for pain responses by paw pinch and blink tests. No action was taken if responses were seen. The surgical areas were prepared by hair removal (shaving and chemical depilatory) and cleaning after the anesthetic had taken effect. Then, the femoral vein was cannulated using PE50 tubing (Clay Adams, Becton, Dickinson and Company, Franklin Lakes, NJ). The carotid artery was then cannulated with PE90 tubing. The venous cannula was used for glucose, saline, and other injections. A tracheotomy was performed and a tube inserted to enable the attachment of a ventilator to aid respiration. All incisions were shut with surgical clips. Throughout the surgery and the experiment, the rat was warmed using an electric heating pad. A glass window in a metal frame was placed along the midline of the back of the rat. A circular incision was made to all implantations of a DT/MH-functionalized AgFON sensor subcutaneously such that the substrate was in contact with the interstitial fluid and optically accessible through the window. The rat was then positioned within a holder in the conventional sample position on a lab-scale Raman spectroscopy system (Figure15.5). The system consisted of a monochromatic laser, a band-pass filter, a series of steering and focusing optics to deliver the laser light to the sample, and collection optics that convey the scattered light to the detector, a long-pass filter to reject Rayleigh scattering, a 3 m spectrograph for wavelength dispersion, and a CCD detector. Following IACUC protocol, at the conclusion of the experiment the animals were sacrificed by an overdose of anesthetic and bilateral thorachotomy.

Glucose levels in the rat were controlled through intermittent intravenous infusion for 3 h. The glucose was infused over 5–10 min at a concentration of 1 g/mL in sterile phosphate buffered saline via the femoral venous cannula. A small droplet of blood was drawn from the femoral arterial cannula, the glucose concentration was measured with the One Touch II glucometer (Lifescan, Inc.), and corresponding SERS measurements were taken. SERS spectra were acquired through the optical

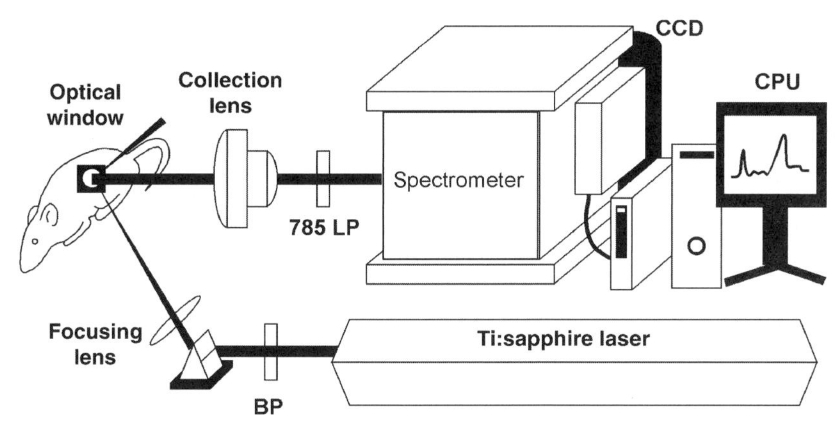

Figure 15.5 Schematic of instrumental apparatus. The DT/MH-functionalized AgFON was surgically implanted into a rat with an optical window and integrated into a conventional laboratory Raman spectroscopy system. The Raman spectroscopy system consists of a Ti: sapphire laser ($\lambda_{ex} = 785$ nm), band-pass filter, beam-steering optics, collection optics, and a long-pass filter to reject Raleigh scattered light. All of the optics fit on a 4 ft $\times$ 10 ft optical table.

window using a Ti:sapphire laser ($\lambda_{ex} = 785$ nm, $P = 50$ mW, and $t = 2$ min). The partial least squares method previously described was used to analyze the data collected.

15.3 RESULTS

The following results show considerable progress toward an implantable, real-time, continuous, SERS-based glucose sensor. Our work demonstrates the ability to detect glucose both *in vitro* and *in vivo* with SERS using a mixed SAM consisting of decanethiol and mercaptohexanol (DT/MH). We demonstrate (1) reversibility of the sensor, (2) long-term stability of the DT/MH-functionalized AgFON surface (3) real-time partitioning and departitioning of the glucose sensor, (4) quantitative measurements, and (5) comparison of the SERS-based sensor to electrochemical measurements.

15.3.1 Reversibility of the DT/MH AgFON

To successfully monitor fluctuations in glucose concentration throughout the day, an implantable glucose sensor must be reversible. The DT/MH-modified AgFON sensor was exposed to cycles of 0 and 100 mM aqueous glucose solutions (pH $\sim$ 7) to demonstrate the reversibility of the sensor (Figure15.6). The sensor was not flushed between measurements to simulate real-time sensing. To minimize effective laser power fluctuations, nitrate was used as an internal standard (1053 cm^{-1} peak) in all the experiments. This band was used to normalize the spectra and corresponds to a symmetric stretching of NO_3^-.[63] SER spectra were collected at each step ($\lambda = 532$ nm, $P = 10$ mW, and $t = 20$ min) (Figure15.6a–c (left)).

The experimental spectra shown in Figure15.6 d–f (right) compare the experimental spectra to the normal Raman spectrum of a saturated aqueous glucose solution. Peaks at 1462, 1365, 1268, 1126, 915, and 850 cm^{-1} correspond to glucose peaks in the saturated solution (Figure15.6d).[64] In all the difference spectra, a sharp peak can be seen at 1053 cm^{-1}, representing imperfect subtraction of the nitrate internal standard. Figure15.6e demonstrates the partitioning of glucose in the DT/MH SAM with features at 1461, 1371, 1269, 1131, 916, and 864 cm^{-1}. The shift in the glucose peaks compared to the saturated solution is due to the fact that SERS bands can shift up to 25 cm^{-1} when compared to normal Raman bands of the same analyte.[65] Complete departitioning of glucose is represented by the absence of glucose spectral features in the difference spectra shown in Figure15.6f. The data demonstrate that the DT/MH-mixed SAM has a reversible sensing surface for optimal partitioning and departitioning of glucose.

15.3.2 Stability of DT/MH AgFON

An implantable glucose sensor must also be stable for a minimum of 3 days.[66] Previously, we demonstrated by electrochemical and SERS measurements that

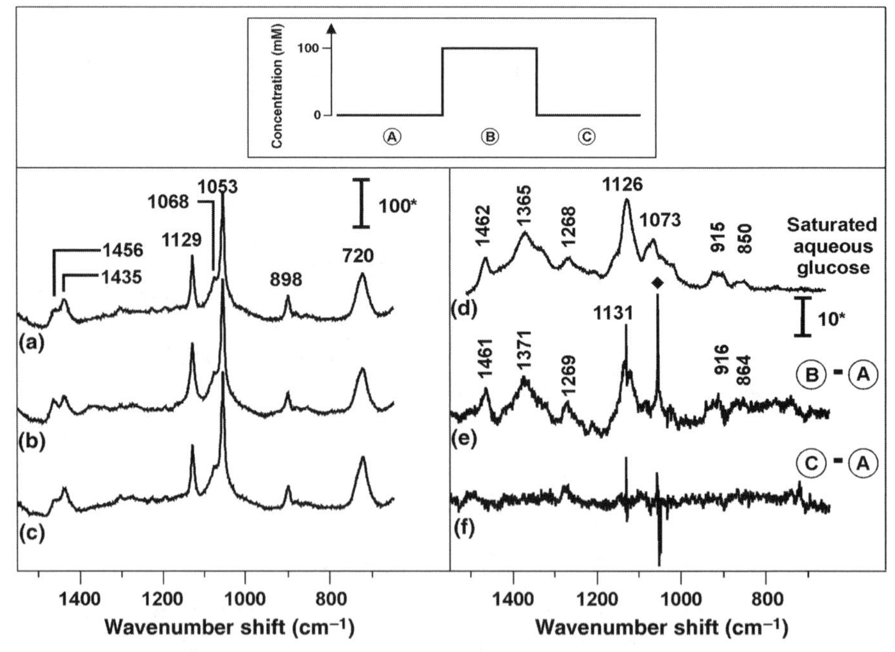

Figure 15.6 Inset shows the step changes in glucose concentration experienced by the sensor. (a, b, and c) SER spectra collected for each step for alternate 0 and 100 mM aqueous glucose solutions. (d) The normal Raman spectrum of aqueous saturated glucose solution. (e and f) Show the differences indicating glucose partitioning and departitioning. Raman bands in the difference spectra showing partitioning of glucose (e) agree well with the reference glucose spectrum (d). The absence of spectral features in (f) indicates complete departioning of glucose. Symbol * denotes analog-to-digital units/mW/s. $\lambda_{ex} = 532$ nm, $P = 10$ mW, $t = 20$ min, and pH ~ 7.

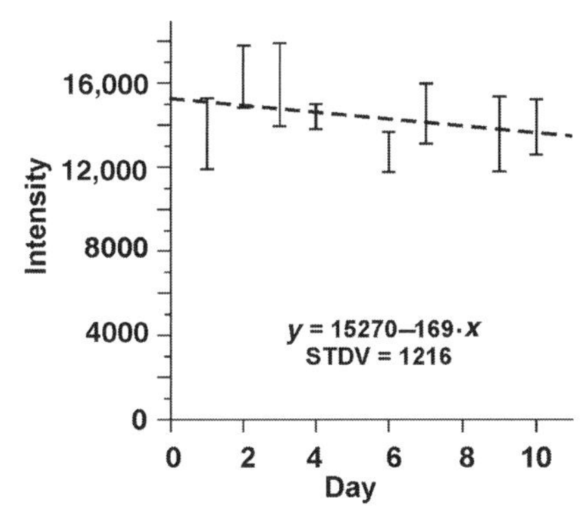

Figure 15.7 Stability of DT/MH-functionalized AgFON in bovine plasma. Plot depicts the intensity of the 1119 cm^{-1} peak versus time. Signal intensity stayed stable over a 10-day period with a 2% change in intensity of the 1119 cm^{-1} peak. STDV = 1216 counts, $\lambda_{ex} = 785$ nm, $P_{laser} = 55$ mW, and $t = 2$ min.

SAM-functionalized AgFON substrates were stable for at least 3 days in phosphate-buffered saline.[67] Here, we show 10 days stability of the DT/MH-functionalized AgFON surface in bovine plasma. SER spectra were taken once every 24 h on three different spots for three different samples ($\lambda = 785$ nm, $t = 2$ min). Figure15.7 gives the plot of the average intensity of the 1119-cm^{-1} peak for DT/MH on the AgFON, which corresponds to the stretching vibration of a C–C bond,[68] for each day as a function of time. The 1119 cm^{-1} peak was chosen because it gives a strong signal for the DT/MH SAM allowing fluctuations in intensity to be easily measured. From the first day to the last, only a 2% change in intensity was observed with a standard deviation (STDV) of 1216 counts. The change was seen as not significant and can be explained by the rearrangement of the SAM during the incubation in bovine plasma.[69] This experiment demonstrates that the DT/MH SAM is intact and well ordered over a 10-day period, making this SAM-functionalized surface a good candidate as an implantable sensor.

15.3.3 Temporal Response

In addition to reversibility and stability, the sensor must be able to partition and departition glucose on a reasonable timescale. The *in vitro* real-time response was examined by using bovine plasma. The DT/MH-functionalized AgFON was placed in bovine plasma for ~5 h. The AgFON substrate was then placed in a flow cell and SER spectra were collected continuously ($\lambda_{ex} = 785$ nm) with a 15 s integration time. Spectra were collected at 785 nm to reduce autoflourescence of proteins from the bovine plasma. To observe partitioning, bovine plasma (used as purchased) spiked with 50 mM glucose was injected at $t = 0$ s. Departitioning was observed by injecting 0 mM glucose in bovine plasma at $t = 225$ s. The amplitude of the 1462 cm^{-1} peak was plotted versus time as shown in Figure15.8a (top). An offset was subtracted from the data such that the first intensity value is zero. The $1/e$ time constant was calculated from an exponential curve fitted to the data points. The fit yields a $1/e$ time constant of 28 s for partitioning and 25 s for departitioning.[58]

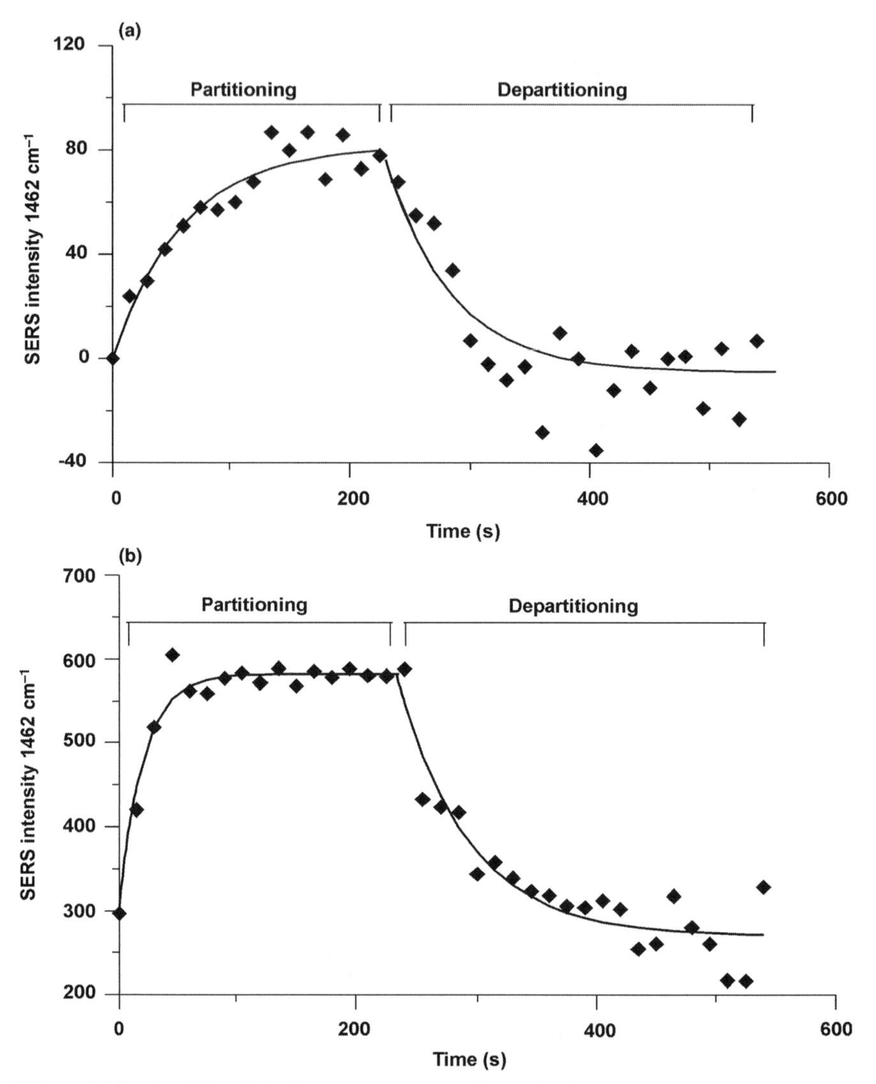

Figure 15.8 Real-time SERS response to a step change in glucose concentration. Glucose was injected at $t = 0$ s, and the cell was flushed with glucose-depleted bovine plasma at $t = 225$ s. ($\lambda_{ex} = 785$ nm, $P = 100$ mW, and $t = 15$ s) (a) Partitioning and departioning of glucose after being incubated in bovine plasma for ~5 h. The 1/*e* time constants were calculated to be 28 s for partitioning and 25 s for departitioning. (b) Partitioning and departitioning of glucose after being implanted in a rat for ~5 h. The 1/*e* time constants were calculated to be 9 s for partitioning and 27 s for departitioning.

A similar experiment was conducted after the sensor was implanted in the rat for ~5 h. This experiment was done to test whether implanting the sensor fouls the surface and alters the partitioning and departitioning of glucose. The sensor was removed from the rat and immediately placed in a flow cell containing bovine plasma to simulate the *in vivo* environment. The 1/*e* time constant was then determined as described above. In this case, the fit yields a $1/e$ time constant of 9 s for partitioning and 27 s for departitioning (Figure15.8b, bottom). These results indicate that changes in glucose concentration can be detected rapidly both *in vitro* and *in vivo*.

15.3.4 *In Vitro* Quantitative Detection

A viable glucose sensor must also be able to detect glucose in the clinically relevant concentration range, 10–450 mg/dL (0.56–25 mM), in the presence of interfering analytes, under physiological pH. Data are often presented on a Clarke error grid as described in Chapter 1. To evaluate the SERS-based glucose sensor against the Clarke error grid, the DT/MH-functionalized AgFON samples were placed in a flow cell containing water (pH ~ 7) with lactate (1 mM) and urea (2.5 mM) in physiological concentrations, which are potential interferences for glucose. Random concentrations of glucose solutions ranging from 10 to 450 mg/dL were then introduced into the cell and allowed to incubate for 2 min to ensure complete partitioning. A near-infrared laser source ($\lambda_{ex} = 785$ nm, $P = 8.4$ mW, and $t = 2$ min) was used to collect SER spectra from multiple locations on two substrates. The calibration model was constructed by partial least squares leave-one-out (PLS-LOO) analysis using 46 randomly chosen independent spectral measurements of known glucose concentrations. Seven latent variables were used as the basis for the calibration model. The variables take into account the environment in the laboratory, variation in laser power, and SERS enhancement at different locations. The number of latent variables that are used should be chosen carefully because using too many latent variables can result in overfitting of the data. Using a cross-validation step in the calibration process allows us to determine the number of latent variables to use. The number of latent variables is chosen by constructing a plot of root mean squared error of cross-validation (RMSECV) versus latent variables and finding the minimum RMSECV value.[61] The PLS analysis resulted in a RMSEC of 9.89 mg/dL (0.549 mM). This RMSEC value is lower than previously reported when using the EG3-modified AgFON. In order to validate the PLS model, 23 independent data points were used, resulting in a root mean square error of prediction (RMSEP) of 92.17 mg/dL (5.12 mM). As seen in Figure15.9, 98% of the calibration set and 87% of the validation set fall within the A and B ranges of the Clarke error grid.

The glucose sensor was then evaluated in bovine plasma to simulate the *in vivo* environment. The bovine plasma was passed through a 0.45 μm diameter pore filter before use, and then spiked with glucose concentrations from 10 to 450 mg/dL. DT/MH-functionalized AgFON substrates were placed in the flow cell and allowed to incubate in the glucose-spiked bovine plasma. SERS spectra were collected from multiple spots and multiple samples at various random glucose concentrations ($\lambda = 785$ nm, $P = 10$–30 mW, and $t = 2$ min). The calibration model was constructed with 92 data points using PLS-LOO with seven latent variables. The analysis resulted in an RMSEC of 34.3 mg/dL (1.90 mM). The model was validated by using 46 data points, resulting in an RMSEP of 83.16 mg/dL (4.62 mM). Plotting the data demonstrated that 98% of the calibration set and 85% of the validation set fall within the A and B ranges of the Clarke error grid (Figure15.10). The results indicate that the DT/MH-modified AgFON glucose sensor can make accurate glucose measurements *in vitro* even in the presence of interfering analytes.

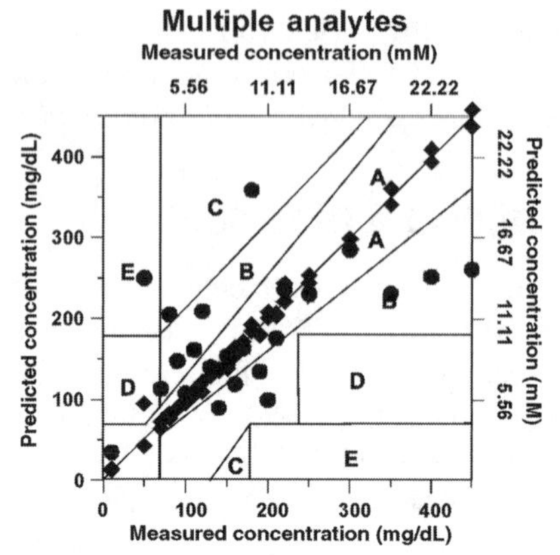

Figure 15.9 Calibration (◆) and validation (●) plot of glucose in an aqueous solution consisting of 1 mM lactate and 2.5 mM urea at pH ~ 7 on a DT/MH-functionalized AgFON surface. Two substrates and multiple spots were used to measure glucose concentrations in the physiological range (10–450 mg/dL). The calibration plot was constructed using 46 data points with PLS-LOO analysis. The validation plot was constructed using 23 data points. RMSEC 9.89 mg/dL (0.55 mM) and RMSEP = 92.17 mg/dL (5.12 mM). $\lambda_{ex} = 785$ nm, $P_{laser} = 8.4$ mW, and $t = 2$ min.

15.3.5 *In Vivo* Detection

Finally, the DT/MH AgFON was evaluated *in vivo*. A representative Clarke error grid analysis of a single rodent is shown in Figure15.11. All measurements were taken from a single spot on the implanted DT/MH-functionalized AgFON surface.

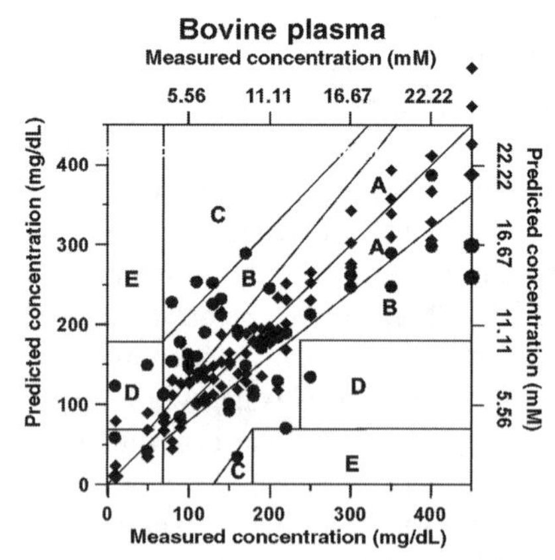

Figure 15.10 Calibration (◆) and validation (●) plot of glucose in bovine plasma on a DT/MH-functionalized AgFON surface. Three substrates and multiple spots were used to measure glucose concentrations in the physiological range (10–450 mg/dL) over the course of 2 days. PLS calibration plot was constructed using 92 data points. The validation plot was constructed using 46 data points. RMSEC = 34.3 mg/dL (1.9 mM) and RMSEP = 83.16 mg/dL (4.62 mM). $\lambda_{ex} = 785$ nm, $P_{laser} = 10–30$ mW, and $t = 2$ min.

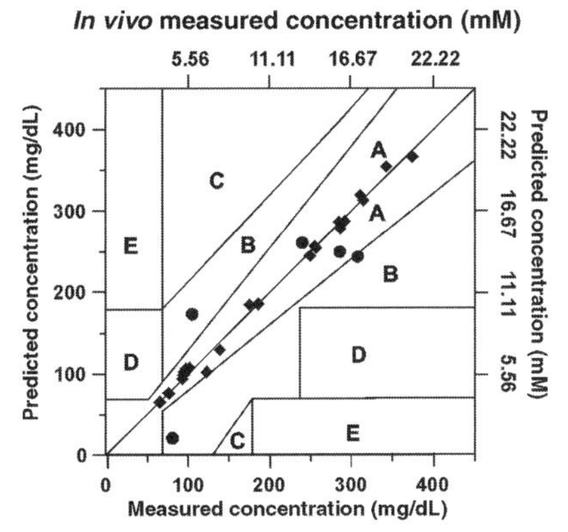

Figure 15.11 Calibration (◆) and validation (●) plot using a single sensor and a single spot on a DT/MH-functionalized AgFON surface *in vivo*. PLS calibration was constructed using 21 data points correlated with the commercial glucometer. The validation plot was constructed using 5 data points. RMSEC = 7.46 mg/dL (0.41 mM) and RMSEP = 53.42 mg/dL (2.97 mM). $\lambda_{ex} = 785$ nm, $P = 50$ mW, and $t = 2$ min.

The calibration set was calculated using 21 data points. The validation set was constructed using five independent data points. The sensor showed a relatively low error (RMSEC = 7.46 mg/dL (0.41 mM) and RMSEP = 53.42 mg/dL (2.97 mM)).[60] The RMSEC and RMSEP can be improved by increasing the number of data points

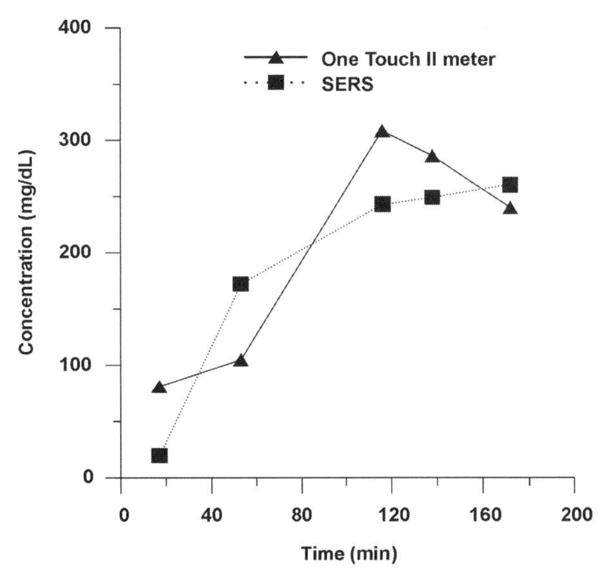

Figure 15.12 Comparison of *in vivo* glucose measurement using SERS and the One Touch II blood glucose meter. The glucose concentration change is plotted over time. Glucose bolus was started at $t = 60$ min. Squares (■■) are measurements made using the SERS sensor and triangles (▲) are measurements made using One Touch II blood glucose meter. $\lambda_{ex} = 785$ nm, $P = 50$ mW, and $t = 2$ min.

in the calibration set and by optimizing the substrate. These data compare favorably with existing detection methods, which have instrument-dependant coefficients of variation of 0.96–26.9% (0.096–2.69 mM, 1.75–49 mg/dL at 10 mM), as well as with our *in vitro* results.

In addition, to plot the data against the Clarke error grid, it is important to directly compare the SERS-based glucose sensor to a commercially available glucose sensor. Glucose concentration was measured in the rat using both the DT/MH AgFON SERS sensor and the One Touch II blood glucose meter. The data were plotted with respect to time and is shown in Figure15.12. Both the SERS-based measurements and the standard glucometer effectively tracked the sharp rise in glucose concentration after the start of the glucose infusion ($t = 60$ min) as well as subsequent changes in glucose concentration.[60]

15.4 CONCLUSIONS

The automatic and continuous monitoring of glucose levels is critically important for the care and treatment of millions of diabetics, particularly juvenile cases where observance of a strict medical regimen is often problematic. With further refinement of the system (e.g., instrumentation, optics, and surface), the SERS-based glucose sensor has the potential to replace conventional personal and point-of-care systems. The Raman spectrometer currently used for developing this sensor consists of a number of discrete components (laser, delivery optics, collection optics, spectrograph, and detector) that fit on 4 ft × 10 ft optical table and weighs a few hundred pounds. In order for the sensor to be used routinely in a clinical setting, the current spectrometer needs to be miniaturized. Several smaller Raman spectrometers have been developed that are paving the way for substantial miniaturization in the future. Small Raman systems that are now commercially available include Inspector Raman, a handheld system from DeltaNu with dimensions (6.5 × 4.5 × 2.5 in.), and a miniature 785 nm optimized spectrograph developed by Acton Instruments with dimensions 14 × 9.2 × 5.3 in. DeltaNu has also developed a second-generation handheld Raman spectrometer known as the ReporteR, which is palm-sized spectrometer (6 × 3 × 1.75 in.) and only weighs 11 oz. In addition, the Van Duyne group has made initial steps toward a miniature Raman system by replacing components of a laboratory-scale Raman system with smaller commercially available components. This system successfully collected Raman spectra using a laser pointer as the excitation source and a reflectance probe fiber-optic cable for laser delivery and collection.[70] Finally, Ondax, Inc. has fabricated a specialized 1 in. multiline filter that only allows certain wavelengths of light to pass through. This reduces the amount of noise that is collected and increases efficiency by giving the same signal to noise with a factor of 1.8 less collection time.[71] Combining this multiline filter with a notch filter to reject the laser light would produce an analyte-specific sensor. In addition, the development of an implantable sensor usable transdermally, where both the excitation and Stokes-shifted photons pass directly through the skin, will minimize the invasiveness of the procedure. Looking to the future, we predict that the functionalized

SERS sensor approach will have important applications in both the treatment and the care of diabetes and will open up new areas of fundamental research.

ACKNOWLEDGMENTS

Funding for this work was provided by the NIH (5 R56 DK078691-02) and the U.S. Army Medical Research and Material Command's Military Operational Medical Research Program/Julia Weaver Fund (W81XWH-04-1-0630), and the NSF (CHE0414554).

REFERENCES

1. Cameron BD, Gorde HW, Satheesan B, Cote GL. The use of polarized laser light through the eye for noninvasive glucose monitoring. *Diabetes Technology & Therapeutics* 1999, 1, 125–143.
2. Malin SF, Ruchti TL, Blank TB, Thennadil SN, Monfre SL. Noninvasive prediction of glucose by near-infrared diffuse reflectance spectroscopy. *Clinical Chemistry* 1999, 45, 1651–1658.
3. Arnold MA. Progress in noninvasive glucose monitoring with near infrared transmission spectroscopy. *Abstracts of Papers of the American Chemical Society* 2002, 224, U114–U114.
4. Arnold FH, Zheng W, Michaels AS. A membrane-moderated, conductimetric sensor for the detection and measurement of specific organic solutes in aqueous solutions. *Journal of Membrane Science* 2000, 167, 227–239.
5. Asher SA, Alexeev VL, Goponenko AV, Sharma AC, Lednev IK, Wilcox CS, Finegold DN. Photonic crystal carbohydrate sensors: low ionic strength sugar sensing. *Journal of the American Chemical Society* 2003, 125, 3322–3329.
6. Cao YC, Jin RC, Nam JM, Thaxton CS, Mirkin CA. Raman dye-labeled nanoparticle probes for proteins. *Journal of the American Chemical Society* 2003, 125, 14676–14677.
7. Doering WE, Nie S. Single-molecule and single-nanoparticle SERS: examining the roles of surface active sites and chemical enhancement. *Journal of Physical Chemistry B* 2001, 106, 311–317.
8. Ni J, Lipert RJ, Dawson GB, Porter MD. Immunoassay readout method using extrinsic Raman labels adsorbed on immunogold colloids. *Analytical Chemistry* 1999, 71, 4903–4908.
9. McCreery RL. *Raman Spectroscopy for Chemical Analysis*, Vol. 157. John Wiley & Sons, Inc., New York, 2000, p. 420.
10. Lambert J, Storrie-Lombardi M, Borchert M. Measurement of physiologic glucose levels using Raman spectroscopy on a rabbit aqueous humor model. *IEEE LEOS Newsletter* 1998, 12, 19–22.
11. Berger AJ, Koo TW, Itzkan I, Horowitz G, Feld MS. Multicomponent blood analysis by near-infrared Raman spectroscopy. *Applied Optics* 1999, 38, 2916–2926.
12. American National Standard for the Safe Use of Lasers: ANSI Z-136. http://www.inform.umd.edu/PRES/policies/vi1600a.html.
13. Bell AF, Barron LD, Hecht L. Vibrational Raman optical activity study of D-Glucose. *Carbohydrate Research* 1994, 257, 11–24.

14. Chaiken J, Finney WF, Yang X, Knudson PE, Peterson KP, Peterson CM, Weinstock RS, Hagrman D. Progress in the noninvasive, *in-vivo*, tissue modulated Raman spectroscopy of human blood. *Proceedings of SPIE* 2001, 4254, 216–227.
15. Van Duyne RP. Applications of Raman spectroscopy in electrochemistry. *Journal of Physique* 1977, 38, (C5)239–(C5)252.
16. Campion A, Kambhampati P. Surface-enhanced Raman scattering. *Chemical Society Reviews* 1998, 27, 241–250.
17. Kneipp K, Wang Y, Kneipp H, Perelman LT, Itzkan I, Dasari RR, Feld MS. Single molecule detection using surface-enhanced Raman scattering (SERS). *Physical Review Letters* 1997, 78, 1667–1670.
18. Nie S, Emory SR. Probing single molecules and single nanoparticles by surface-enhanced Raman scattering. *Science* 1997, 275, 1102–1106.
19. Haynes CL, Van Duyne RP. Plasmon-sampled surface-enhanced Raman excitation spectroscopy. *Journal of Physical Chemistry B* 2003, 107, 7426–7433.
20. Schatz GC, Van Duyne RP. Electromagnetic mechanism of surface-enhanced spectroscopy. In: Chalmers JM, Griffiths PR (Eds), *Handbook of Vibrational Spectroscopy*, Vol. 1. John Wiley & Sons, Ltd, Chichester, 2002, pp. 759–774.
21. Flaugh PL, O'Donnell SE, Asher SA. Development of a new optical wavelength rejection filter: demonstration of its utility in Raman spectroscopy. *Applied Spectroscopy* 1984, 38, 847–850.
22. Munro CH, Pajcini V, Asher SA. Dielectric stack filters for *ex situ* and *in situ* UV optical-fiber probe Raman spectroscopic measurements. *Applied Spectroscopy* 1997, 51, 1722–1729.
23. Emory SR, Haskins WE, Nie S. Direct observation of size-dependent optical enhancement in single metal nanoparticles. *Journal of the American Chemical Society* 1998, 120, 8009–8010.
24. Emory SR, Nie S. Screening and enrichment of metal nanoparticles with novel optical properties. *Journal of Physical Chemistry B* 1998, 102, 493–497.
25. Kahl M, Voges E, Kostrewa S, Viets C, Hill W. Periodically structured metallic substrates for SERS. *Sensors and Actuators B: Chemical* 1998, 51, 285–291.
26. Yang WH, Hulteen JC, Schatz GC, Van Duyne RP. A surface-enhanced hyper-Raman and surface-enhanced Raman scattering study of *trans*-1,2-bis(4-pyridyl)ethylene adsorbed onto silver film over nanosphere electrodes. Vibrational assignments: experiment and theory. *Journal of Chemical Physics* 1996, 104, 4313–4323.
27. Pipino ACR, Schatz GC, Van Duyne RP. Surface-enhanced second-harmonic diffraction: experimental investigation of selective enhancement. *Physical Review B* 1996, 53, 4162–4169.
28. Zhu J, Xu F, Schofer SJ, Mirkin CA. The first Raman spectrum of an organic monolayer on a high-temperature superconductor: direct spectroscopic evidence for a chemical interaction between an amine and $YBa_2Cu_3O_{7-\delta}$. *Journal of the American Chemical Society* 1997, 119, 235–236.
29. Freeman RG, Grabar KC, Allison KJ, Bright RM, Davis JA, Guthrie AP, Hommer MB, Jackson MA, Smith PC, Walter DG, Natan MJ. Self-assembled metal colloid monolayers: an approach to SERS substrates. *Science* 1995, 267, 1629–1632.
30. Caldwell WB, Chen K, Herr BR, Mirkin CA, Hulteen JC, Van Duyne RP. Self-assembled monolayers of ferrocenylazobenzenes on Au(111)/mica films: surface-enhanced Raman scattering (SERS) response vs. surface morphology. *Langmuir* 1994, 10, 4109–4115.

31. Van Duyne RP, Hulteen JC, Treichel DA. Atomic force microscopy and surface-enhanced Raman spectroscopy. I. Ag island films and Ag film over polymer nanosphere surfaces supported on glass. *Journal of Chemical Physics* 1993, 99, 2101–2115.
32. Liao PF. Silver structures produced by microlithography. In: Chang RK, Furtak TE (Eds), *Surface Enhanced Raman Scattering*. Plenum Press, New York, 1982, pp. 379–390.
33. Liao PF, Bergman JG, Chemla DS, Wokaun A, Melngailis J, Hawryluk AM, Economou NP. Surface-enhanced Raman scattering from microlithographic silver particle surfaces. *Chemical Physics Letters* 1981, 81, 355–359.
34. Howard RE, Liao PF, Skocpol WJ, Jackel LD, Craighead HG. Microfabrication as a scientific tool. *Science* 1983, 221, 117–121.
35. Elghanian R, Storhoff JJ, Mucic RC, Letsinger RL, Mirkin CA. Selective colorimetric detection of polynucleotides based on the distance-dependent optical properties of gold nanoparticles. *Science* 1997, 277, 1078–1081.
36. Mucic RC, Storhoff JJ, Letsinger RL, Mirkin CA. DNA-induced assembly of gold nanoparticles: a method for rationally organizing colloidal particles into ordered macroscopic materials. *Nature* 1996, 382, 607–609.
37. Storhoff JJ, Elghanian R, Mucic RC, Mirkin CA, Letsinger RL. One-pot colorimetric differentiation of polynucleotides with single base imperfections using gold nanoparticle probes. *Journal of the American Chemical Society* 1998, 120, 1959–1964.
38. BruchezJrM, Moronne M, Gin P, Weiss S, Alivisatos AP. Semiconductor nanocrystals as fluorescent biological labels. *Science* 1998, 281, 2013–2018.
39. Chan WCW, Nie S. Quantum dot bioconjugates for ultrasensitive nonisotopic detection. *Science* 1998, 281, 2016–2018.
40. Pan G, Kesavamoorthy R, Asher SA. Nanosecond switchable polymerized crystalline colloidal array Bragg diffracting materials. *Journal of the American Chemical Society* 1998, 120, 6525–6530.
41. Weissman JM, Sunkara HB, Tse AS, Asher SA. Thermally switchable periodicities and diffraction from mesoscopically ordered materials. *Science* 1996, 274, 959–960.
42. Asher S, Chang S-Y, Tse A, Liu L, Pan G, Wu Z, Li P. Optically nonlinear crystalline colloidal self assembled submicron periodic structures for optical limiters. *Material Research Society Symposium Proceedings* 1995, 374, 305–310.
43. Lidorikis E, Li Q, Soukoulis CM. Optical bistability in colloidal crystals. *Physical Review E* 1997, 55, 3613–3618.
44. Mansour K, Soileau MJ, Van Stryland EW. Nonlinear optical properties of carbon-black suspensions (ink). *Journal of the Optical Society of America B* 1992, 9, 1100–1109.
45. Woileau MJ. Materials for optical switches, isolators, and limiters. *Proceedings of SPIE* 1989, 1105, 187.
46. Haynes CL, Van Duyne RP. Nanosphere lithography: a versatile nanofabrication tool for studies of size-dependent nanoparticle optics. *Journal of Physical Chemistry B* 2001, 105, 5599–5611.
47. McFarland AD, Young MA, Dieringer JA, Van Duyne RP. Wavelength-scanned surface-enhanced Raman excitation spectroscopy. *Journal of Physical Chemistry B* 2005, 109, 11279–11285.
48. Hulteen JC, Van Duyne RP. Nanosphere lithography: a materials general fabrication process for periodic particle array surfaces. *Journal of Vacuum Science and Technology A* 1995, 13, 1553–1558.

49. Haynes CL, Van Duyne R. Plasmon scanned surface-enhanced Raman scattering excitation profiles. *Material Research Society Symposium Proceedings* 2002, 728, S10.7.1–S10.7.6.

50. Smith WE, RodgerC. Surface-enhanced Raman scattering. In: Chalmers JM, Griffiths PR (Eds), *Handbook of Vibrational Spectroscopy*, Vol. 1. John Wiley & Sons, Chichester, UK, 2002, pp. 775–784.

51. Freunscht P, Van Duyne RP, Schneider S. Surface-enhanced Raman spectroscopy of *trans*-stilbene adsorbed on silver film over nanosphere surfaces modified by platinum or alkanethiol deposition. *Chemical Physics Letters* 1997, 281, 372–378.

52. Blanco Gomis D, Tamayo DM, Mangas Alonso J. Determination of monosaccharides in cider by reversed-phase liquid chromatography. *Analytica Chimica Acta* 2001, 436, 173.

53. Yang L, Janle E, Huang T, Gitzen J, Kissinger PT, Vreeke M, Heller A. Applications of "wired" peroxidase electrodes for peroxide determination in liquid chromatography coupled to oxidase immobilized enzyme reactors. *Analytical Chemistry* 1995, 34, 1326–1331.

54. Carron KT, Kennedy BJ. Molecular-specific chromatographic detector using modified SERS substrates. *Analytical Chemistry* 1995, 67, 3353–3356.

55. Deschaines TO, Carron KT. Stability and surface uniformity of selected thiol-coated SERS surfaces. *Applied Spectroscopy* 1997, 51, 1355–1359.

56. Shafer-Peltier KE, Haynes CL, Glucksberg MR, Van Duyne RP. Toward a glucose biosensor on surface-enhanced Raman scattering. *Journal of the American Chemical Society* 2003, 125, 588–593.

57. Yonzon CR, Haynes CL, Zhang XY, Walsh JT, Van Duyne RP. A glucose biosensor based on surface-enhanced Raman scattering: improved partition layer, temporal stability, reversibility, and resistance to serum protein interference. *Analytical Chemistry* 2004, 76, 78–85.

58. Lyandres O, Shah NC, Yonzon CR, Walsh JT, Glucksberg MR, Van Duyne RP. Real-time glucose sensing by surface-enhanced Raman spectroscopy in bovine plasma facilitated by a mixed decanethiol/mercaptohexanol partition layer. *Analytical Chemistry* 2005, 77, 6134–6139.

59. Zhang X, Young MA, Lyandres O, Van Duyne RP. Rapid detection of an anthrax biomarker by surface-enhanced Raman spectroscopy. *Journal of the American Chemical Society* 2005, 127, 4484–4489.

60. Stuart DA, Yuen JM, Shah NC, Lyandres O, Yonzon CR, Glucksberg MR, Walsh JT, Van Duyne RP. *In vivo* glucose measurement by surface-enhanced Raman spectroscopy. *Analytical Chemistry* 2006, 78, 7211–7215.

61. Beebe KR, Pell RJ, Seasholtz MB. *Chemometrics: A Practical Guide*, Wiley Interscience, New York, 1998, p. 348.

62. Junod A, Lambert AE, Stauffac W, Renold AE. Diabetogenic action of streptozotocin: relationship of dose to metabolic response. *Journal of Clinical Investigation* 1969, 48, 2129–2139.

63. Mosier-Boss PA, Lieberman SH. Detection of nitrate and sulfate anions by normal Raman spectroscopy and SERS of cationic-coated, silver substrates. *Applied Spectroscopy* 2000, 54, 1126–1135.

64. Soderholm S, Roos YH, Meinander N, Hotokka M. Raman spectra of fructose and glucose in the amorphous and crystalline states. *Journal of Raman Spectroscopy* 1999, 30, 1009–1018.

65. Stacy AM, Van Duyne RP. Surface enhanced Raman and resonance Raman spectroscopy in a non-aqueous electrochemical environment: tris(2,2′-bipyridine)ruthenium(II) adsorbed on silver from acetonitrile. *Chemical Physics Letters* 1983, 102, 365–370.
66. Kaufman FR, Gibson LC, Halvorson M, Carpenter S, Fisher LK, Pitukcheewanont P. A pilot study of the continuous glucose monitoring system: clinical decisions and glycemic control after its use in pediatric type 1 diabetic subjects. *Diabetes Care* 2001, 24, 2030–2034.
67. Stuart DA, Yonzon CR, Zhang X, Lyandres O, Shah NC, Glucksberg MR, Walsh JT, Van Duyne RP. Glucose sensing using near infrared surface-enhanced Raman spectroscopy: gold surfaces, 10-day stability, and improved accuracy. *Analytical Chemistry* 2005, 77, 4013–4019.
68. Bryant MA, Pemberton JE. Surface Raman-scattering of self-assembled monolayers formed from 1-alkanethiols – behavior of films at Au and comparison to films at Ag. *Journal of the American Chemical Society* 1991, 113, 8284–8293.
69. Biebuyck HA, Bain CD, Whitesides GM. Comparison of organic monolayers on polycrystalline gold spontaneously assembled from solutions containing dialkyl disulfides or alkanethiols. *Langmuir* 1994, 10, 1825–1831.
70. Young MA, Stuart DA, Lyandres O, Glucksberg MR, Van Duyne RP. Surface-enhanced Raman spectroscopy with a laser pointer light source and miniature spectrometer. *Canadian Journal of Chemistry* 2004, 82, 1435–1441.
71. Caltech Astronomy: Holographic Filter Allows Astronomers to Read between the Lines. http://www.astro.caltech.edu/palomar/ohfilter.html (accessed on June 25, 2008).

INDEX

In Vivo Glucose Sensing, Edited by David D. Cunningham and Julie A. Stenken

CHEMICAL ANALYSIS

A SERIES OF MONOGRAPHS ON ANALYTICAL CHEMISTRY
AND ITS APPLICATIONS

Series Editor
J. D. WINEFORDNER

Vol. 1 **The Analytical Chemistry of Industrial Poisons, Hazards, and Solvents**. *Second Edition.* By the late Morris B. Jacobs
Vol. 2 **Chromatographic Adsorption Analysis**. By Harold H. Strain (*out of print*)
Vol. 3 **Photometric Determination of Traces of Metals**. *Fourth Edition*
Part I: General Aspects. By E. B. Sandell and Hiroshi Onishi
Part IIA: Individual Metals, Aluminum to Lithium. By Hiroshi Onishi
Part IIB: Individual Metals, Magnesium to Zirconium. By Hiroshi Onishi
Vol. 4 **Organic Reagents Used in Gravimetric and Volumetric Analysis**. By John F. Flagg (*out of print*)
Vol. 5 **Aquametry: A Treatise on Methods for the Determination of Water**. *Second Edition* (*in three parts*). By John Mitchell, Jr. and Donald Milton Smith
Vol. 6 **Analysis of Insecticides and Acaricides**. By Francis A. Gunther and Roger C. Blinn (*out of print*)
Vol. 7 **Chemical Analysis of Industrial Solvents**. By the late Morris B. Jacobs and Leopold Schetlan
Vol. 8 **Colorimetric Determination of Nonmetals**. *Second Edition.* Edited by the late David F. Boltz and James A. Howell
Vol. 9 **Analytical Chemistry of Titanium Metals and Compounds**. By Maurice Codell
Vol. 10 **The Chemical Analysis of Air Pollutants**. By the late Morris B. Jacobs
Vol. 11 **X-Ray Spectrochemical Analysis**. *Second Edition.* By L. S. Birks
Vol. 12 **Systematic Analysis of Surface-Active Agents**. *Second Edition.* By Milton J. Rosen and Henry A. Goldsmith
Vol. 13 **Alternating Current Polarography and Tensammetry**. By B. Breyer and H.H.Bauer
Vol. 14 **Flame Photometry**. By R. Herrmann and J. Alkemade
Vol. 15 **The Titration of Organic Compounds** (*in two parts*). By M. R. F. Ashworth
Vol. 16 **Complexation in Analytical Chemistry: A Guide for the Critical Selection of Analytical Methods Based on Complexation Reactions**. By the late Anders Ringbom
Vol. 17 **Electron Probe Microanalysis**. *Second Edition.* By L. S. Birks
Vol. 18 **Organic Complexing Reagents: Structure, Behavior, and Application to Inorganic Analysis**. By D. D. Perrin
Vol. 19 **Thermal Analysis**. *Third Edition.* By Wesley Wm.Wendlandt
Vol. 20 **Amperometric Titrations**. By John T. Stock
Vol. 21 **Reflctance Spectroscopy**. By Wesley Wm.Wendlandt and Harry G. Hecht
Vol. 22 **The Analytical Toxicology of Industrial Inorganic Poisons**. By the late Morris B. Jacobs
Vol. 23 **The Formation and Properties of Precipitates**. By Alan G.Walton
Vol. 24 **Kinetics in Analytical Chemistry**. By Harry B. Mark, Jr. and Garry A. Rechnitz
Vol. 25 **Atomic Absorption Spectroscopy**. *Second Edition.* By Morris Slavin
Vol. 26 **Characterization of Organometallic Compounds** (*in two parts*). Edited by Minoru Tsutsui
Vol. 27 **Rock and Mineral Analysis**. *Second Edition.* By Wesley M. Johnson and John A. Maxwell
Vol. 28 **The Analytical Chemistry of Nitrogen and Its Compounds** (*in two parts*). Edited by C. A. Streuli and Philip R.Averell
Vol. 29 **The Analytical Chemistry of Sulfur and Its Compounds** (*in three parts*). By J. H. Karchmer
Vol. 30 **Ultramicro Elemental Analysis**. By Güther Toölg
Vol. 31 **Photometric Organic Analysis** (*in two parts*). By Eugene Sawicki
Vol. 32 **Determination of Organic Compounds: Methods and Procedures**. By Frederick T. Weiss
Vol. 33 **Masking and Demasking of Chemical Reactions**. By D. D. Perrin

Vol. 34. **Neutron Activation Analysis**. By D. De Soete, R. Gijbels, and J. Hoste
Vol. 35 **Laser Raman Spectroscopy**. By Marvin C. Tobin
Vol. 36 **Emission Spectrochemical Analysis**. By Morris Slavin
Vol. 37 **Analytical Chemistry of Phosphorus Compounds**. Edited by M. Halmann
Vol. 38 **Luminescence Spectrometry in Analytical Chemistry**. By J. D.Winefordner, S. G. Schulman, and T. C. O'Haver
Vol. 39. **Activation Analysis with Neutron Generators**. By Sam S. Nargolwalla and Edwin P. Przybylowicz
Vol. 40 **Determination of Gaseous Elements in Metals**. Edited by Lynn L. Lewis, Laben M. Melnick, and Ben D. Holt
Vol. 41 **Analysis of Silicones**. Edited by A. Lee Smith
Vol. 42 **Foundations of Ultracentrifugal Analysis**. By H. Fujita
Vol. 43 **Chemical Infrared Fourier Transform Spectroscopy**. By Peter R. Griffiths
Vol. 44 **Microscale Manipulations in Chemistry**. By T. S. Ma and V. Horak
Vol. 45 **Thermometric Titrations**. By J. Barthel
Vol. 46 **Trace Analysis: Spectroscopic Methods for Elements**. Edited by J. D.Winefordner
Vol. 47 **Contamination Control in Trace Element Analysis**. By Morris Zief and James W. Mitchell
Vol. 48 **Analytical Applications of NMR**. By D. E. Leyden and R. H. Cox
Vol. 49 **Measurement of Dissolved Oxygen**. By Michael L. Hitchman
Vol. 50 **Analytical Laser Spectroscopy**. Edited by Nicolo Omenetto
Vol. 51 **Trace Element Analysis of Geological Materials**. By Roger D. Reeves and Robert R. Brooks
Vol. 52 **Chemical Analysis by Microwave Rotational Spectroscopy**. By Ravi Varma and Lawrence W. Hrubesh
Vol. 53 **Information Theory as Applied to Chemical Analysis**. By Karl Eckschlager and Vladimir Stepanek
Vol. 54 **Applied Infrared Spectroscopy: Fundamentals, Techniques, and Analytical Problemsolving**. By A. Lee Smith
Vol. 55 **Archaeological Chemistry**. By Zvi Goffer
Vol. 56 **Immobilized Enzymes in Analytical and Clinical Chemistry**. By P. W. Carr and L. D. Bowers
Vol. 57 **Photoacoustics and Photoacoustic Spectroscopy**. By Allan Rosencwaig
Vol. 58 **Analysis of Pesticide Residues**. Edited by H. Anson Moye
Vol. 59 **Affity Chromatography**. By William H. Scouten
Vol. 60 **Quality Control in Analytical Chemistry**. *Second Edition*. By G. Kateman and L. Buydens
Vol. 61 **Direct Characterization of Fineparticles**. By Brian H. Kaye
Vol. 62 **Flow Injection Analysis**. By J. Ruzicka and E. H. Hansen
Vol. 63 **Applied Electron Spectroscopy for Chemical Analysis**. Edited by Hassan Windawi and Floyd Ho
Vol. 64 **Analytical Aspects of Environmental Chemistry**. Edited by David F. S. Natusch and Philip K. Hopke
Vol. 65 **The Interpretation of Analytical Chemical Data by the Use of Cluster Analysis**. By D. Luc Massart and Leonard Kaufman
Vol. 66 **Solid Phase Biochemistry: Analytical and Synthetic Aspects**. Edited by William H. Scouten
Vol. 67 **An Introduction to Photoelectron Spectroscopy**. By Pradip K. Ghosh
Vol. 68 **Room Temperature Phosphorimetry for Chemical Analysis**. By Tuan Vo-Dinh
Vol. 69 **Potentiometry and Potentiometric Titrations**. By E. P. Serjeant
Vol. 70 **Design and Application of Process Analyzer Systems**. By Paul E. Mix
Vol. 71 **Analysis of Organic and Biological Surfaces**. Edited by Patrick Echlin
Vol. 72 **Small Bore Liquid Chromatography Columns: Their Properties and Uses**. Edited by Raymond P.W. Scott
Vol. 73 **Modern Methods of Particle Size Analysis**. Edited by Howard G. Barth
Vol. 74 **Auger Electron Spectroscopy**. By Michael Thompson, M. D. Baker, Alec Christie, and J. F. Tyson
Vol. 75 **Spot Test Analysis: Clinical, Environmental, Forensic and Geochemical Applications**. By Ervin Jungreis
Vol. 76 **Receptor Modeling in Environmental Chemistry**. By Philip K. Hopke

Vol. 77 **Molecular Luminescence Spectroscopy: Methods and Applications** (*in three parts*). Edited by Stephen G. Schulman
Vol. 78 **Inorganic Chromatographic Analysis**. Edited by John C. MacDonald
Vol. 79 **Analytical Solution Calorimetry**. Edited by J. K. Grime
Vol. 80 **Selected Methods of Trace Metal Analysis: Biological and Environmental Samples**. By Jon C.VanLoon
Vol. 81 **The Analysis of Extraterrestrial Materials**. By Isidore Adler
Vol. 82 **Chemometrics**. By Muhammad A. Sharaf, Deborah L. Illman, and Bruce R. Kowalski
Vol. 83 **Fourier Transform Infrared Spectrometry**. By Peter R. Griffiths and James A. de Haseth
Vol. 84 **Trace Analysis: Spectroscopic Methods for Molecules**. Edited by Gary Christian and James B. Callis
Vol. 85 **Ultratrace Analysis of Pharmaceuticals and Other Compounds of Interest**. Edited by S. Ahuja
Vol. 86 **Secondary Ion Mass Spectrometry: Basic Concepts, Instrumental Aspects, Applications and Trends**. By A. Benninghoven, F. G. Rüenauer, and H.W.Werner
Vol. 87 **Analytical Applications of Lasers**. Edited by Edward H. Piepmeier
Vol. 88 **Applied Geochemical Analysis**. By C. O. Ingamells and F. F. Pitard
Vol. 89 **Detectors for Liquid Chromatography**. Edited by Edward S.Yeung
Vol. 90 **Inductively Coupled Plasma Emission Spectroscopy: Part 1: Methodology, Instrumentation, and Performance; Part II: Applications and Fundamentals**. Edited by J. M. Boumans
Vol. 91 **Applications of New Mass Spectrometry Techniques in Pesticide Chemistry**. Edited by Joseph Rosen
Vol. 92 **X-Ray Absorption: Principles,Applications,Techniques of EXAFS, SEXAFS, and XANES**. Edited by D. C. Konnigsberger
Vol. 93 **Quantitative Structure-Chromatographic Retention Relationships**. By Roman Kaliszan
Vol. 94 **Laser Remote Chemical Analysis**. Edited by Raymond M. Measures
Vol. 95 **Inorganic Mass Spectrometry**. Edited by F.Adams,R.Gijbels, and R.Van Grieken
Vol. 96 **Kinetic Aspects of Analytical Chemistry**. By Horacio A. Mottola
Vol. 97 **Two-Dimensional NMR Spectroscopy**. By Jan Schraml and Jon M. Bellama
Vol. 98 **High Performance Liquid Chromatography**. Edited by Phyllis R. Brown and Richard A. Hartwick
Vol. 99 **X-Ray Fluorescence Spectrometry**. By Ron Jenkins
Vol. 100 **Analytical Aspects of Drug Testing**. Edited by Dale G. Deustch
Vol. 101 **Chemical Analysis of Polycyclic Aromatic Compounds**. Edited by Tuan Vo-Dinh
Vol. 102 **Quadrupole Storage Mass Spectrometry**. By Raymond E. March and Richard J. Hughes (*out of print: see Vol. 165)*
Vol. 103 **Determination of Molecular Weight**. Edited by Anthony R. Cooper
Vol. 104 **Selectivity and Detectability Optimization in HPLC**. By Satinder Ahuja
Vol. 105 **Laser Microanalysis**. By Lieselotte Moenke-Blankenburg
Vol. 106 **Clinical Chemistry**. Edited by E. Howard Taylor
Vol. 107 **Multielement Detection Systems for Spectrochemical Analysis**. By Kenneth W. Busch and Marianna A. Busch
Vol. 108 **Planar Chromatography in the Life Sciences**. Edited by Joseph C. Touchstone
Vol. 109 **Fluorometric Analysis in Biomedical Chemistry: Trends and Techniques Including HPLC Applications**. By Norio Ichinose, George Schwedt, Frank Michael Schnepel, and Kyoko Adochi
Vol. 110 **An Introduction to Laboratory Automation**. By Victor Cerdá and Guillermo Ramis
Vol. 111 **Gas Chromatography: Biochemical, Biomedical, and Clinical Applications**. Edited by Ray E. Clement
Vol. 112 **The Analytical Chemistry of Silicones**. Edited by A. Lee Smith
Vol. 113 **Modern Methods of Polymer Characterization**. Edited by Howard G. Barth and Jimmy W. Mays
Vol. 114 **Analytical Raman Spectroscopy**. Edited by Jeanette Graselli and Bernard J. Bulkin
Vol. 115 **Trace and Ultratrace Analysis by HPLC**. By Satinder Ahuja
Vol. 116 **Radiochemistry and Nuclear Methods of Analysis**. By William D. Ehmann and Diane E.Vance

Vol. 117 **Applications of Fluorescence in Immunoassays**. By Ilkka Hemmila
Vol. 118 **Principles and Practice of Spectroscopic Calibration**. By Howard Mark
Vol. 119 **Activation Spectrometry in Chemical Analysis**. By S. J. Parry
Vol. 120 **Remote Sensing by Fourier Transform Spectrometry**. By Reinhard Beer
Vol. 121 **Detectors for Capillary Chromatography**. Edited by Herbert H. Hill and Dennis McMinn
Vol. 122 **Photochemical Vapor Deposition**. By J. G. Eden
Vol. 123 **Statistical Methods in Analytical Chemistry**. By Peter C. Meier and Richard Züd
Vol. 124 **Laser Ionization Mass Analysis**. Edited by Akos Vertes, Renaat Gijbels, and Fred Adams
Vol. 125 **Physics and Chemistry of Solid State Sensor Devices**. By Andreas Mandelis and Constantinos Christofides
Vol. 126 **Electroanalytical Stripping Methods**. By Khjena Z. Brainina and E. Neyman
Vol. 127 **Air Monitoring by Spectroscopic Techniques**. Edited by Markus W. Sigrist
Vol. 128 **Information Theory in Analytical Chemistry**. By Karel Eckschlager and Klaus Danzer
Vol. 129 **Flame Chemiluminescence Analysis by Molecular Emission Cavity Detection**. Edited by David Stiles, Anthony Calokerinos, and Alan Townshend
Vol. 130 **Hydride Generation Atomic Absorption Spectrometry**. Edited by Jiri Dedina and Dimiter L. Tsalev
Vol. 131 **Selective Detectors: Environmental, Industrial, and Biomedical Applications**. Edited by Robert E. Sievers
Vol. 132 **High-Speed Countercurrent Chromatography**. Edited by Yoichiro Ito and Walter D. Conway
Vol. 133 **Particle-Induced X-Ray Emission Spectrometry**. By Sven A. E. Johansson, John L. Campbell, and Klas G. Malmqvist
Vol. 134 **Photothermal Spectroscopy Methods for Chemical Analysis**. By Stephen E. Bialkowski
Vol. 135 **Element Speciation in Bioinorganic Chemistry**. Edited by Sergio Caroli
Vol. 136 **Laser-Enhanced Ionization Spectrometry**. Edited by John C. Travis and Gregory C. Turk
Vol. 137 **Fluorescence Imaging Spectroscopy and Microscopy**. Edited by Xue Feng Wang and Brian Herman
Vol. 138 **Introduction to X-Ray Powder Diffractometry**. By Ron Jenkins and Robert L. Snyder
Vol. 139 **Modern Techniques in Electroanalysis**. Edited by Petr Vaňýek
Vol. 140 **Total-Reflction X-Ray Fluorescence Analysis**. By Reinhold Klockenkamper
Vol. 141 **Spot Test Analysis: Clinical, Environmental, Forensic, and Geochemical Applications**. *Second Edition.* By Ervin Jungreis
Vol. 142 **The Impact of Stereochemistry on Drug Development and Use**. Edited by Hassan Y. Aboul-Enein and Irving W.Wainer
Vol. 143 **Macrocyclic Compounds in Analytical Chemistry**. Edited by Yury A. Zolotov
Vol. 144 **Surface-Launched Acoustic Wave Sensors: Chemical Sensing and Thin-Film Characterization**. By Michael Thompson and David Stone
Vol. 145 **Modern Isotope Ratio Mass Spectrometry**. Edited by T. J. Platzner
Vol. 146 **High Performance Capillary Electrophoresis: Theory, Techniques, and Applications**. Edited by Morteza G. Khaledi
Vol. 147 **Solid Phase Extraction: Principles and Practice**. By E. M. Thurman
Vol. 148 **Commercial Biosensors: Applications to Clinical, Bioprocess and Environmental Samples**. Edited by Graham Ramsay
Vol. 149 **A Practical Guide to Graphite Furnace Atomic Absorption Spectrometry**. By David J. Butcher and Joseph Sneddon
Vol. 150 **Principles of Chemical and Biological Sensors**. Edited by Dermot Diamond
Vol. 151 **Pesticide Residue in Foods: Methods, Technologies, and Regulations**. By W. George Fong, H. Anson Moye, James N. Seiber, and John P. Toth
Vol. 152 **X-Ray Fluorescence Spectrometry**. *Second Edition.* By Ron Jenkins
Vol. 153 **Statistical Methods in Analytical Chemistry**. *Second Edition.* By Peter C. Meier and Richard E. Züd
Vol. 154 **Modern Analytical Methodologies in Fat- and Water-Soluble Vitamins**. Edited by Won O. Song, Gary R. Beecher, and Ronald R. Eitenmiller
Vol. 155 **Modern Analytical Methods in Art and Archaeology**. Edited by Enrico Ciliberto and Guiseppe Spoto

Vol. 156 **Shpol'skii Spectroscopy and Other Site Selection Methods: Applications in Environmental Analysis, Bioanalytical Chemistry and Chemical Physics**. Edited by C. Gooijer, F. Ariese and J.W. Hofstraat

Vol. 157 **Raman Spectroscopy for Chemical Analysis**. By Richard L. McCreery

Vol. 158 **Large (C> = 24) Polycyclic Aromatic Hydrocarbons: Chemistry and Analysis**. By John C. Fetzer

Vol. 159 **Handbook of Petroleum Analysis**. By James G. Speight

Vol. 160 **Handbook of Petroleum Product Analysis**. By James G. Speight

Vol. 161 **Photoacoustic Infrared Spectroscopy**. By Kirk H. Michaelian

Vol. 162 **Sample Preparation Techniques in Analytical Chemistry**. Edited by Somenath Mitra

Vol. 163 **Analysis and Purification Methods in Combination Chemistry**. Edited by Bing Yan

Vol. 164 **Chemometrics: From Basics to Wavelet Transform**. By Foo-tim Chau, Yi-Zeng Liang, Junbin Gao, and Xue-guang Shao

Vol. 165 **Quadrupole Ion Trap Mass Spectrometry**. *Second Edition.* By Raymond E. March and John F. J. Todd

Vol. 166 **Handbook of Coal Analysis**. By James G. Speight

Vol. 167 **Introduction to Soil Chemistry: Analysis and Instrumentation**. By Alfred R. Conklin, Jr.

Vol. 168 **Environmental Analysis and Technology for the Refining Industry**. By James G. Speight

Vol. 169 **Identification of Microorganisms by Mass Spectrometry**. Edited by Charles L. Wilkins and Jackson O. Lay, Jr.

Vol. 170 **Archaeological Chemistry**. *Second Edition.* By Zvi Goffer

Vol. 171 **Fourier Transform Infrared Spectrometry**. *Second Edition.* By Peter R. Griffiths and James A. de Haseth

Vol. 172 **New Frontiers in Ultrasensitive Bioanalysis: Advanced Analytical Chemistry Applications in Nanobiotechnology, Single Molecule Detection, and Single Cell Analysis**. Edited by Xiao-Hong Nancy Xu

Vol. 173 **Liquid Chromatography Time-of-Flight Mass Spectrometry: Principles, Tools, and Applications for Accurate Mass Analysis.** Edited by Imma Ferrer and E. Michael Thurman

Vol. 174 **In Vivo Glucose Sensing.** Edited by David O. Cunningham and Julie A. Stenken